Machine Learning for Plant Biology

Edited by Jen-Tsung Chen

National University of Kaohsiung
Taiwan

WILEY

This edition first published 2026

Registered Office(s)
John Wiley & Sons, Inc., 111 River Street, Hoboken, NJ 07030, USA
John Wiley & Sons Ltd, New Era House, 8 Oldlands Way, Bognor Regis, West Sussex, PO22 9NQ, UK

For details of our global editorial offices, customer services, and more information about Wiley products visit us at www.wiley.com.

The manufacturer's authorized representative according to the EU General Product Safety Regulation is Wiley-VCH GmbH, Boschstr. 12, 69469 Weinheim, Germany, e-mail: Product_Safety@wiley.com.

Wiley also publishes its books in a variety of electronic formats and by print-on-demand. Some content that appears in standard print versions of this book may not be available in other formats.

Library of Congress Cataloging-in-Publication Data:

Names: Chen, Jen-Tsung, editor
Title: Machine learning for plant biology / edited by Jen-Tsung Chen.
Description: First edition. | Hoboken, NJ : Wiley, 2026. | Includes bibliographical references and index.
Identifiers: LCCN 2025027934 | ISBN 9781394329618 hardback | ISBN 9781394329632 adobe pdf | ISBN 9781394329625 epub | ISBN 9781394329649
Subjects: LCSH: Botany—Mathematical models | Machine learning
Classification: LCC QK46.5.M3 M33 2026
LC record available at https://lccn.loc.gov/2025027934

Cover Design: Wiley
Cover Image: © piranka/Getty Images

STIX Two Text 9.5/12.5 by Lumina Datamatics

Table of Contents

Preface

For decades, bioinformatics has made a considerable contribution to the progress of biological studies in the post-genomic era, and its technical development has always led to certain approaches for researchers to accelerate knowledge exploration, particularly capable of dealing with big data of omics, modeling complicated biological networks, and predicting crucial players in cellular processes. In recent years, a subtopic of bioinformatics, artificial intelligence (AI), has had a revolutionary technical breakthrough, i.e., machine learning (ML), deep learning, and generative AI, which are expected to make a comprehensive impact on scientific research, including the field of biological science.

Among AI technologies, ML models have been successfully applied to plant biological studies for years, including the use of deep learning in the detection and analysis of plant morphological phenotypes, and their applications have expanded to construct phenomics in recent years and subsequently move forward to integrate with multiple omics approaches. Additionally, ML models are promising to assist in the organization of single-cell resolution and spatial omics for "3D biology" in the future.

This book aims to summarize current technical advancements in AI/ML technologies particularly for applications in plant research including a series of sub-fields, such as species identification, systems biology, functional genomics, genomic selection, phenotyping, synthetic biology, biotechnology, diversity, ecology, spatial omics, plant disease diagnosis and protection, plant stress physiology, genome editing, and plant secondary metabolism. It discusses agricultural applications, such as AI/ML-enabled smart agriculture and speed breeding, and additionally, ethical considerations and regulations of AI/ML applications in plant biology and agriculture.

This book is a pioneer in providing ways to use ML tools in plant biological research and agricultural applications. It is a suitable reference for a wide range of readers, including students, researchers, lecturers, and professors in all sub-fields of plant sciences and agriculture.

This book summarizes current achievements of ML-based technologies, highlighting the applications in investigating complicated biological networks of plants and modeling biotic and abiotic stress responses of major crops for developing sustainable agriculture to face a changing global climate. It presents cutting-edge AI tools, particularly machine and deep learning models, and generative AI, combined with multiple omics, functional genomics, genome editing technologies, and high-throughput technologies, for uncovering interactions between plants and pathogens and environmental stresses. These combinatorial strategies provided by this book can advance the exploration of plant biology, plant stress physiology, plant protection, and speed and smart breeding for future anti-stress and high-yield plants, achieve sustainable agriculture, and additionally, support the "SDGs" by the United Nations.

This book is an ideal reference providing AI/ML technologies by sharing crucial aspects of methods, applications, and future perspectives and it's designed for students and researchers to efficiently overview these critical subtopics of plant biology and biotechnology and thus realize the concept and hopefully, inspire the ideas of future experiments. It may advance the exploration of knowledge and eventually lead to better development of future crops by scientists, plant biologists, and crop breeders.

The Editor appreciates all chapter contributors and the publisher.

List of Contributors

N. Abinesh
Hindusthan College of Arts & Science, Coimbatore, India

Arunachalam Abitha
Department of Human Genetics and Molecular Biology, Bharathiar University, Coimbatore, Tamil Nadu, India

Boris Adam
INRAE Centre Clermont-Auvergne-Rhône-Alpes, Clermont-Ferrand, France

Sarfraz Ahmad
Department of Genetics and Plant Breeding, College of Agriculture Mandawa, Swami Keshwanand Rajasthan Agriculture University, Bikaner, Rajasthan, India

Eman A. Ahmed
Department of Community Organization, Faculty of Social Work, Assiut University, Assiut, Egypt

Arumugam Vijaya Anand
Department of Human Genetics and Molecular Biology, Bharathiar University, Coimbatore, Tamil Nadu, India

Jeyabal Philomenathan Antony Prabhu
Department of Human Genetics and Molecular Biology, Bharathiar University, Coimbatore, Tamil Nadu, India

Esakkimuthu Balaji
Sudha Fertility Hospital, Hyderabad, Telangana, India

Vikas Belwal
Department of Tree Improvement, College of Forestry Ranichauri, VCSG Uttarakhand University of Horticulture and Forestry, Bharsar, Uttarakhand, India

Vadivelu Bharathi
Department of Biochemistry, Thanthai Periyar Government College, Tiruchirappalli, Tamil Nadu, India

Natasha Charaya
Department of Biotechnology, Guru Jambheshwar University of Science & Technology (GJUS&T) Hisar, Haryana, India

Ankan Das
Palli Siksha Bhavana (Institute of Agriculture), Visva-Bharati University, Sriniketan, West Bengal, India

Soumya Shephalika Dash
Institute of Agriculture, Visva-Bharati, Sriniketan, West Bengal, India

Juel Debnath
College of Agriculture, Kerala Agricultural University, Thiruvananthapuram, Kerala, India

Sandip Debnath
Department of Genetics and Plant Breeding, Institute of Agriculture, Visva-Bharati University, Sriniketan, Birbhum,
West Bengal, India

Nathan Drogue
INRAE Centre Clermont-Auvergne-Rhône-Alpes, Clermont-Ferrand, France

Haragopal Dutta
Division of Genetics, ICAR- Indian Agricultural Research Institute, New Delhi, India

Suman Dutta
School of Agriculture and Rural Development, Ramakrishna Mission Vivekananda Educational and Research Institute, Belur Math, Howrah,
West Bengal, India

Atef M. El-Sagheer
Agricultural Zoology and Nematology Department, Faculty of Agriculture, Al-Azhar University, Assiut, Egypt

Julien Garnier
Laboratoire Angevin de Recherche en Ingénierie des Systèmes (LARIS), UMR INRAe-IRHS, Université d'Angers, Angers, France

Shanmugam Gavaskar
Department of Computer Application, Bharathiar University, Coimbatore, Tamil Nadu, India

Subhadwip Ghorai
Department of Agriculture, Brainware University, Barasat, Kolkata, West Bengal, India

Mohamed M.M. Hamd
Foundations of Education, Faculty of Education, Assiut University, Assiut, Egypt

Shanmugasundaram Hariharan
Department of Artificial Intelligence and Data Science, Vardhaman College of Engineering, India

Soham Hazra
Department of Agriculture, Brainware University, Barasat, Kolkata, West Bengal, India

Shimi Jose
Hawkesbury Institute for the Environment, Western Sydney University, Richmond, Australia

Kalaichelvi Kalaignan
Department of Agronomy, Krishi Vigyan Kendra, Virdhachalam, India

Monika Kalia
Department of Entomology, Dr YS Parmar University of Horticulture & Forestry, COH & F, Neri, Hamirpur (HP), India

Sonika Kalia
Department of Biotechnology, School of Applied and Life Sciences, Uttaranchal University, Dehradun, India

Annasamy Kaliyan
Department of Horticulture, Pushkaram College of Agriculture Sciences, Pudukkottai, India

Sahely Kanthal
Department of Agriculture, Brainware University, Barasat, Kolkata, West Bengal, India

Rubina Khan
Department of Genetics and Plant Breeding, Sri Karan Narendra College of Agriculture Jobner, Sri Karan Narendra Agriculture University, Jobner, Rajasthan, India

Pankaj Kumar
Department of Crop Improvement, College of Forestry Ranichauri, VCSG Uttarakhand University of Horticulture and Forestry, Bharsar, Uttarakhand, India

Sudhir Kumar
Division of Plant Physiology, ICAR-IARi, New Delhi, India

Shilpi Kumari
Department of Biotechnology, School of Applied and Life Sciences, Uttaranchal University, Dehradun, India

Karthiba Loganathan
Department of Plant Pathology, Tamil Nadu Agricultural University, Coimbatore, India

Ramasamy Manikandan
Department of Biochemistry, Shrimati Indira Gandhi College, Tiruchirappalli, Tamil Nadu, India

Mani Manoj
Department of Human Genetics and Molecular Biology, Bharathiar University, Coimbatore, Tamil Nadu, India

Jeya Rani Maria Michael
Department of Agronomy, Pushkaram College of Agriculture Sciences, Pudukkottai, India

Sanjay Mochary
Department of Agriculture, Brainware University, Barasat, Kolkata, West Bengal, India

Sneha Mohan
College of Agriculture, Kerala Agricultural University, Thiruvananthapuram, Kerala, India

Rajdeep Mohanta
Department of Agriculture, Brainware University, Barasat, Kolkata, West Bengal, India

Sucharita Mohapatra
College of Agriculture, Odisha University of Agriculture and Technology, Bhubaneswar, Odisha, India

Thamaraiselvan Nandhini Devi
Department of Biomedical Engineering, Selvam College of Technology, Namakkal, Tamil Nadu, India

Gyanisha Nayak
Crop Improvement Division, ICAR-National Rice Research Institute, Cuttack, Odisha, India

Sahfigul Ameed Nihaal Fathima
Department of Human Genetics and Molecular Biology, Bharathiar University, Coimbatore, Tamil Nadu, India

Ehsas Pachauri
Department of Genetics and Plant Breeding, IARI-Central Arid Zone Research Institute, Jodhpur, India

Murali Sankar Perumal
Department of Plant Pathology, Pushkaram College of Agriculture Sciences, Pudukkottai, India

Anantha Raju Pokkaru
Department of Plant Breeding and Genetics, Cotton Research Station, Veppanthattai, India

Sourish Pramanik
Department of Plant Pathology, Punjab Agricultural University, Ludhiana, Punjab, India; Ethophilia Research Foundation, Santiniketan, West Bengal, India

Mohammad Ghani Raghib
Department of Horticulture, Faculty of Agriculture, Nangarhar University, Afghanistan

Mehdi Rahimi
Department of Biotechnology, Institute of Science and High Technology and Environmental Sciences, Graduate University of Advanced Technology, Kerman, Iran

Mehjebin Rahman
College of Agriculture, Assam Agricultural University, Assam, India

Krishna Kumar Rai
Department of Biotechnology, Amity School of Biological Sciences, Amity University Punjab, Mohali, Punjab, India

Pejman Rasti
Laboratoire Angevin de Recherche en Ingénierie des Systèmes (LARIS), UMR INRAe-IRHS, Université d'Angers, Angers, France

Indra Rautela
Department of Biotechnology, School of Applied and Life Sciences, Uttaranchal University, Dehradun, India

Simon Ravé
Laboratoire Angevin de Recherche en Ingénierie des Systèmes (LARIS), UMR INRAe-IRHS, Université d'Angers, Angers, France

Raveena Ravi
Department of Environmental Sciences, Tamil Nadu Agricultural University, Coimbatore, India

Lellapalli Rithesh
College of Agriculture, Kerala Agricultural University, Thiruvananthapuram, Kerala, India

David Rousseau
Laboratoire Angevin de Recherche en Ingénierie des Systèmes (LARIS), UMR INRAe-IRHS, Université d'Angers, Angers, France

Soumik Dey Roy
Department of Agriculture, Brainware University, Barasat, Kolkata, West Bengal, India

Ehab A A Salama
Department of Agricultural Botany (Genetics), Saba Basha (Faculty of Agriculture), Alexandria University, Alexandria, Egypt

Manasa Samuthiravelu
Department of Seed Science and Technology, TNAU, Coimbatore, India

Saniya
Department of Horticulture, ICAR-National Research Centre for Grapes, Pune, Maharashtra, India

Hamdy A. Sayed
Social planning Department, Faculty of Social Work, Assiut University, Assiut, Egypt

A.V. Senthil Kumar
Hindusthan College of Arts & Science, Coimbatore, India

Dibyendu Seth
Department of Genetics and Plant Breeding, Chaudhary Charan Singh Haryana Agriculture University (CCSHAU), Hisar, Haryana, India

Mohammed Jaffer Shakeera Banu
Division of Molecular Biology. Microbiology Laboratory, Coimbatore, Tamil Nadu, India

Anwesha Sharma
College of Agriculture, Assam Agricultural University, Assam, India

Vinay Sharma
Center of Excellence in Genomics and Systems Biology (CEGSB), and Center for Prebreeding Research (CPBR), International Crops Research Institute for the Semi-Arid Tropics (ICRISAT), Patancheru, India

Ramaraj Sivamano
Department of Human Genetics and Molecular Biology, Bharathiar University, Coimbatore, Tamil Nadu, India

Ravichandran Sneha
Department of Human Genetics and Molecular Biology, Bharathiar University, Coimbatore, Tamil Nadu, India

Alagarsamy Sumitrha
Department of Computer Science and Engineering, SNS College of Technology, Coimbatore, Tamil Nadu, India

Kyla L. Tennin
United Nations Peace Ambassador Foundation, Southaven, MS, USA

Thiruveni Thangaraj
Department of Agricultural Entomology, Nammazhvar College of Agriculture and Technology, Kamuthy, India

Sathyalingam Sathya Trisha
Department of Human Genetics and Molecular Biology, Bharathiar University, Coimbatore, Tamil Nadu, India

Sadhana Veeramani
Department of Agricultural Entomology, Aravindhar Agricultural Institute of Technology, Tiruvannamalai, India

Shanmugam Velayuthaprabhu
Department of Biotechnology, Bharathiar University, Bharathiar University, Coimbatore, Tamil Nadu, India

Kannan Vijayarani
Department of Education, Bharathiar University, Coimbatore, Tamil Nadu, India

1

Edge-Based Machine Learning for Computer Vision in Smart Plant Biology Imaging

Julien Garnier[1], Simon Ravé[1], Nathan Drogue[2], Boris Adam[2], Pejman Rasti[1], David Rousseau[1]*

[1] *Laboratoire Angevin de Recherche en Ingénierie des Systèmes (LARIS), UMR INRAe-IRHS, Université d'Angers, Angers, France*
[2] *INRAE Centre Clermont-Auvergne-Rhône-Alpes, Clermont-Ferrand, France*

*Corresponding author: david.rousseau@univ-angers.fr

1.1 Introduction

Ground plant monitoring systems, based on vehicle or connected sticks, deployed in fields increasingly rely on sensor-equipped devices to collect large-scale multimodal data. These data sources include images, temperature, humidity, soil moisture measurements, and other environmental data, all of which are essential for the assessment of plant growth and health (Li et al. 2020). However, most existing systems do not perform real-time processing on-site. Instead, raw sensor data are transmitted to centralized servers or cloud platforms to be processed. Nowadays, these are massively dominated by deep learning models (Chen et al. 2023; Arya et al. 2022). While this approach enables high-accuracy analysis leveraging powerful computational resources, it presents several critical limitations.

First, the transmission of high-resolution image data requires substantial network bandwidth, which can be impractical in remote agricultural environments with limited connectivity. Second, reliance on centralized processing introduces latency, preventing real-time analysis. Finally, many edge devices operate on battery power or other constrained energy sources, making continuous data transmission and computationally intensive deep learning processing infeasible (Fanariotis et al. 2023).

These challenges highlight the need for real-time, energy-efficient plant imaging analysis directly on edge devices while balancing computational cost and model accuracy (Shadrin et al. 2019; Sharma et al. 2022; Mohammad-Razdari et al. 2022). Recent research has explored deploying deep learning models on embedded platforms to eliminate the need for cloud-based inference and improve response times in agricultural applications (Howard et al. 2017; Tan and Le 2019; Zhang et al. 2017). However, standard deep learning architectures are often too computationally demanding for edge devices, necessitating efficient model compression, optimization techniques, and lightweight architectures (Saha et al. 2024; Liang et al. 2021; ExecuTorch 2024; Hinton et al. 2015).

This chapter provides a panorama of strategies to optimize deep learning–based computer vision models at the edge. Our goal is to review key approaches that improve model efficiency while maintaining predictive performance, making embedded AI a viable solution for several applications, including agricultural and biological applications. Such techniques need to be more widespread to further extend the applicability of the works developed in affordable plant phenotyping on low-cost single-board computers (Valle et al. 2017; Tovar et al. 2018; Grindstaff et al. 2019; Garbouge et al. 2021, 2022; Cordier et al. 2024).

This chapter is organized as follows. Section 2 provides an overview of the current electronic devices for embedded AI-driven computer vision, focusing on the hardware components required for efficient processing. In Section 3, we delve into light deep learning strategies that aim to reduce model complexity while preserving performance. Section 4 illustrates these strategies through a practical use case in plant imaging, showcasing the impact of edge-based machine learning on an image segmentation task.

Machine Learning for Plant Biology, First Edition. Edited by Jen-Tsung Chen.

Finally, Section 5 discusses the trade-offs between accuracy, inference speed, and energy consumption, offering insights into the challenges and opportunities for future research in this area.

1.2 Electronic Devices for Embedded AI-driven Computer Vision

This section provides an overview of electronic devices suitable for performing computer vision tasks at the edge. Edge computing devices are designed to integrate compact and energy-efficient processing units, enabling AI-based inference with minimal reliance on cloud resources. Among the most commonly employed solutions are single-board computers (SBCs), which offer a balance between computational power, flexibility, and power efficiency. These platforms, such as the Raspberry Pi or NVIDIA Jetson series (NVIDIA 2024), are widely used for embedded AI applications due to their ability to support various deep learning frameworks while maintaining relatively low power consumption. Table 1.1 presents a selection of widely used SBCs that support embedded AI, illustrating their key specifications and suitability for various edge computing scenarios.

An alternative approach involves leveraging modern smartphones, which integrate a high-performance processor, dedicated AI accelerators, built-in cameras, and a battery-powered system, eliminating the need for additional accessories or external devices. Moreover, many contemporary smartphones are waterproof and incorporate specialized neural processing units (NPUs), significantly enhancing AI inference efficiency (Qualcomm 2023). However, despite these advantages, smartphones generally exhibit higher power consumption than SBCs, which can be a limiting factor for long-term deployment in edge environments.

For applications with stringent power constraints, microcontrollers such as the Arduino (Arduino Hardware) or Espressif ESP32 (ESP32 Wi-Fi & Bluetooth) provide an ultra-low-power alternative. These devices are highly energy-efficient and well-suited for lightweight AI models, such as TinyML applications (Warden and Situnayake 2019). However, since they lack a standard operating system and have limited computational resources, compatibility with common AI libraries is restricted, often requiring model quantization and optimization to function effectively.

SBCs can be enhanced with dedicated AI accelerators to improve computational efficiency for machine learning tasks. These accelerators connect to the SBC through various interfaces, with the most accessible option being a USB port. Among USB-based accelerators, the Google Coral USB Accelerator (Coral USB Accelerator) is one of the most widely adopted solutions. Previously, the Intel Movidius Neural Compute Stick (Intel NCS2) offered a similar alternative, but it is no longer available. However, EyeCloudAI has developed the NCS2 Replacement (2024) as a successor to this device.

Beyond USB, another common interface for AI accelerators is the M.2 connector, which is often interchangeable with PCIe in many devices. The Google Coral accelerator is also available in this format, along with other solutions such as the Hailo-8, which is integrated into the Raspberry Pi AI Hat designed specifically for the Raspberry Pi 5. Additional AI accelerators supporting both M.2 and 2PCIe connectivity include the Axelera METIS (AI acceleration hardware) and the EdgeCortix Sakura II (SAKURA^{TM} Modules and Cards). Meanwhile, the NVIDIA Jetson module, which features an integrated NVIDIA GPU core, is exclusively available with a PCIe connector, offering a high-performance solution for edge AI applications.

Table 1.1 List of some available SBCs.

System	Processor	RAM	CUDA/TPU	Power	Price (€)
Raspberry pi 5	Quad core 2.4 Ghz	2/4/8 Gb	With HAT 13 Tops	3–10 W	90
Raspberry pi zero 2	Quad core 1 Ghz	512 Mb	None	0.5–2.5 W	20
Nvidia Jetson Orin Nano super	Six core 1.7 Ghz	8 Gb	1024 Cuda core 32 tensor core 67 Tops	7–25 W	240
Google coral dev board	Quad core	1/ 4 Gb	Google edge tpu 4 tops	3–10 W	120
Udoo bolt V8	Quad core 2 Ghz	up to 32 Gb	AMD Vega 8 GPU	12–25 W	450
Orange Pi 5	Eight core 2.4 Ghz	2/4/8/16 Gb	Built-in AI accelerator NPU 6 TOPS	3–10 W	110
Latte panda 3 delta	Quad core 2 Ghz	8 Gb	Intel uhd graphic	4–16 W	230
Odroid M1S	Quad core 1.8 Ghz	4/8 Gb	None	5–21 W	120
BeagleBone Ai-64	Dual core 2 Ghz	4 Gb	up to 8 TOPS (8b) at 1.0 GHz	5–20 W	160

Beyond hardware acceleration, efficient embedded AI systems also require software optimizations and model architecture adaptations to operate effectively within low-power environments. This is what we propose in the following section.

1.3 Light Deep Learning Strategies

Deep learning models for embedded AI should operate under resource constraints, which are including limited computational power, memory, and energy efficiency. Standard frameworks such as Torch and TensorFlow are widely used for deep learning applications, but their deployment on edge devices is often complicated due to their high computational requirements.

The first and more accessible option for optimizing embedded AI models is the use of libraries designed for resource-constrained environments. There are two main libraries, ExecuTorch (ExecuTorch), which is the Torch Mobile replacement, and LiteRT (LiteRT overview), which is known as TensorFlow Lite. These frameworks are completely optimized to reduce memory footprint and computational overhead in order to make them optimized for mobile and embedded devices. Furthermore, LiteRT extends its compatibility to microcontrollers, with an extensive list of supported hardware available online (LiteRT for Microcontrollers).

Beyond libraries selection, we can also use model optimizing techniques that play an important role in reducing computational demands while maintaining similar performance. Quantization and pruning are among the most commonly applied techniques (Liang et al. 2021), aiming to reduce the bit precision of model parameters and eliminate redundant connections, respectively. These methods are frequently incorporated during the conversion of models to low-power inference engines, ensuring that computational and memory constraints are met without requiring substantial modifications to the original model architecture. Another widely used optimization technique is knowledge distillation (Hinton et al. 2015), wherein a larger teacher model generates soft targets for training a more compact student model. This approach facilitates knowledge transfer while reducing model complexity and inference time. However, despite its effectiveness in computational efficiency, knowledge distillation does not inherently minimize RAM consumption, which remains a critical factor in real-time embedded applications.

In addition to software-level optimizations, the design of model architectures suited for low-power hardware is crucial to efficient inference on edge devices. Several lightweight deep learning models fall into this category, including ShuffleNet (Zhang et al. 2017), which employs grouped convolutions and channel shuffling, and MobileNet (Howard et al. 2017), which utilizes depth-wise separable convolutions to minimize parameters and floating-point operations. Another approach, EfficientNet (Tan and Le 2019), optimizes model efficiency through compound scaling, adjusting width, depth, and resolution. Additionally, AttendNets (Wong et al. 2020) have emerged as a compelling alternative, further improving performance in resource-constrained environments.

Another approach for reducing inference cost is the use of Kolmogorov–Arnold networks, which introduce learnable activation functions at the edge. Unlike conventional deep neural networks, where activation functions are predefined and fixed, the Kolmogorov–Arnold formulation allows activation functions to adapt dynamically based on the input distribution. This adaptability reduces the number of required layers and computations; however, it also increases training complexity, potentially leading to longer convergence times.

Finally, the last approach that can be named in this section is Neural Architecture Search (NAS), which can enable the automated design of optimized deep learning models. However, standard NAS methods are computationally expensive, which limits their practical applicability for embedded systems.

To overcome these challenges, hardware-aware NAS approaches have been developed to tailor model architectures to specific deployment environments. As an example, TinyTNAS (Saha et al. 2017) use hardware characteristics directly into the search process to optimize energy efficiency, while other approaches, such as (Garavagno et al. 2023), try to prioritize the generation of minimal architectures with the lowest possible computational footprint.

Most of the above light approaches fall under the category of TinyML, which is a domain dedicated to deploying machine learning models on highly resource-constrained hardware (Schizas et al. 2022). Beyond the use of dedicated deep learning frameworks and optimized architectures, alternative solutions such as the Fast Artificial Neural Network (FANN) library (FANN on GitHub) provide efficient neural network training and deployment on low-power systems. A specialized variant of FANN, designed explicitly for microcontrollers, further extends its applicability in extreme low-power environments (Wang et al. 2020). These approaches collectively enable the realization of AI capabilities on embedded platforms while maintaining stringent efficiency requirements.

1.4 Benchmark of Light Embedded Deep Learning on a Plant Imaging Use Case

To better assess the performance of embedded deep learning models in real-world conditions, we propose to test the methods listed in the previous section on a practical plant imaging use case.

Specifically, we address a segmentation task focused on soil coverage estimation.

1.4.1 Image Acquisition and Segmentation

In this study, images are acquired using the PHACC stick, a system developed by INRAE Clermont-Ferrand, France, as illustrated in Figure 1.1. The system integrates multiple environmental sensors, such as those measuring soil temperature, soil humidity, external temperature, hygrometry, wind speed, and rain fall.

The system is also equipped with two cameras: one positioned at an 45° angle and another mounted to provide top-down views of the scene. Both cameras are connected to a Raspberry Pi, which serves as the central computational unit. The Raspberry Pi handles image processing and data transmission, utilizing a GSM module for communication with a remote server. More details about the PHACC stick is available on (https://ue1375.clermont.hub.inrae.fr/outils/piquets-connectes).

An example of 36 images captured by the PHACC stick is shown in Figure 1.2, where the left image is taken at a 45° angle and the right image presents a top-down view. Since the images were captured on different days, at various times, and from different angles, natural variations in illumination are present.

To mitigate these lighting inconsistencies, an illumination normalization operation was applied. The images were first converted to the YCrCb color space, and a Gaussian kernel with a standard deviation of 5 pixels was used to estimate the image background. This estimated background was then subtracted from the Y channel before converting the images back to the RGB color space. An example of these normalized images is shown in Figure 1.3.

Figure 1.1 PHACC stick deployed in a field.

Following this normalization, plant segmentation is performed on all 36 images using the VegAnn pre-trained model (Madec et al. 2023), which is based on a U-Net architecture (Ronneberger et al. 2015) with a ResNet backbone, implemented using the Torch framework. The segmentation generated by the model on normalized images is shown in Figure 1.4.

1.4.2 Model Adaptation

In the initial stage, the model is transformed into three distinct versions. The first version is created through quantization of the original model, which is referred to as Torch Quantized. The second version is obtained by converting the model to the Torch Mobile format. The last version is obtained by converting the model to the TensorFlow Lite format through an ONNX conversion, to be executed on both single-threaded and multi-threaded configurations. Following these conversions, the models are tested on three distinct devices: a Raspberry Pi 4 with 4 GB of RAM, a Jetson Nano with 2 GB of RAM, and a Raspberry Pi 4 with 2 GB of RAM. The average time required to process an image, the Dice score, and the average energy consumption of all models and systems are evaluated and listed in the Table 1.2.

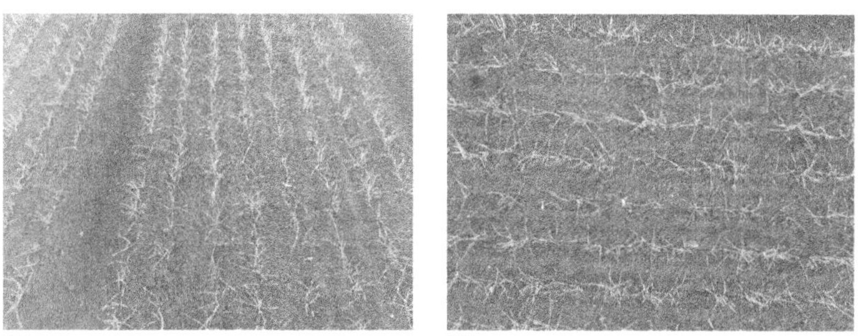

(a) (b)

Figure 1.2 Examples of images captured: (a) image captured by the camera at a 45° angle and (b) image captured by a camera with a top-down view.

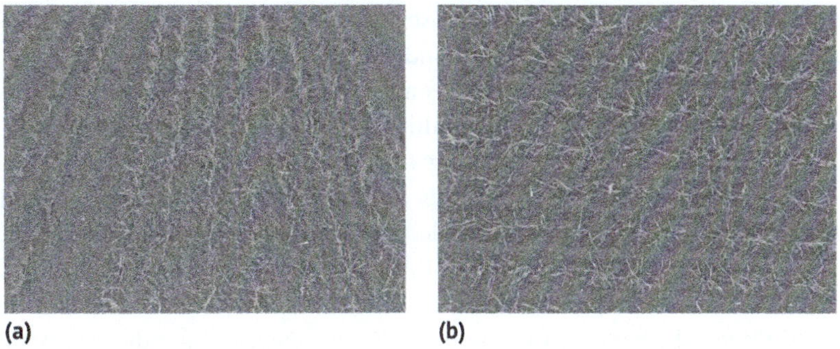

Figure 1.3 Examples of normalized images: (a) image captured by the camera at a 45° angle after normalization and (b) image captured by a camera with a top-down view after normalization.

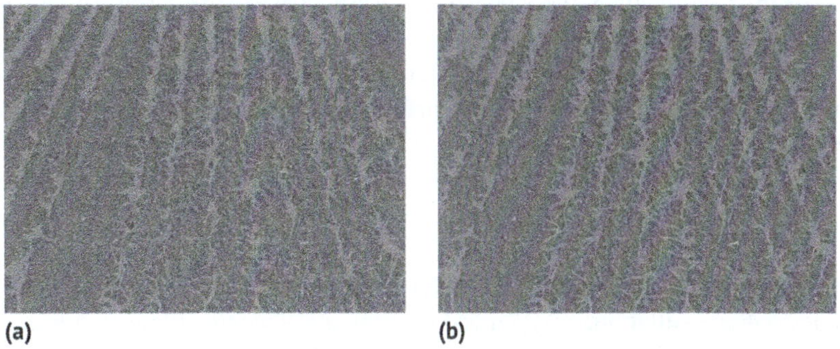

Figure 1.4 Example of the segmentation results from the VegAnn model on two types of images: (a) segmentation on the image taken by the camera at a 45° angle and (b) segmentation on the image taken by the top-down view camera.

Table 1.2 Results of different model formats on multiple single-board computers.

Model Format	System	Average Time/Image (s)	Dice Score	Average Consumption (mAh)/Image
Torch	Rasp Pi 4 4GB	22	0.99	4.20
Torch Quantized	Rasp Pi 4 4GB	20	0.99	3.92
Torch Mobile	Rasp Pi 4 4GB	16	0.56	2.88
Tf Lite Single Thread	Rasp Pi 4 4GB	25	0.84	3.78
Tf Lite Mutli Thread	Rasp Pi 4 4GB	12	0.84	2.88
Torch	Jetson Nano 2GB	420	0.99	55.81
Torch Quantized	Jetson Nano 2GB	420	0.99	56.01
Torch Mobile	Jetson Nano 2GB	15	0.56	6.31
Tf Lite Single Thread	Jetson Nano 2GB	31	0.84	7.48
Tf Lite Multi Thread	Jetson Nano 2GB	10	0.84	4.22
Torch	Rasp Pi 4 2GB	118	0.99	17.97
Torch Quantized	Rasp Pi 4 2GB	127	0.99	17.69
Torch Mobile	Rasp Pi 4 2GB	19	0.56	4.42
Tf Lite Single Thread	Rasp Pi 4 2GB	27	0.84	7.62
Tf Lite Multi Thread	Rasp Pi 4 2GB	14.75	0.84	5.1

Upon examining the results in Table 1.2, it is clear that the Torch Mobile model has a significantly lower Dice score (0.56) on all three systems, compared to the original Torch model and other lightweight models, which achieve 0.99 (Torch and Torch Quantized) or 0.84 (TensorFlow Lite models). This indicates that while Torch Mobile improves speed and reduces power consumption, it does so at the cost of segmentation performance.

The processing time varies across models and hardware. On the Raspberry Pi 4 with 4 GB of RAM, the original Torch model takes 22 seconds per image, and the quantized version slightly reduces this to 20 seconds. Since Torch runs on multiple threads by default, the TensorFlow Lite model in multi-threaded mode is the fastest at 12 seconds, while its single-threaded version takes 25 seconds.

The Torch Mobile model processes an image in 16 seconds, making it faster than the original Torch models but still slower than TensorFlow Lite with multiple threads. On the Jetson Nano with 2 GB of RAM, the Torch and Torch Quantized models are extremely slow, requiring 420 seconds per image, making them unsuitable for real-time use in this scenario. The Torch Mobile model is much faster at 15 seconds, while the TensorFlow Lite model takes 31 seconds in single-threaded mode and 10 seconds in multi-threaded mode, showing a major speed improvement when using multiple threads.

The Raspberry Pi 4 with 2 GB of RAM struggles with the standard Torch models, taking 118 seconds for Torch and 127 seconds for the quantized version. However, Torch Mobile reduces the time to 19 seconds, while TensorFlow Lite multi-threaded completes the task in 14.75 seconds, making it the best-performing option on this system.

Regarding energy consumption, the Jetson Nano is the most power-hungry device. The standard Torch model consumes 55.81 mAh, while the same model on the Raspberry Pi 4 (4 GB) only uses 4.20 mAh. TensorFlow Lite multi-threaded mode is the most efficient across platforms, consuming 2.88 mAh on the Raspberry Pi 4 (4 GB) and 4.22 mAh on the Jetson Nano.

Since Torch runs on multiple threads by default, it is important to compare it fairly with TensorFlow Lite in multi-threaded mode. The results suggest that TensorFlow Lite multi-threaded mode is the best choice for balancing speed and energy efficiency, particularly on lower-power devices like the Raspberry Pi.

1.4.3 Knowledge Distillation

One approach for adapting the model for deployment on edge devices is the creation of an additional model through knowledge distillation, as discussed in Section 3. This technique involves training a student model using the knowledge extracted from a larger, pre-trained teacher model. The goal is to achieve a balance between computational efficiency and predictive accuracy, ensuring that the model remains suitable for edge deployment.

In this case, the original VegAnn model serves as the teacher, while two student models are trained with encoder sizes of three and four. The choice of these encoder sizes is motivated by the need to reduce model complexity while preserving sufficient feature extraction capacity. An encoder size of three provides a more compact model, optimizing efficiency for constrained hardware, while an encoder size of four maintains a higher level of expressiveness, helping to retain accuracy. Larger encoders would introduce additional computational overhead with diminishing compression benefits, whereas a significantly smaller encoder could lead to excessive loss of information, negatively impacting segmentation performance.

The qualitative differences between the student models and our best-performing model, the TensorFlow Lite model from Table 1.2, are visually illustrated in Figure 1.5, which shows the segmentation on a representative image. Additionally, a quantitative comparison of the models' performance is provided in Table 1.3.

(a) (b) (c)

Figure 1.5 Example of the prediction of the TensorFlow lite model and the model obtained by knowledge distillation: (a) segmentation results of TensorFlow lite model, (b) segmentation results of the 3-depth model obtained by knowledge distillation, and (c) segmentation results of the 4-depth model obtained by knowledge distillation.

Table 1.3 Results of models obtained by knowledge distillation on a Raspberry Pi 4 4GB.

Model	Average Time/Image (s)	Dice Score	Average Consumption (mAh)/Image
TensorFlow Lite multi-threaded	12	0.84	2.88
3-Depth	7.8	0.85	3.02
4-Depth	11	0.86	4.57

A comparison of the results obtained from the two student models reveals that their segmentation performance remains close to that of the original TensorFlow Lite VegAnn model, with Dice scores of 0.85 and 0.86, respectively. The original model achieves a Dice score of 0.84, indicating that knowledge distillation preserves the segmentation quality while offering potential efficiency gains. One notable observation is that while all models tend to segment similar regions, the student models consistently predict slightly larger areas, particularly in the upper portion of the images.

When evaluating computational efficiency, the model with a three-depth encoder demonstrates a significant reduction in inference time, cutting prediction time by approximately 50% compared to the TensorFlow Lite model. This reduction directly contributes to lower energy consumption, as the shorter processing time leads to reduced power usage. The three-depth model is the most energy-efficient, consuming 3.02 mAh per image, compared to 5.1 mAh for the original model. Similarly, the four-depth model achieves a balance between accuracy (0.86 Dice score) and energy consumption (4.57 mAh per image).

1.4.4 Kolmogorov–Arnold Network

The final approach for optimizing predictions on embedded devices involves training a model with an alternative architecture. In this case, we employ a Kolmogorov–Arnold Network (KAN)-based architecture, specifically a variant known as U-KAN (Li et al. 2024) for segmentation tasks. The U-KAN architecture is implemented using the Torch library and trained on the VegAnn dataset (Madec et al. 2023), the same dataset used for training the VegAnn model. Two versions of the model are trained: one using raw images without preprocessing and the other incorporating normalization by mean and standard deviation before the training process. In both cases, illumination normalization is applied during inference before making predictions on the test images. The segmentation results of these models are presented in Figure 1.6 and Table 1.4.

The Dice scores for the two KAN-based models, one trained with preprocessing and the other without, were 0.68 and 0.67, respectively. This shows that the pre-processing step, which involves normalizing images, does not significantly enhance performance in this case. Visually, both models show similar segmentation patterns, with the primary differences being the detection of a smaller area at the bottom of the image and a larger area at the top, which is composed of ground. This can indicate that the models may be slightly sensitive to variations in image content, especially in regions with sparse vegetation.

In terms of computational performance, both KAN models were deployed on a Raspberry Pi 4 with 4 GB of RAM and compared to the TensorFlow Lite VegAnn model. Unlike the VegAnn model, both KAN models were run using the Torch

(a) (b) (c)

Figure 1.6 Example of the prediction of the TensorFlow lite model and the KAN model: (a) prediction on image by the original model converted to TensorFlow lite, (b) prediction on image by the KAN model without preprocessing, and (c) prediction on image by the KAN model with preprocessing.

Table 1.4 Results of Kolmogorov–Arnold Network models on a Raspberry Pi 4 4G.

Model	Time/Image (s)	Dice Score	Consumption (in mAh)
VegAnn (Tflite)	12	0.84	2.88
KAN without preprocessing	17	0.68	2.87
KAN with preprocessing	16	0.67	2.72

library without converting them into a lightweight format. Both KAN models required approximately 16 and 17 seconds per image, which is comparable to the segmentation time of the TensorFlow Lite VegAnn model of 12 seconds.

For energy consumption, both KAN models consume 2.87 mAh (for the model without preprocessing) and 2.72 mAh (for the model with preprocessing) per image, which is lower than the 2.88 mAh consumed by the TensorFlow Lite VegAnn model. Additionally, the KAN models exhibit a lower peak RAM usage, stabilizing at around 800 MB after an initial spike to 1.3 GB at the start of the segmentation process.

1.5 Discussion

The findings of the experiments in the previous section emphasize the fine balance between model complexity, inference time, and energy efficiency when deploying deep learning–based segmentation models on embedded systems. While methods such as knowledge distillation and alternative architectures like U-KAN have improved efficiency, further exploration is necessary to ensure these models meet real-time demands.

Among the techniques evaluated, knowledge distillation emerged as the most effective approach for maintaining segmentation performance while reducing model size and inference time. By transferring knowledge from a larger teacher model to a compact student model, this method achieved a good balance between efficiency and performance. At inference, these distilled models significantly reduced computational load, making them particularly well suited for edge deployment.

Alternative architectures such as U-KAN offered additional improvements by introducing learnable activation functions tailored for segmentation tasks. While this approach demonstrated promising results in optimizing model expressiveness under resource constraints, it required fine-tuning to maintain generalization across diverse datasets. In contrast, traditional deep learning models without compression or architectural adaptations exhibited the highest accuracy but imposed excessive computational costs, making them impractical for embedded systems.

Despite these optimizations, the need for continuous monitoring in real-time applications introduces additional challenges. Tasks such as agricultural monitoring and plant health assessment require models capable of processing spatio-temporal data rather than still images. The shift from single-frame analysis to sequential data introduces the need for architectures that capture temporal dependencies while remaining computationally efficient. Traditional 2D convolutional networks perform well on static images but struggle with movement patterns over time. To address this, 3D Convolutional Neural Networks (C3D) (Köpüklü et al. 2021) have been developed, extending standard convolutions into the temporal dimension. While effective, their computational demands limit their applicability to embedded devices.

Reducing computational overhead without losing temporal modeling has led to approaches such as two-stream networks (Simonyan and Zisserman 2014). These models separate spatial and motion processing by analyzing RGB frames for appearance while using optical flow to track motion patterns. Despite reducing computational load compared to full 3D convolutions, computing optical flow remains resource-intensive. A more efficient alternative is Temporal Shift Modules (TSM) (Lin et al. 2019), which modify 2D CNNs to capture temporal dependencies by shifting feature channels along the time axis. This method significantly reduces computational requirements and enables real-time processing on embedded hardware. However, optimizing these models for plant imaging scenarios still requires further refinements.

Beyond video processing, the increasing demand for generative AI and large language models (LLMs) presents additional challenges for edge deployment. Vision models must be optimized for inference efficiency, but generative models, such as diffusion-based architectures, introduce even greater computational complexity. Unlike discriminative models, these networks rely on iterative refinement steps, requiring high computational power. Model compression techniques, including knowledge distillation and structured pruning (He and Xiao 2024), have been applied to generative models to reduce size while maintaining performance. Parameter-efficient fine-tuning (PEFT) (Mangrulkar et al. 2022) has also shown promise in adapting large models to specific tasks without excessive resource consumption.

Quantization plays a crucial role in making large-scale models feasible for edge deployment. Methods like BitNet (Wang et al. 2023) and LLM.int8() (Dettmers et al. 2022) reduce precision while preserving model accuracy. However, the effectiveness of these techniques depends on specialized hardware accelerators, which are not yet standard in embedded computing. Step distillation (Hsieh et al. 2023) introduces a structured approach by using LLMs to generate rationale-based labels, improving knowledge transfer to smaller models. Speculative decoding further accelerates inference by leveraging secondary models to predict intermediate outputs before validation, reducing redundant computations. While promising, these advancements require adaptation to edge hardware constraints before becoming widely applicable.

The interplay between hardware and software is central to optimizing edge AI deployment. Model efficiency depends not only on algorithmic improvements but also on the capabilities of the underlying hardware. AI processors designed for bitwise operations, such as those optimized for BitNet architectures, offer substantial efficiency gains. However, fragmentation in hardware design limits the widespread adoption of such solutions. Optimized software frameworks must complement these advancements, ensuring compatibility with specialized hardware. For example, deep learning models must transition from floating-point operations to binary operations without sacrificing accuracy, requiring efficient compilation strategies tailored to low-power devices.

Despite these advancements, several challenges remain in making deep learning models more practical for edge applications. Neural architecture search (NAS) (White et al. 2023) automates the design of optimized networks, and approaches like TinyTNAS (Saha et al. 2024) have demonstrated success in identifying architectures suited for resource-limited environments. However, NAS methods must be extended to multispectral imaging and real-time video processing to maximize their impact. Another pressing issue is the need for dynamic adaptation, where models can adjust their complexity based on available resources. Edge devices operate in environments where power availability fluctuates, requiring models that can scale performance dynamically while maintaining robustness.

Standardization is another significant hurdle. The diversity of hardware accelerators and software frameworks leads to inconsistencies in model deployment. Establishing common benchmarks and deployment protocols will be essential for broader adoption. Addressing these challenges will be critical for advancing edge AI, enabling efficient and scalable deep learning applications for plant biology imaging and beyond.

1.6 Conclusion

This chapter provided a panorama of deep learning optimization techniques for edge deployment. We emphasized the trade-offs between model complexity, inference speed, and resource efficiency on embedded devices. We compared the main methods of the state-of-the-art on a single real-world use case. A U-Net-like model was used for illustration to estimate plant surface areas from images captured by an open source and open hardware connected stick empowered with a simple Raspberry-pi single-board computer (https://ue1375.clermont.hub.inrae.fr/outils/piquets-connectes).

To enhance deployment feasibility on edge devices, the model was first converted to TensorFlow Lite, significantly reducing computational demands while maintaining accuracy. This conversion enabled inference on single-board computers with low energy consumption, demonstrating the practical benefits of lightweight model formats.

Beyond format conversion, knowledge distillation was applied to train two new models based on the original large U-Net network. The distilled models closely replicated the performance of the original while achieving faster inference times, without increasing RAM consumption. This technique proved effective in optimizing model efficiency for embedded environments where computational resources are constrained.

Additionally, two Kolmogorov–Arnold Network (KAN) models were trained and evaluated. These models achieved comparable predictive accuracy; however, they exhibited higher RAM usage. Their inference speed was similar to that of the TensorFlow Lite model, but they operated without requiring conversion to a lightweight framework. This suggests that while KANs offer an alternative optimization strategy, they introduce memory overhead that may limit their suitability for resource-constrained edge devices.

By comparing these approaches, this study highlights the trade-offs involved in deploying deep learning models on embedded platforms. The results provide concrete performance benchmarks, enabling informed decisions on model selection based on specific deployment constraints. Future extension of such a panorama and benchmark could explore further optimizations, including adaptive model architectures and hardware-aware quantization strategies, to enhance the efficiency of edge AI in plant biology applications.

References

AI acceleration hardware. https://www.axelera.ai/ai-acceleration-hardware-products. Accessed 15 November 2024.

Arduino Hardware. https://www.arduino.cc/en/hardware. Accessed 27 November 2024.

Arya, S., Sandhu, K.S., Singh, J., et al. (2022). Deep learning: As the new frontier in high-throughput plant phenotyping. *Euphytica* 218 (4): 47.

Chen, Y., Huang, Y., Zhang, Z., et al. (2023). Plant image recognition with deep learning: A review. *Computers and Electronics in Agriculture* 212: 108072.

Coral USB Accelerator. https://coral.ai/products/accelerator/. Accessed 15 November 11 2024.

Cordier, M., Rasti, P., Torres, C., et al. (2024). Affordable phenotyping at the edge for high-throughput detection of hypersensitive reaction involving cotyledon loss. *Plant Phenomics* 6: 0204.

Dettmers, T., Lewis, M., Belkada, Y., et al. (2022). Gpt3. int8 (): 8-bit matrix multiplication for transformers at scale. *Advances in Neural Information Processing Systems* 35: 30318–30332.

ESP32 Wi-Fi & Bluetooth SoC | Espressif Systems. https://www.espressif.com/en/products/socs/esp32. Accessed 10 December 2024.

ExecuTorch. https://pytorch.org/executorch-overview. Accessed 18 November 2024.

Fanariotis, A., Orphanoudakis, T., Kotrotsios, K., et al. (2023). Power efficient machine learning models deployment on edge IoT devices. *Sensors* 23 (3): 1595.

FANN on GitHub. https://github.com/libfann/fann. Accessed: 21 November 2024.

Garavagno, A.M., Ragusa, E., Frisoli, A., et al. (2023). A hardware-aware neural architecture search algorithm targeting low-end microcontrollers. In *2023 18th Conference on Ph.D. Research in Microelectronics and Electronics (PRIME)*, 281–284. IEEE.

Garbouge, H., Rasti, P., and Rousseau, D. (2021). Enhancing the tracking of seedling growth using RGB-depth fusion and deep learning. *Sensors* 21 (24): 8425.

Garbouge, H., Sapoukhina, N., Rasti, P., et al. (2022). Deep learning based detection of seedling development from controlled environment to field. In *XXXI International Horticultural Congress (IHC2022): III International Symposium on Mechanization, Precision Horticulture, and 1360*, 237–244.

Grindstaff, B., Mabry, M.E., Blischak, P.D., et al. (2019). Affordable remote monitoring of plant growth in facilities using raspberry Pi computers. *Applications in Plant Sciences* 7 (8): e11280.

He, Y., and Xiao, L. (2024). Structured pruning for deep convolutional neural networks: A survey. *IEEE Transactions on Pattern Analysis and Machine Intelligence* 46 (5): 2900–2919.

Hinton, G., Vinyals, O., and Dean, J. (2015). Distilling the knowledge in a neural network.

Howard, A.G., Zhu, M., Chen, B., et al. (2017). MobileNets: Efficient convolutional neural networks for mobile vision applications. *arXiv preprint arXiv:1704.04861.*

Hsieh, C.Y., Li, C.L., Yeh, C.K., et al. (2023). Distilling step-by-step! Outperforming larger language models with less training data and smaller model sizes. *arXiv preprint arXiv:2305.02301.*

Intel® Neural Compute Stick 2 (Intel NCS2). https://www.intel.com/content/www/us/en/developer/articles/tool/neural-compute-stick.html. Accessed 15 November 2024.

Köpüklü, O., Kose, N., Gunduz, A., et al. (2021). Resource efficient 3D convolutional neural networks. *In Proceedings of the IEEE/CVF International Conference on Computer Vision Workshops,* 1910–1919.

Li, C., Liu, X., Li, W., et al. (2024). U-Kan makes strong backbone for medical image segmentation and generation. *preprint arXiv:2406.02918.*

Li, Z., Guo, R., Li, M., et al. (2020). A review of computer vision technologies for plant phenotyping. *Computers and Electronics in Agriculture* 176: 105672.

Liang, T., Glossner, J., Wang, L., et al. (2021). Pruning and quantization for deep neural network acceleration: A survey. *Neurocomputing* 461: 370–403.

Lin, J., Gan, C., and Han, S. (2019). TSM: Temporal shift module for efficient video understanding. In *Proceedings of the IEEE/CVF International Conference on Computer Vision*, 7083–7093.

LiteRT for Microcontrollers. https://ai.google.dev/edge/litert/microcontrollers/overview. Accessed 20 November 2024.

LiteRT overview. https://ai.google.dev/edge/litert. Accessed 18 November 2024.

Madec, S., Irfan, K., Velumani, K., et al. (2023). VegAnn: Vegetation annotation of multi-crop RGB images acquired under diverse conditions for segmentation.

Mangrulkar, S., Gugger, S., Debut, L., et al. (2022). PEFT: State-of-the-art parameter-efficient fine-tuning methods. https://github.com/huggingface/peft.

Mohammad-Razdari, A., Rousseau, D., Bakhshipour, A., et al. (2022). Recent advances in e-monitoring of plant diseases. *Biosensors and Bioelectronics* 201: 113953.

NCS2 Replacement | Tiny Accelerator Card based on Intel Movidius Myriad X VPU. https://www.eyecloudai.com/product-page/tiny-ml-accelerator-card-on-intel-movidius-vpu-usb-ncs. Accessed 15 November 2024.

NVIDIA. (2024). Jetson AI platform. https://developer.nvidia.com/embedded-computing.

Qualcomm. (2023). AI on mobile: How neural processing units enhance performance. https://www.qualcomm.com/news/onq/2023/ai-on-mobile-how-npu-enhances-performance.

Ronneberger, O., Fischer, P., and Brox, T. (2015). U-Net: Convolutional networks for biomedical image segmentation. In *Medical Image Computing and Computer-Assisted Intervention–MICCAI 2015: 18th International Conference*, October 5–9, 2015, proceedings, part III 18, 234–241. Springer.

Saha, B., Samanta, R., Ghosh, S.K., et al. (2024). TinyINAS: GPU-free, time-bound, hardware-aware neural architecture search for TinyML time series classification. *arXiv preprint arXiv:2408.16535.*

SAKURA™ Modules and Cards. https://www.edgecortix.com/en/products/sakura-modules-and-cards. Accessed 15 November 2024.

Schizas, N., Karras, A., Karras, C., et al. (2022). TinyML for ultra-low power AI and large scale IoT deployments: A systematic review. *Future Internet* 14 (12): 363.

Shadrin, D., Menshchikov, A., Somov, A., et al. (2019). Enabling precision agriculture through embedded sensing with artificial intelligence. *IEEE Transactions on Instrumentation and Measurement* 69 (7):4103–4113.

Sharma, A., Georgi, M., Tregubenko, M., et al. (2022). Enabling smart agriculture by implementing artificial intelligence and embedded sensing. *Computers & Industrial Engineering* 165: 107936.

Simonyan, K., and Zisserman, A. (2014). Two-stream convolutional networks for action recognition in videos. *Advances in Neural Information Processing Systems* 27: 568–576.

Tan, M., and Le, Q.V. (2019). EfficientNet: Rethinking model scaling for convolutional neural networks. In *International Conference on Machine Learning*, 6105–6114. PMLR.

Tovar, J.C., Hoyer, J.S., Lin, A., et al. (2018). Raspberry pi powered imaging for plant phenotyping. *Applications in Plant Sciences* 6 (3): e1031.

Valle, B., Simonneau, T., Boulord, R., et al. (2017). PYM: A new, affordable, image-based method using a raspberry Pi to phenotype plant leaf area in a wide diversity of environments. *Plant Methods* 13: 1–17.

Wang, H., Ma, S., Dong, L., et al. (2023). Bitnet: Scaling 1-bit transformers for large language models. *arXiv preprint arXiv:2310.11453.*

Wang, X., Magno, M., Cavigelli, L., et al. (2020). FANN-on-MCU: An open-source toolkit for energy-efficient neural network inference at the edge of the internet of things. *IEEE Internet of Things Journal* 7 (5): 4403–4417.

Warden, P., and Situnayake, D. (2019). *TinyML: Machine Learning with TensorFlow Lite on Arduino and Ultra-Low-Power Microcontrollers.* O'Reilly Media.

White, C., Safari, M., Sukthanker, R., et al. (2023). Neural architecture search: Insights from 1000 papers. *arXiv preprint arXiv:2301.08727.*

Wong, A., Famouri, M., and Shafiee, M.J. (2020). Attendnets: Tiny deep image recognition neural networks for the edge via visual attention condensers. *CoRR*, abs/2009.14385.

Zhang, X., Zhou, X., Lin, M., et al. (2017). Shufflenet: An extremely efficient convolutional neural network for mobile devices. In *Proceedings of the IEEE Conference on Computer Vision and Pattern Recognition*, 6848–6856.

2

Machine Learning for Studying Plant Evolutionary Developmental Biology

Mani Manoj[1], Ramaraj Sivamano[1], Arunachalam Abitha[1], Mohammed Jaffer Shakeera Banu[2], Shanmugam Velayuthaprabhu[3], Kannan Vijayarani[4], Jeyabal Philomenathan Antony Prabhu[1], Arumugam Vijaya Anand[1]**

[1] *Department of Human Genetics and Molecular Biology, Bharathiar University, Coimbatore, Tamil Nadu, India*
[2] *Division of Molecular Biology, Microbiology Laboratory, Coimbatore, Tamil Nadu, India*
[3] *Department of Biotechnology, Bharathiar University, Coimbatore, Tamil Nadu, India*
[4] *Department of Education, Bharathiar University, Coimbatore, Tamil Nadu, India*

*Corresponding authors: avahgmb@buc.edu.in; jtonybio@gmail.com

2.1 Introduction to Plant Evolutionary Developmental Biology

2.1.1 Overview of Plant Evolutionary Developmental Biology

Plant evolutionary developmental biology (Evo-Devo) is a discipline that integrates evolutionary and developmental procedures to understand plant morphology and diversity. This technique has been instrumental in elucidating the roles of key transcription elements, consisting of PLETHORA (PLT) in root improvement, which adjust cellular proliferation and differentiation throughout various plant species, highlighting their evolutionary conservation and practical redundancy in angiosperms (Rodríguez-Herrera et al. 2024). The Evo-Devo attitude also challenges conventional phylogenetic interpretations, specifically within the context of Ordovician land vegetation, by focusing on developmental pathways and gene regulatory networks instead of fossil species taxa, thus offering new insights into plant evolution (Strother 2023). In the evolution and development of plants, concepts such as continuum and method morphology provide a deeper understanding of the diversity of plant forms. These perspectives go beyond traditional morphological classifications, recognizing the fluid and interconnected nature of plant organs (Sattler and Rutishauser 2022). An essential aspect of floral development is to understand the origin and evolution of carpels. Regulatory genes such as AGGLOMOTOS and CRABS CLAW play crucial roles in defining the identity and function of the carpel, suggesting that the evolution of these structures occurred gradually through changes in the regulatory genes (Liu et al. 2023). In addition, research in Ecological Evolutionary Developmental Biology (Eco-Evo-Devo) revealed significant diversity in stomatic development in plant species. This variation is shaped by ecological and evolutionary factors. Alos recent advances have made it possible to study stomatic development beyond traditional model plants, leading to a broader understanding of its ecological meaning (Doll et al. 2022).

2.1.2 Key Concepts in Plant Evolution and Development

Plant Evo-Devo is a burgeoning field that integrates concepts from development, genetics, and morphology to understand the evolution of plant forms and functions. Key concepts in plant Evo-Devo include the evolution of developmental processes and the developmental basis of evolutionary changes, which are central to understanding how plants have adapted and diversified over time (Love 2024). Morphological studies, such as continuum and process morphology, challenge classical views by suggesting that plant organs like roots, stems, and leaves exist on a continuum, with homology being a matter of degree rather than strict categorization (Sattler and Rutishauser 2022). This approach is complemented by genetic and molecular analyses that explore the complexity of plant structures, such as leaf complexity and flower organ evolution,

Machine Learning for Plant Biology, First Edition. Edited by Jen-Tsung Chen.

highlighting the role of genetic pathways and environmental interactions in shaping plant phenotypes (Minelli 2018). The field is expanding to include comparative studies and interdisciplinary approaches, such as epigenetics and maternal effects, which provide insights into ecological and evolutionary dynamics (Kramer et al. 2017). The evolution of flowers, a pivotal innovation in plant history, is particularly significant, with genomic studies revealing the genetic and morphological changes that led to the first flowers and the role of gene duplications in floral diversification (Chanderbali et al. 2016).

2.1.3 Importance of Evo-Devo in Understanding Plant Adaptations

Evo-Devo significantly enhances our understanding of plant adaptations to changing environmental conditions by elucidating the genetic and developmental mechanisms underlying these adaptations. Plants have evolved a variety of sophisticated strategies to cope with environmental stresses, such as drought, temperature extremes, and nutrient limitations, which are increasingly relevant in the context of climate change. For instance, the evolution of vascular systems and seeds was crucial for plant colonization of terrestrial environments, allowing them to adapt to diverse ecological niches (Cohen et al. 2023). Modern plants continue to exhibit remarkable adaptive responses, such as xerophytic traits in arid environments and mutualistic relationships with fungi and insects, which are essential for survival in extreme conditions (Neves and Costa 2020). The study of plant responses to combined and sequential abiotic stresses, such as those caused by climate change, highlights the importance of understanding these adaptive mechanisms to improve plant resilience and performance (Alptekin and Kunkowska 2025). Additionally, the evolutionary history of plants, such as the genus Cochlearia, demonstrates how past climatic events have shaped current adaptations, providing insights into how plants might respond to future environmental changes (Wolf et al. 2021). Furthermore, evolutionary approaches in crop adaptation, including genomic selection and genome editing, offer promising strategies for enhancing the resilience of crops to climate change, ensuring food security in the face of global warming (Gao et al. 2023).

2.1.4 Role of Computational and AI Tools in Evo-Devo Studies

Computational and artificial intelligence (AI) tools play a pivotal role in advancing plant Evo-Devo studies by enhancing the understanding of complex developmental processes and improving phenotyping efficiency. The integration of computational approaches, such as the study of PLETHORA transcription factors (TFs), has elucidated their roles in root development across various plant species, highlighting the conservation of gene regulatory networks and their context-specific regulation (Rodríguez-Herrera et al. 2024). AI and machine learning (AI/ML) have revolutionized plant phenotyping by enabling the automatic analysis of large image datasets, such as those used for stomatal patterning, achieving human-level performance in trait quantification with superhuman speed (Tan et al. 2024). AI's potential extends to plant breeding, where it aids in data collection, genetic diversity analysis, and precision breeding, thereby optimizing crop traits and enhancing productivity (Farooq et al. 2024). Furthermore, AI's integration with multi-omics data offers a high-throughput approach to plant science research, facilitating a comprehensive understanding of plant biology and agriculture (Aliyu 2024). The application of AI in plant phenomics has improved data collection and analysis through noninvasive imaging techniques, fostering community-driven research and data-sharing, which are crucial for accurate phenotype studies (Nabwire et al. 2021).

2.2 Basics of ML in Biological Research

2.2.1 Fundamentals of ML in Biology

Plant biology is undergoing a revolution due to ML, which offers strong tools for analyzing large, complicated datasets and revealing information on plant phenomics, disease resistance, and genomics. In the field of genomics, ML has proved crucial in examining transposable elements (TEs) in 352 plant species, demonstrating their important role in functional innovation and genome evolution. This includes gene duplications, noncoding RNA production, and gene domestication all of which are essential for crop improvement and plant evolution (Liu et al. 2024). By predicting gene expression, chromatin interactions, and epigenetic properties, deep learning, a subset of ML, improves genomic research and makes it easier to build and synthesize functional genomic components (Wang et al. 2024). When paired with high-throughput photography, ML in plant phenomics automates the study of plant attributes, increasing the precision and efficiency of phenotypic data collecting. This advancement accelerates breeding processes and provides insights into plant growth, development, and resistance mechanisms (Wu et al. 2024). Additionally, ML models have been developed to predict plant disease development by analyzing transcriptional responses, offering a novel approach to understanding plant immunity and resistance across different pathogens (Sia et al. 2025). In phenotyping, ML-enabled computer vision is used to analyze stomatal traits from images, achieving human-level performance in trait quantification, although challenges remain in applying these models across diverse phenotypic variations (Tan et al. 2024).

2.2.2 Supervised, Unsupervised, and Reinforcement Learning

ML methods, such as supervised, unsupervised, and reinforcement learning, are essential for deciphering complicated biological data and identifying evolutionary patterns in plant Evo-Devo investigations. By analyzing gene expression profiles using techniques like False Positive–Random Forest (FP-RF), Random Forest–Recursive Feature Elimination (RF-RFE), and Support Vector Machine-Recursive Feature Elimination (SVM-RFE), which produce high classification performance and biological relevance, supervised learning is used to choose pertinent genes for plant development, as shown in the study of *Arabidopsis thaliana* (Saura Sánchez 2019). Unsupervised learning, on the other hand, is critical in situations when labeled data is limited, allowing for clustering and dimensionality reduction to detect trends in plant systems biology and phenotyping studies (Yan and Wang 2022). This approach is particularly useful in analyzing TEs across plant genomes, where ML frameworks have uncovered lineage-specific functional gene families influenced by these elements, highlighting their role in genome evolution and functional innovation (Liu et al. 2024). Although it isn't specifically addressed in the papers that are provided, reinforcement learning may be used to model evolutionary scenarios. For example, in the stochastic modeling of flower organ development, regulatory protein interactions are tested for resilience to poor cell fate decisions (Casci 2009). Plant genome dynamics and developmental processes are better understood when these ML paradigms are included into Evo-Devo investigations, providing important information for synthetic biology and crop enhancement. All things considered, these approaches offer an extensive arsenal for investigating the complex genetic linkages supporting developmental robustness as well as the evolvability of plant phenotypes (Minelli 2018).

2.2.3 Deep Learning and Neural Networks in Evo-Devo

Plant Evo-Devo research has undergone a revolution due to the use of deep learning and neural networks, which allow scientists to evaluate intricate biological data, find hidden patterns, and forecast evolutionary processes. To position ancient specimens inside phylogenies, researchers have created neural networks that can categorize pollen specimens and forecast phylogenetic distances. Because of this method, evolutionary links that were previously obscured by the low resolution of ancient pollen records may now be identified. For example, research on the *Podocarpus* species showed that morphological differences can be quantified and ancient specimens may be placed in a phylogeny using Bayesian inference employing deep-learned characteristics from pollen morphology (Adaïmé et al. 2024). Evo-Devo methods have also been employed by researchers to examine how developmental variation contributes to speciation. For instance, research on the variety of leaf shapes in tomatoes and similar species found that the BLADE-ON-PETIOLE (BOP) TF controls a gene regulatory network module. By modifying the expression of important homeobox genes, this module was discovered to regulate leaf shape (Ichihashi et al. 2014). Researchers also recreated the history of the MIR164 family in angiosperms and examined the evolution of microRNAs (miRNAs) in plants using a two-step miRNA identification approach. This study shed light on the possible contribution of *MIR164* genes to the development of carpel closure and leaf morphology (Jasinski et al. 2010).

2.2.4 ML Workflow: Data Collection, Processing, Model Selection, and Interpretation

Data collection, processing, model selection, and interpretation are some of the crucial steps in integrating ML workflows in plant studies, all of which increase agricultural research and practice. Imaging sensors and open-source repositories like PlantVillage are frequently used in plant research for data collecting. These tools make it easier to obtain the sizable datasets required for deep learning model training, especially for disease and stress detection (Walsh et al. 2024). Data cleaning, segmentation, and feature extraction are all part of the preprocessing step, which is essential for improving ML models' accuracy. Because of their capacity to process complicated picture data, techniques like Convolutional Neural Networks (CNNs) are widely used; designs like ResNet and MobileNet are especially well-liked because of their efficiency and capacity for deep feature extraction (Rodríguez-Lira et al. 2024). Plant scientists without a lot of programming experience may now use ML more easily due to tools like AutoML, which automate model selection, hyperparameter optimization, and feature engineering (Shoaib et al. 2023). Interpretation of ML models, especially in the context of plant breeding and stress detection, involves understanding genotype-by-environment interactions and leveraging big data to improve predictive accuracy and breeding strategies (Crossa et al. 2025). Despite these developments, there are still issues with standardizing data gathering procedures and guaranteeing the interpretability of intricate models, both of which are necessary for real-world uses like environmental management and precision agriculture. Overall, the integration of ML in plant studies is revolutionizing the field by enhancing the efficiency and accuracy of plant classification, disease detection, and stress response prediction, thereby supporting sustainable agricultural practices (Karnan and Ragupathy 2023).

2.2.5 Challenges of Applying ML to Evo-Devo Research

Although ML and Evo-Devo have advanced, a number of methodological and technological constraints still exist. For instance, current techniques might not be tailored for plant species, and the reconstruction of Gene Regulatory Networks (GRNs) from single-cell data is still in its infancy (Ferrari et al. 2022). The availability and caliber of data needed for ML algorithms is one of the main problems with ML. Non-model species, which might not have a wealth of genomic or transcriptome data, are frequently used in plant evolution-devo investigations (Vialette-Guiraud et al. 2016). Large-scale plant Evo-Devo datasets might provide a challenge since ML approaches frequently need substantial computer resources. For instance, processing large datasets for genome-wide, organ-delimited GRNs (OD-GRNs) requires a lot of computing power (Ranjan et al. 2023). Although TF activities and regulatory interactions may be predicted by ML models, it is still difficult to understand these models in a biological setting. For example, thorough validation is necessary to assure biological relevance when employing ML models to predict transcriptional responses to environmental perturbations. The integration of ML with Evo-Devo investigations is made more difficult by the absence of experimental confirmation for predicted TF regulators (Smet et al. 2023). ML models often simplify complex biological systems to make predictions. However, biological precision may be lost as a result of this simplification. For instance, Boolean models of GRNs may not fully reflect the intricacy of gene interactions, but they are helpful for qualitative descriptions (Chairez-Veloz et al. 2023).

2.3 ML Applications in Plant Morphological Evolution

2.3.1 ML for Analyzing Fossilized Plant Structures

By providing fresh approaches to comprehending evolutionary connections and paleoenvironmental circumstances, ML has greatly improved the interpretation of fossilized plant structures. The evolutionary history encoded in pollen morphology has been interpreted using deep learning, namely neural networks. As shown with *Podocarpus* species and fossil *Podocarpidites* specimens, ML algorithms can estimate phylogenetic distances and locate fossil specimens across a phylogeny using Bayesian inference by examining the shape, internal structure, and texture of pollen grains (Adaïmé et al. 2024). In order to infer paleoclimatic conditions, ML approaches have also been used to analyze the morphology of leaf epidermal cells, such as those seen in ginkgo leaves. When epidermal cells were segmented using the U-Net architecture, correlations between climatic factors and cell shape were shown to be similar between ancient and present ginkgoales (Zhang et al. 2023). Furthermore, ML-based pipelines like ML-morph facilitate high-throughput morphometric data collection, significantly increasing the speed and accuracy of landmarking biological structures, which is crucial for large-scale phenomic studies (Porto and Voje 2020). Despite these advancements, challenges remain, such as the need for diverse, high-quality image datasets and the complexity of fossil morphologies, which require further development and integration of advanced neural network architectures (Ansari et al. 2024).

2.3.2 Shape and Trait Evolution Using CNNs and Autoencoders

CNNs and autoencoders have demonstrated encouraging progress in automating and improving the precision of plant phenotyping and categorization in the study of plant form and trait development. By utilizing picture datasets and advanced preprocessing techniques, CNNs have been successfully used for leaf morphology classification, attaining high accuracy rates that expedite the identification process in agriculture and botany (Reddy and Iyer 2023). By extracting strong descriptors, CNNs' combination of shape and texture information improves plant species categorization even further. They perform exceptionally well across a range of picture datasets (Atiampo et al. 2022). CNNs have played a key role in high-throughput applications related to plant phenotyping, allowing for the quick and accurate extraction of phenotypic traits which is a critical component of breeding programs and the comprehension of plant–environment interactions (Jiang and Li 2020). Meanwhile, autoencoders, particularly generative recursive autoencoders, have been instrumental in encoding and synthesizing 3D shapes, capturing hierarchical structures that reflect intra-shape relationships. This approach supports applications such as shape classification and synthesis, offering a comprehensive generative pipeline for 3D shapes (Li et al. 2017). Additionally, deep learning techniques have been applied to study pollen morphology in Podocarpus, linking morphological traits to environmental adaptations and evolutionary processes, thereby enhancing our understanding of plant adaptation over time (Adaimé et al. 2025).

2.3.3 3D Reconstruction of Plant Organs Through ML-based Image Processing

The subject of 3D reconstruction of plant organs using ML-based image processing is developing quickly, utilizing a variety of methods to improve speed and accuracy. In this field, deep learning algorithms are essential, especially those made for point cloud segmentation. For example, although it comes at a significant computational cost, the use of transformer designs, such the Stratified Transformer (ST), has demonstrated greater performance in segmenting plant organs from 3D point clouds (Xie et al. 2024). High semantic segmentation accuracy and thorough morphological parameter extraction are made possible by the MVS-Pheno platform, which is an example of the use of point cloud data-driven techniques for reconstructing the 3D morphological structure of wheat (Li et al. 2024). Additionally, the Neural Radiance Field (NeRF) approach, improved with instant NGP and multi-resolution hash coding, offers a fast and high-quality 3D modeling solution by utilizing multi-view images and pose information (Wang and Shen 2024). UAV-based methods further contribute by efficiently generating 3D point clouds from aerial images, significantly reducing processing time while maintaining quality (Huang et al. 2024). Moreover, PointNet-like architectures have been employed for segmenting plant organelles, demonstrating the potential of deep neural networks in classifying and reconstructing detailed plant structures from 3D data (Dahdal and Hament 2024).

2.3.4 Quantitative Trait Analysis Using SVM

The accuracy of quantitative trait analysis in plants is improved by combining SVM with other ML techniques. This is done by utilizing the advantages of several models to tackle intricate data problems. Known for its ability to handle high-dimensional data with resilience, SVM may be further enhanced by combining it with additional algorithms, such as mixed kernel functions, which have been demonstrated to perform noticeably better in genomic selection tasks than conventional parametric models like Genomic Best Linear Unbiased Prediction (GBLUP) and BayesB. In wheat datasets, for example, mixed kernel SVM models, such as Support Vector Machine Regression–Genomic Selection (SVR-GS), have shown up to a 13.3% improvement in prediction accuracy over GBLUP, indicating their potential for enhancing genomic predictions (Wang et al. 2024). Furthermore, it has been demonstrated that multi-trait ML models, which incorporate spectral information, improve the prediction of complex traits such as wheat protein content and grain yield, outperforming traditional models and highlighting the value of combining various data sources and algorithms (Sandhu et al. 2021). SVM and other ML techniques are used in plant omics to analyze vast, diverse datasets and extract useful information for plant development under a range of environmental stressors. Even though SVM performs better in some situations than others, as shown by its higher accuracy in classifying plant growth when compared to Naive Bayes and ELM, more development and integration with other models are required to fully realize its potential in agricultural applications (Isewon et al. 2022).

2.3.5 Integrating Phylogenetics and ML for Morphological Adaptation Studies

Integrating phylogenetics and ML for studying morphological adaptation involves leveraging advanced computational techniques to enhance the understanding of evolutionary processes. Deep learning, a subset of ML, has been successfully used to derive morphological traits from images, which, when combined with molecular data, can improve phylogenetic analyses. This approach, however, faces challenges such as the strength of the phylogenetic signal and resource demands, suggesting a need for improved trait extraction methods and disentangled networks for better interpretation of derived traits (Hunt et al. 2025). Phylogenetic codon models, which assess adaptation through the ratio of non-synonymous to synonymous substitution rates, can be integrated with population-genetic data to provide a comprehensive view of adaptation across evolutionary scales. This integration reconciles phylogenetic and population-genetic approaches, offering mutual confirmation and paving the way for more robust models (Latrille et al. 2023). Morphological studies, such as those on snake feeding systems, demonstrate the importance of phylogenetic frameworks in testing adaptive hypotheses, revealing significant relationships between morphological traits and ecological functions (Vincent et al. 2006). Furthermore, the use of phylogenetic comparative methods in morphometrics allows for modeling evolutionary scenarios and analyzing trait evolution patterns, although challenges remain in handling the high dimensionality of morphometric data (Monteiro 2013). An integrative approach that combines biomechanical analyses with phylogenetic data can elucidate the evolutionary pathways of morphological traits, as demonstrated in studies of carnivorous mammals' feeding biomechanics (Tseng and Flynn 2015).

2.4 Genomic and Transcriptomic Insights Through ML

2.4.1 Evolutionary Genomics: Identifying Selection Signatures with ML

The integration of ML in evolutionary genomics has significantly advanced the identification of selection signatures in plants, offering a robust alternative to traditional statistical models. ML approaches, such as those described by Kumar et al., treat the detection of selection signatures as a classification problem, enhancing predictive accuracy and resolution in population genomics data analysis (Kumar et al. 2022). Deep learning methods, exemplified by the DeepGenomeScan framework, have been particularly effective in identifying loci under complex spatial selection patterns, outperforming traditional genome scan methods like principal component analysis. This approach has been successfully applied to human genetic datasets, identifying loci that contribute to spatial adaptation, which could be translated to plant genomics for similar insights (Qin et al. 2022). In plant breeding, ML and pangenomics have accelerated genomic selection, crucial for improving crop yields under climate change pressures. These methods incorporate nonadditive effects and leverage the entire genome content of a species, enhancing the precision of breeding programs (Bayer et al. 2021). In soybean, for instance, genome-wide selection signatures have been identified across different breeding populations, revealing common and unique haplotypes that inform targeted breeding strategies (Viana et al. 2024).

2.4.2 GRN Prediction Using Graph Neural Networks

GRN prediction using Graph Neural Networks (GNNs) in plants is a burgeoning area of research that leverages advanced computational models to unravel complex gene interactions. The Cross-Attention Complex Dual Graph Embedding Model (XATGRN) addresses the skewed degree distribution in GRNs by employing a cross-attention mechanism and dual complex graph embedding, which enhances the prediction of regulatory relationships and their directionality, outperforming existing methods across various datasets (Xiong et al. 2024). Similarly, the use of Graph Attention Network v2 (GATv2) in GRN analysis highlights the potential of GNNs to overcome traditional limitations by accurately predicting regulatory interactions and identifying key regulators, thus offering deeper biological insights and paving the way for advancements in personalized medicine and drug discovery (Otal et al. 2025). The GRNNLink framework, designed for single-cell RNA-seq data, utilizes a graph recurrent neural network to refine gene features and predict GRNs with high robustness and accuracy, demonstrating its effectiveness across multiple datasets (Bai and Wang 2024). Additionally, a Graph Convolutional Network (GCN) approach guided by causal feature reconstruction enhances GRN inference by mitigating information loss and improving computational efficiency, as evidenced by superior performance on benchmark datasets (Ji et al. 2024).

2.4.3 ML for Comparative Genomics in Evolutionary Studies

ML applications in comparative genomics have significantly advanced evolutionary studies in plant species by providing novel insights and methodologies. One prominent application is the analysis of TEs across 352 plant species, which has illuminated their role in genome evolution. This study utilized a ML framework to identify lineage-specific functional gene families influenced by TEs, highlighting their contribution to key innovations in plant evolution, such as the expansion of TF gene families and potential horizontal gene transfer (HGT; Liu et al. 2024). Another application involves the use of pre-trained DNA language models, such as PlantCaduceus, which have demonstrated high transferability across species, enabling genome-wide deleterious mutation identification and enhancing cross-species prediction without the need for multiple sequence alignment (Zhai et al. 2024). Additionally, ML has been employed to study the evolutionary mechanisms of plant cyclic nucleotide-gated ion channels (CNGCs), focusing on gene family expansion and synteny through comparative genomics (Baloch et al. 2022). In the context of photosynthesis, a ML framework has been developed to distinguish between C3 and C4 photosynthetic pathways in the PACMAD clade, achieving over 99% accuracy in classification and identifying key sequences involved in C4 photosynthesis evolution (Yogadasan et al. 2023). Furthermore, the ISEEML algorithm has been introduced to compare gene expression profiles across species, offering a new metric for evaluating functional similarity and overcoming challenges posed by complex evolutionary histories and genome duplications in plants (Kasianov et al. 2023).

2.4.4 Understanding Non-coding RNA Evolution with NLP-based ML Models

Natural language processing (NLP) and ML models are being used more and more to understand the evolution of non-coding RNAs (ncRNAs) in plants, especially long non-coding RNAs (lncRNAs). Instead of depending just on transcriptome data, which frequently underestimates the quantity of lncRNAs, these algorithms have been modified to detect lncRNAs using genomic sequences. While the quality of genome assembly can influence prediction accuracy, DNA-Bidirectional Encoder Representations from Transformers (DNABERT)-based models, have shown the capacity to predict lncRNAs across a variety of plant species, reaching an accuracy of up to 83.4% for Zea mays (Danilevicz et al. 2022). Furthermore, by including explainable AI into these models, genomic patterns crucial for lncRNA prediction have been discovered, indicating species-wide conserved structures. The Plant-LncPipe pipeline further enhances lncRNA identification by retraining models like CPAT and LncFinder specifically for plants, significantly improving their performance compared to traditional tools developed for animal datasets (Tian et al. 2024). Moreover, the PINC tool, utilizing an automated ML framework, has shown high accuracy in identifying ncRNAs in plants, outperforming existing tools in several evaluation metrics (Zhang et al. 2022). These advancements highlight the potential of NLP-based ML models in uncovering the evolutionary roles of ncRNAs in plant gene regulation, development, and stress responses, providing a more comprehensive understanding of plant genomics (Wrightsman et al. 2024).

2.4.5 Unraveling Epigenetic Modifications in Plant Evolution Using ML

With its ability to provide new insights into plant adaptation and domestication, ML is being used more and more to decipher the complexity of epigenetic changes in plant evolution. The prediction of chromatin remodeling in apple buds using deep learning models, such those created by Saito et al., has shown the significance of epigenetic modifications in cold perception and dormancy, which are essential for plant adaptation to climatic fluctuations (Saito et al. 2024). Similar to this, Korfmann et al. showed how useful methylomes are for detecting evolutionary changes in plant populations. They emphasized how quickly epimutations occur in comparison to genetic mutations and their potential for demographic inference, which genetic variants alone might not be able to provide (Korfmann et al. 2024). Wang et al. (2021) further advanced this field by developing the Smart Model for Epigenetics in Plants (SMEP), a deep-learning tool that predicts various epigenomic modifications with high accuracy, thus facilitating functional research and genome engineering across diverse plant species. These advancements underscore the potential of ML in enhancing our understanding of epigenetic mechanisms, as discussed by Sun et al. (2023), who emphasize the role of ML in deciphering the epigenetic underpinnings of plant adaptation and domestication. Additionally, the study of histone methyltransferases, such as H3K4 and H3K27, reveals the evolutionary dynamics of epigenetic mechanisms, with gene duplications and functional diversification playing key roles in the regulation of new genes and pathways in plants (Casas-Mollano et al. 2023). Collectively, these studies illustrate the transformative impact of ML on epigenetic research, offering new avenues for exploring plant evolution and adaptation.

2.5 Inferring Evolutionary Developmental Pathways Using ML

2.5.1 ML Models for Predicting Gene Expression Patterns

By utilizing extensive transcriptomic and genomic datasets, ML algorithms have demonstrated considerable potential in forecasting gene expression patterns in plants. Using highly variable genes (HVGs) from extensive RNA sequencing data in maize and rice, Wu et al. showed that ML models could accurately predict plant phenotypes, such as tissue types and stress conditions. However, cross-species generalization is still difficult because of expression divergence (Wu et al. 2024). By examining genotype-by-environment interactions and using multi-trait genomic data, Crossa et al. (2025) demonstrated the revolutionary role of ML in plant breeding, where it improves prediction accuracy. By revealing the cis-regulatory code for gene expression, deep learning techniques have made significant progress in this field. Peleke et al. (2024) developed deep learning models that predict gene expression from non-coding regulatory sequences with over 80% accuracy, identifying significant features like untranslated regions of mRNA (UTR) regions that influence expression levels across species such as Arabidopsis and tomato. Tross et al. employed a modified transformer model to predict differential expression across plant organs, identifying regulatory regions in noncoding sequences that correlate with expression patterns, thus offering a scalable approach for genome annotation and targeted genetic editing (Tross et al. 2024).

2.5.2 Identifying Key Developmental Genes via Feature Selection Algorithms

Identifying key developmental genes in plants through feature selection algorithms involves leveraging ML techniques to analyze gene expression data and uncover significant genetic markers. In the study of *Arabidopsis thaliana*, ML methods, such as FP-RF, RF-RFE, and SVM-RFE, have been employed to select relevant genes from a database of over 500 expression profiles, covering various plant tissues like seeds, seedlings, roots, leaves, and flowers. These selected genes demonstrated high classification performance in SVM, RF, and ANN algorithms, indicating their biological relevance to plant development processes (Saura Sánchez 2019). Additionally, the construction of a Feature Importance Network (FIN) using a dataset of 11,801 gene features for 31,522 Arabidopsis genes has revealed novel functional relationships between biological features, providing insights into gene function and facilitating the discovery of new gene interactions (Ng et al. 2022). Furthermore, the identification of circadian genes, which play a crucial role in regulating plant physiological and developmental processes, has been enhanced by employing a support vector machine with a Laplace kernel, achieving notable accuracy in predicting circadian proteins (Meher et al. 2021). Lastly, a dynamic feature choosing algorithm has been proposed to determine major features in gene identification, achieving high accuracy in distinguishing coding from noncoding sequences, as demonstrated in the *S. cerevisiae* genome (Luo et al. 2007). These studies collectively highlight the potential of ML and feature selection algorithms in advancing our understanding of plant developmental genetics and improving gene identification methodologies.

2.5.3 Evolution of Transcription Factor Networks with ML

The evolution of TF networks through ML is a multifaceted process that involves the integration of various computational and experimental approaches to understand and predict GRNs. Recent advancements have leveraged ML to enhance the prediction and analysis of TF binding sites and regulatory networks. For instance, the development of an evolutionary algorithm–based ordinary differential equation (ODE) model has improved the inference of GRN structures by incorporating dynamic transcriptional data, outperforming traditional methods that rely on static data (Li et al. 2023). Additionally, the MATL model employs deep learning techniques, such as multi-scale convolution and transformers, to predict TF binding sites by integrating sequence and shape features, demonstrating superior performance over existing methods (Fan et al. 2024). Furthermore, the MRTLE approach uses phylogenetic information to infer regulatory networks across species, highlighting the role of gene duplication in network evolution (Koch et al. 2017). Ensemble ML methods have also been applied to enhance TF-DNA binding predictions by combining multiple motif exploration algorithms, improving the identification of TF gene targets and binding sites (Fan et al. 2018). Finally, the evolutionary potential of TFs for regulatory rewiring is explored through the analysis of robustness, tunability, and evolvability, providing a mechanistic framework for understanding network evolution (Igler et al. 2018).

2.5.4 Bayesian ML for Inferring Ancestral Gene Interactions

Bayesian ML offers a robust framework for inferring ancestral gene interactions in plants by integrating genetic data with environmental and biological knowledge. The variational Bayes approach, as discussed in the context of genomic prediction models, provides a computationally efficient method for handling genotype–environment (G × E) interactions, which are crucial for understanding plant traits across different environments (Montesinos-López et al. 2017). This method is particularly advantageous in large-scale datasets where traditional Bayesian models become computationally prohibitive

(Yan et al. 2023). Furthermore, Bayesian whole-genome regression models, such as LEMMA, facilitate the joint modeling of genetic effects and G × E interactions, offering insights into the phenotypic variance attributable to these interactions (Kerin and Marchini 2020). The integration of biological knowledge, such as TF binding and protein–protein interactions, into Bayesian frameworks enhances the reconstruction of gene networks, allowing for a more accurate inference of gene regulatory mechanisms (Moon 2013). Additionally, nonparametric Bayesian models, which utilize Dirichlet processes, provide a flexible approach to learning ancestral genetic processes by capturing the shared structural information across populations, thereby improving the inference of population structure and local ancestry estimation (Sohn 2011).

2.5.5 Evolution of Polyploidy and Hybridization Analyzed Through ML Model

The evolution of polyploidy and hybridization in plants, analyzed through ML models, reveals complex interactions and evolutionary dynamics. Deep learning models, such as convolutional neural networks, have been employed to predict subgenome dominance in ancient polyploidized genomes by analyzing DNA sequences and methylation sites, although these models struggle with neo/synthetic polyploidizations (Guo et al. 2024). Polyploidy, a significant evolutionary mechanism, has been linked to speciation and genetic diversity, with events often coinciding with major extinction events, suggesting adaptive potential (Conover 2021). Hybridization, often accompanied by allopolyploidy, can perturb cytonuclear interactions, yet plants exhibit robustness to these perturbations, which may explain the prevalence of allopolyploidy in plant evolution (Sloan et al. 2024). Introgression further complicates polyploid genomes by introducing new alleles from various lineages, as seen in Betula pubescens, where hybridization with multiple species has reshaped its genomic composition, contributing to climate adaptation and meiotic stability (Leal et al. 2024). Phylogenomic approaches, such as Paralog PhyloGenomics, have been developed to unravel the evolutionary history of complex polyploid and hybrid taxa, allowing for the assessment of polyploidization modes and hybridization events, even involving ghost lineages (Walden et al. 2024).

2.6 Phylogenetics and Evolutionary Tree Reconstruction Using ML

2.6.1 Phylogenetic Tree Prediction via Deep Learning

Deep learning has emerged as a promising tool for phylogenetic tree prediction, offering advantages in computational efficiency and accuracy over traditional methods like Maximum Likelihood. PhyloDeep, a deep learning inference tool, has shown potential in enhancing phylodynamic analysis by integrating contact tracing data, which refines poorly resolved phylogenies and improves predictions of epidemiological parameters, such as superspreading potential in outbreaks like SARS-CoV-2 (Xie et al. 2024). Additionally, deep neural networks have demonstrated superior performance in estimating tree branch lengths, particularly in challenging regions of the parameter space, outperforming Maximum Likelihood methods in terms of efficiency and accuracy, especially for long branches associated with distantly related taxa (Suvorov and Schrider 2024). A semi-supervised learning approach has also been proposed, combining deep learning estimation of substitution rates with Bayesian inference, which enhances the accuracy of phylogenetic inference, particularly in complex evolutionary scenarios (Silvestro et al. 2024). Furthermore, DeepDynaForecast utilizes phylogenetic-informed graph deep learning to predict epidemic transmission dynamics, proving effective in identifying high-risk groups and forecasting transmission patterns using large-scale data from epidemics like HIV (Sun et al. 2024). Neural networks have been shown to rival Maximum Likelihood methods in reconstructing phylogenetic trees and selecting evolutionary models, particularly for quartet trees, offering faster and often equally accurate predictions (Figure 2.1; Kulikov et al. 2024).

2.6.2 Bayesian ML for Inferring Evolutionary Relationships

Bayesian ML algorithms significantly enhance the accuracy of inferring evolutionary relationships in plant species by addressing the complexities inherent in phylogenetic analysis. The VBPI-Mixtures algorithm, for instance, improves upon traditional Variational Bayesian Phylogenetic Inference (VBPI) by effectively capturing the multimodal distributions over tree topologies, which are often challenging for sampling-based methods like Black-Box Variational Inference (BBVI) to model accurately. This is achieved through the integration of advanced mixture learning techniques, resulting in superior performance on complex phylogenetic datasets (Kviman et al. 2023). Additionally, the use of Bayesian inference with Markov Chain Monte Carlo (MCMC) methods, as demonstrated in the phylogenetic analysis of the *Dipterocarpaceae*

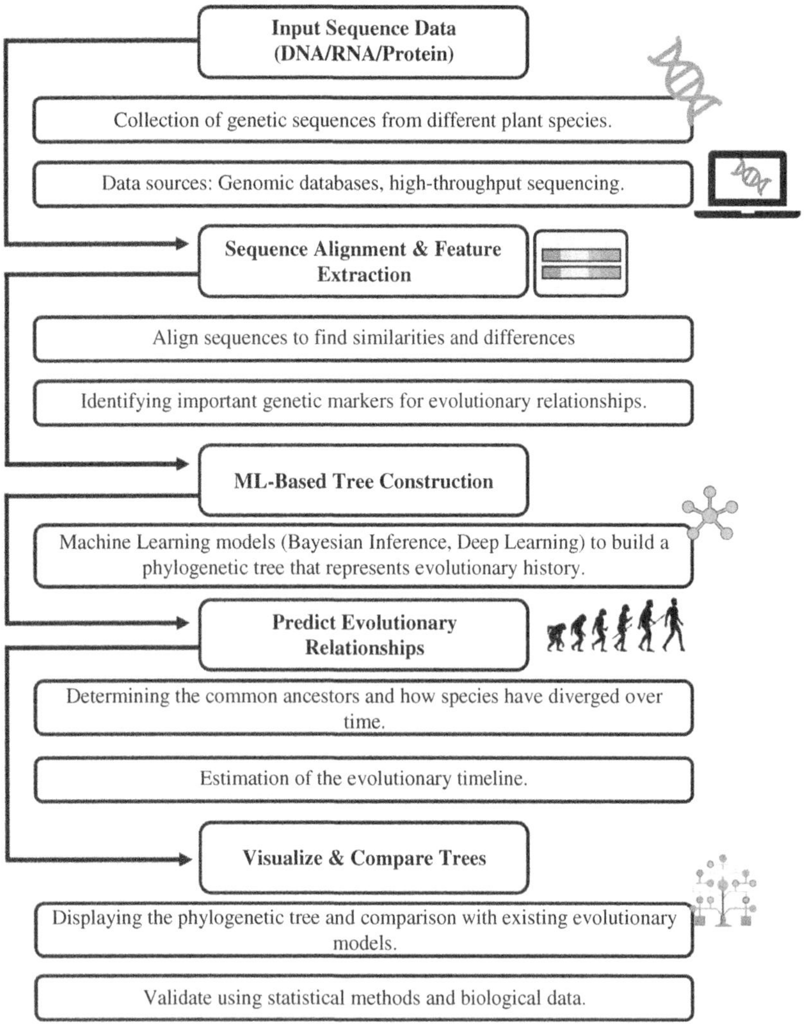

Figure 2.1 ML-driven phylogenetic tree construction.

family, allows for more accurate classification based on genetic data rather than morphological characteristics, which can be misleading due to phenotypic similarities. This approach revealed discrepancies between morphological and genetic classifications, highlighting the precision of Bayesian methods in resolving such inconsistencies (Yunita et al. 2023). Furthermore, advances in variational inference, particularly through the use of flexible prior models and neural network-based parameterization, enhance the estimation of branch lengths and evolutionary parameters. This flexibility in modeling prior densities reduces bias in posterior approximations, offering a more accurate reflection of the current data distribution and improving the overall phylogenetic inference process (Remita et al. 2023).

2.6.3 Predicting Adaptive Radiation Events with ML Models

Predicting adaptive radiation events in plants using ML models involves leveraging genomic prediction techniques that have been increasingly refined through the integration of big data and advanced computational methods. Adaptive radiation, characterized by the rapid evolution of ecological and phenotypic diversity within a lineage, often follows the colonization of new environments or the emergence of key innovations that open new ecological niches (Gavrilets and Vose 2005). ML models, particularly ensemble methods like RF and Extreme Gradient Boosting, have shown promise in genomic prediction by effectively handling complex, nonlinear phenotypes and large datasets, which are crucial for understanding the genetic architecture underlying adaptive radiation. These models can automatically identify relevant features from extensive genomic, phenotypic, and environmental data, enhancing prediction accuracy and providing insights into genotype-by-environment interactions, which are pivotal in adaptive radiation scenarios (Crossa et al. 2025). However, the effectiveness of ML models can be influenced by factors such as sample size, population structure, and genetic architecture, necessitating careful consideration of these elements to optimize predictive performance. Despite challenges like susceptibility to confounding due to population structure, ML methods are less sensitive to low linkage disequilibrium compared to linear parametric methods, making them suitable for predicting complex adaptive radiation events in plants (Farooq et al. 2022). Overall, the integration of ML in genomic prediction offers a robust framework for anticipating adaptive radiation events by analyzing diverse and extensive datasets, thereby advancing our understanding of evolutionary dynamics in plant lineages.

2.6.4 Network-based Approaches for Studying Horizontal Gene Transfer

Network-based approaches for studying HGT in plants have evolved significantly, leveraging both character-based methods and genomic analyses. Traditional sequence-based methods often struggle to detect ancient HGT events due to the erasure of evidence over time through mutations. To address this, character-based methods, such as perfect transfer networks, have been proposed. These networks utilize phylogenetic models to explain character diversity among taxa, assuming unique character births and rare losses, which can indicate HGT even when sequence similarity is low (López Sánchez and Lafond 2024). In plants, HGT is increasingly recognized as a common phenomenon, facilitated by next-generation sequencing technologies that enable comparative genomic approaches to identify such transfers across reproductive barriers (Aubin et al. 2021). Recent studies have employed genome-wide analyses in non-model wild species, revealing HGT events, particularly involving low copy number LTR-retrotransposons from the Copia superfamily, suggesting new routes for genetic exchange in natural ecosystems (Aubin et al. 2023). Additionally, plant–bacteria interactions have been identified as significant contributors to HGT in plants, with bacterial symbionts acting as potential donors. This interaction has been shown to facilitate the transfer of genes involved in metabolism and genetic information processing, highlighting the role of bacterial–plant relationships in plant evolution (Cai et al. 2021). These network-based and genomic approaches provide a comprehensive framework for understanding the mechanisms and implications of HGT in plant genomes, offering insights into evolutionary processes and genetic diversity.

2.6.5 Automating Phylogenomic Inference Using AI-based Pipelines

Automating phylogenomic inference in plants using AI-based pipelines is a rapidly advancing field, leveraging diverse computational tools to handle the complexity and volume of genomic data. The Auto-phylo v2 and auto-phylo-pipeliner provide a comprehensive framework for phylogenetic analysis, incorporating modules for gene annotation, sequence alignment, and the identification of positively selected amino acid sites, which are crucial for understanding evolutionary processes such as S-RNase specificity in *Prunus* species (López-Fernández et al. 2024). CAPTUS, another innovative pipeline, efficiently integrates mixed high-throughput sequencing data types, such as RNA-Seq and genome skimming, to produce phylogenomic datasets. This pipeline has been successfully applied to the Cucurbitales order, revealing complex reticulate evolution patterns and resolving previous phylogenetic conflicts (Ortiz et al. 2023). The SyntenyLink algorithm automates the reconstruction of subgenomes in polyploid plants by analyzing synteny blocks, offering significant improvements in accuracy and efficiency over traditional methods. This is particularly useful for studying the evolutionary history of polyploid species like Brassica (Hewavithana et al. 2023). Additionally, Seqping enhances gene prediction accuracy in plant genomes by using self-training Hidden Markov Models (HMMs) and transcriptomic data, outperforming existing HMM-based tools in species like *Oryza sativa* and *Arabidopsis thaliana* (Chan et al. 2017).

2.7 ML for Studying Developmental Plasticity and Environmental Adaptation

2.7.1 Predicting Phenotypic Plasticity with ML Algorithms

ML algorithms have shown significant potential in predicting phenotypic plasticity in plants by leveraging large-scale datasets and advanced computational techniques. The integration of transcriptomic data, as demonstrated in maize and rice, allows for the accurate prediction of plant phenotypes by focusing on HVGs, which are crucial for identifying tissue types, developmental stages, and stress conditions (Wu et al. 2024). This approach highlights the ability of ML models to generalize across species, although challenges remain due to expression divergence between species (Wu et al. 2024). Additionally, ML techniques have been employed to identify environmental variables influencing phenotypic plasticity, such as temperature and solar radiation, which are critical for breeding crops suited to future climates. The use of genetic algorithms in this context has successfully pinpointed known environmental determinants, underscoring the utility of ML in understanding genotype-by-environment interactions (Kusmec et al. 2024). Furthermore, ML's capacity to handle complex, multi-dimensional datasets, including genomic and environmental data, enhances prediction accuracy and offers insights into G × E interactions, thereby optimizing breeding strategies (Crossa et al. 2025). In the realm of plant phenotyping, ML applications extend to crop management, plant health assessment, and disease detection, with artificial neural networks and stacked models proving particularly effective (Centorame et al. 2024). Despite the challenges of data quality and model explainability, ML's ability to autonomously extract and analyze data features positions it as a superior tool for genotype to phenotype prediction, surpassing traditional statistical methods (Danilevicz et al. 2022).

2.7.2 Climate-responsive Developmental Evolution Using ML

ML is increasingly being leveraged to address the challenges posed by climate change on plant development and agriculture, offering innovative solutions for climate-responsive developmental evolution in plants. The integration of ML with plant genetic resources (PGRs) is crucial for developing climate-resilient crops, as ML algorithms can analyze complex, multi-dimensional datasets to identify genetic traits that enhance adaptability to changing climates (Yoosefzadeh-Najafabadi et al. 2024). This approach has accelerated the discovery of high-yielding crop varieties with traits that mitigate and adapt to climate change, such as heat tolerance, which were previously difficult to identify using traditional methods (Bari et al. 2021). In the context of specific crops like Chinese cabbage, ML models have been successfully employed to predict abiotic stress-responsive genes, demonstrating high accuracy in identifying genes that confer resistance to stresses like cold, heat, drought, and salt (You et al. 2025). Furthermore, ML models have been used to project the effects of climate change on global vegetation growth, revealing potential increases in vegetation under certain climate scenarios, but also highlighting the risk of decline due to rapid temperature rises (Nguyen et al. 2023). These insights are critical as they inform strategies for biodiversity preservation and environmental management. However, the variability in plant developmental responses to climate factors such as CO_2 levels, temperature, and drought stress underscores the need for further research into the molecular mechanisms underlying these responses, particularly the interactive effects of multiple climate change factors (Gray and Brady 2016).

2.7.3 Adaptive Traits Discovery Using ML in Dynamic Environments

The discovery of adaptive traits in plants using ML in dynamic environments involves integrating advanced phenotyping techniques and adaptive ML models to predict and understand plant responses to varying conditions. The AMULET system exemplifies this by employing high-throughput imaging and ML to predict plant traits before they become visible, thus enhancing phenotyping efficiency and aiding in agricultural management (Zdrazil et al. 2024). In dynamic environments, traditional ML models often struggle due to nonstationary data distributions. To address this, adaptive ML frameworks incorporating anomaly detection, transfer learning, and reinforcement learning have been developed, offering improved accuracy and adaptability for real-time decision-making (Raja and Kung 2025). Specifically, adaptive transfer learning has shown promise in plant phenotyping by leveraging knowledge from related tasks to improve prediction accuracy, especially when data is limited. This approach is particularly useful in environments where hyperspectral reflectance data varies, allowing for efficient model retraining across different plants or conditions (Wu et al. 2022). In rapeseed, ML models like RFs have identified dynamic i-traits that contribute to ecotype divergence, highlighting the role of genetic differentiation in environmental adaptation. These traits, with high heritability, serve as biomarkers for predicting ecotype and understanding plant architecture changes during growth stages, thus facilitating yield improvement and adaptation strategies (Feng et al. 2022).

2.7.4 ML-based Prediction of Plant Evolution Under Climate Change

ML has emerged as a powerful tool for predicting plant evolution under climate change, offering insights into species distribution, growth patterns, and biodiversity. In the study of endangered plant species in Egypt, ML algorithms were used to predict the future distribution of *Silene leucophylla* and *Silene schimperiana*, revealing potential range contractions and shifts due to climate change, with a complete loss projected by 2080 under high emission scenarios (Refaat et al. 2024). Similarly, the integration of genomic selection with ML in corn breeding has shown promise in enhancing yield resilience by predicting genotype-by-environment interactions, crucial for developing climate-resilient crops (Saimon et al. 2023). The INTRAGRO approach, which combines dendrometer data with climate models, has successfully predicted intra-annual tree growth dynamics, indicating enhanced growth despite a shortened growing season due to altered precipitation patterns (Aryal et al. 2023). ML models have also been applied to a comprehensive plant database in the United States, which have predicted declines in native phylogenetic diversity, highlighting the complex interplay between taxonomic and phylogenetic diversity under changing environmental conditions (Park et al. 2020). Lastly, ML methods have demonstrated high accuracy in predicting vegetation growth through the growing season using meteorological forecasts, underscoring their potential in improving crop yield predictions (Peng et al. 2022).

2.8 High-throughput Image-based ML Approaches in Evo-Devo

2.8.1 ML for Automated Plant Organ Recognition and Classification

The application of ML, particularly deep learning, in automated plant organ recognition and classification has shown significant advancements, as evidenced by recent research. CNNs have been pivotal in automating feature extraction for plant leaf classification, enhancing accuracy and reducing manual intervention, which is crucial for reliable species identification (Ganesh et al. 2025). However, the reliance on large datasets and computational resources remains a challenge. To address this, a novel approach combining Wavelet Scattering Networks (WSNs) and MobileNetV2 has been proposed, which effectively captures both texture and high-level features, achieving high accuracy even with smaller datasets (Gowthaman and Das 2025). The integration of multimodal deep learning techniques has also been explored to capture the full biological diversity of plant species by fusing data from multiple plant organs, such as flowers, leaves, fruits, and stems. This approach has demonstrated superior performance, achieving an accuracy of 83.48% on a large dataset, surpassing traditional methods (Lapkovskis et al. 2024). In the realm of plant pathology, CNNs have also been employed to classify plant diseases, with models like VGG16 achieving high accuracy in identifying various plant illnesses (Kumar et al. 2024). Additionally, for flower classification, a deep learning framework utilizing a Sequential Fast Forward Feed Perceptron (SFFFP) neural network, has been developed, which automates the measurement of floral dimensions, thus overcoming the limitations of manual methods and achieving a precision accuracy of 96.67% (Kundu et al. 2024).

2.8.2 Deep Learning for Leaf, Flower, and Root Morphological Evolution

Deep learning is increasingly being applied to the study of morphological evolution, including the analysis of leaves, flowers, and roots, by leveraging large datasets and advanced AI techniques. The application of deep learning in phylogenetic placement, as demonstrated by Adaïmé et al. (2024), shows that neural networks can effectively interpret the evolutionary history embedded in pollen morphology, which can be generalized to other plant structures such as leaves and flowers. This approach uses neural networks to analyze the shape, structure, and texture of biological specimens, transforming these features into phylogenetic distances that can be used for evolutionary inference. Similarly, the development of platforms like PlantAI aims to enhance the analysis of herbarium images by employing deep learning models to identify and classify plant morphological traits, including leaves and flowers, thus facilitating biodiversity studies (Sklab et al. 2024). The integration of AI in evolutionary morphology is further supported by the work of He et al. (2024), who highlight the transformative potential of AI in this field, particularly through the use of deep learning for image capture, segmentation, and feature recognition, which are crucial for understanding morphological evolution. These advancements suggest that deep learning not only aids in the classification and analysis of plant morphology but also opens new avenues for exploring evolutionary patterns and relationships, thereby aligning morphological studies with genomic and bioinformatics research.

2.8.3 CNNs for Large-scale Evolutionary Trait Analysis

CNNs have emerged as powerful tools for large-scale evolutionary trait analysis in plants, offering significant advancements in phenotyping efficiency and accuracy. In the context of ornamental plant breeding, CNNs have been utilized to classify novel varieties, such as the radiation mutant BG Regale, achieving high accuracy and precision with optimization techniques like RMSProp, which demonstrated a model accuracy of 98.44% (Delos Reyes et al. 2022). Similarly, in

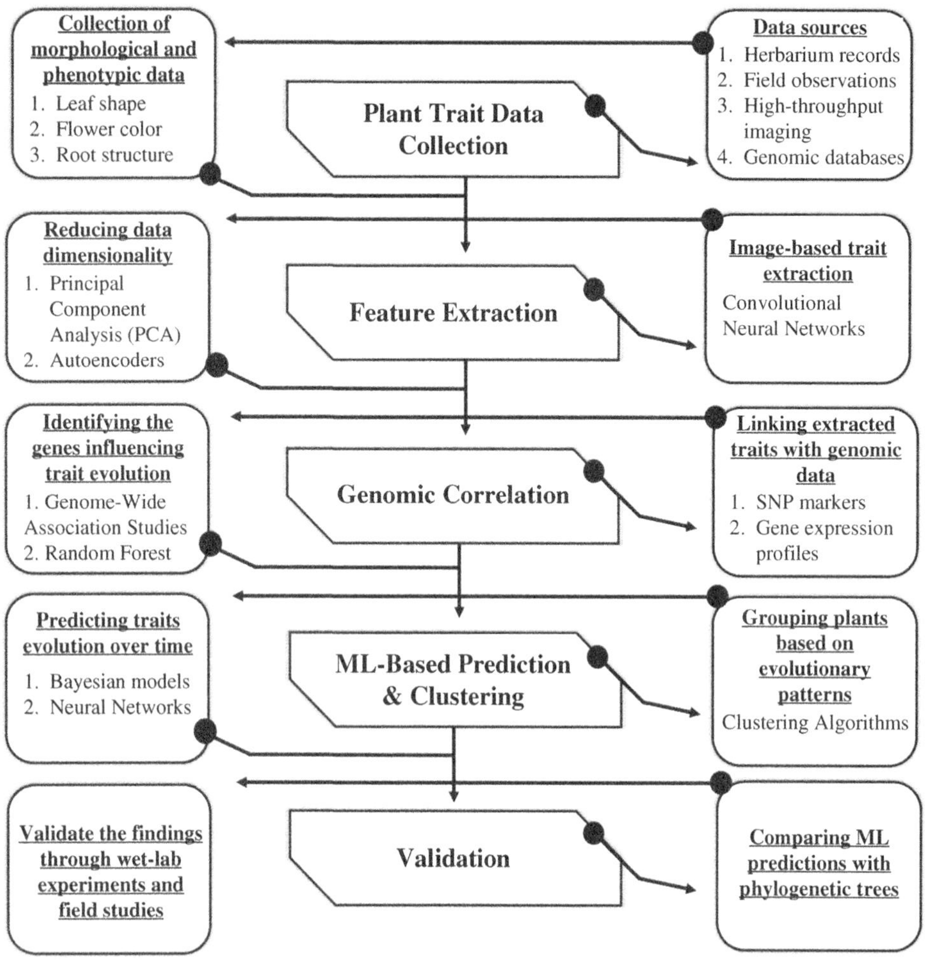

Figure 2.2 ML-based evolutionary trait analysis.

the bioenergy crop Miscanthus, CNNs have been employed to phenotype key traits such as flowering time, culm length, and biomass yield using high-spatial- and temporal-resolution aerial imagery. The integration of 3D-spatiotemporal (ST) CNN architectures with unmanned aerial vehicle (UAV)-based remote sensing has shown substantial improvements in trait assessment, with up to 23% enhancements in R^2 and 20% reductions in mean absolute error compared to 2D-spatial architectures (Varela et al. 2022). Furthermore, the use of CNNs in conjunction with citizen science data has facilitated the prediction of plant functional traits across diverse taxa and biomes, underscoring the potential of CNNs to automate and enhance the assessment of plant functional diversity on a global scale (Schiller et al. 2021). Figure 2.2 outlines how ML is used to analyze evolutionary traits in plants.

2.8.4 Time-series ML for Tracking Developmental Transitions

Time-series ML approaches have shown significant potential in tracking developmental transitions in plants by leveraging various data sources and methodologies. The use of spatiotemporal long short-term memory (ST-LSTM) and memory in memory network (MIM) models has been effective in predicting plant growth and development through time-series images, achieving high structural similarity index measure (SSIM) values and demonstrating the feasibility of predicting future plant phenotypes (Wang et al. 2022). Similarly, the application of deep learning networks like U-Net, combined with ML algorithms such as XGBoost, has enabled the prediction of developmental stages in Arabidopsis, highlighting the importance of selecting essential time intervals for accurate yield prediction (Chang et al. 2021). Furthermore, single-plant-omics has been utilized to resolve transcriptional changes during rapid developmental transitions, such as the vegetative-to-reproductive transition in Arabidopsis, by exploiting developmental asynchrony to order plants by their developmental trajectory (Redmond et al. 2024). In addition, contrastive learning methods, such as the Priori Distance Embedding (PDE), have been employed for pre-training plant time-series images, facilitating tasks like semantic segmentation and anomaly detection in cherry growth modeling (Xu et al. 2024). Lastly, remote sensing–based time-series analysis using satellite observations has been applied to retrieve plant functional traits, offering a nondestructive means to monitor plant conditions across different phenological phases and sites (Švik et al. 2023).

2.8.5 Integrating ML with Phenotyping Platforms for Evo-Devo Research

Integrating ML with phenotyping platforms significantly enhances plant Evo-Devo research by automating and refining the analysis of complex plant traits. High-throughput phenotyping (HTP) technologies, combined with ML, allow for the efficient measurement and analysis of plant phenotypes, facilitating insights into genetic, environmental, and developmental interactions (Wu et al. 2024; Kaya 2025). As previously mentioned, ML models, particularly when applied to large-scale transcriptomic datasets, can predict plant phenotypes with high accuracy by focusing on HVGs, as demonstrated in studies involving maize and rice (Wu et al. 2024). This approach not only aids in understanding plant development but also in precision crop management, enhancing yield and stress resistance. The integration of multi-omics data with AI further enriches phenotyping by providing a comprehensive framework to handle the complexity of plant biology, thus overcoming the limitations of traditional methods (Cembrowska-Lech et al. 2023). In horticulture, this integration supports precision breeding and sustainable crop management, offering solutions to challenges like disease and stress response management. Additionally, ML applications in plant phenotyping, such as in sunflower studies, highlight its role in crop management, plant health assessment, and early disease detection, using algorithms like artificial neural networks to analyze diverse data sources (Centorame et al. 2024).

2.9 Single-cell and Multi-omics ML Integration in Plant Evo-Devo

2.9.1 ML for Single-cell RNA Sequencing Data in Evolutionary Studies

ML has become an invaluable tool in analyzing single-cell RNA sequencing (scRNA-seq) data for evolutionary studies in plants, offering insights into cell identity, differentiation, and evolutionary conservation. In *Arabidopsis thaliana*, the AtML system has been developed to accurately identify root cell stages, achieving high accuracy and recall rates and identifying 160 important marker genes, which enhances understanding of cellular heterogeneity in root development (Yu et al. 2024). Similarly, the SPmarker pipeline uses interpretable ML to identify novel cell-type marker genes, facilitating cross-species mapping by discovering hundreds of new markers, including those with orthologs in rice and other species, thus expanding the known repertoire of marker genes significantly (Yan et al. 2022). The Orthologous Marker Gene Groups (OMGs) method further extends this by enabling the identification and comparison of cell types across diverse plant species without requiring cross-species data integration, revealing 14 conserved cell-type groups across monocots and dicots (Chau et al. 2025). These advancements in ML applications not only improve cell-type annotation but also contribute to understanding evolutionary patterns, such as the developmental hourglass model, which highlights the conservation of mid-embryonic stages across species. Single-cell phylotranscriptomics, another emerging approach, provides insights into the evolutionary history of cell types and the selective forces acting on tissues during adaptation, thereby elucidating the molecular and cellular mechanisms underlying plant evolution (Ma and Zheng 2024).

2.9.2 Integrating Proteomics, Transcriptomics, and Metabolomics with ML

Integrating proteomics, transcriptomics, and metabolomics with ML in plant research offers a comprehensive approach to understanding complex biological systems and enhancing agricultural practices. The integration of multi-omics data allows for a holistic view of plant systems, revealing interactions between gene expression, protein levels, and metabolites, which are crucial for stress response and crop improvement. Strategies for integrating these omics data include co-expression analysis, metabolite–gene networks, and pathway enrichment analysis, which, when combined with ML techniques, can uncover complex patterns and interactions that single-omics analyses might miss (Sanches et al. 2024). ML models, such as AutoGluon-Tabular, have been successfully used to predict genes involved in the biosynthesis of plant specialized metabolites, demonstrating the importance of genomics and proteomics data in enhancing model performance across different plant species (Bai et al. 2024). The application of ML in conjunction with omics data has also been pivotal in exploring plant bioactive pathways, offering new opportunities for the discovery of plant natural products (Durand et al. 2024). Furthermore, the integration of multi-omics and artificial intelligence (AI) in horticultural research is transforming plant phenotyping, enabling precision horticulture and predictive breeding by effectively managing the complexity and volume of data (Cembrowska-Lech et al. 2023).

2.9.3 Deep Learning for Cell Fate and Differentiation Analysis in Evo-Devo

Deep learning has emerged as a powerful tool for analyzing cell fate and differentiation in Evo-Devo in plants, offering novel insights into complex biological processes. FateNet, for instance, integrates dynamical systems theory with deep learning to predict cell fate decisions using scRNA-seq data, providing qualitative insights into biological state transitions and identifying key genes and pathways in differentiation processes such as hematopoiesis (Sadria and Bury 2024). Similarly, the approach of combining deep learning with mathematical modeling has been applied to recognize and reconstruct spatial cell fate patterns in organoids, using synthetic data to train models like graph neural networks and multilayer perceptrons. This method achieves a 70% accuracy in predicting cell fate based on neighboring cells, linking spatial patterns with underlying mechanisms (Dirk et al. 2023). Furthermore, Deep Lineage leverages deep learning to analyze time-series scRNA-seq data with lineage tracing, accurately predicting early cell fate biases and gene expression profiles and identifying key genes and pathways in cellular reprogramming and hematopoiesis (Sadria et al. 2024).

2.9.4 Predicting Evo-Devo Pathways Through Multi-omics Data Fusion

Predicting Evo-Devo pathways in plants through multi-omics data fusion involves integrating diverse datasets to understand the genetic and molecular mechanisms underlying plant development and evolution. Multi-omics integration (MOI) is crucial for this task, as it allows for the comprehensive assimilation of transcriptomics, proteomics, and metabolomics data, which are essential for modeling complex biological systems in plants (Jamil et al. 2020). The integration of these datasets can be enhanced by ML techniques, which have been shown to predict genes involved in specialized metabolite biosynthesis by focusing on key genomic and proteomic features, as demonstrated in *Arabidopsis* and validated in other species like maize and tomato (Bai et al. 2024). This approach is particularly valuable in Evo-Devo studies, where understanding the genetic basis of morphological diversity, such as leaf shape, requires constructing GRNs and identifying key TFs that regulate developmental pathways (Ichihashi et al. 2014). The use of comparative genomics and natural variation among species can further elucidate how developmental pathways are modified during evolution, providing insights into adaptive traits (Irish and Benfey 2004). Recent advancements in omics technologies and ML have opened new avenues for exploring plant bioactive pathways, highlighting the potential of these tools in unraveling the complexities of plant evolution and development (Durand et al. 2024). Figure 2.3 illustrates the integration of multiple biological datasets using ML to study Plant Evo-Devo. It also highlights how different omics data can be used to uncover evolutionary patterns and plant adaptation mechanisms.

2.9.5 ML for Analyzing Spatial and Temporal Omics Data

ML plays a pivotal role in analyzing spatial and temporal omics data in plants, offering innovative solutions to complex challenges in plant research. The integration of ML with high-throughput omics technologies enables the effective handling of large, heterogeneous datasets, crucial for improving agronomic traits and plant resilience against environmental stresses (Isewon et al. 2022). In the realm of plant hormonomics, advancements in mass spectrometry and data analysis methods, including multivariate approaches, have facilitated the ST quantification of phytohormones, enhancing our understanding of plant hormone biosynthesis and signaling at high resolutions (Petřík et al. 2024). Furthermore, ML models like the ST constrained light gradient boosting machine (ST-LGBM), have been developed to reconstruct solar-induced chlorophyll fluorescence (SIF) data, addressing the challenges of uneven spatial and temporal data distribution, thereby improving the accuracy of photosynthesis estimation (Shen et al. 2022). In agricultural robotics, ML techniques such as Hausdorff distance–based skeletonization are employed for tracking plant organ growth, enabling automated monitoring and analysis of plant development through ST point cloud data (Pandey et al. 2024). Additionally, a one-stage ST P-spline-based hierarchical approach has been proposed for analyzing HTP data, which models genetic effects over time while accounting for spatial and temporal noise, thus aiding in genomic selection and plant breeding efforts (Pérez-Valencia et al. 2024).

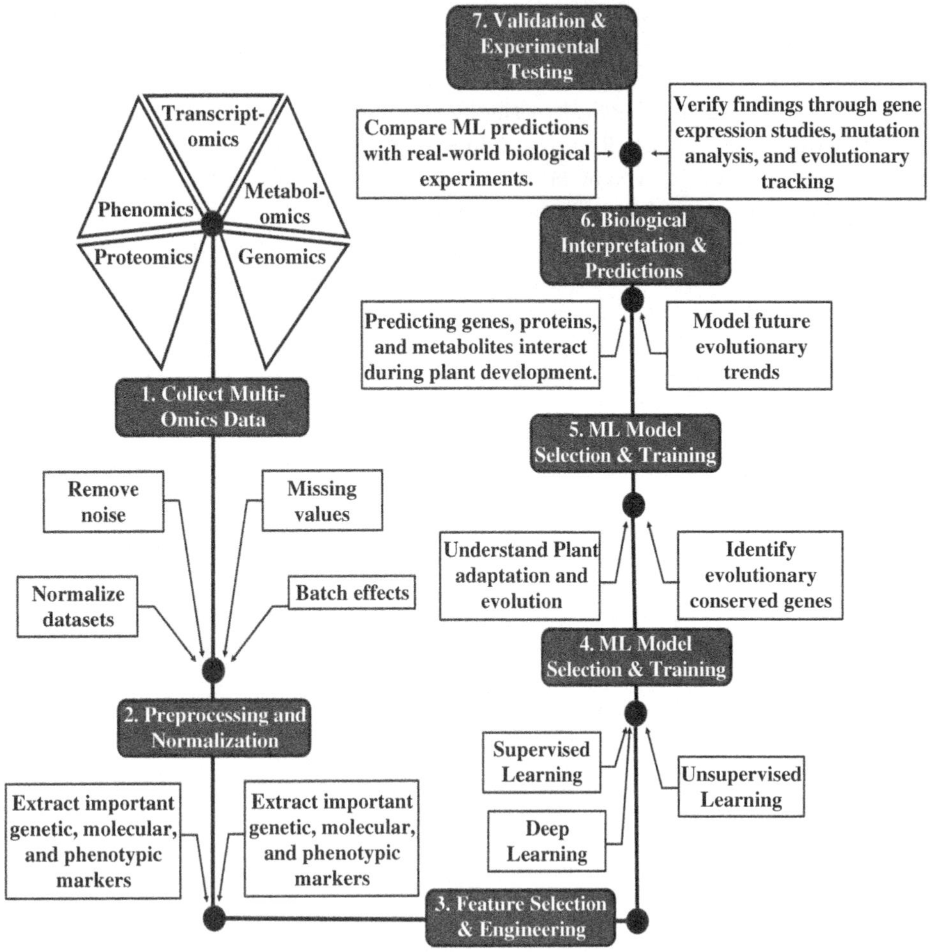

Figure 2.3 Multi-omics data integration using ML in Evo-Devo.

2.10 Ethical, Computational, and Experimental Challenges

The use of ML in studying plant evolutionary developmental biology raises several key ethical considerations. One primary concern is data privacy and security, as ML systems often handle sensitive genetic information, necessitating robust regulations to protect individual rights and maintain public trust. Algorithmic bias is another critical issue, as biases present in training data can lead to inequitable outcomes, potentially skewing research findings and applications in plant science. Transparency in AI decision-making is essential, as "black box" models can hinder trust and understanding, which is particularly important in scientific research where reproducibility and clarity are valued. Furthermore, the ethical implications of genetic manipulation, a potential outcome of ML applications in plant biology, require careful scrutiny to define the limits of human intervention. Engaging diverse stakeholders, including ethicists and policymakers, is crucial to align these technologies with societal values and ensure equitable distribution of benefits, preventing the exacerbation of disparities. Additionally, the rapid progress of AI in plant science necessitates a principled approach to regulation, ethics review, and monitoring to navigate towards ethical and socially responsible AI applications (Wang et al. 2025; Silvia et al. 2024).

2.11 Conclusion

Plant Evolutionary Developmental Biology (Evo-Devo) is being revolutionized by ML, which offers data-driven, high-throughput techniques for researching phylogenetics, environmental adaptations, and trait evolution. Key ML applications have been described in this chapter, such as automated trait analysis, evolutionary tree construction, and predictive modeling of phenotypic plasticity. Traditional methods of phylogenetic classification and trait evolution analysis relied heavily on manual observations and statistical models. But by gleaning intricate traits from transcriptomic, morphological, and genomic data, ML improves these investigations and yields more accurate and extensive evolutionary insights. Image-based trait extraction is made easier by Deep Learning models like CNNs, whereas evolutionary connections are more effectively classified by clustering techniques like K-means and Hierarchical Clustering. Our knowledge of species divergence is further enhanced by ML-driven phylogenetic tree construction, which combines sequence alignment, evolutionary distance calculation, and ancestral trait prediction. ML offers a comprehensive understanding of plant growth and adaption mechanisms by integrating multi-omics data (genomics, transcriptomics, and proteomics). Notwithstanding these developments, issues with experimental validation, dataset biases, and model interpretability still exist. Future studies should concentrate on strengthening cross-species comparisons, including real-time environmental data, and refining AI models for evolutionary biology. Applications of ML in Evo-Devo will transform evolutionary biology as it develops further, allowing for a better comprehension of genetic inheritance, adaption, and plant diversification.

Acknowledgement

All the authors are thankful to their respective Universities and Institutions for their valuable supports.

Data Availability

The authors confirm that the data supporting the findings of this study are available within the book/chapter.

References

Adaïmé, M. É., Kong, S., and Punyasena, S. W. (2024). Deep learning approaches to the phylogenetic placement of extinct pollen morphotypes. *PNAS Nexus* 3 (1): pgad419.

Adaimé, M. É., Urban, M. A., Kong, S. et al. (2025). Pollen morphology, deep learning, phylogenetics, and the evolution of environmental adaptations in *Podocarpus*. *New Phytol* 247: 1460–1473.

Aliyu, B. S. (2024). Artificial intelligence (AI): A powerful tool for advancing plant study and research. *Nigerian Journal of Botany* 37 (1): 61–68.

Alptekin, B. and Kunkowska, A. B. (2025). How plants adapt to combined and sequential abiotic stresses: A transcriptomics approach. *Plant physiology* 197 (1): kiaf006. Ansari, Y., Ishaq, M., Yusuf, M. et al. (2024). Advancing paleontology: A survey on deep learning methodologies in fossil image analysis.

Aryal, S., Grießinger, J., Dyola, N. et al. (2023). INTRAGRO: A machine learning approach to predict future growth of trees under climate change. *Ecology and Evolution* 13 (10): e10626.

Atiampo, A. K., N'Goran, K. A. C. and Trey, Z. F. O. (2022). Global approach of shape and texture features fusion in convolutional neural network for automatic classification of plant species based on leaves images. In *Congress on Intelligent Systems* (pp. 655–673). Springer Nature Singapore.

Aubin, E., El Baidouri, M., and Panaud, O. (2021). Horizontal gene transfers in plants. *Life*. 11 (8): 857.

Aubin, E., Llauro, C., Garrigue, J. et al. (2023). Genome-wide analysis of horizontal transfer in non-model wild species from a natural ecosystem reveals new insights into genetic exchange in plants. *PLoS Genetics* 19 (10): e1010964.

Bai, L., and Wang, S. (2024). GRNNLink: Predicting gene regulatory links from single-cell RNA-seq data using graph recurrent neural network. In *2024 IEEE International Conference on Bioinformatics and Biomedicine (BIBM)* (pp. 491–496). IEEE.

Bai, W., Li, C., Li, W. et al. (2024). Machine learning assists prediction of genes responsible for plant specialized metabolite biosynthesis by integrating multi-omics data. *BMC Genomics* 25 (1): 418.

Baloch, A. A., Kakar, K. U., Nawaz, Z. et al. (2022). Comparative genomics and evolutionary analysis of plant CNGCs. *Biology Methods and Protocols* 7 (1): bpac018.

Bari, A., Ouabbou, H., Jilal, A. et al. (2021). Machine learning speeding up the development of portfolio of new crop varieties to adapt to and mitigate climate change. *bioRxiv* 2021–10.

Bayer, P. E., Petereit, J., Danilevicz, M. F. et al. (2021). The application of pangenomics and machine learning in genomic selection in plants. *The Plant Genome* 14 (3): e20112.

Cai, H., Liu, Y., and Guo, C. (2021). Contribution of plant–bacteria interactions to horizontal gene transfer in plants. *Biotechnology and Biotechnological Equipment* 35 (1): 1587–1592.

Casas-Mollano, J. A., Zacarias, E. and Almeida, J. (2023). Evolution of epigenetic mechanisms in plants: Insights from H3K4 and H3K27 methyltransferases. In *Handbook of Epigenetics* (pp. 499–519). Academic Press.

Casci, T. (2009). Failsafe flowers. *Nature Reviews Genetics* 10 (3): 152–152.

Cembrowska-Lech, D., Krzemińska, A., Miller, T. et al. (2023). An integrated multi-omics and artificial intelligence framework for advance plant phenotyping in horticulture. *biology* 12 (10): 1298.

Centorame, L., Gasperini, T., Ilari, A. et al. (2024). An overview of machine learning applications on plant phenotyping, with a focus on sunflower. *Agronomy* 14 (4): 719.

Chairez-Veloz, J. E., Chávez-Hernández, E. C., Dávila-Velderrain, J. et al. (2023). Uncovering the role that specific genes play in reachability properties of complex biological Gene Regulatory Networks (GRNs): The Arabidopsis thaliana flower development case. *bioRxiv*, 2023–05.

Chan, K. L., Rosli, R., Tatarinova, T. V. et al. (2017). Seqping: Gene prediction pipeline for plant genomes using self-training gene models and transcriptomic data. *BMC Bioinformatics* 18: 1–7.

Chanderbali, A. S., Berger, B. A., Howarth, D. G. (et al). (2016). Evolving ideas on the origin and evolution of flowers: New perspectives in the genomic era. *Genetics* 202 (4): 1255–1265.

Chang, S., Lee, U., Hong, M. J. et al. (2021). Time-series growth prediction model based on U-net and machine learning in Arabidopsis. *Frontiers in Plant Science* 12: 721512.

Chau, T. N., Timilsena, P. R., Bathala, S. P. et al. (2025). Orthologous marker groups reveal broad cell identity conservation across plant single-cell transcriptomes. *Nature Communications* 16 (1): 201.

Cohen, M., Wohlmuth, H., Williams, C. et al. (2023). The evolutionary pathways and ecological adaptations of plants: A comprehensive analysis of survival strategies over geological timescales. *Australian Herbal Insight* 6 (1): 1–5.

Conover, J. L. (2021). *Molecular evolution following Allopolyploidization in Gossypium* (Doctoral dissertation, Iowa State University).

Crossa, J., Montesinos-Lopez, O. A., Costa-Neto, G. et al. (2025). Machine learning algorithms translate big data into predictive breeding accuracy. *Trends in Plant Science*, 30 (2): 167–184.

Dahdal, R. and Hament, B. (2024). Plant Organelle segmentation using 3D point clouds and PointNet. In *2024 Systems and Information Engineering Design Symposium (SIEDS)* (pp. 206–210). IEEE.

Danilevicz, M. F., Gill, M., Anderson, R. et al. (2022). Plant genotype to phenotype prediction using machine learning. *Frontiers in Genetics* 13: 822173.

Delos Reyes, I. V. P., Deocaris, C. C., Sahagun, J. R. et al. (2022). Convolutional neural network analysis of BG regale, the Sansevieria rorida radiation mutant. *Philippine Journal of Science* 151.

Dirk, R., Fischer, J. L., Schardt, S. et al. (2023). Recognition and reconstruction of cell differentiation patterns with deep learning. *PLOS Computational Biology* 19 (10): e1011582.

Doll, Y., Koga, H., and Tsukaya, H. (2022). Eco-evo-devo of plant stomata: The diversity of stomatal development and its evolution. *Plant Morphology* 34 (1): 53–62.

Durand, M., Besseau, S., Papon, N. et al. (2024). Unlocking plant bioactive pathways: Omics data harnessing and machine learning assisting. *Current Opinion in Biotechnology* 87: 103135.

Fan, J., Shen, R., Huang, X. et al. (2024). MATL: A deep neural network using multi-scale convolutions and transformer for transcription factor binding site prediction. In *2024 IEEE International Conference on Bioinformatics and Biomedicine (BIBM)* (pp. 613–618). IEEE.

Fan, Y., Kon, M., and DeLisi, C. (2018). Transcription factor-DNA binding via machine learning ensembles. *arXiv preprint arXiv:1805.03771.*

Farooq, M., van Dijk, A. D., Nijveen, H. et al. (2023). Genomic prediction in plants: Opportunities for ensemble machine learning based approaches. *F1000Research* 11: 802.

Farooq, M.A., Gao, S., Hassan, M.A. et al. (2024). Artificial intelligence in plant breeding. *Trends in Genetics*, 40 (10): 891–908.

Feng, H., Guo, C., Li, Z. et al. (2022). Machine learning assisted dynamic phenotypes and genomic variants help understand the ecotype divergence in rapeseed. *Frontiers in Plant Science* 13: 1028779.

Ferrari, C., Pérez, N. M., and Vandepoele, K. (2022). MINI-EX: Integrative inference of single-cell gene regulatory networks in plants. *Molecular Plant* 15 (11): 1807–1824.

Ganesh, C., Harshavardhan, G., Keerthi, N. R. S. et al. (2025). Enhancing plant leaf classification with deep learning: Automating feature extraction for accurate species identification. *SCT Proceedings in Interdisciplinary Insights and Innovations* 3: 513.

Gao, L., Kantar, M. B., Moxley, D. et al. (2023). Crop adaptation to climate change: An evolutionary perspective. *Molecular Plant* 16 (10): 1518–1546.

Gavrilets, S., and Vose, A. (2005). Dynamic patterns of adaptive radiation. *Proceedings of the National Academy of Sciences* 102 (50): 18040–18045.

Gowthaman, S., and Das, A. (2025). Plant leaf identification using feature fusion of wavelet scattering network and CNN with PCA classifier. *IEEE Access*, 13: 11594–11608.

Gray, S. B., and Brady, S. M. (2016). Plant developmental responses to climate change. *Developmental biology* 419 (1): 64–77.

Guo, Z., Zhang, K., Cai, C. et al. (2024). Deep learning can predict subgenome dominance in ancient but not in neo/synthetic polyploidized genomes. *The Plant Journal* 120 (1): 174–186.

He, Y., Mulqueeney, J. M., Watt, E. C. et al. (2024). Opportunities and challenges in applying AI to evolutionary morphology. *Integrative Organismal Biology* 6 (1): obae036.

Hewavithana, T., Koh, C. S., Kaur, A. et al. (2023). Inference of subgenomes resulting from polyploid events using synteny based dynamic linking and maximum neighbourhood. In *2023 IEEE International Conference on Bioinformatics and Biomedicine (BIBM)* (pp. 104–109). IEEE.

Huang, H., Wang, Z., and Zhao, G. (2024). Efficient 3D reconstruction of multiple plants from UAV images with deep learning. In *Proceedings of the 2024 16th International Conference on Machine Learning and Computing* (pp. 369–375).

Hunt, R., Reyes-Hernández, J. L., Shaw, J. J. et al. (2025). Integrating deep learning derived morphological traits and molecular data for total-evidence phylogenetics: Lessons from digitized collections. *Systematic Biology* 74 (3): syae072.

Ichihashi, Y., Aguilar-Martínez, J. A., Farhi, M. et al. (2014). Evolutionary developmental transcriptomics reveals a gene network module regulating interspecific diversity in plant leaf shape. *Proceedings of the National Academy of Sciences* 111 (25): E2616–E2621.

Igler, C., Lagator, M., Tkačik, G. et al. (2018). Evolutionary potential of transcription factors for gene regulatory rewiring. *Nature Ecology and Evolution* 2 (10): 1633–1643.

Irish, V. F., and Benfey, P. N. (2004). Beyond Arabidopsis. Translational biology meets evolutionary developmental biology. *Plant Physiology* 135 (2): 611–614.

Isewon, I., Apata, O., Oluwamuyiwa, F et al. (2022). Machine learning algorithms: Their applications in plant omics and agronomic traits' improvement. *F1000Research* 11: 1256.

Jamil, I. N., Remali, J., Azizan, K. A. et al. (2020). Systematic multi-omics integration (MOI) approach in plant systems biology. *Frontiers in Plant Science* 11: 944.

Jasinski, S., Vialette-Guiraud, A. C., and Scutt, C. P. (2010). The evolutionary-developmental analysis of plant microRNAs. *Philosophical Transactions of the Royal Society B: Biological Sciences* 365 (1539): 469–476.

Ji, R., Geng, Y., and Quan, X. (2024). Inferring gene regulatory networks with graph convolutional network based on causal feature reconstruction. *Scientific Reports* 14 (1): 21342.

Jiang, Y., and Li, C. (2020). Convolutional neural networks for image-based high-throughput plant phenotyping: A review. *Plant Phenomics* 2020.

Karnan, A., and Ragupathy, R. (2023). A comprehensive study on plant classification using machine learning models. In *International Conference on Information and Communication Technology for Competitive Strategies* (pp. 187–199). Springer Nature Singapore.

Kasianov, A. S., Klepikova, A. V., Mayorov, A. V. et al. (2023). Interspecific comparison of gene expression profiles using machine learning. *PLOS Computational Biology* 19 (1): e1010743.

Kaya, C. (2025). Optimizing crop production with plant phenomics through high-throughput phenotyping and AI in controlled environments. *Food and Energy Security* 14 (1): e70050.

Kerin, M., and Marchini, J. (2020). Inferring gene-by-environment interactions with a Bayesian whole-genome regression model. *The American Journal of Human Genetics* 107 (4): 698–713.

Koch, C., Konieczka, J., Delorey, T. et al. (2017). Inference and evolutionary analysis of genome-scale regulatory networks in large phylogenies. *Cell Systems* 4 (5): 543–558.

Korfmann, K., Zauchner, A., Huo, B. et al. (2024). Methylomes reveal recent evolutionary changes in populations of two plant species. *bioRxiv*: 2024–09.

Kramer, E. M., Kong, H., and Rausher, M. D. (2017). Plant evolutionary developmental biology. Introduction to a special issue. *New Phytologist* 216 (2): 335–336.

Kulikov, N., Derakhshandeh, F., and Mayer, C. (2024). Machine learning can be as good as maximum likelihood when reconstructing phylogenetic trees and determining the best evolutionary model on four taxon alignments. *Molecular Phylogenetics and Evolution* 200: 108181.

Kumar, H., Panigrahi, M., Panwar, A. et al. (2022). Machine-learning prospects for detecting selection signatures using population genomics data. *Journal of Computational Biology* 29 (9): 943–960.

Kumar, M., Gahtori, R., Bisht, S. et al. (2024). Convolutional neural network-based plant pathology recognition. In *2024 Second International Conference on Advances in Information Technology (ICAIT)* (Vol. 1, pp. 1–5). IEEE.

Kundu, R., Gupta, S., Dehury, M. K. et al. (2024). Image-based dimension extraction of flowers' and classification with SFFFP neural network. In *2024 2nd International Conference on Sustainable Computing and Smart Systems (ICSCSS)* (pp. 1298–1303). IEEE.

Kusmec, A., Yeh, C. T. E., Genomes to Fields Initiative, Bohn, M. O. et al. (2024). Data-driven identification of environmental variables influencing phenotypic plasticity to facilitate breeding for future climates. *New Phytologist* 244 (2): 618–634.

Kviman, O., Molén, R., and Lagergren, J. (2023). Improved variational Bayesian phylogenetic inference using mixtures. *arXiv preprint arXiv:2310.00941*.

Lapkovskis, A., Nefedova, N., and Beikmohammadi, A. (2024). Automatic fused multimodal deep learning for plant identification. *arXiv preprint arXiv:2406.01455*.

Latrille, T., Rodrigue, N., and Lartillot, N. (2023). Genes and sites under adaptation at the phylogenetic scale also exhibit adaptation at the population-genetic scale. *Proceedings of the National Academy of Sciences* 120 (11): e2214977120.

Leal, J. L., Milesi, P., Hodková, E. et al. (2024). Complex polyploids: Origins, genomic composition, and role of Introgressed alleles. *Systematic Biology* 73 (2); 392–418.

Li, J., Xu, K., Chaudhuri, S. et al. (2017). Grass: Generative recursive autoencoders for shape structures. *ACM Transactions on Graphics (TOG)* 36 (4): 1–14.

Li, R., Rozum, J. C., Quail, M. M. et al. (2023). Inferring gene regulatory networks using transcriptional profiles as dynamical attractors. *PLoS Computational Biology* 19 (8): e1010991.

Li, W., Wu, S., Wen, W. et al. (2024). Using high-throughput phenotype platform MVS-Pheno to reconstruct the 3D morphological structure of wheat. *AoB Plants* 16 (2): plae019.

Liu, H., Li, J., Gong, P. et al. (2023). The origin and evolution of carpels and fruits from an evo-devo perspective. *Journal of Integrative Plant Biology* 65 (2): 283–298.

Liu, X., Huang, Y., Sahu, S. et al. (2024). Machine learning based pan-plant analyses of transposable elements across 352 species illuminates genome evolution.

López Sánchez, A., and Lafond, M. (2024). Predicting horizontal gene transfers with perfect transfer networks. *Algorithms for Molecular Biology* 19 (1): 6.

López-Fernández, H., Pinto, M., Vieira, C. P. et al. (2024). Auto-phylo v2 and auto-phylo-pipeliner: Building advanced, flexible, and reusable pipelines for phylogenetic inferences, estimation of variability levels and identification of positively selected amino acid sites. *Journal of Integrative Bioinformatics* 21 (2): 20230046.

Love, A. C. (2024). *Evolution and Development: Conceptual Issues*. Cambridge University Press.

Luo, J. W., Yang, L., and Zhang, X. Z. (2007). A method for gene identification by dynamic feature choosing. In *Digital Human Modeling: First International Conference on Digital Human Modeling, ICDHM 2007, Held as Part of HCI International 2007*, Beijing, China, *(July 22–27, 2007)* (pp. 678–683). Springer Berlin Heidelberg.

Ma, F., and Zheng, C. (2024). Single-cell phylotranscriptomics of developmental and cell type evolution. *Trends in Genetics* 40 (6): 495–510.

Meher, P. K., Mohapatra, A., Satpathy, S. et al. (2021). PredCRG: A computational method for recognition of plant circadian genes by employing support vector machine with Laplace kernel. *Plant Methods* 17 (1): 46.

Minelli, A. (2018). *Plant Evolutionary Developmental Biology: The Evolvability of the Phenotype*. Cambridge University Press.

Monteiro, L. R. (2013). Morphometrics and the comparative method: Studying the evolution of biological shape. *Hystrix* 24 (1): 25.

Montesinos-López, O. A., Montesinos-López, A., Crossa, J. et al. (2017). A variational Bayes genomic-enabled prediction model with genotype× environment interaction. *G3: Genes, Genomes, Genetics* 7 (6): 1833–1853.

Moon, J. Y. (2013). *A causal gene network with genetic variations incorporating biological knowledge and latent variables* (Doctoral dissertation, The University of Wisconsin-Madison).

Nabwire, S., Suh, H. K., Kim, M. S. et al. (2021). Application of artificial intelligence in phenomics. *Sensors* 21 (13): 4363.

Neves, R.T., and Costa, F.A. (2020). Unique plant adaptations to varied environments: A comprehensive review of evolutionary mechanisms shaping plant survival across diverse habitats. *Australian Herbal Insight* 3 (1): 1–5.

Ng, J.W.X., Chua, S.K., and Mutwil, M. (2022). Feature importance network reveals novel functional relationships between biological features in Arabidopsis thaliana. *Frontiers in Plant Science* 13: 944992.

Nguyen, K. A., Seeboonruang, U., and Chen, W. (2023). Projected climate change effects on global vegetation growth: A machine learning approach. *Environments* 10 (12): 204.

Ortiz, E. M., Höwener, A., Shigita, G. et al. (2023). A novel phylogenomics pipeline reveals complex pattern of reticulate evolution in Cucurbitales. *BioRxiv*: 2023–10.

Otal, H. T., Subasi, A., Kurt, F. et al. (2025). Analysis of gene regulatory networks from gene expression using graph neural networks. In *Digital Healthcare, Digital Transformation and Citizen Empowerment in Asia-Pacific and Europe for a Healthier Society* (pp. 249–270). Academic Press.

Pandey, S. B., Colliaux, D., and Chaudhury, A. (2024). Spatio-temporal correspondence estimation of growing plants by Hausdorff distance based skeletonization for organ tracking. In *2024 IEEE International Conference on Robotics and Automation (ICRA)* (pp. 13625–13631). IEEE.

Park, D. S., Willis, C. G., Xi, Z. et al. (2020). Machine learning predicts large scale declines in native plant phylogenetic diversity. *New Phytologist* 227 (5): 1544–1556.

Peleke, F. F., Zumkeller, S. M., Gültas, M. et al. (2024). Deep learning the cis-regulatory code for gene expression in selected model plants. *Nature Communications* 15 (1): 3488.

Peng, Q., Li, X., Shen, R. et al. (2022). How well can we predict vegetation growth through the coming growing season?. *Science of Remote Sensing* 5: 100043.

Pérez-Valencia, D. M., Rodríguez-Álvarez, M. X., Boer, M. P. et al. (2024). A one-stage approach for the spatio-temporal analysis of high-throughput phenotyping data. *Journal of Agricultural, Biological and Environmental Statistics*: 1–23.

Petřík, I., Hladík, P., Zhang, C. et al. (2024). Spatio-temporal plant hormonomics: From tissue to subcellular resolution. *Journal of Experimental Botany* 75 (17): 5295–5311.

Porto, A., and Voje, K. L. (2020). ML-morph: A fast, accurate and general approach for automated detection and landmarking of biological structures in images. *Methods in Ecology and Evolution* 11 (4): 500–512.

Qin, X., Chiang, C. W., and Gaggiotti, O. E. (2022). Deciphering signatures of natural selection via deep learning. *Briefings in Bioinformatics* 23 (5): bbac354.

Raja, V., and Kung, J. (2025). Adaptive machine learning models for dynamic environments: A mathematical framework for real-time decision making.

Ranjan, R., Srijan, S., Balekuttira, S. et al. (2023). Genome-wide, Organ-delimited gene regulatory networks (OD-GRNs) provide high accuracy in candidate TF selection across diverse processes. *bioRxiv*: 2023–06.

Reddy, B. D. K., and Iyer, L. S. (2023). Classification of diseased leaves in plants using convolutional neural networks. In *International Conference on Data Science, Computation and Security* (pp. 139–149). Springer Nature Singapore.

Redmond, E. J., Ronald, J., Davis, S. J. et al. (2024). Single-plant-omics reveals the cascade of transcriptional changes during the vegetative-to-reproductive transition. *The Plant Cell* 36 (10): 4594–4606.

Refaat, A. M., Youssef, A. M., Mosallam, H. A. A. et al. (2024). Predicting the effect of climate change on the spatiotemporal distribution of two endangered plant species, Silene leucophylla Boiss. and Silene schimperiana Boiss, using machine learning, in Saint Catherine Protected Area, Egypt. *Beni-Suef University Journal of Basic and Applied Sciences* 13 (1): 98.

Remita, A. M., Vitae, G., and Diallo, A. B. (2023). Prior density learning in variational Bayesian phylogenetic parameters inference. In *RECOMB International Workshop on Comparative Genomics* (pp. 112–130). Springer Nature Switzerland.

Rodríguez Herrera, J., Galván Alcaraz, K. A., Albarrán Hernández, R. U. et al. (2024). A computational Evo-Devo approach for elucidating the roles of PLETHORA transcription factors in regulating root development. *bioRxiv*: 2024–12.

Rodríguez-Lira, D. C., Córdova-Esparza, D. M., Álvarez-Alvarado, J. M. et al. (2024). Trends in machine and deep learning techniques for plant disease identification: A systematic review. *Agriculture* 14 (12): 2188.

Sadria, M., and Bury, T. M. (2024). FateNet: An integration of dynamical systems and deep learning for cell fate prediction. *Bioinformatics* 40 (9): btae525.

Sadria, M., Zhang, A., and Bader, G. D. (2024). Deep lineage: Single-cell lineage tracing and fate inference using deep learning. *bioRxiv*: 2024–04.

Saimon, A. S. M., Moniruzzaman, M., Islam, M. S. et al. (2023). Integrating genomic selection and machine learning: A data-driven approach to enhance corn yield resilience under climate change. *Journal of Environmental and Agricultural Studies* 4 (2): 20–27.

Saito, T., Wang, S., Ohkawa, K. et al. (2024). Deep learning with a small dataset predicts chromatin remodelling contribution to winter dormancy of apple axillary buds. *Tree Physiology* 44 (7): tpae072.

Sanches, P. H. G., de Melo, N. C., Porcari, A. M. et al. (2024). Integrating molecular perspectives: Strategies for comprehensive multi-omics integrative data analysis and machine learning applications in transcriptomics, proteomics, and metabolomics. *Biology* 13 (11): 848.

Sandhu, K., Patil, S. S., Pumphrey, M. et al. (2021). Multitrait machine-and deep-learning models for genomic selection using spectral information in a wheat breeding program. *The Plant Genome* 14 (3): e20119.

Sattler, R., and Rutishauser, R. (2022). Fundamentals of plant morphology and plant Evo-Devo (evolutionary developmental morphology). *Plants* 12 (1): 118.

Saura Sánchez, M. T. (2019). Machine learning para la selección de genes implicados en el desarrollo de Arabidopsis thaliana utilizando datos de expresión génica.

Schiller, C., Schmidtlein, S., Boonman, C. et al. (2021). Deep learning and citizen science enable automated plant trait predictions from photographs. *Scientific Reports* 11 (1): 16395.

Shen, H., Wang, Y., Guan, X. et al. (2022). A spatiotemporal constrained machine learning method for OCO-2 solar-induced chlorophyll fluorescence (SIF) reconstruction. *IEEE Transactions on Geoscience and Remote Sensing* 60: 1–17.

Shoaib, M., Sharma, N., Kotthoff, L. et al. (2023). AutoML: Advanced tool for mining multivariate plant traits. *Trends in Plant Science* 28 (12): 1451–1452.

Sia, J., Zhang, W., Cheng, Mand. et al. (2025). Machine learning-based identification of general transcriptional predictors for plant disease. *New Phytologist* 245 (2): 785–806.

Silvestro, D., Latrille, T., and Salamin, N. (2024). Toward a semi-supervised learning approach to phylogenetic estimation. *Systematic Biology* 73 (5): 789–806.

Silvia, R., Haldi, B., Usman, S. et al. (2024). Comparison of SVM, Naive Bayes, and ELM Models in Plant Growth Classification. *Journal of Data Science* 2024 (56): 1–6.

Sklab, Y., Ariouat, H., Boujydah, Y. et al. (2024). Towards a deep learning-powered herbarium image analysis platform. *Biodiversity Information Science and Standards* 8.

Sloan, D. B., Conover, J. L., Grover, C. E. et al. (2024). Polyploid plants take cytonuclear perturbations in stride. *The Plant Cell* 36 (4): 829–839.

Smet, D., Opdebeeck, H., and Vandepoele, K. (2023). Predicting transcriptional responses to heat and drought stress from genomic features using a machine learning approach in rice. *Frontiers in Plant Science* 14: 1212073.

Sohn, K. A. (2011). *Learning Ancestral Genetic Processes using Nonparametric Bayesian Models* (Doctoral dissertation, University of Chicago).

Strother, P. K. (2023). An evo-devo perspective on no Ordovician land plants. *Estonian Journal of Earth Sciences* 72 (1): 102–105.

Sun, C., Fang, R., Salemi, M. et al. (2024). DeepDynaForecast: Phylogenetic-informed graph deep learning for epidemic transmission dynamic prediction. *PLOS Computational Biology* 20 (4): e1011351.

Sun, S., Ye, X., and Zou, Q. (2023). Machine learning on understanding the epigenetic mechanisms underlying plant adaptation and domestication. *Frontiers in Plant Science* 14: 1236787.

Suvorov, A., and Schrider, D. R. (2024). Reliable estimation of tree branch lengths using deep neural networks. *PLOS Computational Biology* 20 (8): e1012337.

Švik, M., Lukeš, P., Lhotáková, Z. et al. (2023). Retrieving plant functional traits through time series analysis of satellite observations using machine learning methods. *International Journal of Remote Sensing* 44 (10): 3083–3105.

Tan, G. D., Chaudhuri, U., Varela, S. et al. (2024). Machine learning-enabled computer vision for plant phenotyping: A primer on AI/ML and a case study on stomatal patterning. *Journal of Experimental Botany* 75 (21): 6683–6703.

Tian, X. C., Chen, Z. Y., Nie, S. et al. (2024). Plant-LncPipe: A computational pipeline providing significant improvement in plant lncRNA identification. *Horticulture Research* 11 (4): uhae041.

Tross, M. C., Duggan, G., Shrestha, N. et al. (2024). Models trained to predict differential expression across plant organs identify distal and proximal regulatory regions. *bioRxiv*: 2024–06.

Tseng, Z. J., and Flynn, J. J. (2015). An integrative method for testing form–function linkages and reconstructed evolutionary pathways of masticatory specialization. *Journal of The Royal Society Interface* 12 (107): 20150184.

Varela, S., Zheng, X., Njuguna, J. N. et al. (2022). Deep convolutional neural networks exploit high-spatial-and-temporal-resolution aerial imagery to phenotype key traits in miscanthus. *Remote Sensing* 14 (21): 5333.

Vialette-Guiraud, A. C., Andres-Robin, A., Chambrier, P. et al. (2016). The analysis of Gene Regulatory Networks in plant evo-devo. *Journal of Experimental Botany* 67 (9): 2549–2563.

Viana, J. P. G., Avalos, A., Zhang, Z. et al. (2024). Common signatures of selection reveal target loci for breeding across soybean populations. *The Plant Genome* 17 (1): e20426.

Vincent, S. E., Dang, P. D., Herrel, A. et al. (2006). Morphological integration and adaptation in the snake feeding system: A comparative phylogenetic study. *Journal of Evolutionary Biology* 19 (5): 1545–1554.

Walden, N., Kiefer, C., and Koch, M. A. (2024). Unravelling complex hybrid and polyploid evolutionary relationships using phylogenetic placement of paralogs from target enrichment data. *bioRxiv*: 2024–06.

Walsh, J. J., Mangina, E., and Negrão, S. (2024). Advancements in imaging sensors and AI for plant stress detection: A systematic literature review. *Plant Phenomics* 6: 0153.

Wang, B., and Shen, X. (2024). 3D reconstruction of plant based on NeRF. In *2024 IEEE 3rd International Conference on Electrical Engineering, Big Data and Algorithms (EEBDA)* (pp. 345–348). IEEE.

Wang, C., Pan, W., Song, X. et al. (2022). Predicting plant growth and development using time-series images. *Agronomy* 12 (9): 2213.

Wang, J., Zong, W., Shi, L. et al. (2024). Using mixed kernel support vector machine to improve the predictive accuracy of genome selection1. *Journal of Integrative Agriculture*.

Wang, Y., Zhang, P., Guo, W. et al. (2021). A deep learning approach to automate whole-genome prediction of diverse epigenomic modifications in plants. *New Phytologist* 232 (2): 880–897.

Wang, Z., Yuan, H., Yan, J. et al. (2025). Identification, characterization, and design of plant genome sequences using deep learning. *The Plant Journal* 121 (1): e17190.

Wolf, E., Gaquerel, E., Scharmann, M. et al. (2021). Evolutionary footprints of a cold relic in a rapidly warming world. *Elife* 10: e71572.

Wrightsman, T., Ferebee, T.H., Romay, M.C. et al. (2024). Current genomic deep learning architectures generalize across grass species but not alleles. *bioRxiv*: 2024–04.

Wu, J., Ainsworth, E. A., Wang, S. et al. (2022). Adaptive transfer learning for plant phenotyping. *arXiv preprint arXiv:2201.05261*.

Wu, S., Shu, M., and Zeng, B. (2024). Application of machine learning to plant phenomics. In *2024 IEEE 7th Information Technology, Networking, Electronic and Automation Control Conference (ITNEC)* (Vol. 7, pp. 106–112). IEEE.

Wu, Z., Sun, Y., Zhao, X. et al. (2024). Phenotype prediction in plants is improved by integrating large-scale transcriptomic datasets. *NAR Genomics and Bioinformatics* 6 (4): lqae184.

Xie, K., Zhu, J., Ren, H. et al. (2024). Delving into the potential of deep learning algorithms for point cloud segmentation at organ level in plant phenotyping. *Remote Sensing* 16 (17): 3290.

Xie, R., Adam, D. C., Hu, S. et al. (2024). Integrating contact tracing data to enhance outbreak phylodynamic inference: A deep learning approach. *Molecular Biology and Evolution* 41 (11): msae232.

Xiong, J., Yin, N., Liang, S. et al. (2024). Cross-attention graph neural networks for inferring gene regulatory networks with skewed degree distribution. *arXiv preprint arXiv:2412.16220*.

Xu, W., Guo, R., Chen, P. et al. (2024). Cherry growth modeling based on prior distance embedding contrastive learning: Pre-training, anomaly detection, semantic segmentation, and temporal modeling. *Computers and Electronics in Agriculture* 221: 108973.

Yan, H., Lee, J., Song, Q. et al. (2022). Identification of new marker genes from plant single-cell RNA-seq data using interpretable machine learning methods. *New Phytologist* 234 (4): 1507–1520.

Yan, J., and Wang, X. (2022). Unsupervised and semi-supervised learning: The next frontier in machine learning for plant systems biology. *The Plant Journal* 111 (6): 1527–1538.

Yan, Q., Fruzangohar, M., Taylor, J. et al. (2023). Improved genomic prediction using machine learning with Variational Bayesian sparsity. *Plant Methods* 19 (1): 96.

Yogadasan, N., Doxey, A. C., and Chuong, S. D. (2023). A machine learning framework identifies plastid-encoded proteins harboring C3 and C4 distinguishing sequence information. *Genome Biology and Evolution* 15 (7): evad129.

Yoosefzadeh-Najafabadi, M., Hesami, M., and Eskandari, M. (2024). Machine learning-enhanced utilization of plant genetic resources. In *Sustainable Utilization and Conservation of Plant Genetic Diversity* (pp. 619–639). Springer Nature Singapore.

You, X., Shu, Y., Ni, X. et al. (2025). MLAS: Machine learning-based approach for predicting abiotic stress-responsive genes in Chinese cabbage. *Horticulturae* 11 (1): 44.

Yu, S., Fu, X., Deng, X. et al. (2024). AtML: Accurate identification of Arabidopsis thaliana root cell identity via single-cell sequencing analysis using interpretable machine learning.

Yunita, M., Muwardi, R., and Iklima, Z. (2023). Implementation of Bayesian inference MCMC algorithm in phylogenetic analysis of Dipterocarpaceae family. *SINERGI* 27 (1): 23–30.

Zdrazil, J., Kong, L., Klimeš, P. et al. (2024). Decrypting the complex phenotyping traits of plants by machine learning. *bioRxiv*: 2024–11.

Zhai, J., Gokaslan, A., Schiff, Y. et al. (2024). Cross-species modeling of plant genomes at single nucleotide resolution using a pre-trained DNA language model. *bioRxiv*.

Zhang, L., Wang, Y., Ruhl, M. et al. (2023). Machine-learning-based morphological analyses of leaf epidermal cells in modern and fossil ginkgo and their implications for palaeoclimate studies. *Palaeontology* 66 (6): e12684.

Zhang, X., Zhou, X., Wan, M. et al. (2022). PINC: A tool for non-coding RNA identification in plants based on an automated machine learning framework. *International Journal of Molecular Sciences* 23 (19): 11825.

3

Machine Learning for Plant High-Throughput Phenotyping

Dibyendu Seth[1], Sourish Pramanik[2], Ehsas Pachauri[3], Ankan Das[4], Sandip Debnath[5], Mehdi Rahimi[6]*

[1] *Department of Genetics and Plant Breeding, Chaudhary Charan Singh Haryana Agriculture University (CCSHAU), Hisar, Haryana, India*
[2] *Department of Plant Pathology, Punjab Agricultural University, Ludhiana, India*
[3] *Department of Genetics and Plant Breeding, IARI-Central Arid Zone Research Institute, Jodhpur, India*
[4] *Palli Siksha Bhavana (Institute of Agriculture), Visva-Bharati University, Sriniketan, West Bengal, India*
[5] *Department of Genetics and Plant Breeding, Institute of Agriculture, Visva-Bharati University, Sriniketan, West Bengal, India*
[6] *Department of Biotechnology, Institute of Science and High Technology and Environmental Sciences, Graduate University of Advanced Technology, Kerman, Iran*

*Corresponding author: sandip.debnath@visva-bharati.ac.in

3.1 Introduction

By 2050, it is anticipated that there will be between 9 and 10 billion people on the planet, which will require a 25–70% increase in agricultural output to keep up with the rising demand (Hunter et al. 2017). However, a number of biotic and abiotic stresses, including salinity, drought, diseases, and pests, drastically lower crop yields, endangering the world's food security. The substantial drop in crop productivity brought on by stress may endanger the world's food supply. Therefore, enhancing yield protection for sustainable production systems requires a thorough understanding of plant stress. Crop phenotyping, or the methodical measurement and study of observable plant features, is used by plant scientists to address these issues and enhance sustainable production systems and yield protection. Throughout the crop cycle, breeding efforts often seek to phenotype large populations for a variety of attributes (Sandhu et al. 2021a,b). Although useful, traditional phenotyping techniques are frequently time-consuming, expensive, and damaging, which restricts their accuracy and scalability.

High-throughput phenotyping (HTP) has revolutionized contemporary plant science by integrating advanced technologies such as unmanned aerial vehicles (UAVs), automated robotic systems, and multispectral sensors to enable rapid, precise, and noninvasive acquisition of phenotypic data. While controlled environments facilitate standardized HTP workflows, field-based applications face significant challenges due to variable environmental factors that introduce measurement instability and data heterogeneity. The development of many plant phenotyping research institutions and the creation of the International Plant Phenotyping Network (IPPN), which promotes international cooperation in this area, have fueled the transition from manual to automated phenotyping (Table 3.1).

Machine learning (ML) and deep learning (DL) methodologies have emerged as pivotal frameworks in contemporary data analytics, employing principles from probability theory, statistical inference, classification algorithms, regression models, decision-theoretic paradigms, and data visualization strategies. These techniques enable the systematic alignment of extracted information with empirically observed phenotypic characteristics, thereby advancing analytical precision. By integrating neural network architectures with computational models, such approaches facilitate robust correlations between heterogeneous datasets and phenotypic manifestations, solidifying their role as indispensable tools in modern scientific and data-driven investigations. These methods enable researchers to extract significant patterns and insights from complex data, making it easier to identify traits linked to stress tolerance, yield potential, and other agronomically significant characteristics. These methodologies accomplish this through the application of feature extraction, discriminative identification, classificatory frameworks, and predictive criteria to isolate and prioritize pertinent data subsets. By operationalizing structured analytical processes within computational pipelines, such techniques enable the systematic

Machine Learning for Plant Biology, First Edition. Edited by Jen-Tsung Chen.

Table 3.1 Plant phenotyping research institutes established in the past 10 years.

Research Centre	Country	Website
Australian Plant Phenomics Facility (APPF)	Australia	https://www.plantphenomics.org.au
Jülich Plant Phenotyping Centre (JPPC)	Germany	https://www.fz-juelich.de/ibg/ibg-2/EN/Research/ResearchGroups/JPPC/JPPC_node.html
National Plant Phenomics Centre (NPPC)	United Kingdom	https://www.plant-phenomics.ac.uk
Plant Phenotyping and Imaging Research Centre (P2IRC)	Canada	https://p2irc.usask.ca
Plant Phenomics Research Centre (PPRC)	China	http://pprcen.njau.edu.cn
International Plant Phenotyping Network (IPPN)	International	https://www.plant-phenotyping.org

Table 3.2 Types of phenotypic traits measured in HTP.

Platform Used	Trait Recorded
3D Imaging	Shoot mass
	Leaf number, shape, and angle
	Leaf color, leaf health
Optical flow method	Leaf area
Far infrared imaging	Photosynthetic activity
	Effective WUE
Near-infrared imaging	Water content and its movement within leaves and soil
	Carbohydrate content of leaves
Fluorescence imaging and SPAD	Biotic and abiotic stress response
	Chlorophyll content
Visible light scanning	Degree of senescence
Root phenotyping (2D/3D)	Analysis of root morphology
Hyperspectral imaging	Analysis of shoot system
Liquid and gas chromatography	Estimation of chemical content
Mobile NIR	Soil water content
NIR spectroscopy	Seed and fruit phenotypic traits

discernment of informative patterns that align with empirical objectives, thereby enhancing the fidelity of data-driven insights. Plant breeders can address the dual issues of food security and climate change by combining ML and HTP to create resilient and high-yielding crop varieties more quickly. With an emphasis on its uses, difficulties, and prospects, this chapter investigates the function of ML in plant phenotyping.

3.2 Overview of HTP

Phenotyping refers to the measurement and analysis of observable traits in organisms, which result from the interaction between their genetic makeup (genotype) and environmental factors. In modern world, where research work is going on such a massive scale, phenotypic data gathering remains a challenge for crop breeding and functional genomics research (Deery et al. 2016). In agriculture, phenotyping is essential for understanding plant responses to genetic variation, environmental conditions, and management practices. However, traditional phenotyping methods are often limited by their scalability, precision, and cost. HTP addresses these limitations by automating the collection and analysis of phenotypic data. This method is specifically designed to measure a wide range of characteristics (Table 3.2), including growth patterns, yield, biomass, and stress tolerance, with the primary goal of understanding plant responses to genetic variation, environmental conditions, and management practices. HTP platforms utilize advanced technologies such as UAVs, ground-based robots, and multispectral sensors to capture high-resolution images and other data types. These platforms enable researchers to measure a wide range of traits, including plant height, canopy coverage, biomass, and stress tolerance, across large populations in a relatively short time.

The primary objectives of HTP in plant research and agriculture include accelerating plant breeding, understanding genotype-phenotype relationships, enhancing crop resilience and productivity, and optimizing resource use. By enabling the rapid evaluation of phenotypic traits across large populations, HTP speeds up the selection of superior genotypes, which is particularly important for developing crops with increased yield, disease resistance, and stress tolerance (Sandhu et al. 2022). Furthermore, HTP helps dissect the relationship between genetic makeup and observable traits, identifying genetic markers linked to important traits and improving the efficiency of breeding programs (Deery et al. 2016). The automation of data collection in HTP systems reduces manual labor and human error, making it possible to assess thousands of plants simultaneously. This scalability is crucial for identifying superior genotypes with desirable traits such as drought tolerance or disease resistance. For example, drones equipped with RGB cameras can capture images of entire fields, while ground-based robots can measure plant height and leaf area with high precision (Shakoor et al. 2017). Additionally, HTP systems incorporate noninvasive imaging modalities, such as hyperspectral and thermal imaging, to capture quantitative measurements of plant traits. These technologies provide detailed insights into plant physiology and stress responses, improving the reliability of phenotypic data (Tardieu et al. 2017).

3.3 ML for Plant Phenotyping

The capacity to collect enormous volumes of data on crops, soil, and environmental variables has been transformed in agriculture science with the introduction of big data and HTP. However, there are several difficulties in handling such large datasets. Drones, sensors, satellite imaging, and ground-based devices are just a few examples of diverse sources of agricultural data that can provide intricate data streams that require effective integration and processing. Due to the large amount, diversity, and velocity of these datasets, conventional data management techniques are insufficient for prompt analysis and interpretation (Li et al. 2014). Furthermore, HTP produces massive volumes of phenotypic data that need to be processed quickly and accurately to connect with environmental variables and genotypic data, making it challenging to find significant patterns and insights manually.

Addressing these issues requires the use of ML. Large datasets may be automatically analyzed by researchers using ML algorithms, which facilitates more effective data processing, pattern detection, and decision-making. DL, computer vision, and predictive modeling are examples of ML approaches that assist in gleaning valuable insights from complicated datasets for purposes including agricultural trait identification, yield prediction, disease detection, and resource optimization. Furthermore, supervised learning models may be trained to forecast results based on past data, but unsupervised learning techniques can categorize phenotypic data without human involvement. In the end, ML-driven technologies improve crop yield and sustainability by streamlining the data flow and guaranteeing that the vast volumes of data produced by HTP can be efficiently managed, examined, and utilized to make data-driven decisions in real time (Singh et al. 2016).

3.3.1 What Is ML?

ML, a subfield of artificial intelligence (AI), involves the design of computational models and algorithms that autonomously derive insights from data via statistical methodologies. This approach enables predictive analytics and adaptive decision-making without task-specific programming. Central to its framework is the application of statistical analysis to discern patterns and relationships within large-scale datasets, driving progressive performance optimization through iterative data processing. Core methodologies include: supervised learning, which trains models on annotated data for outcome prediction; unsupervised learning, which identifies intrinsic structures in unannotated datasets; and reinforcement learning, where agents refine decision policies through environmental interactions and reward-based feedback (Jordan and Mitchell 2015). These are its three main learning paradigms. Because these techniques are adept at managing complicated datasets, they have been effectively used in a variety of fields, such as image recognition, natural language processing, robotics, and medical diagnostics.

ML has become a useful tool for gleaning insights from nonlinear and high-dimensional data, enhancing conventional analytical techniques. To model complicated relationships in data and enable predictive analytics and automation of jobs that require a great deal of human skill, ML systems employ techniques like neural networks, decision trees, and support vector machines (SVMs; Murphy et al. 2012). ML offers a transformative advantage to plant scientists, including plant breeders, pathologists, physiologists, and agronomists, by enabling multivariate trait analysis through simultaneous parameter extraction, contrasting with conventional phenotypic approaches that isolate individual features. The other important advantage with ML is directly linking the variables extracted from HTP data to plant stresses (Zhang et al. 2019), biomass accumulation (Busemeyer et al. 2013b), grain yield (Crain et al. 2018), and soil characteristics. Models may be continuously improved with more data because of the iterative nature of ML algorithms, which increases efficiency and accuracy. Hence, ML is an essential tool in domains like genetics, climate modeling, and agricultural phenotyping, where innovation and discovery are fueled by the analysis and interpretation of large amounts of data.

3.3.2 ML in Handling Big Data

A core component of ML in big data involves data preprocessing and dimensionality reduction, employing techniques like Principal Component Analysis (PCA), a linear transformation method, and autoencoders, neural network-based architectures, to extract and retain the most discriminative features from high-dimensional datasets. This process minimizes redundancy and enhances data manageability by focusing on essential information, thus allowing researchers to concentrate on core patterns without being overwhelmed by the volume of data (Rathore et al. 2021). Given the enormous amount of data produced, ML models offer scalability through their use in distributed computing frameworks such as TensorFlow, Apache Spark, and Apache Hadoop, which permit parallel processing and the distribution of jobs across several nodes. In big data scenarios, this feature is essential since it greatly speeds up model training and analysis, particularly when working with large datasets that need a lot of processing power. By ensuring scalability, ML frameworks reduce the amount of time and resources required for data processing, hence increasing overall efficiency (L'heureux et al. 2017). Figure 3.1 shows the various applications of ML in bid data.

ML provides automation of data insights by leveraging pattern recognition capabilities, ML algorithms can detect patterns that would be challenging or impossible to uncover with traditional statistical methods (Zhang et al. 2020).

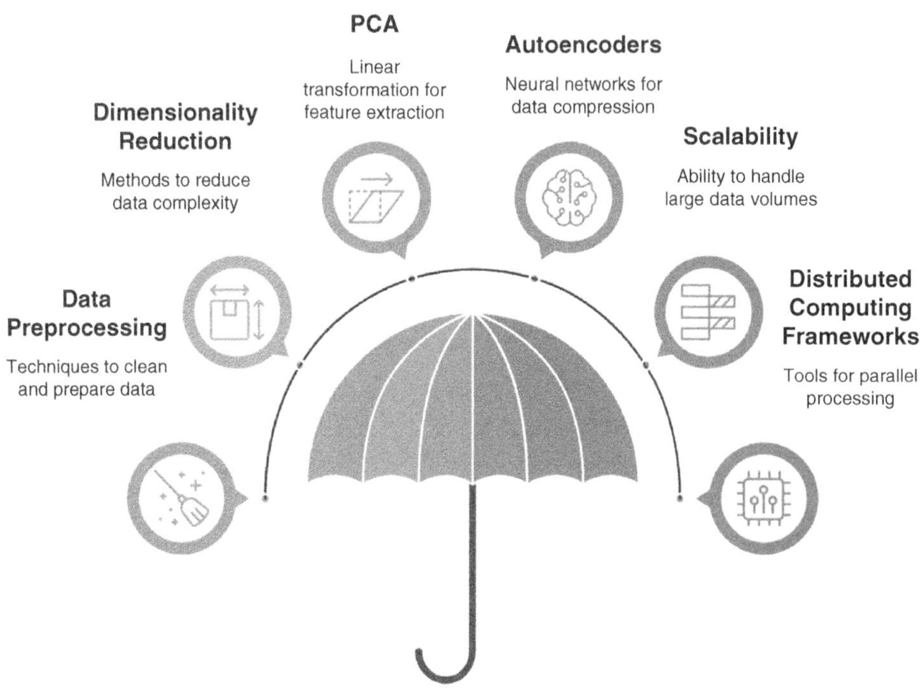

Figure 3.1 Role of ML in handling big data (created with Napkin.ai and the final layout was edited by the authors).

Since real-world datasets frequently include a mix of structured, semi-structured, and unstructured data types, ML's ability to handle data diversity is another strength in big data scenarios. By processing various formats with appropriate algorithms, ML models may adjust to this diversity.

Finally, in large data situations, ML improves anomaly detection, customization, and real-time data analysis. ML models are constantly updated with fresh data for applications that require real-time insights, including fraud detection and recommendation systems, guaranteeing prompt and pertinent solutions (L'heureux et al. 2017). By spotting departures from typical patterns, unsupervised learning methods like isolation forests and autoencoders may spot abnormalities like fraud or equipment failure. Additionally, recommendation engines that customize content recommendations depending on user interactions demonstrate how ML models facilitate customization by learning from individual tastes. By facilitating data-driven decision-making, improving user experience, and optimizing operational results across a range of sectors, ML eventually turns large data from unprocessed information into knowledge that can be put to use (Zhou et al. 2017).

3.4 Overview of ML Algorithms in Phenotyping

ML and DL have become indispensable tools in modern agriculture, particularly in plant phenotyping and stress monitoring. These advanced computational techniques enable precise and automated identification of plant stresses, significantly enhancing agricultural productivity. Supervised learning, a key ML approach, facilitates data-driven decision-making by establishing relationships between input and output variables. The primary objective is to create a predictive model that generalizes well to new data. Classification tasks involve categorizing outputs into predefined labels, while regression problems predict continuous values based on training data. These methods have been widely utilized in agricultural phenotyping, enabling automated stress detection and trait analysis. Numerous studies highlight ML applications in detecting various stress conditions (Table 3.3), such as powdery mildew in cucumber (*Cucumis sativus L.*) (Lin et al. 2019), aflatoxin contamination in maize (Yao et al. 2013), leaf rust in wheat (Ashourloo et al. 2014), and salinity stress in chickpea (*Cicer arietinum L.*) (Atieno et al. 2017).

Table 3.3 Machine learning models in agricultural and biological applications.

Supervised Learning Models		
Model	**Description**	**Applications**
Linear Discriminant Analysis (LDA)	This method employs a linear combination of features to classify observations into discrete categories, particularly suited for scenarios with continuous predictors and a categorical response variable.	Used for early-stage identification of pathological conditions in almond (Prunus dulcis) red leaf blotch (Peña et al. 2015). Applied to identify Huanglongbing disease in citrus orchards using visible-near infrared spectroscopy (Sankaran et al. 2011).
Support Vector Machine (SVM)	Constructs a hyperplane that maximizes separation between different classes in a dataset. The goal is to create the widest possible margin between classes for better classification accuracy.	Used to analyze canopy responses to soil water deficit in spinach (Spinacia oleracea) (Raza et al. 2014). Applied to identify drought stress in barley (Hordeum vulgare) through hyperspectral imaging (Behmann et al. 2014).
Logistic Regression	Uses a logistic function to predict a binary outcome based on predictor variables, often employing odds ratios for classification.	Utilized to distinguish between infected and noninfected plants by selecting relevant hyperspectral bands (Delalieux et al. 2007).
Random Forest	Employs a tree-based classification technique where multiple decision trees are used to improve prediction accuracy while reducing overfitting. It is robust against noise and can handle complex datasets.	Applied to extract relevant features from spectroradiometer data for identifying *phaeosphaeria* leaf spot in maize (Adam et al. 2017).

(Continued)

Table 3.3 (Continued)

Model	Description	Applications
Supervised Learning Models		
Linear Regression	A simple yet widely used model that quantifies the relationship between independent and dependent variables, often used for predictive analysis in plant sciences.	Used in maize to assess water stress by linking multispectral image data with vegetation indices (VI) and crop water stress index (CWSI) (Zhang et al. 2019). Leveraging high-throughput phenotyping methodologies, this study analyzed the relationship between projected shoot area and shoot fresh weight to evaluate salinity stress tolerance in rice cultivars (Hairmansis et al. 2014).
Multiple Linear Regression	Extends linear regression by incorporating multiple independent variables to improve the prediction of the outcome variable.	Developed predictive models for bacterial spot incidence (Jones et al. 2010).
Partial Least Squares Regression (PLSR)	A statistical approach used for modeling complex datasets with numerous variables, effectively handling collinearity among predictors.	This approach was applied to quantify foliar nitrogen levels in *Oryza sativa* using ground-based hyperspectral imaging techniques, correlating reflectance with nitrogen levels (Onoyama et al. 2013). Successfully predicted drought stress in soybean, achieving accuracy scores of 0.973 and 0.969 for different experimental groups (Mo et al. 2015).
Unsupervised Learning Models		
K-means Clustering	A clustering algorithm that groups similar data points based on feature similarity. In the context of plant sciences, it is often used to classify different physiological conditions.	Applied to extract RGB color values from plant surfaces to evaluate salinity tolerance in Arabidopsis thaliana (Awlia et al. 2016).
Dimensionality Reduction (Principal Component Analysis [PCA])	A technique that condenses large datasets into fewer variables while preserving essential information, making it useful for analyzing hyperspectral data.	Used to detect aflatoxin B1 contamination in maize (Zea mays) using hyperspectral imaging (Wang et al. 2011). This approach utilized hyperspectral imaging (400–1000 nm spectral range) to detect toxigenic fungal contamination in maize, coupled with discriminant analysis for categorizing fungal growth stages (Del Fiore et al. 2010).

3.4.1 DL in Plant Phenotyping

A sophisticated subset of ML, DL was first proposed in the 1950s (Samuel 1959) and uses enormous datasets and processing capacity to build hierarchical data representations (Min et al. 2017). As models automatically extract pertinent features from input data, DL does away with the need for manual feature engineering and reduces the amount of time spent on feature selection, which is frequently error prone (Sandhu et al. 2021b). By using deep neural networks for intricate pattern recognition tasks, DL simulates human cognitive processes (Figure 3.2). Convolutional neural networks (CNNs), generative adversarial networks (GANs), recurrent neural networks (RNNs), and multilayer perceptrons (MLP) are notable DL architectures (LeCun et al. 2015). These models are very useful for classification and prediction analysis because they use nonlinear activation functions to find complex patterns. It achieves state-of-the-art performance in phenotyping investigations and performs exceptionally well in applications including object recognition, picture segmentation, and image classification (Pérez-Enciso and Zingaretti 2019). MLP, GANs, CNNs, and RNNs are among the key DL models used for phenomics, however they are not the only ones.

DL models necessitate systematic hyperparameter optimization, encompassing adjustments to neural layer dimensions, activation functions, training iteration parameters, convolutional architecture configurations, feature extraction filters, dropout rates, and regularization methods. The sheer number of possible hyperparameter combinations imposes significant computational demands and necessitates advanced programming expertise. Additionally, overfitting remains a challenge, requiring appropriate mitigation strategies (Sandhu et al. 2021b).

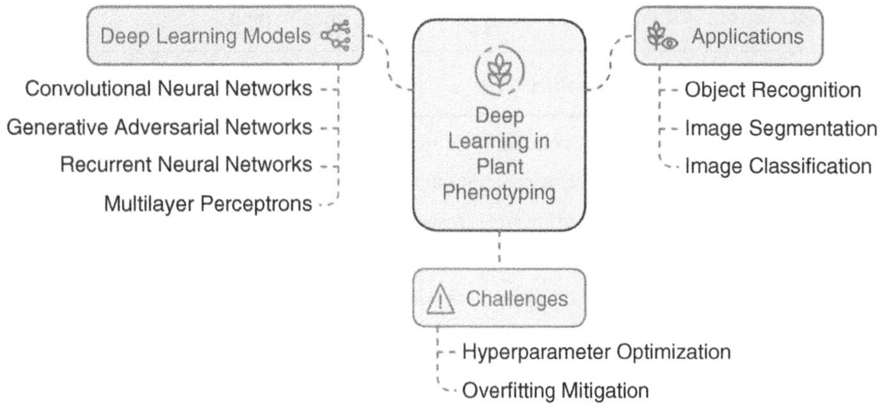

Figure 3.2 Several aspects related to deep learning in plant phenotyping.

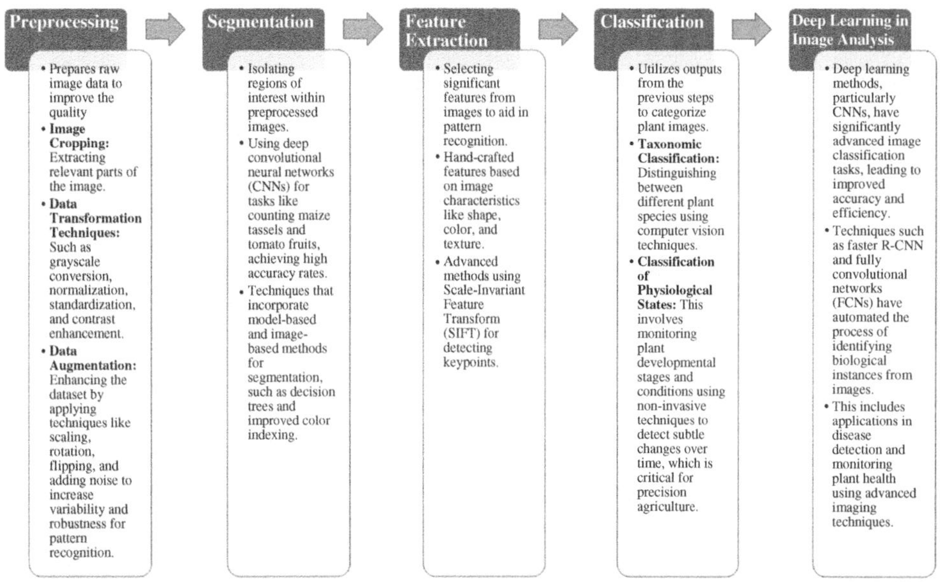

Figure 3.3 Image processing pipeline for precision agriculture: from preprocessing to deep learning analysis.

3.4.2 Computer Vision in Phenotyping

Computer vision is revolutionizing plant phenotyping by enabling high-throughput, noninvasive measurement of key plant traits. The convergence of ML, DL, and computer vision offers a scalable and accurate solutions for stress detection and trait analysis. Utilizing advanced image analysis and ML algorithms, computer vision techniques can rapidly assess parameters such as plant height, leaf area, canopy coverage, and stress indicators (Mochida et al. 2019). CNNs and other ML models facilitate image segmentation, feature extraction, and classification, replacing labor-intensive manual analysis (Figure 3.3). A wide array of imaging sensors, including RGB cameras, multispectral and hyperspectral imaging, thermal infrared sensors, and LiDAR, provide detailed spatial, spectral, and morphological data (Table 3.4). These technologies capture subtle variations in plant responses to environmental conditions, aiding in early stress detection. Furthermore, platforms such as UAVs, ground-based robots, and motorized gantries are increasingly deployed for large-scale phenotyping, enabling continuous crop monitoring throughout growth cycles (Table 3.5). The integration of computer vision with plant phenotyping has significantly advanced genotype-phenotype mapping, establishing critical links between genetic traits and observable characteristics. This capability is particularly vital in precision agriculture and genomic selection, where data-driven decision-making optimizes crop management and breeding programs. However, widespread implementation requires standardized technologies and user-friendly tools. Applications such as EasyPCC and StomataCounter have simplified image segmentation and trait measurement, but further advancements in data integration and accessibility are necessary for broader adoption.

Table 3.4 Various imaging sensors widely used in plant phenotyping.

Sensor Type	Application	Reference
RGB Camera	Quick color and texture-based phenotyping, plant morphology analysis	Mochida et al. (2019)
Multispectral Camera	Capturing spectral information for leaf area index and canopy coverage	Potgieter et al. (2017)
Hyperspectral Camera	Monitoring stress responses, disease detection, and trait analysis	Wahabzada et al. (2016)
Thermal Infrared	Early detection of stress indicators, especially water stress	Poblete et al. (2018)
LiDAR	3D plant structure modeling, biomass estimation, and height measurement	Guo et al. (2018)
Unmanned Aerial Vehicles (UAVs)	Large-scale, field-based phenotyping for traits like canopy coverage and plant height	Yang et al. (2017)
3D Imaging Systems	Detailed 3D reconstruction of plant organs for morphological studies	Montagnoli et al. (2016)

Table 3.5 Various platforms used in plant phenotyping.

Platform	Application	Reference
Unmanned Aerial Vehicles (UAVs)	Large-scale field phenotyping for traits like canopy coverage, plant height, and leaf area	Potgieter et al. (2017)
Ground-Based Robots	In-field high-throughput phenotyping for row crops and detection of stress responses	Underwood et al. (2017)
Tray Conveyor Systems	Controlled environment phenotyping for small plants, such as seedlings and young plants	Montagnoli et al. (2016)
Motorized Gantries	Automated data capture across greenhouse or field trials for growth monitoring	Virlet et al. (2017)
Phenome High-throughput Investigator (PHI)	Automated imaging and growth tracking in controlled environments	Lyu et al. (2017)
PhenoTrac 4 Mobile Platform	Canopy-level phenotyping in field conditions, particularly for cereal crops	Barmeier and Schmidhalter (2017)
Phenomobile	Field-based phenotyping for various traits, including stay-green and canopy cover	Rebetzke et al. (2016)
RIKEN Integrated Plant Phenotyping System	Indoor phenotyping system for stress response and water use efficiency	Fujita et al. (2018)

3.5 Applications of ML in Plant Phenotyping

Smart agriculture relies on Internet of Things (IoT) technologies for improved plant sensing and monitoring. Early detection of plant stress is critical for sustaining agricultural productivity. Delayed identification can lead to severe diseases, significantly impacting crop yields. Compared to traditional manual monitoring, remote sensing methods offer higher accuracy in stress detection. Researchers have proposed automated frameworks that utilize ML to identify and classify plant health conditions in real-time. These frameworks integrate vision-based sensors and environmental monitoring to enhance the precision of stress detection (Figure 3.4).

To minimize damage, visual and environmental sensors detect early-stage leaf discoloration, facilitating proactive interventions. ML-based stress detection frameworks employ vision-based sensors to capture leaf images, extract features, and analyze the data using cloud computing platforms. Techniques such as SVMs are commonly employed for classification tasks, reducing hardware complexity by focusing on diseased leaf portions. By accurately localizing affected plants, this approach enables precision agriculture, optimizing pesticide use and mitigating disease spread. Image processing plays a fundamental role in plant disease detection by enabling preliminary data segmentation and classification, leading to accurate identification of afflicted regions.

Recent advancements in image analysis have significantly transformed plant phenotyping. However, a major challenge persists due to the lack of robust analytical methods for processing complex plant images. Addressing this issue requires

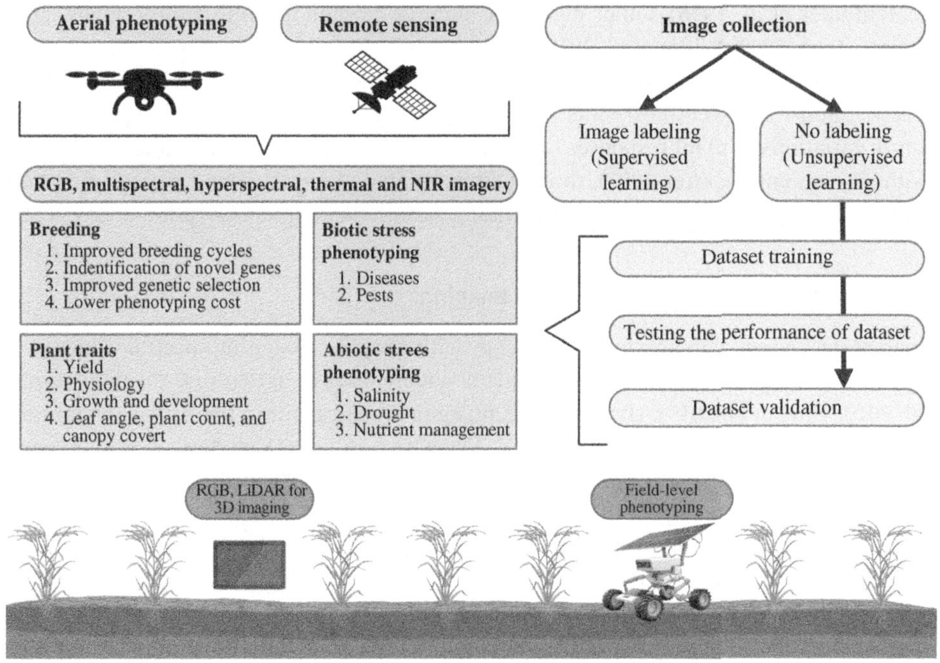

Figure 3.4 Strategies to phenotyping of crops under stress conditions.

refining both imaging techniques and analytical models. HTP begins with appropriate imaging methodologies, which may involve capturing images of multiple plants simultaneously or focusing on individual plants. Dedicated systems, though costly, help isolate plants, thereby simplifying subsequent analysis. Conversely, when plants are not isolated, segmentation—distinguishing the plant from its background—becomes a complex task due to occlusion and overlapping structures. Therefore, accurate feature extraction and image processing are crucial for leveraging ML in phenotyping. Current phenotyping methodologies predominantly rely on domain expertise-driven feature extraction, necessitating specialized proficiency in image analysis. Feature selection serves as a critical determinant of ML model efficacy, underscoring the need for innovations in automated feature engineering and deep neural architectures to elevate phenotyping precision. The adoption of advanced algorithmic frameworks can enhance precision in trait quantification while mitigating dependence on manual annotation. Future efforts should prioritize the refinement of feature selection protocols, optimization of DL architectures, and harmonization of imaging systems to bolster scalability and reproducibility in agricultural phenomics, thereby advancing sustainable agricultural innovation.

3.5.1 ML for Plant Recognition and Disease Detection

Plant recognition and disease identification traditionally rely on visual, chemical, and manual inspection methods. Visual assessment depends on experienced farmers' observations, while chemical methods utilize reagents for disease identification. However, these conventional approaches are time-consuming, labor-intensive, and prone to subjectivity, often resulting in decreased agricultural efficiency. ML, automation, and image processing offer faster, more accurate alternatives for plant disease diagnosis.

Contemporary optical recognition systems integrate preprocessing, image segmentation, discriminative feature extraction, feature correlation analysis, and automated classification algorithms. Digital image processing techniques classify plant species and detect stress conditions by analyzing structural and spectral characteristics of leaves. Studies indicate that green leaves signify healthy plants, whereas stressed plants exhibit changes such as deformation, yellowing, or browning (AlShahrani et al. 2018). The severity of plant diseases can be determined using ML models, which leverage high-resolution imaging data to classify plant health.

3.5.2 Spectral Analysis and Optical Imaging for Stress Detection

The physiological and chemical properties of plants influence light reflectance and absorption. Stress conditions alter these properties, leading to significant spectral variations. For instance, chlorophyll degradation due to stress is associated with increased reflectance around 700 nm (Zhu et al. 2020) and decreased reflectance in the 530–630 nm range

(Lichtenthaler et al. 1996). Other pigments, such as carotenoids and xanthophylls, also impact reflectance properties (Gitelson et al. 2002; Vilfan et al. 2018). Stomatal regulation under stress can further alter leaf temperature, which can be detected using infrared imaging techniques (Sawinski et al. 2013). Optical sensors such as thermal, fluorescence, multispectral, and hyperspectral devices facilitate plant health assessment by capturing reflectance spectra. These sensors analyze variations in plant responses to biotic and abiotic stressors. The visible spectrum is particularly effective for plant health assessments (Carter 1993), though stress-related changes extend beyond this range. Spectroscopic and imaging techniques generate spectral graphs, illustrating variations in light absorption and reflectance patterns.

3.5.3 Hyperspectral and Multispectral Imaging

Hyperspectral imaging integrates spectroscopy with imaging, assigning spectral values to each pixel across multiple wavelengths (Pandey et al. 2017). This technique is widely used in agriculture for phenotyping and stress detection (Rumpf et al. 2010; Yang et al. 2019). Hyperspectral imaging operates within the 250–2500 nm range, covering ultraviolet, visible, and near-infrared spectra (Lowe et al. 2017). Vegetation indices (VIs) derived from hyperspectral data aid in stress monitoring, while spectral disease indices (SDIs) distinguish between specific plant diseases (Meng et al. 2020).

Multispectral imaging, in contrast, captures data within specific wavelength bands rather than across a continuous spectrum. These methods rely on designated wavebands to collect spectral information. While multispectral imaging uses cameras to capture structured data, multispectral spectroscopy generates spectral data for predefined wavebands. RGB imaging, a common multispectral technique, relies on red, green, and blue spectral components to generate visual data (Lowe et al. 2017). Digital cameras and smartphones facilitate RGB imaging, making it more accessible than specialized hyperspectral methods.

3.5.4 Thermal and Fluorescence Imaging for Stress Analysis

Unlike reflectance-based methods, thermal imaging detects radiation emissions, measuring temperature variations associated with plant stress responses (Mastrodimos et al. 2019). Infrared thermography visualizes surface temperature differences, helping to detect stomatal closure, a key stress indicator (Jones 1999). Thermal imaging thus provides valuable insights into plant transpiration and water-use efficiency.

Fluorescence imaging captures emissions from plant tissues excited by specific wavelengths. Chlorophyll fluorescence (650–800 nm) and blue-green fluorescence (400–600 nm) are widely analyzed for stress detection (Krause and Weis 1984; Lang et al. 1992). Fluorescence-based methods distinguish between healthy and stressed plants by analyzing emitted light ratios (Buschmann and Lichtenthaler 1998). Multicolor fluorescence imaging enhances this approach by capturing fluorescence signals across multiple bands, facilitating precise stress assessment (Buschmann and Lichtenthaler 1998).

3.5.5 ML Approaches for Plant Stress Classification

ML models, including SVM, neural networks, and clustering algorithms, enhance plant disease classification and stress prediction. Supervised learning methods such as SVM classify plant health based on predefined training data, while unsupervised clustering techniques identify patterns within unlabeled datasets (Dey 2016). The cosine similarity, Euclidean distance, and Manhattan distance metrics are commonly employed for clustering-based stress classification.

Artificial neural networks (ANNs) have been extensively applied in agricultural applications (Tang et al. 2003). Two major ANN architectures include generalized regression neural networks (GRNN) and feedforward backpropagation neural networks (BPNN). GRNN employs a probabilistic density function to estimate relevant model parameters, while BPNN utilizes multiple hidden layers for feature extraction and pattern recognition. Deep convolutional networks, a subset of ANN models, achieve high accuracy in stress classification, leveraging hierarchical feature extraction techniques. The dropout technique mitigates overfitting in neural networks by preventing complex co-adaptation in training data (AlShahrani et al. 2017).

3.5.6 Automated Image Analysis for Trait Extraction

3.5.6.1 Morphological Trait Measurement

Automated image analysis has significantly enhanced the measurement of morphological traits in plant phenotyping (Table 3.6), addressing challenges associated with real-time assessment of plant height, canopy coverage, and leaf area.

Table 3.6 Advancements in image-based plant phenotyping.

Key Focus	Technologies/Methods	Applications	References
Morphological Trait Measurement: Canopy Coverage and Height	UAVs equipped with RGB cameras, LiDAR, and Structure from Motion-Multi-View Stereo (SfM-MVS); ACE R-CNN for segmentation	Assessment of canopy cover and species identification in complex forest environments using UAV and LiDAR data	Li et al. (2022)
Leaf Area Index (LAI) and Biomass Estimation	Hyperspectral imaging, vegetation indices, red-edge parameters, partial least squares regression (PLSR)	Estimation of leaf area index (LAI) and above-ground biomass (AGB) in winter wheat using UAV-based hyperspectral remote sensing	Tao et al. (2020)
Organ Detection and Counting	Deep learning models (Mask R-CNN, Faster R-CNN), UAV-based imagery for segmentation	Automated detection of plant organs (e.g., grapevine pruning regions, wheat leaf tips) to enhance precision agriculture	Guadagna et al. (2023), Li et al. (2023)
Image-Based Feature Extraction: Object Counting and Yield Estimation	CNN architectures (YOLO, ResNet-50), UAV imaging, deep learning-based segmentation	Automated yield estimation in cotton fields, orchard fruit counting, and phenotyping using UAV imagery	Khaki and Wang (2019), Li et al. (2022)
Domain Adaptation for Cross-Crop Analysis	CycleGAN and domain adaptation techniques for deep learning model generalization	Enhancing plant phenotyping model performance by adapting models trained on one crop to another	Li et al. (2023)
Data Augmentation and Synthetic Data for Model Training	Synthetic image generation, domain adaptation, CNN-based augmentation	Improving robustness of plant stress detection and disease identification models through augmented datasets	Saleem et al. (2022), Shorten and Khoshgoftaar, (2019)

Advances in imaging technologies, including UAVs and ground-based sensors, enable high-resolution data collection, allowing precise measurement of canopy cover and senescence. For example, UAVs equipped with RGB cameras have been successfully deployed in maize trials to assess canopy traits, correlating them with yield potential and stress tolerance. Moreover, recent methodologies facilitate the segmentation of plant organs from 3D images, improving trait quantification even in densely populated crop canopies. The integration of CNNs has further refined morphological trait analysis. Neural networks, such as Mask R-CNN, have proven effective in detecting panicles and spikes, enhancing accuracy even when occlusion is present. These advancements reduce the labor and time needed for field measurements, making HTP platforms increasingly viable for large-scale agricultural research.

3.5.6.2 Image-based Feature Extraction

ML models and image processing algorithms play a crucial role in feature extraction, improving the accuracy of trait analysis. CNN architectures such as ResNet-50 and LedNet have been employed to detect and segment yield-related traits, including fruit and grain heads (Table 3.6). Object counting techniques, utilized in soybean pod and cotton boll detection, leverage RGB and multispectral imagery to quantify key parameters with precision. To enhance model adaptability across different environments, domain adaptation techniques like CycleGAN facilitate model transferability across crop types. Synthetic data generation and geometric transformations further bolster dataset diversity, ensuring robustness against varying field conditions. These advancements position image-based feature extraction as a cornerstone of high-throughput field phenotyping.

3.5.7 Prediction Models for Yield Forecasting

Yield forecasting is crucial for optimizing agricultural management, market strategy, and resource allocation. Accurate predictions enable farmers, policymakers, and agribusinesses to mitigate risks, optimize input usage, and ensure food security. ML's scalability allows it to process diverse data sources, including drone imagery, high-resolution satellite data, and IoT sensor inputs. These models continuously refine their accuracy over time, providing real-time, data-driven insights for optimizing resource use, increasing productivity, and mitigating climate-related uncertainties. Table 3.7 mentions commonly used prediction models employed for yield forecasting.

Table 3.7 Commonly used prediction models for yield forecasting.

Prediction Model	Description	Applications	Limitations	References
Linear Regression	Establishes a relationship between independent and dependent variables	Models the linear link between predictors and yield	Unable to model complex, nonlinear relationships	Shi et al. (2022)
Decision Trees	Splits data at nodes for yield predictions	Can handle nonlinear data and smaller datasets	Prone to overfitting	Nevavuori et al. (2020)
Random Forest	Combines multiple decision trees to improve accuracy	Robust, flexible, models nonlinear relationships, reduces overfitting	More complex	Kuradusenge et al. (2023)
Support Vector Machines (SVM)	Finds the optimal hyperplane to predict outcomes	Suitable for complex and nonlinear relationships	Computationally expensive	Teja et al. (2022)
Convolutional Neural Networks (CNNs)	Analyzes satellite images to detect patterns related to crop health and predict yields	Effective for spatial data	Requires a lot of data and computing power	Russello (2018)
Recurrent Neural Networks (RNNs) and Long Short-Term Memory (LSTM)	Designed for sequential data, well-suited for time series forecasting	Capable of learning long-term dependencies in temporal data	Requires substantial computational resources	Nejad et al. (2023)
Gradient Boosting Machines (GBM)	Builds models sequentially	Highly accurate and handles missing data well	Slow but effective	Dhillon et al. (2023)
K-Nearest Neighbors (KNN)	A nonparametric model, predicts the outcome by finding the nearest neighbor K	Better outcomes	Unable to handle large datasets, expensive	Raman et al. (2024)
Time Series Models (ARIMA, SARIMA)	Forecasts based on past observations and temporal trends	Handles historical yield data and climate patterns	Highly complex	Saini et al. (2020)
Geospatial Models (GIS-based)	Uses remote sensing data to map and predict yield	Handles regional variability and spatial patterns	Data-intensive	Omeka et al. (2024)
Hybrid Models	Combine multiple models to leverage the strengths of each	More robust and adaptable	Complex and requires expertise	Oikonomidis et al. (2022)
Mechanistic Crop Models (DSSAT, APSIM)	Simulation-based models	Makes long-term predictions	Complex to set up	Feleke et al. (2021)

3.6 Integration of ML with Emerging Technologies

The integration of ML with emerging technologies is transforming various industries, including agriculture. Research highlights how robotics, robotic process automation (RPA), and cognitive automation are reshaping traditional agricultural practices, improving efficiency and sustainability (Liakos et al. 2018). Advances in sensor technology, particularly in HTP, have been pivotal in modernizing crop monitoring and assessment. The deployment of UAVs equipped with RGB, multispectral, hyperspectral, and thermal cameras has significantly enhanced agricultural data collection and precision farming (Nduku et al. 2023).

Robotic systems are increasingly automating labor-intensive agricultural tasks, while RPA utilizes software-driven automation to streamline repetitive processes. Cognitive automation, powered by AI and large-scale data analytics, enhances decision-making accuracy and operational efficiency. The IoT has also facilitated smart farming, with IoT-enabled environmental sensors improving data collection on genotype-by-environment (G × E) interactions (Ouafiq et al. 2021).

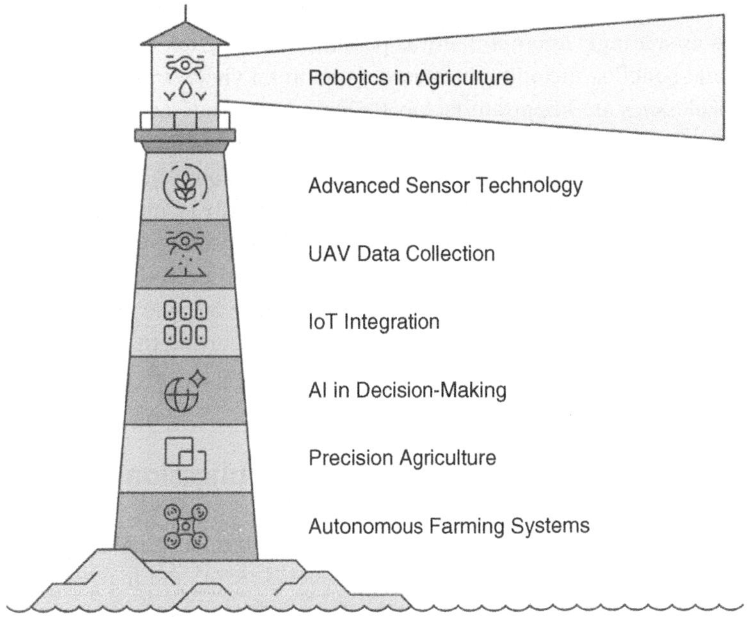

Figure 3.5 Path to autonomous farming.

Innovations such as autonomous tractors, robotic milking systems, and UAV-based crop monitoring are revolutionizing agriculture, with projects like Hands-Free Hectare demonstrating the feasibility of fully autonomous farming.

Precision agriculture utilizes AI and ML to analyze large datasets from sensors, satellite imagery, and historical records, optimizing resource allocation and decision-making. Variable-rate technology (VRT), supported by AI, enables site-specific application of water, fertilizers, and pesticides, maximizing crop yield while minimizing environmental impact (Benos et al. 2021). AI-driven predictive maintenance further enhances farming efficiency by identifying machinery failures before they occur, reducing downtime and repair costs.

ML models serve a critical function in real-time crop health assessment by analyzing multispectral data acquired via UAVs, satellite remote sensing, and IoT-enabled ground sensors. These technologies systematically monitor environmental parameters to identify biotic stressors (e.g., pest infestations) and abiotic constraints (e.g., nutrient deficiencies), enabling timely agricultural interventions. AI-powered forecasting models improve pest management by predicting outbreaks, allowing for targeted interventions that reduce chemical use and environmental harm (Attri et al. 2024). Additionally, AI enhances harvesting efficiency by optimizing harvest times and automating fruit and vegetable picking with robotic systems.

Post-harvest processing benefits significantly from AI integration. ML algorithms improve sorting accuracy, storage conditions, and shelf-life predictions, ensuring higher-quality produce reaches consumers. The convergence of AI, robotics, IoT, and advanced imaging technologies is driving agriculture toward a more data-driven, sustainable, and efficient future, paving the way for fully autonomous farming systems (Figure 3.5).

3.7 Future Directions and Potential of ML in Agriculture

In order to handle issues on the ground, agriculture is a multidisciplinary profession that accepts applications from other sectors. Accurate crop disease diagnosis and biomass estimate are made possible by the development of ML techniques like CNNs and LiDAR. Despite these developments, usage in breeding programs is still restricted because of issues with usability and data integration. Process simplification is the goal of programs like EasyIDP, but wider adoption necessitates cooperation between researchers and breeders to standardize HTP-derived metrics and make sure the tools are useful for crop improvement applications and breeder-oriented. The selection process for stress resilience can be accelerated by combining omic methods with ML-based methodologies. In particular, genomic data may be used to test for ailments on the field and forecast superior phenotypes in real time. These pipelines may aid in the production of new varieties and/or the identification of better lines for future breeding initiatives. Furthermore, precision agriculture promises to advance agriculture in a sustainable manner by offering a number of alternatives to the limited use of pesticides as well as field techniques that can point the way. Utilizing digital cameras and UAVs, a series of programs may be executed to provide the software with accurate data that should be used realistically in the field. This will guarantee that modifications are used as sparingly and optimally as possible, avoiding toxicity or shortage. The use of ML applications in agriculture has several benefits, including increased agricultural yields and better resource management. Nonetheless, data privacy and ethical issues are important factors that require consideration. First of all, questions of ownership and consent are raised by the use of farmers' personal data, such as their financial information and farming methods. Farmers must be given the power to manage their data and be informed about how it will be gathered, kept, and used. Furthermore, bias and fairness concerns are crucial for ML systems. Minority farmers may suffer from distorted predictions made by the resultant models if training datasets are not representative of various farming practices or populations. To reduce this risk, algorithm development must be transparent. Furthermore, to prevent breaches of sensitive data, secure data transfer and storage techniques must be used. In addition to protecting farmers' data, implementing strong cybersecurity measures increases stakeholder's confidence. All things considered, resolving these ethical issues and data security concerns will allow for the proper application of ML in agriculture, encouraging creativity while protecting the rights and interests of those engaged.

3.8 Case Studies and Real-world Applications

ML has held up lot many examples for its contribution in identification of plant responses to abiotic and biotic stresses, such as drought stress in barley (Behmann et al. 2014) and maize (Asaari et al. 2019), yellow rust (Zhang et al. 2014) and powdery mildew (Cao et al. 2013) in wheat, salt stress in okra (feng et al. 2020), and Black Sigatoka disease in banana plants (Ochoa et al. 2016), have been identified using hyperspectral imaging, which has grown in popularity for plant phenotyping and stress detection in agriculture (Rumpf et al. 2010; Zovko et al. 2019) To distinguish between distinct plant diseases, SDIs have been created using hyperspectral data (Meng et al. 2020). Examples include indices for identifying myrtle rust (Heim et al. 2019), Cercospora leaf spot in sugar beet (Mahlein et al. 2013), leaf rust in wheat (Ashourloo et al. 2014), and powdery mildew in wheat (Huang et al. 2014) and sugar beet (Mahlein et al. 2013). The normalized difference vegetation index (NDVI) (Rouse et al. 1974), water index (WI) (Penuelas et al. 1997), and photochemical reflectance index (PRI) (Gamon et al. 1992) are notable vegetation indicators. Developing novel VIs and SDIs for the identification of very precise plant stressors is made possible by the enormous quantity of spectral data that is gathered through hyperspectral imaging. Additionally, multispectral spectroscopy and multispectral imaging have been both effectively employed for recognizing plant stresses. For instance, multispectral spectroscopy has been utilized to identify nitrogen deficiency stress in maize (Kitić et al. 2019), drought stress in tomato plants (Veys et al. 2017), and nitrogen deficiency in canola plants (Habibullah et al. 2020), while multispectral imaging has been employed to recognize leaf spot disease in oilseed rape (Veys et al. 2017), gray mold in tomato leaves (Fahrentrapp et al. 2019), and nutrient deficiencies in tomato plants (Cardim Ferreira Lima et al. 2020). Compared to their hyperspectral equivalents, multispectral methods provide more cheap sensors; nevertheless, because of their wider wavebands, they do not give sufficient knowledge regarding the plant and its surroundings. However, multispectral techniques also have the benefit of being portable and adaptable, which can help with the development of customized devices. RGB imaging's primary benefits are its low cost and compact, portable sensor size. Numerous plant stresses, including iron deficiency chlorosis in soybeans (Naik et al. 2017), multiple nutrient deficiencies in black gram (Watchareeruetai et al. 2018), early and late blight in potato plants (Islam et al. 2017), and biotic stresses in wheat (Xie et al. 2016), have been successfully evaluated using RGB image sensors, which are already found on smartphones. Additionally, a number of biotic and abiotic stressors, including *Aspergillus carbonarius* infection in grapes (Mastrodimos et al. 2019), apple scab disease (Oerke et al. 2011), drought stress in maize (Casari et al. 2019), and drought stress in sesame plants (Khorsandi et al. 2018), have been identified using thermography. Numerous studies have employed fluorescence imaging to identify biotic and abiotic stressors, including cold stress in tomato seedlings (Dong et al. 2019), herbicide stress in soybeans (Li et al. 2017), and biotic and abiotic stressors in barley, grapevine, and sugar beet (Konanz et al. 2014). A smartphone with band-pass filters could represent a reasonably easy and portable way to get fluorescence images; nevertheless, methods using this kind of equipment for plant stress applications are still hard to come across.

3.9 Conclusion

ML has become a transformative tool for interpreting large-scale datasets derived from HTP, offering significant potential to advance the development of climate-resilient, high-yield crop varieties. This integration addresses critical agricultural challenges, including food security and climate adaptation. ML applications span real-time plant health diagnostics, yield prediction, and trait analysis, yet challenges persist in data heterogeneity, algorithmic biases, and ethical considerations surrounding data privacy and model transparency. The synergistic integration of ML with robotics, UAVs, and IoT technologies enhances automation, precision data acquisition, and resource-use optimization in farming systems. By resolving existing limitations and capitalizing on technological advancements, ML-driven innovations can promote sustainable agricultural practices. In conclusion, ML holds a promising trajectory in agriculture, with continued refinement of algorithms and interdisciplinary collaboration poised to support resilient crop development and equitable food systems in response to global demographic demands.

References

Adam, E., Deng, H., Odindi, J. et al. (2017). Detecting the early stage of phaeosphaeria leaf spot infestations in maize crop using in situ hyperspectral data and guided regularized random forest algorithm. *J Spectrosc* 2017: 1–8. https://doi.org/10.1155/2017/6961387

AlShahrani, A.M., Al-Abadi, M.A., Al-Malki, A.S. et al. (2018). Automated system for crops recognition and classification. In *Computer Vision: Concepts, Methodologies, Tools, and Applications* (pp. 1208–1223). IGI Global.

Asaari, M.S.M., Mertens, S., Dhondt, S. et al. (2019). Analysis of hyperspectral images for detection of drought stress and recovery in maize plants in a high-throughput phenotyping platform. *Computers and Electronics in Agriculture* 162: 749–758.

Ashourloo, D., Mobasheri, M., and Huete, A. (2014). Evaluating the effect of different wheat rust disease symptoms on vegetation indices using hyperspectral measurements. *Remote Sens (basel)* 6: 5107–5123. https://doi.org/10.3390/rs6065107

Atieno, J., Li, Y., Langridge, P. et al. (2017). Exploring genetic variation for salinity tolerance in chickpea using image-based phenotyping. *Sci Rep* 7: 1300. https://doi.org/10.1038/s41598-017-01211-7

Attri, I., Awasthi, L. K., and Sharma, T. P. (2024). Machine learning in agriculture: A review of crop management applications. *Multimedia Tools and Applications* 83 (5): 12875–12915.

Awlia, M, Nigro, A, Fajkus, J. et al. (2016). High-throughput non-destructive phenotyping of traits that contribute to salinity tolerance in Arabidopsis thaliana. *Front Plant Sci* 7: 1414. https://doi.org/10.3389/fpls.2016.01414

Barmeier, G. and Schmidhalter, U. (2017). High-throughput field phenotyping of leaves, leaf sheaths, culms and ears of spring barley cultivars at anthesis and dough ripeness. *Front Plant Sci* 8: 1920.

Behmann, J., Schmitter, P., Steinrücken, J., et al. (2014). Ordinal classification for efficient plant stress prediction in hyperspectral data. *Int Arch Photogramm Remote Sens Spatial Inf Sci* XL–7:29–36. https://doi.org/10.5194/isprs archi ves-XL-7-29-2014

Behmann, J., Steinrücken, J., and Plümer, L. (2014). Detection of early plant stress responses in hyperspectral images. *ISPRS Journal of Photogrammetry and Remote Sensing* 93: 98–111.

Benos, L., Tagarakis, A. C., Dolias, G. et al. (2021). Machine learning in agriculture: A comprehensive updated review. *Sensors* 21 (11): 3758.

Buschmann, C., and Lichtenthaler, H.K. (1998). Principles and characteristics of multi-colour fluorescence imaging of plants. *Journal of Plant Physiology* 152 (2–3): 297–314.

Busemeyer, L., Ruckelshausen, A., Möller, K. et al. (2013b). Precision phenotyping of biomass accumulation in triticale reveals temporal genetic patterns of regulation. *Sci Rep* 3: 2442. https://doi.org/10.1038/srep0 2442

Cao, X., Luo, Y., Zhou, Y. et al. (2013). Detection of powdery mildew in two winter wheat cultivars using canopy hyperspectral reflectance. *Crop Protection* 45: 124–131.

Cardim Ferreira Lima, M., Krus, A., Valero, C. et al. (2020). Monitoring plant status and fertilization strategy through multispectral images. *Sensors* 20 (2): 435.

Carter, G. A. (1993). Responses of leaf spectral reflectance to plant stress. *American Journal of Botany* 80 (3): 239–243.

Casari, R. A., Paiva, D. S., Silva, V. N. et al. (2019). Using thermography to confirm genotypic variation for drought response in maize. *International Journal of Molecular Sciences* 20 (9): 2273.

Crain, J., Mondal, S., Rutkoski, J. et al. (2018). Combining high-throughput phenotyping and genomic information to increase prediction and selection accuracy in wheat breeding. *Plant Genome* 11 (1): 170043. https://doi.org/10.3835/plantgenome2017.05.0043

Deery, D. M., Rebetzke, G. J., Jimenez-Berni, J. A. et al. (2016). Methodology for high-throughput field phenotyping of canopy temperature using airborne thermography. *Frontiers in Plant Science* 7: 1808.

Del Fiore, A., Reverberi, M., Ricelli, A. et al. (2010). Early detection of toxigenic fungi on maize by hyperspectral imaging analysis. *Int J Food Microbiol* 144: 64–71. https://doi.org/10.1016/j.ijfoodmicro.2010.08.001

Delalieux, S., van Aardt, J., Keulemans, W. et al. (2007). Detection of biotic stress (Venturia inaequalis) in apple trees using hyperspectral data: Non-parametric statistical approaches and physiological implications. *Eur J Agron* 27: 130–143. https:// doi.org/10.1016/j.eja.2007.02.005

Dey, N. (ed.). (2016). *Classification and Clustering in Biomedical Signal Processing*. IGI global.

Dhillon, M. S., Dahms, T., Kuebert-Flock, C. et al. (2023). Integrating random forest and crop modeling improves the crop yield prediction of winter wheat and oil seed rape. *Frontiers in Remote Sensing* 3: 1010978.

Dong, Z., Men, Y., Li, Z. et al. (2019). Chlorophyll fluorescence imaging as a tool for analyzing the effects of chilling injury on tomato seedlings. *Scientia Horticulturae* 246: 490–497.

Fahrentrapp, J., Ria, F., Geilhausen, M. et al. (2019). Detection of gray mold leaf infections prior to visual symptom appearance using a five-band multispectral sensor. *Frontiers in Plant Science* 10: 628.

Feleke, H. G., Savage, M. J., and Tesfaye, K. (2021). Calibration and validation of APSIM–Maize, DSSAT CERES–Maize and AquaCrop models for Ethiopian tropical environments. *South African Journal of Plant and Soil* 38 (1): 36–51.

Feng, X., Zhan, Y., Wang, Q. et al. (2020). Hyperspectral imaging combined with machine learning as a tool to obtain high-throughput plant salt-stress phenotyping. *The Plant Journal* 101 (6): 1448–1461.

Fujita, M., Tanabata, T., Urano, K. et al. (2018). RIPPS: A plant phenotyping system for quantitative evaluation of growth under controlled environmental stress conditions. *Plant Cell Physiol* 59 (10): 2030–8.

Gamon, J. A., Penuelas, J., and Field, C. B. (1992). A narrow-waveband spectral index that tracks diurnal changes in photosynthetic efficiency. *Remote Sensing of Environment* 41 (1): 35–44.

Gitelson, A. A., Zur, Y., Chivkunova, O. B., and Merzlyak, M. N. (2002). Assessing carotenoid content in plant leaves with reflectance spectroscopy. *Photochemistry and Photobiology* 75 (3): 272–281.

Guadagna, P., Fernandes, M., Chen, F. et al. (2023). Using deep learning for pruning region detection and plant organ segmentation in dormant spur-pruned grapevines. *Precision Agriculture* 24 (4): 1547–1569.

Guo, Q., Wu, F., Pang, S. et al. (2018). Crop 3D-a LiDAR based platform for 3D high-throughput crop phenotyping. *Sci China Life Sci* 61 (3): 328–39.

Habibullah, M., Mohebian, M. R., Soolanayakanahally, R. et al. (2020). Low-cost multispectral sensor array for determining leaf nitrogen status. *Nitrogen* 1 (1): 67–80.

Hairmansis, A., Berger, B., Tester, M. et al. (2014). Image-based phenotyping for non-destructive screening of different salinity tolerance traits in rice. *Rice (NY)* 7: 16. https://doi.org/10.1186/s12284-014-0016-3

Heim, R.H.J., Wright, I.J., Allen, A.P. et al. (2019). Developing a spectral disease index for myrtle rust (*Austropuccinia psidii*). *Plant Pathology* 68 (4): 738–745.

Hunter, M.C., Smith, R.G., Schipanski, M.E. et al. (2017). Agriculture in 2050: Recalibrating targets for sustainable intensification. *Bioscience* 67: 386–391. https://doi.org/10.1093/biosci/bix010

Islam, M., Dinh, A., Wahid, K. et al. (2017). Detection of potato diseases using image segmentation and multiclass support vector machine. In *2017 IEEE 30th Canadian Conference on Electrical and Computer Engineering (CCECE)* (pp. 1–4). IEEE.

Jones, C. D, Jones, J. B., and Lee, W. S. (2010). Diagnosis of bacterial spot of tomato using spectral signatures. *Comput Electron Agric* 74: 329–335. https://doi.org/10.1016/j.compag.2010.09.008

Jones, H. G. (1999). Use of infrared thermometry for estimation of stomatal conductance as a possible aid to irrigation scheduling. *Agricultural and Forest Meteorology* 95 (3): 139–149.

Jordan, M. I., and Mitchell, T. M. (2015). Machine learning: Trends, perspectives, and prospects. *Science* 349 (6245): 255–260.

Khaki, S., and Wang, L. (2019). "Crop yield prediction using deep neural networks." *Frontiers in Plant Science* 10: 621.

Khorsandi, A., Hemmat, A., Mireei, S. A. et al. (2018). Plant temperature-based indices using infrared thermography for detecting water status in sesame under greenhouse conditions. *Agricultural Water Management* 204: 222–233.

Kitić, G., Tagarakis, A., Cselyuszka, N. et al. (2019). A new low-cost portable multispectral optical device for precise plant status assessment. *Computers and Electronics in Agriculture* 162: 300–308.

Konanz, S., Kocsányi, L., and Buschmann, C. (2014). Advanced multi-color fluorescence imaging system for detection of biotic and abiotic stresses in leaves. *Agriculture* 4 (2): 79–95.

Krause, G. H. and Weis, E. (1984). Chlorophyll fluorescence as a tool in plant physiology: II. Interpretation of fluorescence signals. *Photosynthesis Research* 5: 139–157.

Kuradusenge, M., Hitimana, E., Hanyurwimfura, D. et al. (2023). Crop yield prediction using machine learning models: Case of Irish potato and maize. *Agriculture* 13 (1): 225.

L'heureux, A., Grolinger, K., Elyamany, H.F. et al. (2017). Machine learning with big data: Challenges and approaches. *IEEE Access* 5: 7776–7797.

Lang, M., Siffel, P., Braunova, Z., et al. (1992). Investigations of the blue-green fluorescence emission of plant leaves. *Botanica Acta* 105 (6): 435–440.

LeCun, Y., Bengio, Y., and Hinton, G. (2015). Deep learning. *Nature* 521: 436–444. https://doi.org/10.1038/nature14539

Li, F., Bai, J., Zhang, M. et al. (2022). Yield estimation of high-density cotton fields using low-altitude UAV imaging and deep learning. *Plant Methods* 18 (1): 55.

Li, H., Wang, P., Weber, J. F. et al. (2017). Early identification of herbicide stress in soybean (Glycine max (L.) Merr.) using chlorophyll fluorescence imaging technology. *Sensors* 18 (1): 21.

Li, L., Zhang, Q., and Huang, D. (2014). A review of imaging techniques for plant phenotyping. *Sensors* 14 (11): 20078–20111.

Li, Y., Chai, G., Wang, Y. et al. (2022). ACE R-CNN: An attention complementary and edge detection-based instance segmentation algorithm for individual tree species identification using UAV RGB images and LiDAR data. *Remote Sensing* 14 (13): 3035.

Li, Y., Zhan, X., Liu, S. et al. (2023). Self-supervised plant phenotyping by combining domain adaptation with 3D plant model simulations: Application to wheat leaf counting at seedling stage. *Plant Phenomics* 5: 0041.

Liakos, K.G., Busato, P., Moshou, D. et al. (2018). Machine learning in agriculture: A review. *Sensors* 18 (8): 2674.

Lichtenthaler, H.K., Gitelson, A., and Lang, M. (1996). Non-destructive determination of chlorophyll content of leaves of a green and an aurea mutant of tobacco by reflectance measurements. *Journal of Plant Physiology* 148 (3–4): 483–493.

Lin, K., Gong, L., Huang, Y. et al. (2019). Deep learning-based segmentation and quantification of cucumber powdery mildew using convolutional neural network. *Front Plant Sci* 10: 155. https://doi.org/10.3389/fpls.2019.00155

Lowe, A., Harrison, N., and French, A. P. (2017). Hyperspectral image analysis techniques for the detection and classification of the early onset of plant disease and stress. *Plant Methods* 13 (1): 80.

Lyu, J.I, Baek, S.H, Jung, S. et al. (2017). High-throughput and computational study of leaf senescence through a phenomic approach. *Front Plant Sci* 8: 1–8.

Mahlein, A. K., Rumpf, T., Welke, P. et al. (2013). Development of spectral indices for detecting and identifying plant diseases. *Remote Sensing of Environment* 128: 21–30.

Mastrodimos, N., Lentzou, D., Templalexis, C. et al. (2019). Development of thermography methodology for early diagnosis of fungal infection in table grapes: The case of Aspergillus carbonarius. *Computers and Electronics in Agriculture* 165: 104972.

Meng, R., Lv, Z., Yan, J. et al. (2020). Development of spectral disease indices for southern corn rust detection and severity classification. *Remote Sensing* 12 (19): 3233.

Min, S., Lee, B., and Yoon, S. (2017). Deep learning in bioinformatics. *Brief Bioinformatics* 18: 851–869. https://doi.org/10.1093/bib/bbw068

Mo, C., Kim, M.S., Kim, G. et al. (2015). Detecting drought stress in soybean plants using hyperspectral fluorescence imaging. *J Biosyst Eng* 40: 335–344. https://doi.org/10.5307/JBE.2015.40.4.335

Mochida, K., Koda, S., Inoue, K. et al. (2019). Computer vision-based phenotyping for improvement of plant productivity: A machine learning perspective. *GigaScience* 8 (1): giy153.

Montagnoli, A., Terzaghi, M., Fulgaro, N. et al. (2016). Non-destructive phenotypic analysis of early stage tree seedling growth using an automated stereovision imaging method. *Front Plant Sci* 7: 1644.

Murphy, K.P. (2012). *Machine Learning: A Probabilistic Perspective*. MIT press.

Naik, H. S., Zhang, J., Lofquist, A. et al. (2017). A real-time phenotyping framework using machine learning for plant stress severity rating in soybean. *Plant Methods* 13: 1–12.

Nduku, L., Munghemezulu, C., Mashaba-Munghemezulu, Z. et al. (2023). Global research trends for unmanned aerial vehicle remote sensing application in wheat crop monitoring. *Geomatics* 3 (1): 115–136.

Nejad, S.M.M., Abbasi-Moghadam, D., Sharifi, A. et al. (2022). Multispectral crop yield prediction using 3D-convolutional neural networks and attention convolutional LSTM approaches. *IEEE Journal of Selected Topics in Applied Earth Observations and Remote Sensing* 16: 254–266.

Nevavuori, P., Narra, N., and Lipping, T. (2019). Crop yield prediction with deep convolutional neural networks. *Computers and Electronics in Agriculture* 163: 104859.

Nevavuori, P., Narra, N., Linna, P. et al. (2020). Crop yield prediction using multitemporal UAV data and spatio-temporal deep learning models. *Remote Sensing* 12 (23): 4000.

Ochoa, D., Cevallos, J., Vargas, G. et al. (2016). Hyperspectral imaging system for disease scanning on banana plants. In *Sensing for Agriculture and Food Quality and Safety VIII* (Vol. 9864, pp. 85–90). SPIE.

Oerke, E. C., Fröhling, P., and Steiner, U. (2011). Thermographic assessment of scab disease on apple leaves. *Precision Agriculture* 12: 699–715.

Oikonomidis, A., Catal, C., and Kassahun, A. (2022). Hybrid deep learning-based models for crop yield prediction. *Applied Artificial Intelligence* 36 (1): 2031822.

Omeka, M.E., Igwe, O., Onwuka, O.S. et al. (2024). Efficacy of GIS-based AHP and data-driven intelligent machine learning algorithms for irrigation water quality prediction in an agricultural-mine district within the Lower Benue Trough, Nigeria. *Environmental Science and Pollution Research* 31 (41): 54204–54233.

Onoyama, H, Ryu, C, Suguri, M. et al. (2013). Potential of hyperspectral imaging for constructing a year-invariant model to estimate the nitrogen content of rice plants at the panicle initiation stage. *IFAC Proc* 46: 219–224. https://doi.org/10.3182/20130828-2-SF-3019.00054

Ouafiq, E. M., Elrharras, A., Mehdary, A. et al. (2021). IoT in smart farming analytics, big data based architecture. In *Human Centred Intelligent Systems: Proceedings of KES-HCIS 2020 Conference* (pp. 269–279). Springer Singapore.

Pandey, P., Ge, Y., Stoerger, V. et al. (2017). High throughput in vivo analysis of plant leaf chemical properties using hyperspectral imaging. *Frontiers in Plant Science* 8: 1348.

Peña, J.M., Torres-Sánchez, J., Serrano-Pérez, A. et al. (2015). Quantifying efficacy and limits of unmanned aerial vehicle (UAV) technology for weed seedling detection as affected by sensor resolution. *Sensors* 15: 5609–5626. https://doi.org/10.3390/s150305609

Peñuelas, J., Pinol, J., Ogaya, R.. et al. (1997). Estimation of plant water concentration by the reflectance water index WI (R900/R970). *International Journal of Remote Sensing* 18 (13): 2869–2875.

Pérez-Enciso, M. and Zingaretti, L.M. (2019). A guide for using deep learning for complex trait genomic prediction. *Genes (basel)*. https://doi.org/10.3390/genes10070553

Poblete, T., Ortega-Farias, S., and Ryu, D. (2018). Automatic coregistration algorithm to remove canopy shaded pixels in UAV-borne thermal images to improve the estimation of crop water stress index of a drip-irrigated cabernet sauvignon vineyard. *Sensors (Basel)* 18 (2): E397.

Potgieter, A.B, George-Jaeggli, B., Chapman, S.C. et al. (2017). Multispectral imaging from an unmanned aerial vehicle enables the assessment of seasonal leaf area dynamics of sorghum breeding lines. *Front Plant Sci* 8: 1532.

Raman, R., Kantari, H., Gokhale, A.A. et al. (2024). Agriculture Yield Estimation Using Machine Learning Algorithms. In *2024 International Conference on Automation and Computation (AUTOCOM)* (pp. 187–191). IEEE.

Rathore, M. M., Shah, S. A., Shukla, D. et al. (2021). The role of AI, machine learning, and big data in digital twinning: A systematic literature review, challenges, and opportunities. *IEEE Access* 9: 32030–32052.

Raza, S.A., Smith, H.K., Clarkson, G.J.J. et al. (2014). Automatic detection of regions in spinach canopies responding to soil moisture deficit using combined visible and thermal imagery. *PLoS ONE* 9: e97612. https://doi.org/10.1371/journal.pone.0097612

Rebetzke, G.J., Jimenez-Berni, J.A., Bovill, W.D. et al. (2016). High throughput phenotyping technologies allow accurate selection of stay-green. *J Exp Bot* 67 (17): 4919–24.

Rouse, J.W., Haas, R.H., Schell, J.A. et al. (1974). Monitoring vegetation systems in the Great Plains with ERTS. *NASA Spec. Publ* 351 (1): 309.

Rumpf, T., Mahlein, A. K., Steiner, U. et al. (2010). Early detection and classification of plant diseases with support vector machines based on hyperspectral reflectance. *Computers and Electronics in Agriculture* 74 (1): 91–99.

Russello, H. (2018). Convolutional neural networks for crop yield prediction using satellite images. *IBM Center for Advanced Studies*.

Ryu, J.H., Jeong, H., and Cho, J. (2020). Performances of vegetation indices on paddy rice at elevated air temperature, heat stress, and herbicide damage. *Remote Sensing* 12 (16): 2654.

Saini, U., Kumar, R., Jain, V. et al. (2020). Univariant time series forecasting of agriculture load by using LSTM and GRU RNNs. In *2020 IEEE Students Conference on Engineering & Systems (SCES)* (pp. 1–6). IEEE.

Saleem, M.H., Potgieter, J., and Arif, K.M. (2022). A performance-optimized deep learning-based plant disease detection approach for horticultural crops of New Zealand. *IEEE Access* 10: 89798–89822.

Samuel, A.L. (1959). Some studies in machine learning using the game of checkers. *IBM J Res Dev* 3: 210–229. https://doi.org/10.1147/rd.33.0210

Sandhu, K.S., Mihalyov, P.D., Lewien, M.J. et al. (2021a). Genomic selection and genome-wide association studies for grain protein content stability in a nested association mapping population of wheat. *Agronomy* 11 (12): 2528.

Sandhu, K.S., Patil, S.S., Pumphrey, M.O. et al. (2021b). Multi-trait machine and deep learning models for genomic selection using spectral information in a wheat breeding program. *The Plant Genome* 14 (3): e20119. https://doi.org/10.1002/TPG2.20119

Sandhu, K.S., Shiv, A., Kaur, G. et al. (2022). Integrated approach in genomic selection to accelerate genetic gain in sugarcane. *Plants* 11 (16): 2139.

Sankaran, S., Mishra, A., Maja, J.M. et al. (2011). Visible-near infrared spectroscopy for detection of Huanglongbing in citrus orchards. *Comput Electron Agric* 77: 127–134. https://doi.org/10.1016/j.compag.2011.03.004

Sawinski, K., Mersmann, S., Robatzek, S. et al. (2013). Guarding the green: Pathways to stomatal immunity. *Molecular Plant-Microbe Interactions* 26 (6): 626–632.

Shakoor, N., Lee, S., and Mockler, T.C. (2017). High throughput phenotyping to accelerate crop breeding and monitoring of diseases in the field. *Current Opinion in Plant Biology* 38: 184–192.

Shi, G., Du, X., Du, M. et al. (2022). Cotton yield estimation using the remotely sensed cotton boll index from UAV images. *Drones* 6 (9): 254.

Shorten, C., and Khoshgoftaar, T.M. (2019). A survey on image data augmentation for deep learning. *Journal of Big Data* 6 (1): 1–48.

Singh, A., Ganapathysubramanian, B., Singh, A.K. et al. (2016). Machine learning for high-throughput stress phenotyping in plants. *Trends in Plant Science* 21 (2): 110–124.

Tang, L., Tian, L., and Steward, B.L. (2003). Classification of broadleaf and grass weeds using Gabor wavelets and an artificial neural network. *Transactions of the ASAE* 46 (4): 1247.

Tao, H., Feng, H., Xu, L. et al. (2020). Estimation of crop growth parameters using UAV-based hyperspectral remote sensing data. *Sensors* 20 (5): 1296.

Tardieu, F., Cabrera-Bosquet, L., Pridmore, T. et al. (2017). Plant phenomics, from sensors to knowledge. *Current Biology* 27 (15): R770–R783.

Teja, M.S., Preetham, T.S., Sujihelen, L. et al. (2022). Crop recommendation and yield production using SVM algorithm. In *2022 6th International Conference on Intelligent Computing and Control Systems (ICICCS)* (pp. 1768–1771). IEEE.

Underwood, J., Wendel, A., Schofield, B., et al. (2017). Efficient infield plant phenomics for row-crops with an autonomous ground vehicle. *J Field Robot* 34 (6): 1061–83.

Veys, C., Hibbert, J., Davis, P. et al. (2017). An ultra-low-cost active multispectral crop diagnostics device. In *2017 IEEE SENSORS* (pp. 1–3). IEEE.

Vilfan, N., Van der Tol, C., Yang, P. et al. (2018). Extending Fluspect to simulate xanthophyll driven leaf reflectance dynamics. *Remote Sensing of Environment* 211: 345–356.

Virlet, N., Sabermanesh, K., Sadeghi-Tehran, P. et al. (2017). Field scanalyzer: An automated robotic field phenotyping platform for detailed crop monitoring. *Funct Plant Biol* 44 (1): 143–153.

Wahabzada, M., Mahlein, A.K., Bauckhage, C. et al. (2016). Plant phenotyping using probabilistic topic models: Uncovering the hyperspectral language of plants. *Sci Rep* 6: 22482.

Wang, J., Nakano, K., and Ohashi, S. (2011). Nondestructive detection of internal insect infestation in jujubes using visible and near-infrared spectroscopy. *Postharvest Biol Technol* 59: 272–279. https://doi.org/10.1016/j.postharvbio.2010.09.017

Watchareeruetai, U., Noinongyao, P., Wattanapaiboonsuk, C. et al. (2018). Identification of plant nutrient deficiencies using convolutional neural networks. In *2018 International Electrical Engineering Congress (iEECON)* (pp. 1–4). IEEE.

Xie, X., Zhang, X., He, B. et al. (2016). A system for diagnosis of wheat leaf diseases based on Android smartphone. In *Optical Measurement Technology and Instrumentation* (Vol. 10155, pp. 572–580). SPIE.

Yang, G., Liu, J., Zhao, C. et al. (2017). Unmanned aerial vehicle remote sensing for field-based crop phenotyping: Current status and perspectives. *Front Plant Sci* 8: 1111.

Yang, W., Yang, C., Hao, Z. et al. (2019). Diagnosis of plant cold damage based on hyperspectral imaging and convolutional neural network. *IEEE Access* 7: 118239–118248.

Yao, H., Hruska, Z., Kincaid, R. et al. (2013). Detecting maize inoculated with toxigenic and atoxigenic fungal strains with fluorescence hyperspectral imagery. *Biosys Eng* 115: 125–135. https://doi.org/10.1016/j.biosystemseng.2013.03.006.

Zhang, J., Pu, R., Loraamm, R.W. et al. (2014). Comparison between wavelet spectral features and conventional spectral features in detecting yellow rust for winter wheat. *Computers and Electronics in Agriculture* 100: 79–87.

Zhang, L., Zhang, H., Niu, Y. et al. (2019) Mapping maize water stress based on UAV multispectral remote sensing. *Remote Sens (basel)* 11: 605. https://doi.org/10.3390/rs11060605

Zhang, Z. (2020). Big data analysis with artificial intelligence technology based on machine learning algorithm. *Journal of Intelligent & Fuzzy Systems* 39 (5): 6733–6740.

Zhou, L., Pan, S., and Wang, J. et al. (2017). Machine learning on big data: Opportunities and challenges. *Neurocomputing* 237: 350–361.

Zhu, J., He, W., and Yao, J. et al. (2020). Spectral reflectance characteristics and chlorophyll content estimation model of Quercus aquifolioides leaves at different altitudes in Sejila Mountain. *Applied Sciences* 10 (10): 3636.

Zovko, M., Žibrat, U., and Knapič, M. et al. (2019). Hyperspectral remote sensing of grapevine drought stress. *Precision Agriculture* 20: 335–347.

4

Machine Learning for Studying Plant Secondary Metabolites

Saniya[1], Sarfraz Ahmad[2], Mohammad Ghani Raghib[3]*

[1]Department of Horticulture, ICAR-National Research Centre for Grapes, Pune, Maharashtra, India
[2]Department of Genetics and Plant Breeding, College of Agriculture, Mandawa, Swami Keshwanand Rajasthan Agriculture University, Bikaner, Rajasthan, India
[3]Department of Horticulture, Faculty of Agriculture, Nangarhar University, Afghanistan

*Corresponding author: sarfraz_firoz@rediffmail.com

4.1 Introduction

Plant secondary metabolites (PSMs) are a diverse group of organic compounds that play essential roles in plant physiology, defense, and ecological interactions. Unlike primary metabolites such as carbohydrates, proteins, and nucleic acids, PSMs are not directly involved in basic metabolic functions but contribute significantly to plants' adaptation to environmental stressors and pathogenic threats (Kumar et al. 2023). These compounds have gained considerable attention due to their remarkable therapeutic, industrial, and agricultural potential.

PSMs are broadly categorized into three major classes: phenolics, alkaloids, and terpenoids. Each of these groups encompasses thousands of bioactive compounds with unique structural features and biological activities. Phenolics, including flavonoids, tannins, and lignins, are known for their antioxidant, antimicrobial, and ultraviolet (UV)-protective properties (Fernandes et al. 2021; Ahmad et al. 2023). Alkaloids, such as morphine, caffeine, and quinine, exhibit potent pharmacological effects, making them valuable in modern medicine (Bhambhani et al. 2021). Terpenoids, a large and structurally diverse class, are crucial in essential oil production and play vital roles in plant–insect communication, defense mechanisms, and human health (Toffolatti et al. 2021).

The study of PSM biosynthesis is challenging due to the complexity of metabolic pathways and the diverse environmental conditions that influence metabolite production. Advances in multi-omics technologies, including genomics, transcriptomics, proteomics, and metabolomics, have significantly enhanced our understanding of these biosynthetic pathways. However, the sheer volume and complexity of the data generated necessitate advanced computational tools to extract meaningful insights.

Machine learning (ML) has emerged as a transformative tool in biological research, offering robust methods for analyzing large datasets, identifying patterns, and predicting biological outcomes. In recent years, ML techniques have shown exceptional promise in PSM research, enabling accurate identification of metabolites, discovery of biosynthetic pathways, and prediction of compound bioactivity (Behsaz et al. 2021). By leveraging powerful algorithms such as Random Forest (RF), Support Vector Machines (SVMs), and Deep Learning models (Chi et al. 2024), researchers have gained unprecedented insights into the complex interplay of genes, enzymes, and metabolites. One of the most notable applications of ML in PSM research is in metabolite identification. Spectral data obtained from techniques such as Liquid Chromatography–Mass Spectrometry (LC-MS) and Gas Chromatography–Mass Spectrometry (GC-MS) often contain extensive noise and overlapping peaks, making manual interpretation challenging. ML algorithms efficiently extract relevant features from these datasets, improving metabolite identification accuracy and reducing analysis time (Bilbao et al. 2023; Halderet al. 2024).

Machine Learning for Plant Biology, First Edition. Edited by Jen-Tsung Chen.

In addition to metabolite identification, ML models have proven valuable in biosynthetic pathway prediction. Traditional pathway discovery relies heavily on experimental approaches, which are time-consuming and resource-intensive. ML models trained on annotated pathway data can predict unknown steps in biosynthetic networks by inferring gene–metabolite relationships (Pearcy et al. 2021). Such predictive models have significantly enhanced the identification of novel enzymes and regulatory elements involved in PSM biosynthesis.

Another emerging area is bioactivity prediction for pharmaceutical and agricultural applications. ML models trained on chemical structure data combined with biological activity databases have successfully predicted promising bioactive compounds. These approaches have led to the discovery of PSM-based therapeutics with antimicrobial, anticancer, and anti-inflammatory properties (Sigler et al. 2025). Furthermore, crop improvement strategies have benefited greatly from ML applications. By identifying genetic markers linked to secondary metabolite production, researchers can develop breeding programs aimed at enhancing crop resistance to pests, improving nutritional value, and increasing yields (Sun et al. 2024).

Despite the growing success of ML in PSM research, several challenges remain. One major limitation is data integration, as PSM research often involves integrating heterogeneous data types such as genomic sequences, transcriptomics profiles, and metabolomics data. Aligning these diverse datasets requires sophisticated data preprocessing and feature selection strategies (Cembrowska-Lech et al. 2023). Another critical concern is model interpretability. While deep learning models excel in predictive accuracy, their “black-box” nature makes biological interpretation difficult (Murmu et al. 2024). Developing interpretable ML frameworks is essential for improving our understanding of the biological mechanisms underlying PSM production. Lastly, the availability of annotated datasets presents a significant bottleneck. As many PSMs remain unidentified or poorly characterized, supervised learning models may lack sufficient labeled data for training (Tariq et al. 2020). Techniques, such as transfer learning, semi-supervised learning, and active learning, are being explored to mitigate this challenge (Hesami et al. 2022).

In this chapter, we explore how ML techniques are transforming the study of PSMs. We discuss key ML approaches; highlight their applications in metabolite prediction, biosynthetic pathway analysis, and bioactivity discovery; and address challenges associated with data integration and model interpretation. By combining ML methods with experimental plant biology techniques, researchers can unlock the full potential of PSMs for diverse applications in medicine, agriculture, and industry.

4.2 ML Techniques in Metabolite Research

ML techniques play a pivotal role in understanding and predicting PSM profiles. These techniques offer powerful tools for analyzing complex metabolomics data, identifying biomarkers, and uncovering biosynthetic pathways. Figure 4.1 shows key ML techniques commonly utilized in PSMs research.

4.2.1 Supervised Learning Techniques

Supervised learning is widely used in metabolomics for classification, regression, and predictive modeling.

- **RF:** RF is an ensemble-based algorithm known for handling large datasets with high-dimensional features. In plant metabolomics, RF has successfully predicted metabolite profiles in crops such as *Solanum lycopersicum* (tomato) and *Arabidopsis thaliana* by analyzing gene expression patterns linked to metabolite production (Tinte et al. 2021).
- **SVM:** SVM excels in distinguishing metabolic profiles across different plant species. For example, SVM models have classified flavonoid content in *Camellia sinensis* (tea) based on leaf metabolite data (Zhao et al. 2022).
- **Artificial Neural Networks (ANNs):** ANNs are effective in identifying nonlinear relationships between metabolites and gene expressions. Studies in *Glycine max* (soybean) have used ANNs to predict isoflavone biosynthesis with high accuracy (Bernardy et al. 2024).

4.2.2 Unsupervised Learning Techniques

Unsupervised learning is crucial for discovering hidden patterns, clustering metabolites, and reducing data dimensionality.

- **Principal Component Analysis (PCA):** PCA is widely applied for dimensionality reduction and visualizing metabolite clustering. In metabolomic profiling of *Brassica oleracea* (broccoli), PCA effectively separated samples based on glucosinolate content (Fiehn et al. 2020).

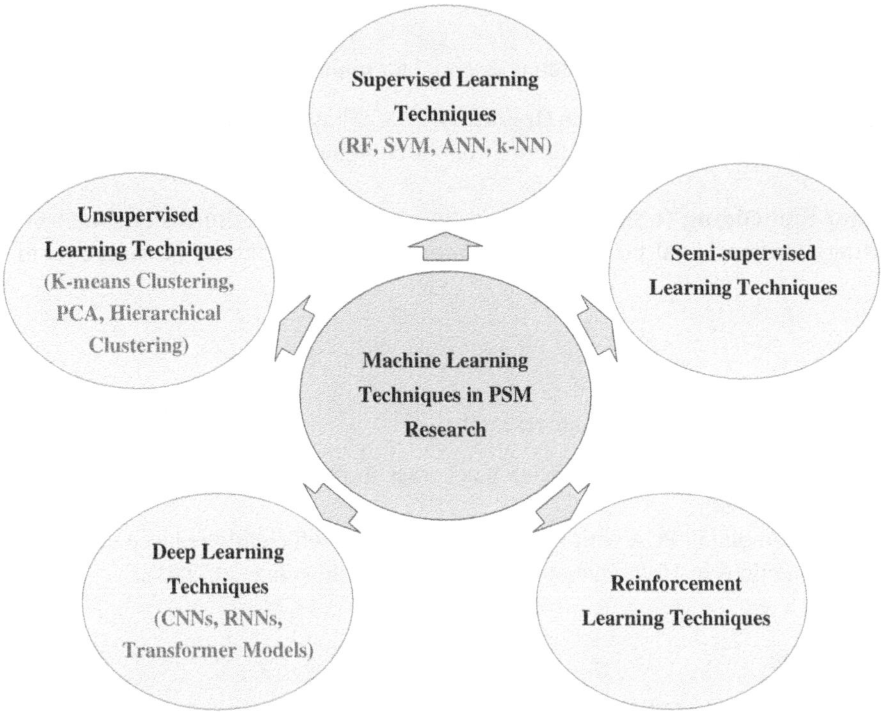

Figure 4.1 Key machine learning techniques in plant secondary metabolites' research.

- **K-means clustering:** K-means has been applied to group metabolites in *Oryza sativa* (rice) based on biochemical pathways, aiding in stress resistance research (Debnath et al. 2023).

4.2.3 Deep Learning Techniques

Deep learning techniques, particularly convolutional neural networks (CNNs) and recurrent neural networks (RNNs), offer advanced pattern recognition capabilities in metabolomics.

- **CNNs:** CNNs have been employed to identify metabolic fingerprints from spectral data. In grapevine (*Vitis vinifera*), CNN models successfully predicted phenolic compound levels, improving wine quality assessment (Carneiro et al. 2024).
- **RNNs:** RNNs excel in processing sequential data. In metabolic pathway prediction, RNN models have been applied to infer time-series metabolite changes in *Arabidopsis thaliana* during stress responses (Szeptycki 2023).

4.2.4 Ensemble Learning Techniques

Ensemble models combine multiple algorithms to improve prediction accuracy and model robustness.

- **XGBoost (Extreme Gradient Boosting):** XGBoost models have demonstrated high accuracy in predicting alkaloid biosynthesis patterns in *Papaver somniferum* (opium poppy) (Bai et al. 2024).
- **Stacked Models:** Combining RF, SVM, and neural networks has improved metabolite classification accuracy in multi-omics studies of medicinal plants like *Withania somnifera* (ashwagandha) (Abraham and Kellogg 2021).

4.2.5 Reinforcement Learning in Metabolomics

Reinforcement learning (RL) is emerging in metabolite research for optimizing experimental design and bioactive compound discovery.

- **RL for pathway optimization:** RL models have been applied to enhance metabolic engineering strategies in *Saccharomyces cerevisiae* for improved alkaloid synthesis (Cao et al. 2024).

4.2.6 Feature Selection and Dimensionality Reduction

Since metabolomics datasets are often large and noisy, feature selection is essential for improving model accuracy.

- **LASSO regression (Least Absolute Shrinkage and Selection Operator):** LASSO has effectively identified key biomarkers in plant species like *Capsicum annuum* (chili pepper), linking volatile compounds to fruit aroma (Velázquez et al. 2023).
- **t-Distributed Stochastic Neighbor Embedding (t-SNE):** t-SNE aids in visualizing high-dimensional metabolite profiles, revealing distinct patterns in medicinal plants such as *Curcuma longa* (turmeric) (Cashman et al. 2025).

4.2.7 Hybrid Models

Combining ML techniques often enhances performance in metabolite research.

- **Hybrid RF-SVM models:** These models have improved classification accuracy in predicting secondary metabolite content in *Citrus sinensis* (orange) (Liu et al. 2023).
- **Deep learning integrated with PCA:** Combining PCA with CNN architectures has efficiently reduced data noise while improving phenolic compound predictions in *Malus domestica* (apple) (Houetohossou et al. 2023).

4.2.8 Emerging Techniques in Metabolomics

- **Graph Neural Networks:** Graph Neural Networks (GNNs) are gaining attention for their ability to analyze complex biochemical pathways in plants.
- **Autoencoders:** These unsupervised neural networks efficiently compress metabolomic data for improved downstream analysis.

ML techniques are revolutionizing plant metabolomics research by providing powerful tools for data analysis, biomarker discovery, and pathway prediction. Integrating robust ML models with accurate data preprocessing strategies will further enhance our understanding of PSMs, driving advancements in agriculture, medicine, and biotechnology.

4.3 Applications of ML in PSM Research

ML has emerged as a transformative tool in studying PSMs, providing robust methods for data analysis, predictive modeling, and biological insight generation. PSMs, including alkaloids, flavonoids, terpenoids, and phenolic compounds, are vital for plant defense, environmental adaptation, and human applications such as pharmaceuticals and nutraceuticals (Ahmad et al. 2023). ML techniques have demonstrated significant success in predicting metabolite biosynthesis, identifying key regulatory genes, and optimizing metabolic engineering strategies (Table 4.1). This section explores the diverse applications of ML in PSM research, highlighting key advancements, methodologies, and case studies.

4.3.1 Predicting Metabolic Pathways

Predicting biosynthetic pathways is crucial for understanding how plants synthesize complex secondary metabolites.

- **Bayesian networks and decision trees:** These models predict metabolic pathway structures by integrating genomic, transcriptomic, and metabolomic data. For example, Bayesian models successfully identified key genes involved in glucosinolate biosynthesis in *Arabidopsis thaliana* (Kooke et al. 2021).
- **SVM:** SVM has been applied to predict alkaloid biosynthetic pathways in *Papaver somniferum* (opium poppy), improving knowledge of morphinan alkaloid synthesis (Mukherjee and Mukherjee 2024).
- **Deep Neural Networks:** Deep neural networks (DNN) models trained on gene expression data have accurately reconstructed flavonoid biosynthetic pathways in *Glycine max* (soybean) (Gai et al. 2025).

Table 4.1 Applications of machine learning in plant secondary metabolite research.

Application	Machine Learning Technique	Plant Species	Metabolite Class	Key Findings	Reference
Biosynthetic pathway prediction	Bayesian Networks	*Arabidopsis thaliana*	Glucosinolates	Identified regulatory genes in glucosinolate biosynthesis.	Kooke et al. (2021)
	Support Vector Machines (SVM)	*Papaver somniferum* (opium poppy)	Alkaloids	Predicted morphinan alkaloid biosynthetic pathways.	Mukherjee and Mukherjee (2024)
	Deep Neural Networks (DNN)	*Glycine max* (soybean)	Flavonoids	Accurately reconstructed flavonoid biosynthesis pathways.	Gai et al. (2025)
Identification of biosynthetic genes	Random Forest	*Withania somnifera* (ashwagandha)	Withanolides	Identified key genes involved in withanolide synthesis.	Ahmad et al. (2024)
	LASSO Regression	*Vitis vinifera* (grapevine)	Anthocyanins	Discovered key genes controlling grape color traits.	Beaumont et al. (2025)
	XGBoost	*Camellia sinensis* (tea)	Catechins	Identified genes regulating catechin biosynthesis.	Modak et al. (2024)
Metabolite profiling and classification	K-means Clustering	*Rauwolfia serpentina*	Alkaloids	Classified alkaloid profiles across chemotypes.	Pandita et al. (2021)
	Principal Component Analysis (PCA)	*Curcuma longa* (turmeric)	Curcuminoids	Distinguished curcuminoid profiles among cultivars.	Salem et al. (2022)
	Convolutional Neural Networks (CNN)	*Capsicum annuum* (chili pepper)	Carotenoids	Detected pigment variations to enhance breeding.	Mahlein et al. (2018)
Enhancing plant stress response	Recurrent Neural Networks (RNN)	*Oryza sativa* (rice)	Drought-induced metabolites	Modeled drought-responsive metabolite changes.	Anand et al. (2024)
	Autoencoders	*Brassica napus* (canola)	Oxidative stress markers	Identified metabolites linked to stress resilience.	Hussain et al. (2022)
Chemotaxonomy and species identification	Random Forest	*Echinacea purpurea* (purple coneflower)	Phenolic acids	Classified phenolic profiles across plant populations.	Park et al. (2024)
	t-SNE (t-Distributed Stochastic Neighbor Embedding)	*Mentha arvensis* (wild mint)	Essential oils	Distinguished essential oil profiles for species identification.	Cavalloro et al. (2022)
Pharmacological and nutraceutical prediction	QSAR (Quantitative Structure–Activity Relationship)	*Catharanthus roseus* (Madagascar periwinkle)	Alkaloids	Predicted anticancer activity of alkaloid derivatives.	Pushpalatha et al. (2022)
	Graph Neural Networks (GNN)	*Ginkgo biloba*	Flavonoids	Predicted bioactive potential of ginkgo-derived compounds.	Spanakis et al. (2025)
Metabolic engineering for enhanced metabolite production	Genetic Algorithms (GA)	*Datura stramonium* (thorn apple)	Tropane Alkaloids	Optimized gene-editing strategies to increase alkaloid yield.	De-la-Cruz et al. (2021)
	Reinforcement Learning (RL)	Engineered Yeast Systems	Secondary Metabolites	Improved biosynthetic efficiency in synthetic biology frameworks.	Naseri et al. (2023)

(Continued)

Table 4.1 (Continued)

Application	Machine Learning Technique	Plant Species	Metabolite Class	Key Findings	Reference
Environmental monitoring using metabolites	Random Forest	*Taraxacum officinale* (dandelion)	Phenolic Compounds	Predicted pollution-induced changes in phenolic profiles.	Kumar et al. (2024)
	Support Vector Machines (SVM)	*Digitalis purpurea* (foxglove)	Alkaloids	Detected soil contamination effects on alkaloid content.	Sotenjwa (2020)
Predictive modeling for agricultural improvement	Deep Learning Models (CNN, LSTM)	*Lavandula angustifolia* (lavender)	Essential Oils	Predicted optimal conditions for maximizing essential oil production.	Aslan and Polat (2024)
	Bayesian Optimization	*Panax ginseng* (ginseng)	Saponins	Tuned nutrient conditions to enhance saponin production.	Kim et al. (2023)

4.3.2 Identification of Key Biosynthetic Genes

Identifying genes responsible for PSM production is critical for metabolic engineering.

- **RF:** RF models have been used to rank genes by importance in metabolite biosynthesis. In *Withania somnifera* (ashwagandha), RF models identified genes linked to withanolide biosynthesis (Ahmad et al. 2024).
- **LASSO Regression:** This technique has efficiently identified candidate genes for anthocyanin biosynthesis in *Vitis vinifera* (grapevine), aiding in grape coloration studies (Beaumont et al. 2025).
- **XGBoost Models:** In *Camellia sinensis* (tea), XGBoost identified genes regulating catechin biosynthesis, improving understanding of tea polyphenols (Modak et al. 2024).

4.3.3 Metabolite Profiling and Classification

ML models excel in profiling and classifying metabolite data from complex plant systems.

- **K-means clustering:** This unsupervised method successfully grouped alkaloid profiles in *Rauwolfia serpentina* (Indian snakeroot), distinguishing between different chemotypes (Pandita et al. 2021).
- **PCA:** PCA has been widely used in *Curcuma longa* (turmeric) to classify curcuminoid profiles across different cultivars, enhancing medicinal plant breeding efforts (Salem et al. 2022).
- **CNNs:** CNN models analyzing spectral data efficiently identified carotenoid content in *Capsicum annuum* (chili pepper), optimizing breeding for improved pigment production (Mahlein et al. 2018).

4.3.4 Enhancing Plant Stress Response Through Metabolite Analysis

PSMs are vital in improving plant resilience against abiotic and biotic stresses. ML models are increasingly used to explore metabolite-driven stress responses.

- **RNNs:** RNNs have effectively modeled time-series metabolite changes in *Oryza sativa* (rice) during drought stress, revealing critical metabolites associated with stress tolerance (Anand et al. 2024).
- **Autoencoders:** These deep learning models efficiently identified biomarkers linked to oxidative stress resistance in *Brassica napus* (canola), guiding selective breeding strategies (Hussain et al. 2022).

4.3.5 Chemotaxonomy and Species Identification

ML algorithms facilitate plant species identification and classification based on their metabolite profiles.

- **RF:** RF models have accurately classified medicinal plants like *Echinacea purpurea* based on phenolic acid content (Park et al. 2024).
- **t-SNE:** This technique effectively distinguished essential oil profiles in *Mentha arvensis* (wild mint) populations (Cavalloro 2022).

4.3.6 Pharmacological and Nutraceutical Research

ML techniques are increasingly applied to predict the pharmacological properties of PSMs and optimize their bioactivity.

- **Quantitative Structure–Activity Relationship models: Quantitative Structure–Activity Relationship** (QSAR) models trained on chemical structure data have successfully predicted the anticancer potential of alkaloids from *Catharanthus roseus* (Madagascar periwinkle) (Pushpalatha et al. 2022).
- **GNNs:** GNNs have been applied to predict bioactive properties of flavonoids derived from *Ginkgo biloba*, improving functional food formulation (Spanakis et al. 2025).

4.3.7 Metabolic Engineering for Enhanced PSM Production

ML-driven strategies are advancing metabolic engineering approaches to increase PSM yields.

- **Genetic Algorithm (GA):** GA models have optimized gene-editing strategies to enhance tropane alkaloid production in *Datura stramonium* (thorn apple) (De-la-Cruz et al. 2021).
- **RL:** RL has been used to optimize synthetic biology frameworks for improving secondary metabolite synthesis in engineered yeast systems (Naseri 2023).

4.3.8 PSM-based Environmental Monitoring

ML algorithms are being applied to monitor environmental stress impacts on PSM profiles.

- **Random forest and gradient boosting:** These techniques have been applied to predict shifts in phenolic compound concentrations in response to pollution in *Taraxacum officinale* (dandelion) (Kumar et al. 2024).
- **SVM models:** SVM has been used to detect soil contamination effects on alkaloid content in *Digitalis purpurea* (foxglove) (Sotenjwa 2020).

4.3.9 Predictive Modeling for Agricultural Improvement

ML models are applied to predict optimal growth conditions for enhancing PSM content in crops.

- **Deep learning models:** CNN and LSTM models have successfully predicted environmental conditions for maximizing essential oil production in *Lavandula angustifolia* (lavender) (Aslan and Polat 2024).
- **Bayesian optimization:** This method efficiently tuned nutrient conditions to boost saponin content in *Panax ginseng* (ginseng) (Kim et al. 2023).

4.4 Challenges and Future Directions

The application of ML in PSM research presents exciting opportunities for advancing agricultural productivity, improving plant resilience, and discovering novel bioactive compounds. However, several challenges must be addressed to fully harness the potential of ML in this field. Furthermore, future research directions should focus on enhancing model interpretability, improving data quality, and integrating multi-omics datasets for comprehensive insights.

4.4.1 Challenges in ML for PSM Research

Despite promising advancements, several challenges limit the full-scale application of ML in metabolomics research.

4.4.1.1 Data-related Challenges

The complexity and variability of PSM data pose significant obstacles to ML applications.

- **Data scarcity and imbalance:**
  PSM datasets are often limited, particularly for rare or under-researched plant species. The imbalance in metabolite concentration data, where certain compounds are highly abundant while others are scarcely detected can hinder

model performance. For instance, rare alkaloid biosynthesis in *Papaver somniferum* (opium poppy) requires extensive data augmentation to improve ML model accuracy (Salam et al. 2023).

- **Data noise and variability:**
  Metabolomics data is prone to variability due to differences in extraction methods, environmental factors, and genetic backgrounds. This inconsistency reduces ML model reproducibility. Advanced noise-reduction strategies like autoencoders and denoising algorithms have been employed to improve data quality in *Glycine max* (soybean) metabolite profiling (Zhou et al. 2024).
- **Lack of standardized datasets:**
  Unlike genomics data, which benefits from established repositories (e.g., NCBI, Ensembl), metabolomics lacks comprehensive standardized databases. Efforts to integrate resources such as METLIN, KEGG, and PlantCyc are crucial for advancing ML models in PSM research (Persaud et al. 2025).

4.4.1.2 Algorithmic Challenges

While ML algorithms provide powerful predictive capabilities, certain technical limitations exist.

- **Overfitting and model robustness:**
  Due to high data dimensionality and small sample sizes, ML models in PSM research are prone to overfitting. Techniques such as dropout regularization, cross-validation, and ensemble learning are crucial for enhancing model robustness in metabolomics data analysis (Murmu et al. 2024).
- **Feature selection complexity:**
  Identifying the most relevant metabolic features from thousands of detected compounds remains challenging. Feature selection techniques like LASSO regression, Boruta algorithms, and SHAP (SHapley Additive exPlanations) models are gaining attention for improving model interpretability (Zhang and Wang et al. 2023).

4.4.1.3 Biological Complexity and Unknown Pathways

- **Incomplete biosynthetic pathway knowledge:**
  Despite growing metabolomics data, many PSM biosynthetic pathways remain poorly characterized. For example, the complete biosynthesis of certain terpenoids in *Artemisia annua* (sweet wormwood) remains unclear, limiting ML model predictions (Wan et al. 2024).
- **Diverse environmental interactions:**
  Environmental factors, such as drought, temperature, and soil nutrients, significantly alter metabolite profiles. Developing ML models capable of integrating environmental variables remains a major research challenge.

4.4.1.4 Computational Challenges

- **High computational demand:**
  Advanced ML techniques like DNN and GNN require significant computational resources. Researchers are increasingly adopting cloud computing frameworks and distributed systems to manage large-scale metabolomics datasets.

4.4.2 Future Directions in ML for PSM Research

To address these challenges and unlock the full potential of ML in PSM research, future research efforts should focus on the following key directions:

4.4.2.1 Integrating Multi-omics Data

Integrating genomics, transcriptomics, proteomics, and metabolomics data can enhance ML model accuracy in predicting biosynthetic pathways.

- **Multi-omics data fusion:**
  Techniques like tensor decomposition, matrix factorization, and graph-based models are being developed to integrate multi-omics data. For example, integrative analysis of RNA-Seq and metabolomics data in *Brassica napus* has identified novel genes linked to glucosinolate biosynthesis (Lei et al. 2024).

- **Knowledge graphs:**
  Constructing plant-specific knowledge graphs using ML algorithms can improve pathway reconstruction and metabolite function prediction.

4.4.2.2 Advancing Explainable Artificial Intelligence

Interpretable ML models are critical for gaining biological insights from complex metabolomics data.

- **SHAP and LIME models:**
  These methods provide visual explanations of model predictions, enabling researchers to identify key metabolites and genes influencing specific traits. For example, SHAP analysis in *Nicotiana tabacum* (tobacco) has improved the identification of nicotine biosynthetic genes (Parthiban et al. 2023).
- **Causal inference models:**
  Causal ML models can help identify cause-and-effect relationships in metabolic pathways, reducing false-positive associations.

4.4.2.3 Improving Data Augmentation Strategies

Developing synthetic data generation methods can address data scarcity issues.

- **Generative Adversarial Networks (GANs):**
  GANs have shown promise in generating realistic metabolite profiles for rare plant species, improving model training for datasets with limited observations.

4.4.2.4 Automation and High-throughput Analysis

Automating metabolomics workflows using ML can accelerate research progress.

- **Automated feature engineering:**
  Automated ML (AutoML) frameworks are being explored to streamline feature selection, hyperparameter tuning, and model selection in metabolomics research (Bifarin and Fernández 2024).

4.4.2.5 Advancing Computational Infrastructure

- **Cloud computing and federated learning:**
  Cloud-based frameworks can efficiently manage large-scale metabolomics datasets. Federated learning approaches are emerging to enable collaborative research while preserving data privacy.

4.4.2.6 Developing Crop-specific ML Models

Since PSM biosynthesis varies widely across species, customized ML models tailored for specific crops will improve predictive accuracy.

- **Species-specific genomic data integration:**
  Developing models that incorporate crop-specific genomic, transcriptomic, and metabolic profiles will enhance the precision of ML predictions in plant breeding programs.

4.4.3 Ethical Considerations and Regulatory Frameworks

- **Data privacy and intellectual property:**
  As ML models rely heavily on data sharing, ensuring data privacy and protecting indigenous plant knowledge are critical ethical concerns.
- **Regulatory compliance:**
  Establishing guidelines for ML-driven metabolic engineering and PSM research will promote the safe adoption of these technologies.

  Future research should focus on developing explainable AI (XAI) models, expanding high-quality data repositories, and improving model robustness for real-world applications.

4.5 Conclusion

ML has emerged as a transformative tool in the study of PSMs, enabling advanced data analysis, predictive modeling, and biological discovery. By leveraging techniques such as deep learning, random forests, and support vector machines, researchers have decoded complex biosynthetic pathways, identified key regulatory genes, and enhanced metabolite profiling. These advancements have significantly contributed to crop resilience, metabolic engineering, and the discovery of bioactive compounds for pharmaceuticals and nutraceuticals. Despite these successes, challenges persist, including data variability, scarcity, and the environmental sensitivity of metabolite production. Addressing these issues requires improved data standardization, multi-omics integration, and XAI techniques. Future research should focus on automated ML frameworks, data augmentation, and species-specific models to enhance predictive accuracy. Additionally, integrating ML with emerging approaches such as knowledge graphs and federated learning may further improve PSM research. By overcoming these challenges and embracing interdisciplinary collaboration, ML will continue to drive innovation in PSM biosynthesis, supporting advancements in sustainable agriculture, plant-based medicine, and environmental monitoring. Transparent data sharing, ethical considerations, and collective research efforts will be essential to maximize the scientific and societal impact of ML-driven PSM studies.

References

Abraham, E.J., and Kellogg, J.J. (2021). Chemometric-guided approaches for profiling and authenticating botanical materials. *Frontiers in Nutrition* 8: 780228.

Ahmad, S., Belwal, V., and Punia, S.S. et al. (2023). Role of plant secondary metabolites and phytohormones in drought tolerance: A review. *Gesunde Pflanzen* 75 (4): 729–746.

Ahmad, Z., Shareen, and Ganie, I.B. et al. (2024). Enhancing Withanolide production in the Withania species: Advances in in vitro culture and synthetic biology approaches. *Plants* 13 (15): 2171.

Anand, S., Visakh, R.L., and Nalishma, R. et al. (2024). High throughput phenomics in elucidating drought stress responses in rice (Oryza sativa L.). *Journal of Plant Biochemistry and Biotechnology*: 1–14.

Aslan, İ., and Polat, N. (2024). Deep learning-based classification of mature and immature lavender plants using UAV orthophotos and a hybrid CNN approach. *Earth Science Informatics* 17 (2): 1713–1727.

Bai, W., Li, C., and Li, W. et al. (2024). Machine learning assists prediction of genes responsible for plant specialized metabolite biosynthesis by integrating multi-omics data. *BMC Genomics* 25 (1): 418.

Beaumont, C., Prigent, S., and Mori, K. et al. (2025). Interspecies predictions of growth traits from quantitative transcriptome data acquired during fruit development. *Journal of Experimental Botany*: eraf122.

Behsaz, B., Bode, E., and Gurevich, A. et al. (2021). Integrating genomics and metabolomics for scalable non-ribosomal peptide discovery. *Nature Communications* 12 (1): 3225.

Bernardy, R., Cañizares, L.D.C.C., and Meza, S.L.R. et al. (2024). Identifying soybean genotypes with artificial intelligence and near infrared reflectance. *Ciência e Agrotecnologia* 48: e005224.

Bhambhani, S., Kondhare, K.R., and Giri, A.P. (2021). Diversity in chemical structures and biological properties of plant alkaloids. *Molecules* 26 (11): 3374.

Bifarin, O.O., and Fernández, F.M. (2024). Automated machine learning and explainable AI (AutoML-XAI) for metabolomics: Improving cancer diagnostics. *Journal of the American Society for Mass Spectrometry* 35 (6): 1089–1100.

Bilbao, A., Munoz, N., and Kim, J. et al. (2023). PeakDecoder enables machine learning-based metabolite annotation and accurate profiling in multidimensional mass spectrometry measurements. *Nature Communications* 14 (1): 2461.

Cao, L., Teo, D., and Wang, Y. et al. (2024). Advancements in microbial cell engineering for benzylisoquinoline alkaloid production. *ACS Synthetic Biology* 13 (12): 3842–3856.

Carneiro, G.A., Cunha, A., and Aubry, T.J. et al. (2024). Advancing grapevine variety identification: A systematic review of deep learning and machine learning approaches. *AgriEngineering* 6 (4): 4851–4888.

Cashman, D., Keller, M., and Jeon, H. et al. (2025). A critical analysis of the usage of dimensionality reduction in four domains. *arXiv preprint arXiv:2503.08836*.

Cavalloro, V. (2022). Exploiting the Pavia territory plant biodiversity for the discovery of proteasome modulators.

Cembrowska-Lech, D., Krzemińska, A., and Miller, T. et al. (2023). An integrated multi-omics and artificial intelligence framework for advance plant phenotyping in horticulture. *Biology* 12 (10): 1298.

Chi, J., Shu, J., and Li, M. et al. (2024). Artificial intelligence in metabolomics: A current review. *TrAC Trends in Analytical Chemistry*: 117852.

Debnath, S., Aisha, S., and Malakar, A. et al. (2023). Understanding the cross-talk of major abiotic-stress-responsive genes in rice: A computational biology approach. *Journal of King Saud University-Science* 35 (7): 102786.

De-la-Cruz, I.M., Hallab, A., and Olivares-Pinto, U. et al. (2021). Genomic signatures of the evolution of defence against its natural enemies in the poisonous and medicinal plant Datura stramonium (Solanaceae). *Scientific Reports* 11 (1): 882.

Fernandes, I.D.A.A., Maciel, G.M., and Ribeiro, V.R. et al. (2021). The role of bacterial cellulose loaded with plant phenolics in prevention of UV-induced skin damage. *Carbohydrate Polymer Technologies and Applications* 2: 100122.

Gai, Y., Liu, S., and Zhang, Z. et al. (2025). Integrative approaches to soybean resilience, productivity, and utility: A review of genomics, computational modeling, and economic viability. *Plants* 14 (5): 671.

Halder, M., Kundu, A., and Jha, S. (2024). Secondary metabolites identification techniques of the current era. In *Plant Specialized Metabolites: Phytochemistry, Ecology and Biotechnology* (pp. 1–41). Springer Nature Switzerland.

Hesami, M., Alizadeh, M., and Jones, A.M.P. et al. (2022). Machine learning: Its challenges and opportunities in plant system biology. *Applied Microbiology and Biotechnology* 106 (9): 3507–3530.

Houetohossou, S.C.A., Houndji, V.R., and Hounmenou, C.G. et al. (2023). Deep learning methods for biotic and abiotic stresses detection and classification in fruits and vegetables: State of the art and perspectives. *Artificial Intelligence in Agriculture* 9: 46–60.

Hussain, M.A., Luo, D., and Zeng, L. et al. (2022). Genome-wide transcriptome profiling revealed biological macromolecules respond to low temperature stress in Brassica napus L. *Frontiers in Plant Science* 13: 1050995.

Kim, T.H., Baek, S., and Kwon, K.H. et al. (2023). Hierarchical machine learning-based growth prediction model of Panax ginseng sprouts in a hydroponic environment. *Plants* 12 (22): 3867.

Kooke, R., Kruijer, W., and van Eekelen, H.D. et al. (2021). Local adaptation shapes metabolic diversity in the global population of Arabidopsis thaliana. *bioRxiv*: 2021–09.

Kumar, S., Chakraborty, S., and Ghosh, S. et al. (2024). Revealing soil microbial ecophysiological indicators in acidic environments laden with heavy metals via predictive modeling: Understanding the impacts of black diamond excavation. *Science of the Total Environment* 923: 171454.

Kumar, S., Korra, T., Thakur, R. et al. (2023). Role of plant secondary metabolites in defence and transcriptional regulation in response to biotic stress. *Plant Stress* 8: 100154.

Lei, J., Zhang, W., and Yu, F. et al. (2024). Integrated analysis of transcriptome and metabolome reveals differential responses to Alternaria brassicicola infection in cabbage (Brassica oleracea var. capitata). *Genes* 15 (5): 545.

Liu, X., Wang, Z., and Gmitter Jr, F.G. et al. (2023). Effects of different rootstocks on the metabolites of Huanglongbing-affected sweet orange juices using a novel combined strategy of untargeted metabolomics and machine learning. *Journal of Agricultural and Food Chemistry* 71 (2): 1246–1257.

Mahlein, A.K., Kuska, M.T., and Behmann, J. et al. (2018). Hyperspectral sensors and imaging technologies in phytopathology: State of the art. *Annual Review of Phytopathology* 56 (1): 535–558.

Modak, A., Das, D., and Chatterjee, T.N. et al. (2024). Regression Based EGCG Detection in Green Tea Employing MIP Electrodes. *IEEE Sensors Journal* 24 (11): 18090–18097.

Mukherjee, D., and Mukherjee, A. (2024). Understanding plant secondary metabolism using bioinformatics tools: Recent advances and prospects. *Bioinformatics for Plant Research and Crop Breeding*: 459–488.

Murmu, S., Sinha, D., and Chaurasia, H. et al. (2024). A review of artificial intelligence-assisted omics techniques in plant defense: Current trends and future directions. *Frontiers in Plant Science 15*: 1292054.

Naseri, G. (2023). A roadmap to establish a comprehensive platform for sustainable manufacturing of natural products in yeast. *Nature Communications* 14 (1): 1916.

Pandita, D., Pandita, A., and Wani, S.H. et al. (2021). Crosstalk of multi-omics platforms with plants of therapeutic importance. *Cells* 10 (6): 1296.

Park, S.J., Yoon, H.I., and Lee, H. et al. (2024). Evaluating the accuracy of machine learning classification models for similar herbal medicine using hyperspectral imaging. *Journal of Biosystems Engineering* 49 (2): 156–166.

Parthiban, S., Vijeesh, T., and Gayathri, T. et al. (2023). Artificial intelligence-driven systems engineering for next-generation plant-derived biopharmaceuticals. *Frontiers in Plant Science* 14: 1252166.

Pearcy, N., Hu, Y., and Baker, M.W. et al. (2021). Genome-scale metabolic models and machine learning reveal genetic determinants of antibiotic resistance in Escherichia coli and unravel the underlying metabolic adaptation mechanisms. *Msystems* 6 (4): 10–1128.

Persaud, M., Lewis, A. and Kisiala, A. et al. (2025). Untargeted metabolomics and targeted phytohormone profiling of sweet aloes (Euphorbia neriifolia) from Guyana: An assessment of asthma therapy potential in leaf extracts and latex. *Metabolites* 15 (3): 177.

Pushpalatha, G., G.L., and Devaraj, A. et al. (2022). Artificial intelligence and machine learning approach based in-silico ADMETox and Pharmacokinetic profile of α-Linolenic acid from Catharanthus roseus (L.) G. Don. *Journal of Drug Delivery & Therapeutics* 12.

Salam, U., Ullah, S., Tang, Z.H. et al. (2023). Plant metabolomics: An overview of the role of primary and secondary metabolites against different environmental stress factors. *Life* 13 (3): 706.

Salem, M.A., El-Shiekh, R.A., Fernie, A.R. (2022). Metabolomics-based profiling for quality assessment and revealing the impact of drying of Turmeric (Curcuma longa L.). *Scientific Reports* 12 (1): 10288.

Sigler, G.I., Murtha, J., and Varley, P.R. (2025). Diagnostic advances and novel therapeutics in peritoneal metastasis. *Surgical Oncology Clinics* 34 (2): 173–194.

Sotenjwa, V.Z. (2020). *Phytochemical Variation and Quality Control of Non-volatile Constituents of Artemisia Afra* (Doctoral dissertation, Tshwane University of Technology).

Spanakis, M., Tzamali, E., and Tzedakis, G. et al. (2025). Artificial intelligence models and tools for the assessment of Drug–Herb interactions. *Pharmaceuticals* 18 (3): 282.

Sun, L., Lai, M., and Ghouri, F. et al. (2024). Modern plant breeding techniques in crop improvement and genetic diversity: From molecular markers and gene editing to artificial intelligence—A critical review. *Plants* 13 (19): 2676.

Szeptycki, B. (2023). Developing a neural network model to predict phenotypic reaction of Arabidopsis thaliana to warm periods in winter.

Tariq, M.U., Haseeb, M., and Aledhari, M. et al. (2020). Methods for proteogenomics data analysis, challenges, and scalability bottlenecks: A survey. *IEEE Access* 9: 5497–5516.

Tinte, M.M., Chele, K.H., and van Der Hooft, J.J. et al. (2021). Metabolomics-guided elucidation of plant abiotic stress responses in the 4IR era: An overview. *Metabolites* 11 (7): 445.

Toffolatti, S.L., Maddalena, G., and Passera, A. et al. (2021). Role of terpenes in plant defense to biotic stress. In *Biocontrol Agents and Secondary Metabolites* (pp. 401–417). Woodhead publishing.

Velázquez, R., Rodríguez, A., and Hernández, A. et al. (2023). Spice and herb frauds: Types, incidence, and detection: The state of the art. *Foods* 12 (18): 3373.

Wan, L., Huo, J., and Huang, Q. et al. (2024). Genetics and metabolic responses of Artemisia annua L to the lake of phosphorus under the sparingly soluble phosphorus fertilizer: Evidence from transcriptomics analysis. *Functional & Integrative Genomics* 24 (1): 26.

Zhang, Y., and Wang, Y. (2023). Recent trends of machine learning applied to multi-source data of medicinal plants. *Journal of Pharmaceutical Analysis* 13 (12): 1388–1407.

Zhao, J., Liu, W., and Chen, Y. et al. (2022). Identification of markers for tea authenticity assessment: Non-targeted metabolomics of highly similar oolong tea cultivars (Camellia sinensis var. sinensis). *Food Control* 142: 109223.

Zhou, W., Yan, Z., and Zhang, L. (2024). A comparative study of 11 non-linear regression models highlighting autoencoder, DBN, and SVR, enhanced by SHAP importance analysis in soybean branching prediction. *Scientific Reports* 14 (1): 5905.

5

Machine Learning for Plant Ecological Research

Mani Manoj[1], Sahfigul Ameed Nihaal Fathima[1], Sathyalingam Sathya Trisha[1], Mohammed Jaffer Shakeera Banu[2], Shanmugam Gavaskar[3], Alagarsamy Sumitrha[4], Jeyabal Philomenathan Antony Prabhu[1], Arumugam Vijaya Anand[1]**

[1] *Department of Human Genetics and Molecular Biology, Bharathiar University, Coimbatore, Tamil Nadu, India*
[2] *Division of Molecular Biology, Microbiology Laboratory, Coimbatore, Tamil Nadu, India*
[3] *Department of Computer Application, Bharathiar University, Coimbatore, Tamil Nadu, India*
[4] *Department of Computer Science and Engineering, SNS College of Technology, Coimbatore, Tamil Nadu, India*

*Corresponding authors: avahgmb@buc.edu.in; jtonybio@gmail.com

5.1 Introduction to Machine Learning in Ecology

5.1.1 Overview of Machine Learning in Ecological Studies

Machine learning (ML) has become even more important within ecological research, providing multiple uses ranging from simulating species abundance to forecasting species relations and effects of bioaccumulation. In modeling for species richness, ML algorithms such as Random Forest, Increased Regression Tree, as well as Extreme Gradient Increasing, have demonstrated superior accuracy along with stability compared to many others like Lasso and Conditional Inference Forest, with the latter being the most stable in terms of model performance consistency (Cao et al. 2024). In predator–prey dynamics, Scientific ML models, especially Universal Differential Equations, look likely to accurately forecast ecological interactions using little data, even if noisy, better than Neural Ordinary Differential Equations (Devgupta et al. 2024). ML models are also important in predicting of the eventual bioaccumulation and subsequent histopathological effects of heavy metal pollutants in aquatic ecosystems, highlighting the importance of incorporating of many diverse species and ecological factors to greatly improve predictive capabilities and to fully inform conservation strategies (Obadimu et al. 2024). Furthermore, the validation of several ML models in ecology and evolution is evolving, with a focus on biological validations that consider the utility of these models beyond customary performance metrics, acknowledging that certain models with lower F1 scores can still be valuable for hypothesis testing and detecting biological patterns (Christensen et al. 2024). ML techniques such as Generalized Linear Models, Classification and Regression Trees, Artificial Neural Networks, and Evolutionary Algorithms are employed to predict species interactions in dynamic ecosystems. Each has special strengths as well as limitations, thereby advancing our comprehension of ecological dynamics (Kalpana et al. 2024).

5.1.2 Importance of ML in Plant Ecological Research

ML is central in plant ecological research development through enhanced accuracy, efficiency, and scope of data analysis in various fields. In invasive plant species research, ML methods such as support vector machines and random forests have immensely improved invasive species detection and control through remote sensing methods, hence enabling enhanced ecological management and conservation (Zaka and Samat 2024). Similarly, ML has revolutionized plant–insect interaction by automating the segmentation and classification of electrical penetration graph signals, which are central to the understanding of insect feeding behavior and ecological relevance. This automation not only enhances research

Machine Learning for Plant Biology, First Edition. Edited by Jen-Tsung Chen.

effectiveness but also reduces errors associated with manual evaluations (Dinh et al. 2024). In plant phenomics, ML enables high-throughput plant phenotypic trait evaluation, which accelerates breeding programs and provides insights into complex biological processes essential in crop management and ecological studies (Wu et al. 2024). Additionally, ML algorithms have been used to assess natural ecosystems by classifying plant species using satellite images, with high classification accuracy and as an effective tool for biodiversity monitoring (Marian et al. 2024). Lastly, ML methods are central in plant functional trait divergence research, allowing researchers to identify key traits that determine phenotypic variation and providing insightful views into evolutionary processes across various species and environmental conditions (Majumder and Mason 2024).

5.1.3 Comparison of Traditional vs. ML-based Ecological Analysis

Traditional ecological analysis methods use physical models and time-consuming approaches, which are designed to be limited by their complexity and specificity to specific services or datasets. Although these methods have commonly served as a basis, they might be hindered by low generalizability and efficiency, particularly in dynamic or large-scale ecosystems (Deng et al. 2024). In contrast, ML-based approaches offer a more data-oriented and adaptable method, capable of handling large-scale interactions and datasets and various ecological environments. For instance, ML models like Random Forest and XGBoost have shown higher accuracy and reproducibility in predicting species richness and modeling ecological interactions and have shown superior performance compared to traditional methods using a variety of metrics, such as R^2 (Kalpana et al. 2024). ML approaches have also been successfully used to estimate high conservation value areas (HCVAs) and define ecological protection red lines, with simpler and more accurate frameworks compared to traditional physical models. Moreover, ML approaches have also been used to predict chronic toxicity in aquatic ecosystems, using molecular descriptors to assess ecological risks better than traditional methods (Lei et al. 2024). In the study of dissolved organic matter (DOM) diversity in riverine ecosystems, ML has been used to uncover complex associations between environmental parameters and DOM composition, with information not uncovered by traditional approaches due to their narrow focus (Müller et al. 2024).

5.2 Data Sources and Preprocessing for Plant Ecology

5.2.1 Types of Ecological Data

Ecological information on plants can be derived from different sources, from satellite imagery, sensor measurements, and field surveys, each carrying different information and challenges. Satellite imagery, for example, from Sentinel-2, is extensively used for monitoring ecological systems and phenology of vegetation, providing high-resolution data that can be paired with ML models to land cover and plant species classification over large areas (Gong et al. 2024). This approach allows for ecological system mapping and monitoring of vegetation change over time, crucial for understanding ecosystem processes and informing conservation efforts (Sunde et al. 2024). Remote sensing is also crucial in the assessment of plant functional diversity (FD), offering a global view that may complement field data; however, it is difficult to correlate these data sources due to differences in scale. Improved sensor technologies, such as optical and radar sensors, are expanding the ability to monitor FD and ecosystem functioning in near real-time, thus evading the limitations of field surveys that lack continuous spatial and temporal coverage (Ma et al. 2020). Field measurements, though time-consuming, are required for the validation of remote sensing data and ensuring the accuracy of ecological assessments. The integration of these diverse data streams, facilitated by normalization techniques that harmonize functional diversity metrics across different resolutions and spectral information, enhances the comparability and robustness of ecological monitoring (Pacheco-Labrador et al. 2023).

5.2.2 Data Collection Techniques

Plant ecological data collection has undergone tremendous changes with the development of remote sensing, IoT, and citizen science, providing various means of inclusive ecosystem monitoring. Remote sensing methods, including Spectral Variation Hypothesis (SVH)-based methods, have been effective in estimating plant diversity and heterogeneity, allowing large-scale biodiversity estimation using spectral diversity and ecosystem spectral heterogeneity indices (Liccari et al. 2022).

These methods, aided by Sentinel-2A-like instruments, allow for the detection of biodiversity hotspots and biological invasion risk areas. Smartphone-based 3D scanning methods, in contrast, have emerged as a new method for nondestructive plant biomass monitoring based on Structure from Motion (SFM) to generate high-resolution ecological data. This method, validated by high correlation with traditional biomass harvesting, is accessible and cost-effective, encouraging citizen science participation (Dietrich et al. 2025). Citizen science, particularly through platforms like iNaturalist and Flora Incognita, has become instrumental in capturing global plant trait patterns and phenological dynamics. These crowd-sourced data collections complement professional efforts by filling gaps in plant trait data and providing insights into macro phenological patterns and climate-induced phenological shifts (Wolf et al. 2023). The integration of citizen science with Earth observation data has enabled the generation of global maps of plant traits, enhancing spatial applicability and accuracy despite inherent noise and biases (Lusk et al. 2024).

5.2.3 Data Cleaning, Normalization, and Feature Engineering

Data cleansing in the field of plant ecology is crucial to the accuracy and reliability of ecological investigations, especially in large databases and biodiversity data. The TRY Plant Trait Database, a leading source of plant trait data, has been identified to have a significant percentage of unusable or untraceable data; i.e., only 22.9% of the specific leaf area data was found usable after rigorous cleansing steps compared to 64.9% that is commonly accepted as usable in standard procedure (Augustine et al. 2024). In response to these difficulties, packages like the datacleanr R package have been created to make data cleansing interactive and reproducible. This package allows defective data structures to be identified and reduces excessive data loss while filtering and thereby enhances reproducibility and confidence in ecological inferences (Hurley et al. 2022). Similarly, the Biodiversity Data Cleaning (BDC) R package is an all-encompassing suite for standardization, integration, and cleansing of biodiversity data. It includes modules focusing on integrating datasets, correction of taxonomic names, and correction of errors in spatial and temporal data, and thereby enhances the fitness-for-use of biodiversity datasets (Ribeiro et al. 2022). Automated cleansing pipelines for data, such as those used for North American Ephedra, have been found to reduce errors in species distribution models conspicuously by removing false positives and unusable data, but still require expert input for the assurance of taxonomic accuracy (Führding-Potschkat et al. 2022).

Normalizing plant ecological data for cross-study comparison involves a few effective methods, each tailored for different data types and study goals. For transcriptional analysis between species, methods like EB, DWD, and XPN have been applied, with XPN optimized for reducing experimental differences and EB for maintaining biological differences. A new proposed method, CSN, is optimized to balance between these two factors and thus is seen to hold promise for use in more inter-species comparisons (Feldman et al. 2024). For functional diversity measures, a global normalization approach has been proposed to cater to heterogeneity in data collected by different sensors, e.g., field surveys and satellite imagery. The approach eliminates the effect of dimensionality, making measures like Rao's quadratic entropy index more comparable among different streams of data (Pacheco-Labrador et al. 2023). For phenomics data generated by high throughput plant phenotyping systems, the Digital Adjustment of Plant Development (DAPD) method normalizes time-series measurements by referencing an early developmental stage, significantly reducing variance and improving measurement accuracy (Lozano-Claros et al. 2020). In ecohydrological research, standardization in isotope-based research is recommended as the high number of method options carries the risk of interpretive vagueness and comparability issues. A standardized approach is recommended to overcome these issues (Millar et al. 2022). Lastly, in RNA sequencing, the use of external RNA spike-ins as normalization controls is highlighted for its ability to account for variations in total RNA abundance, offering a more accurate identification of differentially expressed genes (Laosuntisuk et al. 2024).

Effective feature engineering methods for extracting relevant information from remote sensing data in plant ecology research involve a combination of advanced techniques that leverage multisource data and sophisticated algorithms. One prominent approach is the use of multisource high-resolution remote sensing images, which enhances vegetation extraction through comprehensive multifeature perception. This method employs a random forest model for feature selection and a multifeature synthesis perception convolutional network (MSCIN) to improve multiscale feature extraction and global information interaction, thereby enhancing the differentiation of vegetation from other land features (Li et al. 2024). Another effective technique is the geographic object-based image analysis (GEOBIA), which utilizes data from various sources such as Gaofen-1 satellite, Ziyuan-3 satellite, and airborne LiDAR to extract hierarchical vegetation information. This method demonstrates that LiDAR data, with its high spatial resolution, significantly improves the accuracy of vegetation feature extraction across different levels of detail (Mao et al. 2020). Additionally, Principal Component Analysis

(PCA) has been shown to be a powerful tool for processing multispectral satellite images, achieving high accuracy in vegetation land cover classification by reducing data dimensionality and enhancing feature uniqueness (Parveen et al. 2016). Furthermore, thermal remote sensing offers novel insights into plant ecology by linking surface temperatures to vegetation dynamics, allowing for the evaluation of plant species, traits, and structural characteristics through high spectral resolution thermal infrared emissivity (Farella et al. 2022).

5.3 Machine Learning Techniques for Plant Ecology

5.3.1 Supervised Learning

Supervised learning, which is a core ML branch, is widely utilized across fields such as the identification of plant species and the estimation of biomass. Supervised learning encompasses the training of models on labeled data to acquire the input–output mapping in order to make predictions on unseen data (Nagaraj 2023). In biomass estimation, supervised learning approaches like Gaussian process regression (GPR) have proven to outperform conventional nonlinear regression in performance. For example, in a research work involving cedar and pine trees, GPR performed 1.05 to 1.34 times better than nonlinear regression in accuracy, which indicates its suitability in situations with high flexibility and accuracy in estimating biomass factors (Diamantopoulou et al. 2024). Equally, supervised learning is used in plant species classification via state-of-the-art methods such as Vision Transformers, which use transfer learning to efficiently address multi-label classification problems. This method, used in the PlantCLEF 2024 competition, involves processing large-scale datasets with distributed computing frameworks like Spark, showcasing the integration of supervised learning with modern data processing technologies (Gustineli et al. 2024). In addition, supervised learning's flexibility is realized through its use in various fields like epilepsy research, where it contributes to activities such as seizure detection and EEG analysis, highlighting its extensive viability and capacity to cater to intricate prediction and classification problems (Vankov and Gadhoumi 2024).

5.3.2 Unsupervised Learning

In plant systems biology, unsupervised learning (UL) is crucial due to the difficulty and expense of obtaining labeled data, which is often required for supervised learning. Techniques such as clustering and dimensionality reduction are employed to analyze large datasets generated by high-throughput omics technologies, facilitating insights into plant phenotyping and systems biology without the need for extensive labeled samples (Yan and Wang 2022). For ecological interactions, UL methods such as self-organizing maps (SOM) have been used to predict interactions in ecological networks with reasonable accuracy and very low noise levels, particularly for large networks. This approach bypasses the need for trait or phylogenetic information, relying only on interaction matrices, and has been shown to have potential to fill knowledge gaps in ecological metawebs (Adhurya and Park 2024). Additionally, UL has been applied to improve sward content prediction and herbage mass estimation, using contrastive learning to reduce the dependency on ground-truth images, thus enhancing the generalizability of models across different geographical locations (Albert et al. 2022).

5.3.3 Deep Learning

In plant leaf classification, Convolutional Neural Networks (CNNs) learn to extract features automatically, outdoing conventional methods using manual feature extraction, thus making the species identification more accurate and less prone to human error (Ganesh et al. 2025). Combining deep learning with other methods, including Wavelet Scattering Networks (WSNs) and MobileNetV2, enhances the process further. WSNs extract texture patterns well, even with small datasets, while MobileNetV2 extracts detailed features such as edges and shapes with high accuracy levels of 98.75% and 98.7% on the Flavia and Folio datasets, respectively (Gowthaman and Das 2025). In agriculture, deep learning models such as CNNs and VGG16 streamline vegetable classification by automating sorting, grading, and quality inspection, with higher accuracy and improving retail processes (Jillelamudi et al. 2024). Deep learning also makes it easy to identify plants with diseases, important for crop health and production. Models such as DenseNet-121 and DenseNet169 have high accuracy, 99.2% and 99.61% respectively, in disease classification of different plant species, proving them effective in early detection and prevention (Raghavendran et al. 2024).

5.3.4 Reinforcement Learning

Reinforcement Learning (RL) is now a robust tool in plant ecology studies, yielding new methods for improving conservation and accuracy in a variety of applications. One of these notable applications is the use of Deep Reinforcement Learning (DRL) in planning connectivity conservation, which mitigates the challenge of ecosystem fragmentation by spatially optimizing to achieve maximum habitat connectivity. This method has potential in solving complicated spatial optimization problems, such as maximizing the Integral Index of Connectivity and the Probability of Connectivity Index, which are critical to the sustainability of animal populations and enabling gene flow among fragmented habitats (Equihua et al. 2024). RL is also used in developing adaptive digital twin models of forest ecosystems, from IoT data and spatiotemporal graph modeling to dynamically represent and control forest health and resources. This model enhances the accuracy of forest management by optimizing the trade-off between model accuracy and computational efficiency, thereby enabling sustainable use of resources (Damaševičius and Maskeliūnas 2024). In plant health, Q-learning, a type of DRL, is used in simulating diverse growth conditions, generating customized strategy for optimal plant growth, thereby enabling customized plant growth strategies. This application is founded on a large dataset to enable environmental sustainability and ecological consciousness (Upadhyaya 2024). RL is also used in fisheries management and ecological tipping point environments, showing it can outperform other methods and solve sequential decision-making problems in conservation (Lapeyrolerie et al. 2022).

5.4 Applications of Machine Learning in Plant Ecology

ML has significantly enhanced species distribution modeling (SDM) by addressing challenges such as data quality, scalability, and complex interspecies interactions. Deep learning methods such as neural networks and autoencoders have improved the speed and accuracy of SDMs. For instance, DeepMaxent utilizes neural networks to automatically extract shared features among species, enhancing model performance in regions with uneven sampling and allowing for robust environmental feature prediction across large datasets without increased memory demand (Ryckewaert et al. 2024). Similarly, EcoVAE employs generative models to predict global species distributions, capturing underlying distribution patterns and interspecies interactions through unsupervised learning, and offering new applications for biodiversity monitoring without relying on environmental variables (Yan et al. 2024). MiTREE, a vision-transformer-based model, integrates spatial cross-modal relationships and ecological context, improving predictions of bird species encounter rates by leveraging remote sensing data and location information (Chen and Chiang 2024). Moreover, deep learning models, such as multi-layer perceptron, convolutional neural network, and vision transformer, have demonstrated excellent predictive power for multi-species community data, addressing data imbalance issues and accommodating various data types (Hu et al. 2025). The Bayesian information-theoretic minimum message length (MML) principle offers a robust model comparison framework across various model families, facilitating accurate feature learning and addressing model family misspecification (Wen and Dowe 2024).

ML has significantly advanced vegetation classification and mapping, offering efficient and scalable solutions across various landscapes and data types. In forest environments, the integration of Sentinel-1 and Sentinel-2 data with ML models like XGBoost has proven effective for mapping dominant tree species, achieving high accuracy by utilizing multi-temporal and multi-source data, including spectral indices and texture features (Guo et al. 2024). In agriculture, ML facilitates the identification of crop types and monitoring of vegetation, particularly in arid landscapes. Through the use of Sentinel-1 SAR data and Landsat 8 optical imagery, Random Forest models can efficiently classify crops and monitor vegetation dynamics, enhancing the management of agricultural resources (Mayouf et al. 2024). The use of deep learning, particularly CNNs, further simplifies feature extraction in plant leaf classification, enhancing accuracy in species identification and removing human intervention (Ganesh et al. 2025). Additionally, archived grayscale aerial imagery can be effectively utilized for vegetation classification through ML models, with region-based approaches achieving notable accuracy in distinguishing land cover types (Teeravech et al. 2024). In heterogeneous landscapes, the optimal choice of Sentinel-2 data's spectral variables, processed through ML, optimizes large-scale crop mapping, and the use of Red Edge region variables is significant for accurate land use and land cover classification (de Oliveira Júnior et al. 2024).

ML has emerged as a strong mechanism for forecasting plant growth patterns, providing tremendous improvements in agricultural operations. Different ML models, including Support Vector Machine (SVM), Naive Bayes, and

Extreme Learning Machine (ELM), have been used to forecast plant growth milestones based on environmental and treatment-related data; among them, SVM has been found to provide better accuracy, although it needs further optimization for practical application (Silvia et al. 2024). The incorporation of big data, including genomic, phenotypic, and environmental factors, has significantly improved the predictive accuracy of ML models in plant breeding, enabling a better understanding of genotype-by-environment interaction and better breeding strategies (Crossa et al. 2025). Furthermore, ML algorithms have been employed to forecast crop yields based on the analysis of soil data and imagery, which enables the interpretation of soil characteristics important for crop growth, thus enabling farmers to make highly informed decisions on crop selection and yield forecasting (Venkatesh and Naik 2024). The application of IoT data in greenhouses has also been investigated, where models such as XGBoost perform better than others in forecasting variables associated with plant growth, illustrating the strength of ML in being capable of handling plant growth variables effectively to improve crop yields (Rajathi et al. 2024). Furthermore, ML techniques have been employed for crop yield forecasting in the context of smart agriculture, where models such as XGBoost and Random Forest provide low mean squared errors, indicating their potential in facilitating sustainable crop yield improvement and supporting agricultural planning and decision-making (Nuser et al. 2024).

With its sophisticated prediction skills and insights into ecosystem dynamics, ML has become a crucial tool in evaluating the effects of climate change on plant ecosystems. The integration of ML with ecological models enhances the prediction and mitigation of climate change effects on ecosystems by combining data-driven approaches with ecological expertise, as demonstrated by the use of hybrid modeling and real-time monitoring techniques (Desai and Sharma 2024). ML algorithms and geographic information systems (GIS)-based software have been used to evaluate remote sensing data in the Western Himalayan environment, exposing major vegetation vulnerabilities brought on by climate change (Ahlawat and Roy 2023). Likewise, ML methodologies, including Long-Short Term Memory (LSTM) networks, have been employed to anticipate the effects of climatic conditions on paddy crop yield, underscoring the significance of forecasting in the management of agricultural ecosystems amid fluctuating weather patterns (Thirumal and Latha 2023). Furthermore, ML models, including recurrent deep neural networks, have been applied to predict ecosystem CO_2 assimilation, demonstrating superior performance over traditional models in capturing the spatial and temporal variability of ecosystem photosynthesis across diverse environmental conditions (De Bartolomeis et al. 2023). These applications underscore the potential of ML in enhancing the understanding of climate change impacts on plant ecosystems, facilitating more accurate predictions and effective management strategies. However, challenges such as data standardization and model interpretability remain, necessitating ongoing interdisciplinary collaboration to fully harness ML's potential in climate change research (Hamdan et al. 2024).

ML offers significant potential in developing early warning systems for invasive species in plant ecosystems by enhancing detection, monitoring, and management capabilities. The integration of ML with remote sensing technologies, such as drones and satellites, allows for precise mapping and monitoring of invasive plant species across large and remote areas, significantly reducing fieldwork costs and improving access to data (Fenollosa and Salguero-Gómez 2025). ML algorithms, including support vector machines, random forests, and deep learning models like convolutional neural networks, have demonstrated high accuracy in identifying and classifying invasive species from complex visual data (Zaka and Samat 2024). For instance, the EfficientNetB0 model achieved an 84% accuracy in classifying multiple invasive plant species, highlighting the efficacy of deep learning in this domain (Nunavath and Danilin 2024). Additionally, ML models can predict the phenological stages of invasive species by analyzing environmental conditions, thereby improving the timing and effectiveness of detection efforts. These models can forecast phenological stages at a daily temporal resolution, facilitating the creation of "phenoregions" that map areas with similar phenological dynamics, which is crucial for targeted surveillance and monitoring activities. The use of opportunistic biodiversity observation data further enhances these models' predictive capabilities, underscoring the transformative potential of ML in addressing the challenges posed by biological invasions (Capinha et al. 2024).

5.5 Remote Sensing and AI in Plant Ecology

5.5.1 Use of Satellite and Drone Imagery for Plant Monitoring

The combination of ML models with satellite and drone imagery greatly enhances the accuracy of plant monitoring and decision-making in agriculture. The combination of unmanned aerial vehicles (UAVs) and satellite imagery, as

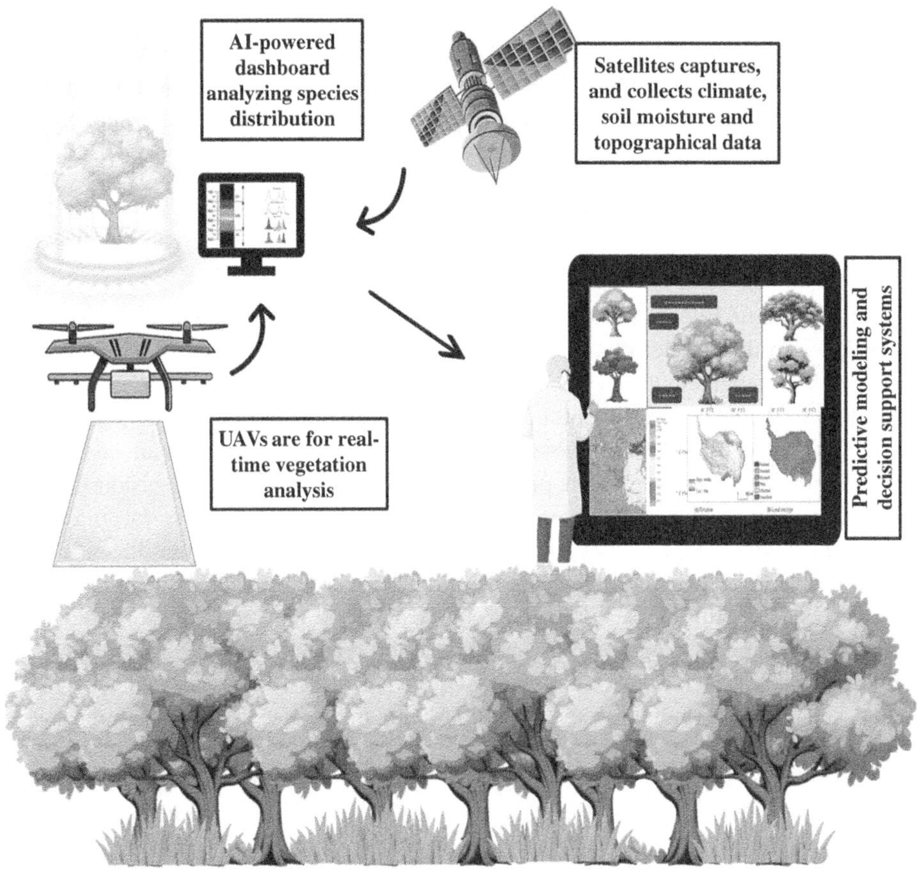

Figure 5.1 AI-driven plant ecology monitoring.

demonstrated by Ogungbuyi et al. greatly enhances the accuracy of predictions regarding pasture biomass, with UAS data reducing the standard error of mean biomass predictions when combined with Sentinel-2 imagery and random forest models illustrated in Figure 5.1 (Ogungbuyi et al. 2024). Deep learning models, particularly CNNs, have been shown to possess a greatly improved ability in the early detection and prediction of crop diseases compared to traditional methods, thus facilitating accurate diagnosis and allowing timely interventions for crop protection (Mudarisov and Miftakhov 2024). The combination of Synthetic Aperture Radar (SAR) data from Sentinel-1 with Landsat 8 optical imagery, as studied by Mayouf et al. greatly enhances crop type classification and vegetation monitoring, with accuracy levels above 85% in differentiating between diverse crop types (Mayouf et al. 2024). Additionally, Dixit highlights the ability of deep learning and image recognition technologies in intelligent farming systems, with high accuracy in crop monitoring and disease detection, thus allowing data-driven decision-making and optimized resource allocation (Dixit 2024). Additionally, ML techniques, including ensemble techniques and CNNs, are employed to process complex data from diverse sources, thus improving predictive accuracy and decision-making capabilities in precision agriculture (Bachu et al. 2024).

5.5.2 Image Segmentation and Object Detection for Vegetation Analysis

Object detection and image segmentation are very important in vegetation analysis, using artificial intelligence (AI) and ML will help in maximizing efficiency and accuracy in plant research. Deep learning models, such as YOLO and Mask R-CNN, have been effectively applied to detect and segment weeds in agricultural crops using UAV-captured images, with YOLOv8s achieving a high mAP50 of 97% and precision of 99.7% in soybean and bean crops (Silva et al. 2024). Multi-scale convolutional neural networks (MSCNN) further advance vegetation detection by integrating multi-scale feature extraction and attention mechanisms, achieving a global accuracy and boundary F1 score of up to 98% across diverse datasets, thus proving robust in complex environments (Albalooshi 2024). The FA-HRNet model introduces a fusion attention approach, enhancing semantic segmentation in remote sensing images by addressing species heterogeneity and seasonal variations, improving MIoU and PA by 2.17% and 4.85%, respectively (He et al. 2024). In handling the problem of missing parts of plant images, a shallow network plan using adversarial learning and fractal dimension estimation greatly improves classification accuracy from 41.16% to 90.51% even when 40–60% of the parts of the plant are missing. Additionally, AI-based apps for plant disease detection utilize CNNs for lesion segmentation, offering comprehensive disease analysis and management recommendations (Batchuluun et al. 2024).

5.5.3 AI Models for Automated Plant Health Assessment

AI model development for plant health assessment automation has come a long way, leveraging deep learning and ML methods to enhance accuracy and efficiency in disease detection. CNNs are extensively used, as seen in a model that leverages image processing combined with data augmentation methods to detect healthy and diseased leaves of plants and achieve high accuracy in real-time detection applications illustrated in Figure 5.2 (Aishwarya et al. 2025). Similarly, a CNN model based on the ResNet architecture, which leverages transfer learning, has been integrated into an easy-to-use interface for real-time plant health classification, providing real-time predictions and enabling informed decision-making for agricultural experts (Shastry et al. 2024). Comparative assessment of state-of-the-art architectures like Inception V3, ResNet-50, and Vision Transformers combined with SHapley Additive exPlanations (SHAP) has shown that Inception V3 exhibits superiority in accuracy and interpretability, especially in the case of Taro disease classification (Maeda-Gutiérrez et al. 2024). Apart from CNNs, other AI methods like Multiple Linear Regression, Naïve Bayes, and Support Vector Machines have been explored, each with varying strengths and weaknesses in the case of plant disease detection (Navaneetha et al. 2024). The integration of AI with satellite imagery, IoT devices, and mobile applications further enhances real-time monitoring and predictive capabilities, addressing issues of scalability and accessibility in large-scale agricultural applications (Manoj and Sasilatha 2024).

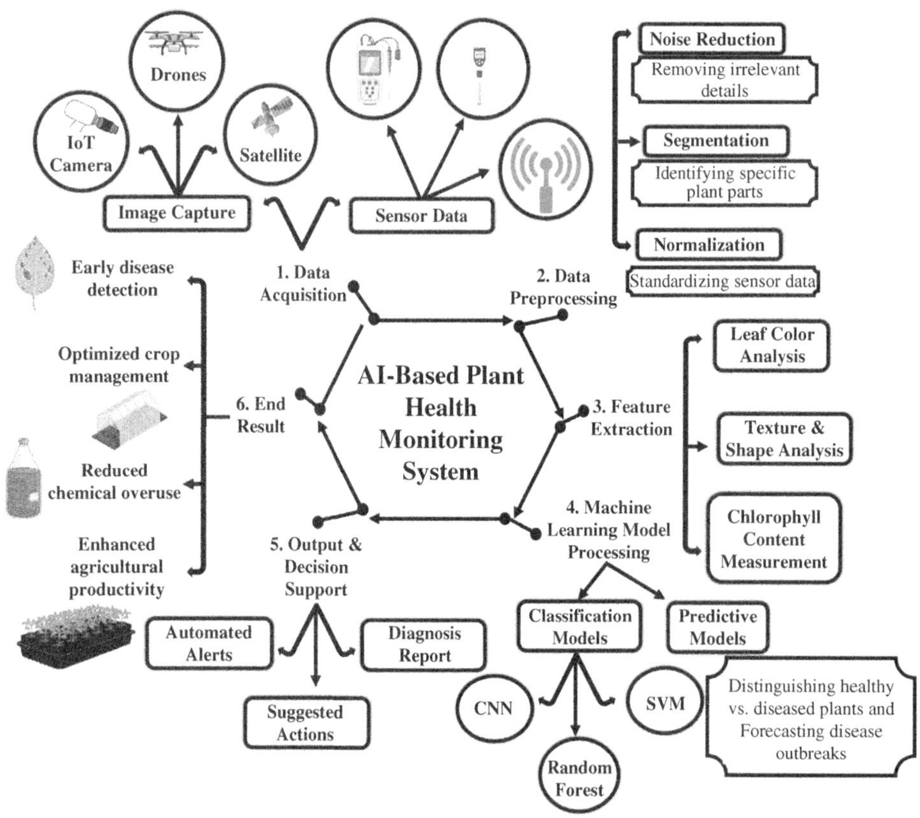

Figure 5.2 AI-based plant health monitoring system.

5.6 Biodiversity Conservation and Ecosystem Monitoring

5.6.1 Machine Learning for Biodiversity Pattern Analysis

ML has emerged as a powerful tool for analyzing patterns of plant biodiversity, leading to immense advances in understanding and forecasting plant diversity across various dimensions. ML algorithms, such as Random Forests and neural networks, has enabled researchers to model species richness and phylogenetic richness with high accuracy, explaining up to 80.9% and 83.3% of these diversities, respectively, by considering current climate and environmental heterogeneity as primary drivers (Hamdan et al. 2024). In natural systems, ML has been used to differentiate morphologically similar plant species based on satellite imagery with over 90% classification accuracy rates (Marian et al. 2024). Besides, deep learning models are in the process of being developed to process herbarium images, thus facilitating the identification of plant morphological features and enhancing the efficiency of biodiversity analyses. These models are integrated into platforms like PlantAI, enabling users to construct personalized datasets and train models for individual traits, thus improving model generalization and usability (Sklab et al. 2024). Additionally, ML has been applied to predict declines in native plant phylogenetic diversity in response to climate change, highlighting the potential for these techniques to forecast future biodiversity patterns and inform conservation efforts (Park et al. 2020).

5.6.2 AI-driven Monitoring of Endangered Plant Species

The use of AI-based monitoring systems for endangered plant species utilizes advanced technologies to support conservation efforts, offering new strategies to the critical challenges of biodiversity loss. The integration of AI in monitoring endangered species is exemplified by the application of ML algorithms, such as Random Forest and CNNs, which are used to predict species that might be under extinction, thereby aiding in ecological conservation illustrated in Figure 5.3 (Rao and Subrahmanyam 2024). In urban areas, AI greatly enhances the identification and monitoring of species, achieving great accuracy in biodiversity assessment and management, an element that is key to the sustainable management of urban ecosystems (Rahmati 2024). The use of UAV in combination with AI-based object detection systems, such as the Maxent predictive model, facilitates efficient monitoring of rare plant species, such as *Geum radiatum*. This approach encourages

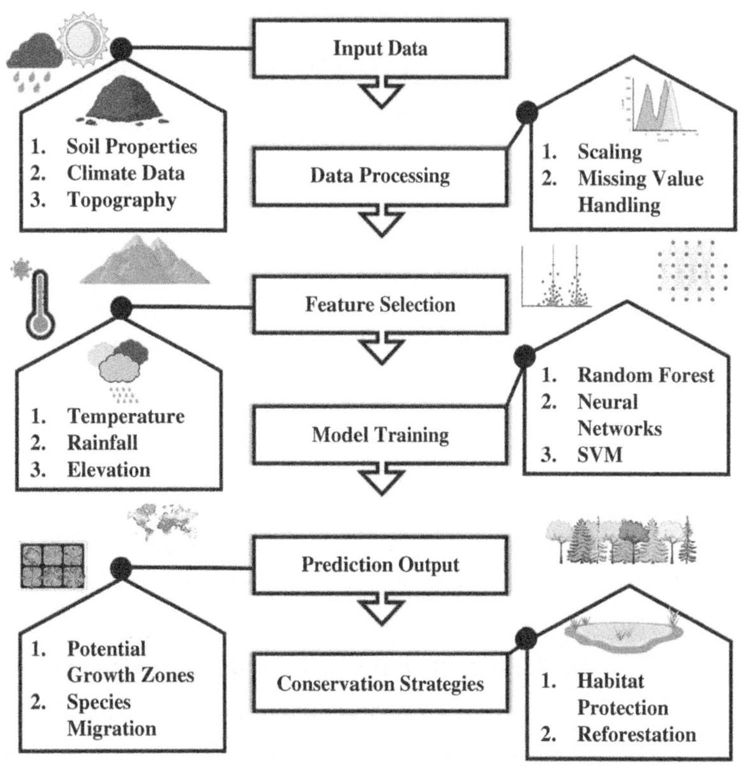

Figure 5.3 Predictive modeling for plant species distribution.

specific data collection and has been effective in established plant habitats and discovering new ones; however, further improvement in data collection strategies and computer vision technology is needed (Reckling et al. 2021). Furthermore, platforms like Conservation AI utilize ML and computer vision to monitor wildlife, including critically endangered species, through real-time and non-real-time analysis, supporting both immediate conservation actions and long-term habitat assessments. These AI-based methods not only enhance the accuracy and efficacy of monitoring endangered plant species but also offer flexible, scalable solutions that can be applied across a wide range of ecological environments, highlighting the ability of AI to make significant inputs to global conservation efforts (Fergus et al. 2024).

5.6.3 Predicting Ecosystem Resilience and Response to Environmental Stressors

Predicting ecosystem resilience and response to environmental stressors involves understanding the complex interplay of various ecological, genomic, and management factors. Ecological network analysis reveals that the diversity of traits within a community, such as those of benthic macroinvertebrates, significantly influences its capacity to withstand and recover from stressors like nutrient enrichment. More complex trait networks tend to exhibit less variability in response to stress, suggesting that maintaining or enhancing baseline network complexity could bolster resilience against environmental changes (Gladstone-Gallagher et al. 2023). Additionally, the concept of "ecosystem response footprints" emphasizes the importance of considering the spatial and temporal scales of ecosystem responses, which often differ from the footprints of individual stressors. This approach can guide environmental management by highlighting the need for holistic strategies that account for cumulative effects and nonlinear ecosystem shifts (Low et al. 2023). The Pressure State Response model offers a framework for assessing and predicting ecological resilience, as demonstrated in China's coastal zones, where resilience has declined over recent decades. This model can be applied to other regions to forecast resilience under various scenarios, aiding in the development of targeted conservation efforts (Wan et al. 2024). At a more granular level, genomic and epigenomic processes play crucial roles in resilience, with changes at these levels potentially cascading to influence broader ecological resilience. Understanding these processes can inform strategies to enhance resistance and recovery in ecosystems facing climate-related stressors (Metzger et al. 2024). Finally, integrating model-driven engineering and AI can address the challenges of predicting ecosystem resilience by synthesizing diverse data and adapting neural network architectures to specific ecosystem scenarios, thus providing a robust framework for future research and management strategies (Sousa 2023).

5.7 Predictive Modeling for Ecological Trends

5.7.1 Time-series Forecasting of Ecological Parameters

Time-series forecasting of ecological parameters in plants is a critical area of research that leverages advanced ML models to predict key vegetation metrics such as Leaf Area Index (LAI), Vegetation Health Index (VHI), Normalized Difference Vegetation Index (NDVI), chlorophyll a, and dissolved oxygen. The LAI, a fundamental structural variable in ecosystem modeling, can be effectively predicted using deep learning models like the attention-enhanced LSTM (AELSTM), which utilizes climate data to achieve high temporal and spatial consistency with satellite-derived LAI, demonstrating R2 values between 0.86 and 0.93 across various biomes (Xiong et al. 2023). Similarly, the ConvLSTM model, combining CNN and LSTM, has been employed to forecast VHI using MODIS time series data, showing superior performance with Root-mean-square deviation (RMSE) values as low as 0.025 for short-term forecasts (Kartal et al. 2024). The NDVI, another crucial indicator of vegetation health, benefits from a novel forecasting model that integrates time series decomposition with CNN-LSTM, achieving high prediction accuracy with RMSE, NSE, and MAE values of 0.0573, 0.9617, and 0.0447, respectively (Gao et al. 2023). Additionally, the forecasting of chlorophyll a and dissolved oxygen in aquatic ecosystems like the Mar Menor has been successfully implemented using neural networks, with models such as Time2Vec and BiLSTM achieving R2 values above 0.92, indicating their efficacy in environmental monitoring (López-Andréu et al. 2023).

5.7.2 ML Models for Predicting Drought and Deforestation Impact

ML models, which use cutting-edge methods to improve accuracy and applicability, are now essential for forecasting the effects of drought and deforestation. By combining meteorological forecasts with spatiotemporal satellite data,

Convolutional Long-Short Term Memory (ConvLSTM) models have shown successful results in predicting vegetation density as indicated by the NDVI index. By using historical and projected climate data, this method seeks to improve predictive power using a guided-prediction technique and offer early warnings of drought consequences in Swiss forests. Similarly, a spatiotemporal deep learning approach has been suggested to estimate drought consequences in forests using Sentinel-2 satellite data and meteorological factors. This model incorporates topographic and environmental data to improve interpretability and predict potential drought occurrences by forecasting NDVI values (Durukan et al. 2023). A review of deep learning models highlights their effectiveness in handling complex datasets for drought forecasting, with models like Deep Neural Networks and CNN being particularly pertinent. However, the availability and quality of the data are necessary for these models to succeed (Gyaneshwar et al. 2023). Furthermore, the effects of drought on agriculture have been investigated using ML methods, such as polynomial regression and artificial neural networks, with a focus on the significance of drought spatial features as input data. Drought vulnerability has been evaluated using ensemble ML approaches, such as M5P with Random SubSpace and Rotation Forest models, which have shown excellent accuracy in mapping drought-prone regions and supporting strategic planning for damage mitigation (Saha et al. 2023).

5.7.3 Climate–Plant Interaction Modeling

The accuracy of climate–plant interaction models in predicting future plant responses is influenced by several key factors, including model complexity, training collinearity, collinearity shift, and predictor novelty. These factors interactively affect model forecasting, with increased predictor novelty and collinearity shift significantly degrading model performance, particularly under high model complexity and training collinearity conditions (Chen et al. 2024). Additionally, the models often fail to incorporate bio physiologically meaningful variables, such as soil water content and specific climatic events during critical growth periods, which are crucial for capturing the physiological responses of plants to climate changes (Gardner et al. 2019). The bidirectional nature of climate–vegetation interactions also plays a critical role, as Earth System Models (ESMs) need to accurately simulate both the impact of climate on vegetation and the feedbacks of vegetation on climate. However, these models often overestimate the climate control on vegetation dynamics, particularly the dominance of precipitation in arid regions and air temperature on forest phenology (Claessen et al. 2019). Furthermore, climate change influences the interactions between abiotic and biotic stresses on plants, altering their primary and secondary metabolism, which in turn affects their adaptive responses (Surówka et al. 2020).

5.8 Challenges and Limitations of Machine Learning in Plant Ecology

Data scarcity and imbalance are significant challenges in ML models applied to plant ecology, impacting the accuracy of predictions, particularly for minority plant species. These issues arise due to the unequal distribution of class instances, where minority classes, often representing rare or newly identified plant species, are underrepresented in datasets. This imbalance leads to biased models that favor majority classes, resulting in poor predictive performance for minority classes (Gonzalez-Villanueva and Carranza-Rojas 2024). Traditional ML methods struggle with these imbalances, as they tend to memorize patterns from the majority class, increasing the risk of overfitting and reducing the model's ability to generalize. Various strategies have been proposed to address these challenges, including data-level methods like SMOTE, which can introduce noise and redundancy, and algorithm-level techniques such as cost-sensitive learning and ensemble methods, which require careful tuning (Altalhan et al. 2025). Hybrid approaches that combine these strategies have shown promise in enhancing accuracy and adaptability. Advanced AI techniques, including deep learning architectures, have also (Apu et al. 2025) been employed, but they are often constrained by computational demands and the risk of overfitting. Moreover, the reliance of ML models on the magnitude of encoded signals during inference can further impair generalization, even with data augmentation (Dablain and Chawla 2024). To improve the balanced accuracy for minority plant species, integrating deep learning techniques with traditional methods has been shown to enhance prediction accuracy without compromising overall performance (Gonzalez-Villanueva and Carranza-Rojas 2024).

Achieving model interpretability in ecological applications using ML presents several challenges, primarily due to the complexity and volume of ecological data and the inherent trade-offs between model accuracy and interpretability. One significant challenge is the information sparsity in ecological datasets, such as those from Passive Acoustic Monitoring (PAM), which often contain vast amounts of non-informative data. This necessitates the development of methods that can filter out noise and focus on ecologically relevant features, as demonstrated by the use of Gaussian Mixture

Models (GMMs) to enhance interpretability in soundscape data analysis (Omprakash et al. 2024). Another challenge is the performance-interpretability trade-off, where models like neural networks offer high accuracy but at the cost of reduced interpretability. This trade-off is evident in ecological studies using sentinel species data, where hybrid models combining neural networks and elastic nets are recommended to balance accuracy and interpretability (McMillan et al. 2024). Additionally, the lack of interpretability in machine-learned acoustic features is compounded by issues such as inductive biases and crude temporal integration methods. Approaches like convolutional Variational Auto-Encoders (VAEs) have been proposed to address these issues by providing interpretable latent spaces and improving temporal aggregation, thus enhancing the understanding of ecological processes through soundscape analysis (Gibb et al. 2024). Furthermore, the misconception that only black-box models can achieve high accuracy is challenged by the development of interpretable models like Generalized Additive Models (GAMs), which can capture complex patterns while remaining interpretable, suggesting that the trade-off between performance and interpretability is not as strict as previously thought (Kruschel et al. 2025).

The primary ethical concerns associated with biases in ecological AI models and their impact on environmental decision-making revolve around algorithmic bias, data discrimination, and environmental inequity. Algorithmic bias in AI models, as highlighted in studies, often stems from nonrepresentative datasets, leading to entrenched stereotypes and skewed decision-making outcomes, which can significantly affect ecological applications such as climate risk modeling and wildlife habitat monitoring (Al-Adwan 2025). These biases can result in data discrimination, where certain demographic or ecological data are underrepresented or misrepresented, potentially leading to inequitable environmental policies and practices. Furthermore, the environmental footprint of AI systems, exacerbated by power-hungry servers, raises concerns about environmental inequity, as the environmental costs of AI deployment can vary significantly across different regions, potentially disadvantaging certain areas (Hajiesmaili et al. 2024). Additionally, the techno-optimism surrounding AI in conservation can distract from addressing these biases and their broader implications, such as changes in staffing and working practices in the conservation sector, which may inadvertently harm biodiversity (Sandbrook 2025). Therefore, it is crucial to develop frameworks that integrate AI's capabilities with sustainable practices, ensuring that AI innovations align with principles of justice, equity, and environmental sustainability to mitigate these ethical concerns (Hajiesmaili et al. 2024).

5.9 Future Perspectives and Emerging Technologies

The future perspectives and emerging technologies of ML in plant ecological research are multifaceted, with significant advancements anticipated across various domains. Remote sensing (RS) technologies, such as hyperspectral imaging and LiDAR, integrated with ML methods like SVM and random forests, are enhancing the detection and management of invasive plant species (IPS), thereby improving ecological management strategies (Zaka and Samat 2024). In plant phenomics, ML is revolutionizing the field by automating the analysis of plant phenotypic traits, which accelerates breeding processes and provides insights into complex biological mechanisms, thus supporting crop management and ecological studies (Wu et al. 2024). In Asia, the adoption of AI and ML has propelled ecological research forward, focusing on applied issues like biodiversity conservation and climate change, with deep learning technologies emerging as pivotal tools for big data collection and analysis (Ryo 2024). Additionally, ML models, particularly random forests, have shown superior performance over traditional models in predicting soil seed bank (SSB) characteristics, demonstrating the potential of ML to streamline labor-intensive ecological research (Rosbakh et al. 2022). In the realm of plant disease management, ML and deep learning approaches are being utilized for automated disease detection and diagnosis, with future directions pointing towards explainable AI, active learning, and edge computing to enhance sustainable crop production and plant health management (Samia et al. 2022).

5.10 Conclusion

AI-based plant health monitoring systems are revolutionizing ecological research and precision agriculture by offering real-time, data-driven insights. The integration of ML models, remote sensing and IoT-based sensors enables early disease detection and improved decision-making. By analyzing plant phenotypes, environmental factors, and soil conditions, AI enhances agricultural productivity while minimizing resource wastage. One of the most significant advantages of AI-based systems is their predictive capability, which can predict possible plant diseases before the occurrence of visible signs. ML algorithms, particularly deep learning algorithms like CNNs, scan high-resolution images to precisely classify the health status of plants. AI-based automated alert systems also provide timely recommendations for irrigation, application of pesticides, and soil nutrients, making sustainable agriculture practices easy. Even with great advances, problems including data heterogeneity, model accuracy, and high computational remain unsolved. Future research should aim to integrate AI, climate modeling, and plant genomics to create holistic plans for preventing disease. AI will most likely remain crucial in maintaining global food security and promoting ecological sustainability by making agriculture more efficient and resilient.

Acknowledgement

All the authors are thankful to their respective Universities and Institutions for their valuable support.

Data Availability

The authors confirm that the data supporting the findings of this study are available within the book/chapter.

References

Adhurya, S., and Park, Y.S. (2024). A novel method for predicting ecological interactions with an unsupervised machine learning algorithm. *Methods in Ecology and Evolution* 15 (7): 1247–1260.

Ahlawat, A., and Roy, A. (2023). Assessing the vulnerability of Western Himalayan ecosystem to climate change using machine learning algorithms. In *Remote Sensing for Agriculture, Ecosystems, and Hydrology XXV* (Vol. 12727, pp. 78–95). SPIE.

Aishwarya, M., Pranathi, K., Vaishnavi, B. et al. (2025). The Future of Plant Health: Deep Learning Solutions. *Journal of Computer Networks and Virtualization* 3 (1): 1–7. Al-Adwan, M.A.S. (2025). Harnessing artificial intelligence for environmental sustainability: Ethical considerations and practical implications in achieving SDG 9 And SDG 16. *Journal of Lifestyle and SDGs Review* 5 (2), e03779–e03779.

Albalooshi, F.A. (2024). Novel approach in vegetation detection using multi-scale convolutional neural network. *Applied Sciences* 14 (22): 10287.

Albert, P., Saadeldin, M., Narayanan, B. et al. (2022). Utilizing unsupervised learning to improve sward content prediction and herbage mass estimation. *arXiv preprint arXiv:2204.09343.*

Altalhan, M., Algarni, A., and Alouane, M.T.H. (2025). Imbalanced data problem in machine learning: A review. *IEEE Access.*

Apu, K.U., Ali, M., Islam, M.F. et al. (2025). A systematic literature review on AI approaches to address data imbalance in machine learning. *Frontiers in Applied Engineering and Technology* 2 (01): 58–77.

Augustine, S.P., Bailey-Marren, I., Charton, K.T. et al. (2024). Improper data practices erode the quality of global ecological databases and impede the progress of ecological research. *Global Change Biology* 30 (1): e17116.

Bachu, L., Kandibanda, A., Grandhi, N. et al. (2024). Machine learning for enhanced crop management and optimization of yield in precision agriculture. In *2024 8th International Conference on I-SMAC (IoT in Social, Mobile, Analytics and Cloud) (I-SMAC)* (pp. 1289–1293). IEEE.

Batchuluun, G., Kim, S.G., Kim, J.S. et al. (2024). Artificial intelligence-based segmentation and classification of plant images with missing parts and fractal dimension estimation. *Fractal and Fractional* 8 (11): 633.

Cao, Y., Schartel, T., Houghton, D. et al. (2024). Evaluating machine leaning algorithms for accuracy, stability, and predictor discriminative ability based on ten datasets. *Stability, and Predictor Discriminative Ability Based on Ten Datasets.*

Capinha, C., Monteiro, A.T., and Ceia-Hasse, A. (2024). Supporting early detection of biological invasions through short-term spatial forecasts of detectability. *bioRxiv*: 2024–06.

Chen, T., and Chiang, Y.Y. (2024). MiTREE: Multi-input transformer ecoregion encoder for species distribution modelling. In *Proceedings of the 7th ACM SIGSPATIAL International Workshop on AI for Geographic Knowledge Discovery* (pp. 110–120). Association for Computing Machinery.

Chen, X., Liang, Y., and Feng, X. (2024). Influence of model complexity, training collinearity, collinearity shift, predictor novelty and their interactions on ecological forecasting. *Global Ecology and Biogeography* 33 (3): 371–384.
Christensen, C., Ferreira, A.C., Cherono, W. et al. (2024). Moving towards more holistic validation of machine learning-based approaches in ecology and evolution. *bioRxiv*.
Claessen, J., Molini, A., Martens, B. et al. (2019). Global biosphere–climate interaction: A causal appraisal of observations and models over multiple temporal scales. *Biogeosciences* 16 (24): 4851–4874.
Crossa, J., Montesinos-Lopez, O.A., Costa-Neto, G. et al. (2025). Machine learning algorithms translate big data into predictive breeding accuracy. *Trends in Plant Science* 30 (2): 167–184.
Dablain, D.A., and Chawla, N.V. (2024). The hidden influence of latent feature magnitude when learning with imbalanced data. In *International Conference on Pattern Recognition and Artificial Intelligence* (pp. 76–91). Springer Nature Singapore.
Damaševičius, R., and Maskeliūnas, R. (2024). A reinforcement learning-based adaptive digital twin model for forests. In *2024 4th International Conference on Applied Artificial Intelligence (ICAPAI)* (pp. 1–7). IEEE.
De Bartolomeis, P., Meterez, A., Shu, Z. et al. (2023). An effective machine learning approach for predicting ecosystem CO_2 assimilation across space and time. *EGUsphere 2023*: 1–31.
de Oliveira Júnior, J.G., Esquerdo, J.C.D.M., and Lamparelli, R.A.C. (2024). Optimum combination of spectral variables for crop mapping in heterogeneous landscapes based on sentinel-2 time series and machine learning. *ISPRS Annals of the Photogrammetry, Remote Sensing and Spatial Information Sciences* 10: 85–92.
Deng, J., Xie, Y., Wei, R. et al. (2024). Application of machine learning in ecological red line identification: A case study of Chengdu–Chongqing urban agglomeration. *Diversity* 16 (5): 300.
Desai, K. and Sharma, N. (2024). Harnessing the power of machine learning to predict and mitigate the impacts of climate change on global ecosystems. In *2024 International Conference on Advances in Computing Research on Science Engineering and Technology (ACROSET)* (pp. 1–6). IEEE.
Devgupta, R., Dandekar, R.A., Dandekar, R. et al. (2024). Scientific machine learning in ecological systems: A study on the predator-prey dynamics. *arXiv preprint arXiv:2411.06858*.
Diamantopoulou, M.J., Kurnaz, E., Kalkanlı Genç, Ş. et al. (2024). Assessment of a probabilistic supervised machine learning method to estimate biomass expansion and conversion factors: A case study on cedar and pine trees. *Canadian Journal of Forest Research*.
Dietrich, P., Elias, M., Harpole, S. et al. (2025). Advancing plant biomass measurements: Integrating smartphone-based 3D scanning techniques for enhanced ecosystem monitoring. *Methods in Ecology and Evolution* 00: 1–10.
Dinh, Q.D., Kunk, D., Hy, T.S. et al. (2024). Machine learning for characterizing plant-insect interactions through electrical penetration graphic signal. *bioRxiv*: 2024–06.
Dixit, R.R. (2024). Integrating deep learning and image recognition into smart farming. In *Smart Agriculture* (pp. 97–110). CRC Press.
Durukan, E.D., Ledain, S., Brunschwiler, T. et al. (2023). Forest drought impact prediction based on spatio-temporal satellite imagery and weather forecasts--A spatio-temporal approach using convolutional LSTM models. In *European Geoscience Union (EGU) Assembly*.
Equihua, J., Beckmann, M., and Seppelt, R. (2024). Connectivity conservation planning through deep reinforcement learning. *Methods in Ecology and Evolution* 15 (4): 779–790.
Farella, M.M., Fisher, J.B., Jiao, W. et al. (2022). Thermal remote sensing for plant ecology from leaf to globe. *Journal of Ecology* 110 (9): 1996–2014.
Feldman, S., Ner-Gaon, H., Treister, E. et al. (2024). Comparison and development of cross-study normalization methods for inter-species transcriptional analysis. *PLoS One* 19 (9): e0307997.
Fenollosa, E., and Salguero-Gomez, R. (2025). AI and Big Data for invasion biology: Finding, modelling and forecasting the population dynamics of invaders.
Fergus, P., Chalmers, C., Longmore, S. et al. (2024). Harnessing artificial intelligence for wildlife conservation. *Conservation* 4 (4): 685–702.
Führding-Potschkat, P., Kreft, H., and Ickert-Bond, S.M. (2022). Influence of different data cleaning solutions of point-occurrence records on downstream macroecological diversity models. *Ecology and Evolution* 12 (8): e9168.
Ganesh, C., Harshavardhan, G., Keerthi, N.R.S. et al. (2025). Enhancing plant leaf classification with deep learning: Automating feature extraction for accurate species identification. *SCT Proceedings in Interdisciplinary Insights and Innovations* 3: 513.
Gao, P., Du, W., Lei, Q. et al. (2023). NDVI forecasting model based on the combination of time series decomposition and CNN–LSTM. *Water Resources Management* 37 (4): 1481–1497.

Gardner, A.S., Maclean, I.M., and Gaston, K.J. (2019). Climatic predictors of species distributions neglect biophysiologically meaningful variables. *Diversity and Distributions* 25 (8): 1318–1333.

Gibb, K.A., Eldridge, A., Sandom, C.J. et al. (2024). Towards interpretable learned representations for ecoacoustics using variational auto-encoding. *Ecological Informatics* 80: 102449.

Gladstone-Gallagher, R.V., Hewitt, J.E., Siwicka, E. et al. (2023). Ecological network analysis of traits reveals variable response capacity to stress. *Proceedings of the Royal Society B* 290 (1998): 20230403.

Gong, Z., Ge, W., Guo, J. et al. (2024). Satellite remote sensing of vegetation phenology: Progress, challenges, and opportunities. *ISPRS Journal of Photogrammetry and Remote Sensing* 217: 149–164.

Gonzalez-Villanueva, R., and Carranza-Rojas, J. (2023). Improving balanced accuracy for minority plant species under data imbalance: A multi-architectural ensemble approach. In *2023 IEEE 5th International Conference on BioInspired Processing (BIP)* (pp. 1–6). IEEE.

Gowthaman, S., and Das, A. (2025). Plant leaf identification using feature fusion of wavelet scattering network and CNN with PCA classifier. *IEEE Access*.

Guo, H., Boonprong, S., Wang, S. et al. (2024). Dominant tree species mapping using machine learning based on multi-temporal and multi-source data. *Remote Sensing* 16 (24): 4674.

Gustineli, M., Miyaguchi, A. and Stalter, I. (2024). Multi-label plant species classification with self-supervised vision transformers. *arXiv preprint arXiv:2407.06298*.

Gyaneshwar, A., Mishra, A., and Chadha, U. et al. (2023). A contemporary review on deep learning models for drought prediction. *Sustainability* 15 (7): 6160.

Hajiesmaili, M., Ren, S., Sitaraman, R.K. et al. (2024). Towards environmentally equitable AI. *arXiv preprint arXiv:2412.16539*.

Hamdan, A., Ibekwe, K.I., Etukudoh, E.A. et al. (2024). AI and machine learning in climate change research: A review of predictive models and environmental impact. *World Journal of Advanced Research and Reviews* 21 (1): 1999–2008.

He, B., Wu, D., Wang, L. et al. (2024). FA-HRNet: A new fusion attention approach for vegetation semantic segmentation and analysis. *Remote Sensing* 16 (22): 4194.

Hu, Y., Si-Moussi, S., and Thuiller, W. (2025). Introduction to deep learning methods for multi-species predictions. *Methods in Ecology and Evolution* 16 (1): 228–246.

Hurley, A.G., Peters, R.L., Pappas, C. et al. (2022). Addressing the need for interactive, efficient, and reproducible data processing in ecology with the datacleanr R package. *PloS one* 17 (5): e0268426.

Jillelamudi, R., Tarimella, S.S., Kotha, I. et al. (2024). Agricultural practices: Deep learning for vegetable classification. In *2024 First International Conference on Innovations in Communications, Electrical and Computer Engineering (ICICEC)* (pp. 1–6). IEEE.

Kalpana, B., Krishnamoorthy, P., Kanageswari, S. et al. (2024). Machine learning approaches for predicting species interactions in dynamic ecosystems. *The Scientific Temper* 15: 2961–2967.

Kartal, S., Iban, M.C., and Sekertekin, A. (2024). Next-level vegetation health index forecasting: A ConvLSTM study using MODIS Time Series. *Environmental Science and Pollution Research* 31 (12): 18932–18948.

Kruschel, S., Hambauer, N., Weinzierl, S. et al. (2025). Challenging the performance-interpretability trade-off: An evaluation of interpretable machine learning models. *Business & Information Systems Engineering*: 1–25.

Laosuntisuk, K., Vennapusa, A., Somayanda, I.M. et al. (2024). A normalization method that controls for total RNA abundance affects the identification of differentially expressed genes, revealing bias toward morning-expressed responses. *The Plant Journal: For Cell and Molecular Biology* 118 (5): 1241–1257.

Lapeyrolerie, M., Chapman, M.S., Norman, K.E. et al. (2022). Deep reinforcement learning for conservation decisions. *Methods in Ecology and Evolution* 13 (11): 2649–2662.

Lei, L., Zhang, L., Han, Z. et al. (2024). Advancing chronic toxicity risk assessment in freshwater ecology by molecular characterization-based machine learning. *Environmental Pollution* 342: 123093.

Li, Y., Min, S., Song, B. et al. (2024). Multisource high-resolution remote sensing image vegetation extraction with comprehensive multifeature perception. *Remote Sensing* 16 (4): 712.

Liccari, F., Sigura, M., Bacaro, G. (2022). Use of remote sensing techniques to estimate plant diversity within ecological networks: A worked example. *Remote Sensing* 14 (19): 4933.

López-Andreu, F.J., López-Morales, J.A., Hernández-Guillen, Z. et al. (2023). Deep learning-based time series forecasting models evaluation for the forecast of chlorophyll a and dissolved oxygen in the Mar Menor. *Journal of Marine Science and Engineering* 11 (7): 1473.

Low, J.M., Gladstone-Gallagher, R.V., Hewitt, J.E. et al. (2023). Using ecosystem response footprints to guide environmental management priorities. *Ecosystem Health and Sustainability* 9: 0115.

Lozano-Claros, D., Meng, X., Custovic, E. et al. (2020). Developmental normalization of phenomics data generated by high throughput plant phenotyping systems. *Plant Methods* 16: 1–11.

Lusk, D., Wolf, S., Moreno Martínez, Á. et al. (2024). Combining citizen science and Earth observation data to produce global maps of 31 plant traits. In *EGU General Assembly Conference Abstracts* (p. 2477).

Ma, X., Migliavacca, M., Wirth, C. et al. (2020). Monitoring plant functional diversity using the reflectance and echo from space. *Remote Sensing* 12 (8): 1248.

Maeda-Gutiérrez, V., Oropeza-Valdez, J.J., Reveles-Gómez, L.C. et al. (2024). AI-driven plant health assessment: A comparative analysis of inception V3, ResNet-50 and ViT with SHAP for accurate disease identification in Taro. *Agronomy* 15 (1): 77.

Majumder, S., and Mason, C.M. (2024). A machine learning approach to study plant functional trait divergence. *Applications in Plant Sciences* 12 (5): e11576.

Manoj, R.K., and Sasilatha, T. (2024). Automated plant disease detection using efficient deep ensemble learning model for smart agriculture. In *Using Traditional Design Methods to Enhance AI-Driven Decision Making* (pp. 318–336). IGI Global Scientific Publishing.

Mao, X., Deng, Y., Zhu, L. et al. (2020). Hierarchical geographic object-based vegetation type extraction based on multi-source remote sensing data. *Forests* 11 (12): 1271.

Marian, M.L., Nasui, D., Ghise, C.R. et al. (2024). Using machine learning algorithms for natural habitats assessment. *Carpathian Journal of Earth and Environmental Sciences* 19 (1): 103–113.

Mayouf, R., Hanafi, M.T., and Mostefaoui, O. (2024). Machine learning-based crop type detection and vegetation monitoring using Sentinel-1 SAR and Landsat 8. *South Florida Journal of Development* 5 (11): e4620–e4620.

McMillan, P.G., Feng, Z.Z., Arciszewski, T.J. et al. (2024). Assessing the interpretability–performance trade-off of artificial neural networks using sentinel fish health data. *Environments* 11 (5): 94.

Metzger, D.C., Earhart, M.L., and Schulte, P.M. (2024). Genomic and epigenomic influences on resilience across scales: Lessons from the responses of fish to environmental stressors. *Integrative and Comparative Biology* 64 (3): 853–866.

Millar, C., Janzen, K., Nehemy, M.F. et al. (2022). On the urgent need for standardization in isotope-based ecohydrological investigations. *Hydrological Processes* 36 (10): e14698.

Mudarisov, S.G., and Miftakhov, I.R. (2024). Deep learning methods and UAV technologies for crop disease detection. *Agricultural Machinery and Technologies* 18 (4): 24–33.

Müller, M., D'Andrilli, J., Silverman, V. et al. (2024). Machine-learning based approach to examine ecological processes influencing the diversity of riverine dissolved organic matter composition. *Frontiers in Water* 6: 1379284.

Nagaraj, A. (2023). *COVID 19–Monitoring with IoT Devices*. Bentham Science Publishers.

Navaneetha, D., Maringanti, T., Repaka, S. et al. (2024). Analysis of automatic health monitoring in agricultural crops using artificial intelligence methods. In *2024 International Conference on Data Science and Network Security (ICDSNS)* (pp. 1–4). IEEE.

Nunavath, V., and Danilin, O. (2024). Deep learning based multi-class classification of invasive plant species. In *2024 Second International Conference on Advances in Information Technology (ICAIT)* (Vol. 1, pp. 1–5). IEEE.

Nuser, M., Alshirah, M., Mashaqbeh, S.A. et al. (2024). Adoption of machine learning methods for crop yield prediction-based smart agriculture and sustainable growth of crop yield production – Case study in Jordan. *Qeios*.

Obadimu, C.O., Shaibu, S.E., Ekwere, I.O. et al. (2024). Machine learning-based forecasting of bioaccumulation and histopathological effects in aquatic organisms. *Fudma Journal of Sciences* 8 (6): 485–496.

Ogungbuyi, M.G., Mohammed, C., Fischer, A.M. et al. (2024). Integration of drone and satellite imagery improves agricultural management agility. *Remote Sensing* 16 (24): 4688.

Omprakash, A., Balakrishnan, R., Ewers, R. et al. (2024). Interpretable and robust machine learning for exploring and classifying soundscape data. *bioRxiv*: 2024–11.

Pacheco-Labrador, J., de Bello, F., Migliavacca, M. et al. (2023). Normalizing functional diversity metrics across heterogeneous data streams: From field to satellite data. In *EGU General Assembly Conference Abstracts* (pp. EGU–10085).

Park, D.S., Willis, C.G., Xi, Z. et al. (2020). Machine learning predicts large scale declines in native plant phylogenetic diversity. *New Phytologist* 227 (5): 1544–1556.

Parveen, R., Kulkarni, S., and Mytri, V.D. (2016). Vegetation extraction from multi sensor IRS-1C LISS III image by principle component analysis and comparative assessment. In *2016 2nd International Conference on Advances in Electrical, Electronics, Information, Communication and Bio-Informatics (AEEICB)* (pp. 368–372). IEEE.

Raghavendran, C.V., Siddhardha, A.S.A., Sameeunnisa, M. et al. (2024). Optimizing leaf disease detection in agriculture using deep learning. In *2024 First International Conference on Software, Systems and Information Technology (SSITCON)* (pp. 1–7). IEEE.

Rahmati, Y. (2024). Artificial intelligence for sustainable urban biodiversity: A framework for monitoring and conservation. *arXiv preprint arXiv:2501.14766*.

Rajathi, K., David, S.A., and Naveena, N.R. (2024). Comparative performance of machine learning models in predicting plant growth using IoT data. In *2024 5th International Conference on Smart Electronics and Communication (ICOSEC)* (pp. 636–640). IEEE.

Rao, Y.P., and Subrahmanyam, M.V.V.S. (2024). Prediction of endangered species 41 using artificial intelligence. In *Algorithms in Advanced Artificial Intelligence* (pp. 267–272). CRC Press.

Reckling, W., Mitasova, H., Wegmann, K. et al. (2021). Efficient drone-based rare plant monitoring using a species distribution model and AI-based object detection. *Drones* 5 (4): 110.

Ribeiro, B.R., Velazco, S.J.E., Guidoni-Martins, K. et al. (2022). BDC: A toolkit for standardizing, integrating and cleaning biodiversity data. *Methods in Ecology and Evolution* 13 (7): 1421–1428.

Rosbakh, S., Pichler, M., and Poschlod, P. (2022). Machine-learning algorithms predict soil seed bank persistence from easily available traits. *Applied Vegetation Science* 25 (2): e12660.

Ryckewaert, M., Marcos, D., Botella, C. et al. (2024). Applying the maximum entropy principle to multi-species neural networks improves species distribution models. *arXiv preprint arXiv:2412.19217*.

Ryo, M. (2024). Ecology with artificial intelligence and machine learning in Asia: A historical perspective and emerging trends. *Ecological Research* 39 (1): 5–14.

Saha, S., Kundu, B., Paul, G.C. et al. (2023). Proposing an ensemble machine learning based drought vulnerability index using M5P, dagging, random sub-space and rotation forest models. *Stochastic Environmental Research and Risk Assessment* 37 (7): 2513–2540.

Samia, I.J., Jahan, M., and Rabeya, T. (2022). A survey on plants leaf diseases using machine learning and deep learning approaches. In *International Joint Conference on Advances in Computational Intelligence* (pp. 471–480). Springer Nature Singapore.

Sandbrook, C. (2025). Beyond the hype: Navigating the conservation implications of artificial intelligence. *Conservation Letters* 18 (1): e13076.

Shastry, K.A., Jagadish, B., Bhat, V.C. et al. (2024). Automated urban infrastructure classification using aerial imagery. In *2024 International Conference on Intelligent Algorithms for Computational Intelligence Systems (IACIS)* (pp. 1–6). IEEE.

Silva, J.A.O.S., de Siqueira, V.S., Mesquita, M. et al. (2024). Deep learning for weed detection and segmentation in agricultural crops using images captured by an unmanned aerial vehicle. *Remote Sensing* 16 (23): 4394.

Silvia, R., Haldi, B., Usman, S. et al. (2024). Comparison of SVM, Naive Bayes, and ELM Models in Plant Growth Classification. *Journal of Data Science* 2024 (56): 1–6.

Sklab, Y., Ariouat, H., Boujydah, Y. et al. (2024). Towards a deep learning-powered herbarium image analysis platform. *Biodiversity Information Science and Standards* 8.

Sousa, T. (2023). Towards modeling and predicting the resilience of ecosystems. In *2023 ACM/IEEE International Conference on Model Driven Engineering Languages and Systems Companion (MODELS-C)* (pp. 159–165). IEEE.

Sunde, M., Diamond, D., and Elliott, L. (2024). Ecological systems classification: Integrating machine learning, ancillary modeling, and Sentinel-2 Satellite imagery. *Remote Sensing* 16 (23): 4440.

Surówka, E., Rapacz, M., and Janowiak, F. (2020). Climate change influences the interactive effects of simultaneous impact of abiotic and biotic stresses on plants. *Plant Ecophysiology and Adaptation under Climate Change: Mechanisms and Perspectives I: General Consequences and Plant Responses*: 1–50.

Yan, Y., Shao, B., and Davis, C.C. (2024). A generative deep learning approach for global species distribution prediction. *bioRxiv*: 2024–12.

Zaka, M.M., and Samat, A. (2024). Advances in remote sensing and machine learning methods for invasive plants study: A comprehensive review. *Remote Sensing* 16 (20): 3781.

6

Machine Learning for Modeling Plant Abiotic Stress Responses

Haragopal Dutta[1], *Suman Dutta*[2]*

[1] *Division of Genetics, ICAR–Indian Agricultural Research Institute, New Delhi, India*
[2] *School of Agriculture and Rural Development, Ramakrishna Mission Vivekananda Educational and Research Institute, Belur Math, Howrah, West Bengal*

*Corresponding author: duttasumansiliguri@gmail.com

6.1 Introduction

Abiotic stress factors, including drought, salinity, and extreme temperatures, significantly impede plant growth and development, thereby threatening global food security. These stresses reduce crop yields and can compromise the economic viability of agricultural systems worldwide (Zhu 2016). Traditional methods for enhancing plant resilience to these stressors, such as selective breeding and genetic engineering, often require extensive time and labor, limiting their applicability in responding to rapidly changing environmental conditions (Seo et al. 2020). In contrast, machine learning (ML) presents a transformative solution for analyzing complex plant–environment interactions, enabling the identification of key stress-responsive genes, traits, and regulatory networks (Kimotho et al. 2019; Dutta et al. 2024). For instance, utilizing ML algorithms to predict gene regulatory networks associated with abiotic stress responses can provide insights into transcription factor binding sites and their target genes, which are crucial for developing stress-tolerant plant varieties (Amin et al. 2019). This high-throughput analysis not only accelerates the discovery of relevant genetic traits but also enhances our understanding of plant signaling pathways under abiotic stress conditions (Aloo et al. 2023). The signaling pathways involved in plant responses to abiotic stresses are complex and involve numerous phytohormones, such as abscisic acid (ABA), which plays a pivotal role in mediating stress responses (Manzoor et al. 2016; Sah et al. 2016). The interaction between stress-responsive transcription factors (TFs) and these hormones is critical; studies have demonstrated that manipulation of TFs associated with ABA signaling enhances drought resistance in various plant species (Yoon et al. 2020). Furthermore, genetic modifications that involve overexpressing these TFs can lead to improved tolerance to adverse environmental conditions (S. Chen et al. 2021; Jha and Mishra 2022). This underscores the potential of combining ML with genetic and genomic approaches to streamline the identification and manipulation of advantageous traits and genes associated with stress resilience. ML not only aids in gene discovery but also facilitates the effective integration of multi-dimensional datasets encompassing genomic, transcriptomic, and phenotypic information (Kimotho et al. 2019; Radwan 2015). By employing ML frameworks, researchers can elucidate the metabolic and physiological responses of plants to abiotic stresses, leading to the identification of novel genetic traits that can be targeted to enhance stress tolerance (Rane et al. 2021). This integrated approach is increasingly vital as climate change exacerbates the frequency and intensity of abiotic stress events, necessitating robust and rapid improvements in crop resilience strategies (Yoon et al. 2020; Zhu 2016). Thus, leveraging ML in the study of plant abiotic stress responses represents a promising frontier in agricultural research, providing a pathway to optimize crop breeding strategies and ensure sustainable food production in the face of environmental challenges. By enhancing our understanding of the underlying genetic and molecular mechanisms, ML can significantly contribute to the development of resilient plant varieties that can thrive under stressful conditions, thereby ultimately supporting global food security.

Machine Learning for Plant Biology, First Edition. Edited by Jen-Tsung Chen.

6.2 Definition of Abiotic Stress

Plant abiotic stress refers to the nonliving environmental factors that adversely affect the growth, development, and productivity of plants (Figure 6.1, Table 6.1). These stresses include a range of stressors such as drought, salinity, extreme temperatures (both hot and cold), flooding, and ultraviolet (UV) radiation, all of which can significantly disrupt the normal biological functions of plants (Ghosh and Xu 2014; Malwattage et al. 2024; Kumar et a. 2024). Understanding abiotic stress is crucial because, with climate change, the intensity and frequency of these environmental conditions are anticipated to increase, posing a greater challenge to agricultural sustainability and food security worldwide (Lee et al. 2016). The mechanisms through which plants respond to abiotic stress are complex and involve various physiological, biochemical, and molecular strategies (Dossa et al. 2019). During abiotic stress events, plants generate reactive oxygen species (ROS) as part of the signaling cascade that initiates adaptive responses (Liu et al. 2019; Xiao et al. 2021). This ROS production is a double-edged sword; while it plays a role in stress signaling and activating defense mechanisms, excessive ROS can lead to oxidative stress, further compromising cellular function and health (Dutta et al. 2025). TFs play a pivotal role in mediating plant responses to abiotic stresses. Various TFs have been shown to regulate gene expression involved in stress tolerance pathways. Transcription factors such as MYB and NAC are instrumental in enhancing plant resilience to abiotic stresses like drought and salinity by modulating the expression of stress-responsive genes (Cheng et al. 2022; Kimotho et al. 2019). The overexpression of certain TFs has been linked to improved stress tolerance; however, it may also come with trade-offs in terms of growth and reproductive fitness in genetically modified plants. The hormone ABA is another critical player in the abiotic stress response. ABA accumulates in response to stress conditions and triggers physiological adjustments, such as stomatal closure to conserve water during drought. The interaction between ABA signaling and other pathways, including those regulated by various TFs and ROS, creates a sophisticated network that plants utilize to optimize their response to abiotic stressors (Schmidt et al. 2024). Moreover, the cumulative effects of multiple abiotic stresses, such as simultaneous drought and heat, complicate plant responses. Research indicates that plants do not necessarily respond to multiple stresses in the same way they do under individual stresses, highlighting the need for integrated approaches in studying stress responses (Guo et al. 2016) such as the role of heat shock proteins (HSPs), which assist in protecting proteins from denaturation under stress (W. Hu et al. 2017).

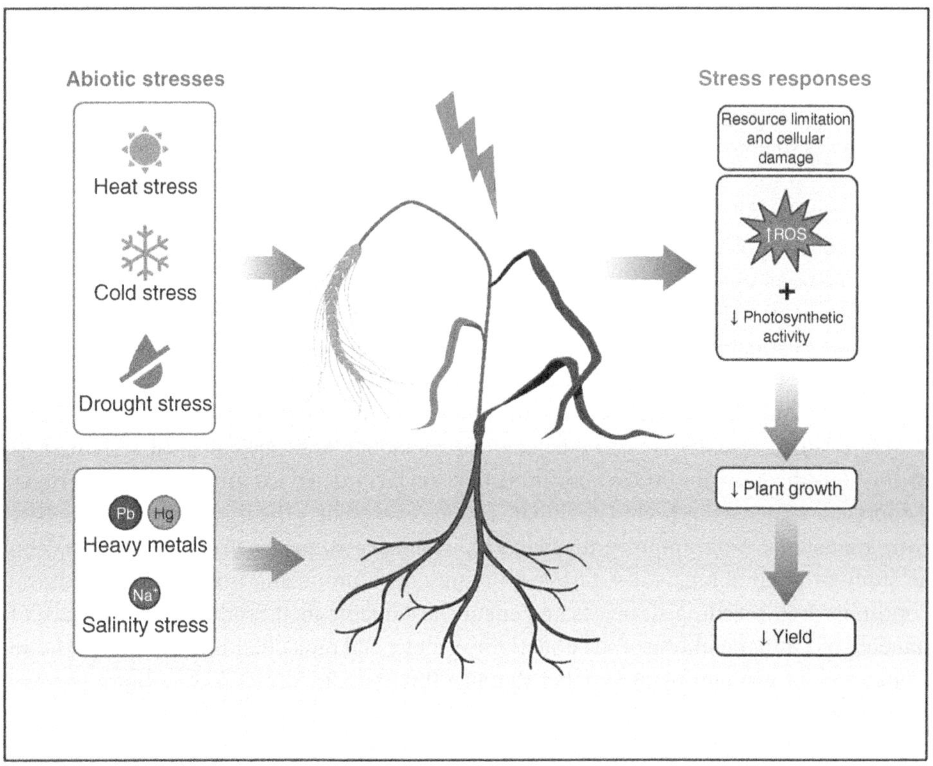

Figure 6.1 Abiotic stress on crop plants.

Table 6.1 Different kinds of abiotic stress and its causes and mitigation strategies.

Stress Type	Stress Range	Mitigation Strategy	References
Drought	Soil moisture < 50% of field capacity	Drought-tolerant varieties (e.g., DT maize), mulching, drip irrigation	Farooq et al. (2009)
Salinity	EC > 4 dS/m (soil), Na^+ accumulation	Use of salt-tolerant cultivars (e.g., CSR-30 rice), gypsum application	Munns and Tester (2008)
Heat stress	Temp > 35 °C during flowering/grain filling	Heat-tolerant genotypes (e.g., HD 2967 wheat), early sowing	Kopecká et al. (2023)
Cold stress	Temp < 10 °C for warm-season crops	Cold-tolerant varieties (e.g., maize hybrids), seed priming	Wang and Frei (2011)
Flooding	Soil saturation > 48 h	Raised bed planting, anaerobic germination traits in rice	Bailey-Serres and Voesenek (2008)
Heavy metals	Cd > 3 mg/kg, Pb > 100 mg/kg in soil	Phytoremediation, biochar amendment, tolerant genotypes	Zhang et al. (2023)
UV-B Radiation	> 10 kJ/m²/day	Application of antioxidants, protective structures	Jeyasri et al. (2021)
Nutrient stress	N < 10 ppm; P < 5 ppm; K < 50 ppm (soil)	Integrated nutrient management, fortified fertilizers, precision agriculture	Wang and Frei (2011)

6.3 Effect of Abiotic Stress on Crops

The effect of abiotic stress on crops is a critical area of agricultural research, especially in the context of increasing environmental pressures such as climate change. Abiotic stresses encompass a range of environmental conditions that negatively impact plant growth and productivity, including drought, salinity, temperature extremes, and heavy metal toxicity (Table 6.2). Among these, salinity stress has emerged as one of the most significant threats to agricultural sustainability, affecting approximately 20% of the world's arable land and further exacerbated by factors such as poor irrigation practices and global warming (Kamran et al. 2020; Ullah et al. 2017). Salinity adversely influences several physiological processes in crops, leading to a decline in growth and yield (Negrão et al. 2017; Okhovatian-Ardakani et al. 2010). High salinity levels disrupt water uptake due to osmotic pressure imbalances and result in ion toxicity, thereby reducing plant productivity (Handayani et al. 2024). Various crops, including wheat, rice, and cotton, have been reported to suffer substantial yield losses when exposed to salt stress (Dey and Raichaudhuri 2022; dos Santos et al. 2022; Saharan et al. 2022). Furthermore, soil salinity impacts nutrient absorption and leads to physiological drought, where plants are unable to extract sufficient water despite its presence in the soil (Baddour 2024; TUĞAÇ 2021). In addition to salinity, drought stress presents another formidable challenge. Drought conditions inhibit essential processes such as photosynthesis, leading to reduced biomass and yield. Moreover, plants exhibit morphological and biochemical changes in response to drought, such as root adaptation and the production of Osmo protectants like proline, which help in maintaining cellular turgor. The capacity of crops to cope with drought is increasingly important, particularly as water scarcity becomes more pronounced in various regions due to climate change. Differential gene expression analysis revealed increased changes of gene expression in *Ain1* in comparison to the freezing adapted *Osl1*. A lower degree of adjustment of transcription supported the notion of *Osl1* adaptability to freezing stress by showing lower levels of response. The overlap between the effects of drought, salinity, heavy metals, UV-irradiation, herbivory, and exposure to chemicals or other pathogens may lead to a dynamic in which a single cause–effect relationship may be very difficult to unravel (Dresselhaus and Hückelhoven 2018; Waseem et al. 2024). Temperature extremes, both high and low, also constitute significant abiotic stressors. Heat stress can cause denaturation of proteins and enzymes, impacting metabolic processes. The early heat stress response in rice seed also triggers protection mechanisms against cell or tissue damage from exposure to high temperatures. For example, *Heat Shock Factor (HSF)* genes were observed to be differentially overexpressed in the study. In comparison to this, *HsfA2, HsfA3, HsfA7, HsfB1*, and *HsfB2* were stimulated by heat stress in *A. thaliana*, wheat (*Triticum aestivum* L.), maize (*Zea mays* L.), and tomato (*Solanum lycopersicum* L.) (Chauhan et al. 2013). *AtHsfA6a* confers salt and dehydration stress resistance in Arabidopsis. A seed preferential *TaHsf* confers abiotic stress tolerance in Arabidopsis, and overexpression

Table 6.2 Different kinds of abiotic stress and its effects on crop plants.

Stress Type	Physiological Effect on Crops	Crops/Plants	Differential Gene Expression	References
Drought	Reduced turgor pressure, photosynthesis, root elongation	Rice, Wheat, Maize, Sorghum	Upregulation of *DREB2A, NCED3, LEA, RD29A, SnRK2s*	Shinozaki and Yamaguchi-Shinozaki (2007)
Salinity	Ionic imbalance, osmotic stress, leaf senescence	Tomato, Barley, Rice, Cotton	Activation of *SOS1, NHX1, HKT1;5, APX, WRKY8, MYB*	Munns and Tester (2008)
Heat Stress	Membrane instability, enzyme inactivation, impaired pollination	Wheat, Tomato, Chickpea, Canola	Induction of *HSP70, HSP101, DREB2, MBF1c, HSFA2*	Kotak et al. (2007)
Cold Stress	Reduced metabolic rate, electrolyte leakage, growth suppression	Maize, Rice, Brassica, Arabidopsis	Upregulation of *CBF/DREB1, COR15a, RD29A, ICE1, KIN1*	Chinnusamy et al. (2007)
Heavy Metal Toxicity	ROS accumulation, growth inhibition, photosynthetic decline	Soybean, Mustard, Rice, Sunflower	Expression of *GST, MT1, PCS1, ABC transporters, ZIP6, HMA4*	Clemens, (2006), Sharma and Dietz (2009)
Flooding/Waterlogging	Anaerobic respiration, leaf epinasty, chlorosis	Rice, Maize, Soybean, Cotton	Upregulation of *ADH1, PDC1, SUS, Sub1A, ERF-VII, RAmy3D*	Bailey-Serres and Voesenek (2008)
UV-B Radiation	DNA damage, ROS production, altered morphology	Arabidopsis, Soybean, Lettuce	Induction of *UVR8, CHS, HY5, ELIP, FAD7*	Jenkins (2009)
Nutrient Deficiency	Chlorosis, poor root architecture, reduced growth	Wheat, Maize, Rice, Tomato	Differential expression of *PHR1, NRT1.1, IRT1, SPX1, ZIP4, NAS3*	López-Arredondo et al. (2014), Marschner (2011)

of *GmHsf-24* increased drought and heat stress resistance in Arabidopsis (Wei et al. 2016). Conversely, cold stress affects membrane stability and enzymatic activity, leading to inhibited growth. The interplay between these abiotic stressors can result in compounded negative effects, leading to a greater-than-additive reduction in crop productivity. To combat these stresses, ongoing research is focused on understanding the genetic and molecular mechanisms that underlie plant tolerance. Advances in crop breeding, particularly with the incorporation of traits that confer resilience to abiotic stresses, have the potential to enhance food security (Dey and Raichaudhuri 2022; Dresselhaus and Hückelhoven 2018). Biotechnology approaches such as gene editing and the application of plant growth–promoting bacteria (PGPB) have shown promise in enhancing crop performance under stress conditions (BRAHIM et al. 2019; Xu et al. 2019).

6.4 Key Applications of Machine Learning in Abiotic Stress Research

6.4.1 Phenotypic Prediction

ML algorithms can predict plant phenotypes under stress by integrating diverse data sources such as genomic, transcriptomic, proteomic, and metabolomic profiles. By training models on these datasets, researchers can predict how plants will respond to different stress conditions, assisting in the identification of traits linked to stress tolerance (Table 6.3). A significant finding in the recent literature indicates that ML approaches, such as support vector machines (SVM) and random forest (RF) models, excel in capturing complex biological interactions and predicting stress responses with high accuracy. For instance, Meher et al. utilized SVM alongside K-mer compositional features to develop ASRpro, a computational model that identifies proteins linked to multiple abiotic stresses in plants, demonstrating consistent accuracy through cross-validation techniques (Meher et al. 2024). Additionally, Meher et al. reported that their methods hold potential for effectively capturing insights into plant stress responses and enhancing the understanding of stress-related proteins in crops (Meher et al. 2022). While Fiesta describes the accuracy of RF algorithms in the context of aquaculture systems,

Table 6.3 Phenotypic prediction for machine learning in abiotic stress research.

AI-Based Tool	Phenotypic Prediction Under Abiotic Stress	Feature	Sensor	Algorithm	Function	Crop Phenotype Assessed	Abiotic Stress Type	Advantages/ Disadvantages	References
GROWSCREEN FLUORO	Chlorophyll fluorescence and leaf growth	Chlorophyll fluorescence	Fluorescence	Not specified	Stress tolerance analysis	Various crops	Multiple	Enables precise collection of stress data; noninvasive; high-throughput.	Danilevicz et al. (2022)
HyperART	Disease severity and chlorophyll content	Hyperspectral imaging	Hyperspectral	Not specified	Nondestructive quantification	Maize, rapeseed, barley, tomato	Multiple	Allows for nondestructive measurement; applicable to various crops.	Zhang et al. (2024)
PHENOVISION	Drought stress assessment	Plant growth parameters	RGB	Not specified	Drought stress measurement	Maize	Drought	Provides detailed plant growth analysis; high-throughput capabilities.	Sheikh et al. (2024)
PlantScreen™ Robotic XYZ System	Drought stress assessment	Physiological traits	Multispectral	Not specified	Drought stress measurement	Rice	Drought	Automated system; precise control and measurement; high-throughput.	Singh et al. (2016)
LemnaTec 3D Scanalyzer	Salinity stress effects	3D plant architecture	3D Laser Scanning	Not specified	Salinity stress measurement	Rice, maize, tobacco	Salinity	Provides comprehensive 3D analysis; noninvasive; high-throughput.	Gill et al. (2022)
PhénoField®	Multiple stress assessments	Various phenotypic traits	Multiple	Not specified	Stress effect measurement	Wheat	Multiple	Field-based platform; allows for real-world condition assessment; high-throughput.	Al-Tamimi et al. (2022)
CropQuant	Various stress assessments	Growth and development metrics	RGB	Not specified	Continuous monitoring	Various crops	Multiple	Real-time data collection; portable; user-friendly interface.	Pradhan et al. (2024)
RootReader3D	Root architecture under stress	Root system architecture	X-ray CT	Not specified	Root phenotyping	Various crops	Multiple	Detailed 3D visualization of root systems; nondestructive; high-resolution.	Paul et al. (2025)
PHENOARCH	Canopy architecture analysis	Canopy structure	LiDAR	Not specified	Structural trait assessment	Various crops	Multiple	High-resolution canopy analysis; applicable to various crops; noninvasive.	Paul et al. (2025)

(Continued)

Table 6.3 (Continued)

AI-Based Tool	Phenotypic Prediction Under Abiotic Stress	Feature	Sensor	Algorithm	Function	Crop Phenotype Assessed	Abiotic Stress Type	Advantages/ Disadvantages	References
MVS-Pheno	Multi-view stereo imaging for stress traits	3D surface models	Stereo Cameras	Not specified	Structural and color trait analysis	Various crops	Multiple	Provides detailed 3D surface models; captures color information; nondestructive.	Singh et al. (2016)
Field Scanalyzer	Field-based phenotyping under stress	Various phenotypic traits	Multiple	Not specified	Large-scale field phenotyping	Various crops	Multiple	Enables large-scale field assessments; integrates multiple sensors; high-throughput.	Farooq et al. (2024)
GROWSCREEN-Rhizo	Root growth analysis under stress	Root growth dynamics	Imaging Sensors	Not specified	Root phenotyping	Various crops	Multiple	Facilitates detailed root growth analysis; non-invasive; high-throughput.	Sheikh et al. (2024)
Unmanned Aerial Systems (UAS)	Prediction of phenotypic traits under drought	Vegetation indices	RGB, Multispectral	Machine Learning Models	Trait prediction and monitoring	Maize	Drought	Allows for rapid, large-scale data collection; effective in monitoring plant responses over time; cost-effective compared to traditional methods.	Al-Tamimi et al. (2022)
Deep Convolutional Neural Networks (DCNN)	Identification and quantification of foliar stresses	Visual symptom analysis	RGB	Deep Learning (DCNN)	Stress identification and quantification	Soybean	Multiple	High accuracy in identifying various stresses; provides explanation mechanisms for predictions; robust to input image variations; potential for deployment in mobile platforms for real-time detection.	Ghosal et al. (2018)

the applicability of this finding to plant phenotyping is not directly addressed (Minerva et al. 2024). Therefore, the specific claim about achieving up to 90% accuracy in predicting phenotypic outcomes under controlled conditions must be contextualized and approached with caution. Another crucial area of focus is the application of ML in miRNA prediction related to abiotic stress responses. Chang et al. and Meher et al. have innovated predictive tools that utilize multisource feature fusion and graph neural networks to identify abiotic stress-responsive microRNAs, essential for understanding plant resilience mechanisms (Chang et al. 2024). These methodologies not only streamline the identification of potential biomarkers for stress tolerance but also demonstrate the feasibility of computational techniques in deciphering genetic regulation under adverse conditions. The necessity of utilizing integrated data approaches is further highlighted by various studies that emphasize the importance of identifying networked genes and regulatory elements that govern plant responses to multiple stresses (Pervaiz et al. 2022; Sarkar et al. 2022). Furthermore, the advent of high-throughput phenotyping (HTP) technologies has transformed the landscape of abiotic stress research, offering a nondestructive means to assess plant health and stress resilience. Advanced imaging techniques, such as hyperspectral imaging and multispectral sensing, combined with ML, can quantify plant responses effectively under stress conditions. Studies have illustrated the ability of hyperspectral analysis to detect early signs of stress in crops, paving the way for timely interventions and improved genetic selections (Al-Tamimi et al. 2022; Feng et al. 2020; Lowe et al. 2017). The use of unmanned aerial systems (UAS) for aerial imaging further enhances the capability to monitor and predict plant phenotypes during critical growth phases, thereby contributing to effective breeding programs focused on stress resilience (Herr et al. 2023).

6.4.2 Gene Expression Modeling

ML can model complex gene regulatory networks that control abiotic stress responses. For instance, supervised learning techniques like RF or SVMs can be trained on expression data to predict gene interactions and identify key regulators of stress response pathways. In recent years, the integration of ML methods into plant research has accelerated the understanding of how gene expression is modulated by abiotic stressors such as drought, salinity, and extreme temperatures (Table 6.4). ML has the capability to model intricate gene regulatory networks, which are critical for elucidating the complexities of stress responses in plants. For instance, supervised learning techniques, including RF and SVM, can be effectively trained on gene expression data derived from plants subjected to various stresses to predict gene interactions and identify key regulatory factors that govern these responses (Cheng et al. 2021; Gupta et al. 2021). Research has demonstrated that TFs, integral to plant stress responses, can be prioritized using ML methods. For example, Cheng et al. utilized a RF-based approach to construct a regulatory network pertinent to nitrogen use efficiency (NUE), revealing rankings of differentially expressed genes (DEGs) in response to treatments. Additionally, studies focused on maize highlight the relevance of WRKY and bHLH TFs in regulating gene expression under abiotic stress (Hu et al. 2021; Wang et al. 2018). Such TFs are often early responders in stress signaling and are vital in modulating downstream gene expression cascades during stress (Kimotho et al. 2019). Furthermore, recent innovations in ML have enabled the prediction of transcriptional responses across various plant species facing abiotic stresses. For instance, models that incorporate genomic features have been shown to effectively identify which genes are activated during stress scenarios by analyzing promoter regions enriched with regulatory motifs (Rico-Chávez et al. 2022; Smet et al. 2023). Specifically, He et al. emphasized how differential expression of genes involved in stress signal transduction pathways can be elucidated using ML algorithms that correlate with environmental stress profiles (Shen et al. 2024). This underscores the ability of ML to classify gene expression responses and delineate the underlying genetic architecture of stress tolerance in crops (Ibrahim et al. 2023). Moreover, studies such as those conducted by Meng et al. (2021) affirmed that both coding and noncoding regulatory regions play integral roles in gene expression, facilitating the characterization of plant stress responses through comprehensive analyses. By employing ML frameworks, researchers can now develop robust models that predict the stress-induced changes in gene expression based on historical data, thereby enhancing crop resilience to climate variability (Gupta et al. 2021). Such advancements exemplify the transformative potential of ML in plant abiotic stress research, allowing for a more nuanced understanding of gene–environment interactions critical for agricultural sustainability.

6.4.3 High-throughput Image Analysis

With the increasing use of sensors and imaging technologies, ML can process large-scale image data to monitor plant health in real time under various stress conditions. Techniques such as convolutional neural networks (CNNs) are effective

Table 6.4 Gene expression modeling in plant abiotic stress research involves analyzing differentially expressed genes.

Abiotic Stress type	DEGs (Upregulated/Downregulated)	Genetic Pathway Involved	Crop	Gene Expression Modeling	ML Algorithms	Reference (s)
Drought	Upregulated: ABA-responsive genes Downregulated: Photosynthesis-related genes	Abscisic Acid (ABA) Signaling Pathway	Maize	RNA-Seq data analyzed to identify DEGs responsive to drought stress	Support Vector Machine (SVM)	Azad et al. (2024)
Salt	Upregulated: SOS1, SOS2, SOS3 Downregulated: Aquaporin genes	Salt Overly Sensitive (SOS) Pathway	Rice	Gene expression profiling under salt stress conditions	Random Forest	Chowdhury et al. (2023)
Cold	Upregulated: CBF/DREB family genes Downregulated: Photosynthetic apparatus genes	CBF/DREB Cold Response Pathway	Arabidopsis	Transcriptome analysis to detect cold-responsive DEGs	Decision Tree	Shaik and Ramakrishna (2014)
Heat	Upregulated: Heat Shock Proteins (HSPs) Downregulated: Rubisco activase	Heat Shock Response Pathway	Wheat	Microarray data used to model gene expression changes due to heat stress	Neural Networks	Shaik and Ramakrishna (2014)
Drought	Upregulated: Genes involved in ABA biosynthesis and signaling Downregulated: Growth-related genes	Abscisic Acid (ABA) Signaling Pathway	Arabidopsis	Integrative approach combining transcriptomics and machine learning to identify core drought-responsive genes	Support Vector Machine (SVM)	Gupta et al. (2021)
Multiple Stresses	Upregulated: Genes in MAPK cascade Downregulated: Photosynthesis-related genes	Mitogen-Activated Protein Kinase (MAPK) Pathway	Rice	Meta-analysis of microarray data to identify genes responsive to multiple abiotic stresses	Random Forest	Panahi et al. (2025)

at analyzing plant morphology and detecting stress symptoms, such as wilting or chlorosis, in a noninvasive manner. CNNs are particularly adept at processing complex image data, enabling the detection of stress indicators in plants, such as wilting and chlorosis, in a noninvasive manner (Vaidya 2023). These systems can automatically learn from pixel data, processing diverse forms of imagery to classify plant species accurately and assess their health (Li et al. 2022). For instance, Vaidya discusses the immense potential of CNNs in biodiversity monitoring and agricultural applications, emphasizing their capacity to cope with complex variations in plant morphology. Similarly, the work of Dominiak et al. highlights the transformative role of hyperspectral imaging in assessing plant health (Dominiak-Świgoń et al. 2019), presenting an overview of how advanced imaging techniques can facilitate effective monitoring over extended periods without harming the plants. The challenge of detecting abiotic stresses in plants—including water deficit, nutrient imbalance, and salinity—has been addressed through various imaging and analytical methodologies. Galieni et al. (2021) highlight the criticality of stress detection to improve crop yields, stating that various abiotic stressors can substantially hinder plant growth. In this context, ML techniques serve as pivotal enablers, harnessing large datasets from remote sensing and imaging to derive actionable insights (Mou et al. 2019; Tang et al. 2023). These insights inform agricultural practices and foster improved crop management strategies. Furthermore, the work of Li et al. (2022) illustrates that CNNs excel in recognizing high-dimensional image data, which is particularly advantageous in analyzing plant species from unstable environments—such as arid lands—thereby reducing the need for field specimens that might be rare or endangered. Such advancements underscore the efficiency of CNNs in boosting identification accuracy and operational workflow in agricultural research. As technology progresses, high-throughput imaging combined with smart algorithms allows researchers to assess plant health in real time. For example, studies indicate that the use of real-time fluorescence imaging for analyzing ABA receptors provides critical insights into plant stress responses under saline conditions, allowing timely interventions (Yang et al. 2022). These innovative strategies, alongside the integration of CNNs in identifying plant diseases (Abade et al. 2021), facilitate a comprehensive approach to detecting and understanding plant health in response to abiotic factors, ultimately leading to optimized agricultural outputs.

6.4.4 Omics Data Integration

Plants' responses to abiotic stress are multifaceted and occur at different biological levels. ML techniques can integrate multi-omics data (genomics, transcriptomics, proteomics, and metabolomics) to create comprehensive models of plant stress responses (Roy et al. 2024). For instance, deep learning methods can uncover hidden relationships between genes, proteins, and metabolites that traditional analysis might overlook. In recent years, the integration of multi-omics data has become a pivotal tool for understanding plant responses to abiotic stress through ML approaches (Table 6.5). Abiotic stresses, such as drought, salinity, and temperature extremes, impose significant challenges on plant growth and development, necessitating robust adaptive mechanisms at various biological levels, including genomic, transcriptomic, proteomic, and metabolomic responses (Hu et al. 2022). Effective ML techniques can holistically analyze these diverse data types, facilitating a comprehensive understanding of stress responses in plants. ML models, particularly deep learning architectures, are advantageous for capturing the intricate relationships within multi-omics data. For instance, autoencoders, a type of deep learning model, have been used to stack omics features, enabling the identification of hidden patterns that traditional analytical methods may overlook (Alhnaity et al. 2020; Chaudhary et al. 2018). This capability is crucial for modeling the interconnected pathways through which plants respond to abiotic stressors. Insights gained from these models can help identify key genetic factors involved in stress tolerance, enabling the selection and engineering of resilient crop varieties. Research has indicated the potential for utilizing TFs associated with specific stress responses, such as those linked to Crassulacean acid metabolism (CAM), to enhance the tolerance of plants, including Arabidopsis. The influence of hormone signaling pathways in conjunction with ROS production during abiotic stress is another area where multi-omics integration and ML can provide significant insights. Studies have shown that hormonal signaling, involving the likes of auxins, ABA, and salicylic acid, is intricately tied to the plant's physiological responses to stress (Devireddy et al. 2021; Imran et al. 2021). The combinatory effects of these signaling networks can be modeled using ML approaches to yield predictions about plant stress responses and adaptations, thus contributing to a proficient understanding of plant resilience mechanisms under varying environmental stressors. Moreover, platforms such as IPOP (Integrative Plant Multi-Omics Platform) emphasize the importance of integrating multi-omics data across various species, enhancing cross-species comparisons and evolutionary studies. Such integrative approaches demonstrate the power of systems biology in elucidating networks of interactions at the multi-omics level, opening doors to advanced breeding programs targeting specific traits associated with stress resistance (Kirongo et al. 2019). Additionally, the deployment of new technologies like hyperspectral

Table 6.5 Omics approach in plant abiotic stress research.

Abiotic Stress	Crop	Genomics Data	Transcriptomics Data	Proteomics Data	Metabolomics Data	References
Drought	Rice (*Oryza sativa*)	SNPs in genes like *OsDREB1* and *OsNAC6* associated with drought tolerance	Upregulation of *OsLEA3*, *OsP5CS*, and *OsRAB16* transcripts under drought stress	Increased levels of dehydrins and late embryogenesis abundant (LEA) proteins	Accumulation of proline, soluble sugars, and aromatic amino acids	Naik et al. (2023)
Heat	Wheat (*Triticum aestivum*)	Identification of heat shock protein genes like *TaHSP17* and *TaHSP101*	Upregulation of *TaHsfA2* and *TaMBF1c* transcripts under heat stress	Elevated levels of small heat shock proteins (sHSPs) and HSP70	Changes in metabolites such as raffinose and galactinol involved in thermotolerance	Dwivedi et al. (2023)
Salinity	Tomato (*Solanum lycopersicum*)	QTL mapping for salt tolerance loci including *Solanum pimpinellifolium* introgressions	Differential expression of genes like *SlSOS1* and *SlNHX1* under salt stress	Proteomic changes indicating upregulation of antioxidant enzymes such as superoxide dismutase (SOD) and catalase (CAT)	Accumulation of metabolites like γ-aminobutyric acid (GABA) and flavonoids	Sun et al. (2024)
Cold	Arabidopsis (*Arabidopsis thaliana*)	Identification of cold-responsive genes such as *CBF1*, *CBF2*, and *CBF3*	Induction of *COR15A* and *KIN1* transcripts during cold exposure	Proteomic shifts indicating accumulation of cold-regulated (COR) proteins	Metabolic adjustments like increased levels of sugars (e.g., sucrose, raffinose) for cryoprotection	Aslam et al. (2024)
Drought and Heat	Arabidopsis (*Arabidopsis thaliana*)	Genomic loci associated with combined stress tolerance, including variations in *HSFA1* and *DREB2A* genes	Differential expression of genes such as *HSP101* and *RD29A* under combined stress	Proteomic changes unique to combined stress conditions, including increased levels of HSPs and dehydrins	Metabolite profiles indicating synergistic stress responses, such as elevated levels of osmoprotectants like trehalose	Xu et al. (2023)

imaging in combination with ML allows for HTP, thus accelerating the selection processes in breeding programs for improved abiotic stress tolerance (Huang et al. 2023).

6.4.5 Stress Response Prediction and Simulation

ML models can be used to predict how plants will react to specific combinations of abiotic stresses over time. This has major implications for precision agriculture, where farmers can make data-driven decisions to optimize crop management under changing environmental conditions. Predictive models may also simulate different stress scenarios to aid in breeding programs aimed at enhancing stress tolerance. Recent studies have highlighted the importance of integrating ML with genomic and physiological data to predict plant reactions to specific combinations of abiotic stresses over time. For example, it has been shown that genetic pathways, such as those involving strigolactones, play crucial roles in stress response mechanisms. The identification of genes associated with strigolactone pathways in sorghum indicates their upregulation during stress conditions like drought and cold, thus highlighting the potential of ML models to probe these pathways for enhanced resilience (An et al. 2016). Moreover, understanding the interplay between different stress-responsive TFs can illuminate the mechanisms underlying plant tolerance, suggesting that ML can enhance predictive accuracy by modeling these complex regulatory interactions (Kaniganti et al. 2024). ML approaches are particularly powerful when used to simulate various stress scenarios, which can provide substantial benefits for breeding programs aimed at developing crop varieties with improved stress tolerance. The integration of mechanistic models of plant responses to abiotic stresses allows for more accurate forecasting of plant responses in variable environments (Verma et al. 2016). These approaches have been substantiated by empirical findings, such as the significant roles of hormones like ABA in mediating drought and salinity responses (Huang et al. 2019). By leveraging these insights, breeders can use predictive models to simulate genetic changes and assess their potential impacts on crop performance under predicted future climate scenarios. Furthermore, studies indicate that plants respond dynamically to combinations of abiotic stresses, necessitating an approach that accommodates this complexity. Research on systemic signaling in plants reveals that concurrent abiotic stresses can trigger intricate signaling networks, which can be modeled through ML to predict plant responses accurately (Zandalinas et al. 2020). These models can be essential for agriculture stakeholders, enabling data-driven decision-making to optimize crop management strategies under changing environmental conditions.

6.5 ML Techniques Commonly Used for Abiotic Stress Modeling

6.5.1 Supervised Learning

In the context of plant stress, supervised learning methods such as decision trees, RF, and SVMs are commonly employed. These models require labeled data to train, which may come from controlled experiments that record plant traits or yield under specific stress conditions. Once trained, these models can predict the performance of plants under similar conditions. Supervised learning techniques are increasingly prevalent in modeling abiotic stress responses in plants, providing valuable insights into plant genetics and environmental adaptations. Among these techniques, decision trees, RF, and SVMs have emerged as dominant tools. These algorithms require high-quality labeled data derived from controlled experiments that document plant responses to specific stress conditions, such as drought, salinity, and extreme temperatures. This framework allows for the creation of predictive models that can anticipate plant performance under similar abiotic stresses. Research has shown that RF are particularly effective in predicting abiotic stress scenarios. For instance, a study reported that using water quality parameters from aquaculture systems, RF achieved an accuracy of around 90% in identifying crucial stress indicators like ammonia levels, showcasing their proficiency in distinguishing stress factors affecting plant health. This finding is echoed in a study focused on common genes regulating multiple abiotic stress responses in rice, suggesting that ML models, including RF, can reveal significant genetic pathways involved in stress responses, with an accuracy around 79% and an ROC curve area of 0.796 (Sarkar et al. 2022). Additionally, the use of SVMs has been illustrated in predicting microRNA involvement in abiotic stress, indicating their versatility and capacity for deriving biological insights from complex datasets (Meher et al. 2022). Moreover, the correlation between ML methodologies and plant response data underscores the crucial role of these models in facilitating advanced phenotyping techniques. For example, imaging technologies have been utilized to quantify plant responses to various abiotic stresses, complementing the data used in supervised learning models (Al-Tamimi et al. 2022). The integration of these technologies with ML approaches, including deep learning techniques, can further enhance predictive accuracy by utilizing nondestructive imaging for stress identification across plant species (Ghosal et al. 2018). The relevance of incorporating ML in abiotic stress modeling is further illustrated in studies utilizing ML for differential network analysis in organisms such as Arabidopsis (Ma et al. 2014). Such methodologies allow researchers to identify critical genes and their interactions under stress-induced conditions, thereby fostering a deeper understanding of the genetic basis for stress tolerance in plants.

6.5.2 Unsupervised Learning

Clustering techniques, such as K-means and hierarchical clustering, can be used to identify patterns in gene expression or metabolite data without prior knowledge of stress categories. This can help uncover novel stress-responsive genes or metabolites that are associated with tolerance mechanisms. Unsupervised learning presents a compelling alternative to traditional supervised learning techniques in the context of abiotic stress modeling in plants. While supervised learning typically relies on labeled datasets, which are often scarce or difficult to generate in the field, unsupervised learning techniques can analyze and interpret high-dimensional data without prior labels, enabling more flexible and scalable approaches to modeling stress responses in plant systems (Iqbal et al. 2022; Yan and Wang 2022). Recent studies have demonstrated the efficacy of unsupervised learning methods in identifying latent patterns relevant to abiotic stress. For instance, Károly et al. highlight that deep learning models can benefit from unsupervised clustering methods, which facilitate faster training and require less memory compared to their supervised counterparts (Károly et al. 2018). These advantages make unsupervised learning particularly beneficial in scenarios where biological data is voluminous but indirectly labeled. Additionally, Liu et al. suggest that unsupervised learning can yield important representations from unlabeled data, enhancing the analysis of plant responses to stresses, although their study focuses primarily on feature learning rather than specific abiotic stress contexts (Liu et al. 2017). In practical applications, unsupervised learning has shown promise in analyzing stress-responsive miRNA profiles in plants. Chang et al. utilized graph neural networks alongside multisource feature fusion to predict miRNA-abiotic stress associations, illustrating the potential for uncovering biologically relevant patterns without labeled examples (Chang et al. 2024). Moreover, unsupervised algorithms can effectively categorize plant stress injuries, as reported by Lu et al. (2023), who emphasize their ease of implementation owing to the lack of requirement for prior knowledge. Furthermore, unsupervised learning not only supports the efficient classification of stress types but can also enhance gene-to-phenotype modeling. For example, it allows for the clustering of gene expression data related to stress responses across different plant species, thereby elucidating shared mechanisms of resilience to abiotic stresses. Although the specific connection to gene–phenotype relationships requires further exploration, unsupervised learning holds potential to enrich our understanding in systems biology, where relationships between genes and phenotypes are complex. Despite the advantages of unsupervised learning, researchers must navigate challenges such as distinguishing between relevant and irrelevant features, as indicated by Ahmad et al. (2023) in their analysis of deep learning models for abiotic stress protein classification. Moreover, the integration of unsupervised learning with supervised counterparts may deliver superior outcomes, blending the robustness of labeled data with the flexibility of unlabeled data approaches, although more empirical evidence is needed to confirm these assertions (Sarkar et al. 2022).

6.5.3 Deep Learning

Neural networks, particularly deep learning models, are well-suited for handling large and complex datasets such as high-throughput phenotypic data or genome-wide association studies (GWAS). CNNs can automatically detect stress-related features from plant images, while recurrent neural networks (RNNs) can model temporal aspects of stress responses, such as the gradual decline in growth under prolonged drought. Deep learning has emerged as a powerful approach for modeling plant responses to abiotic stresses such as drought, salinity, and temperature extremes, particularly through the utilization of neural networks like CNNs and RNNs. These methods facilitate the processing of large and complex datasets, including high-throughput phenotypic data and real-time environmental data, leading to improved crop management and breeding strategies. CNNs are particularly adept at analyzing visual data for detecting and diagnosing stress in plants. For instance, studies have highlighted that CNNs effectively identify drought-stressed plants through image analysis by leveraging visual features such as color, shape, and texture (Kamarudin and Ismail 2022). This ability to extract relevant features automatically from images allows for the efficient detection of stressed states without the need for specialized imaging equipment, which can often limit broader applications in agricultural practices. Additionally, CNNs have been applied successfully in weed detection under drought conditions, emphasizing their utility in agricultural management

(Zhuang et al. 2023). On the other hand, RNNs, especially Long Short-Term Memory networks (LSTMs), are well-suited for modeling temporal relationships in stress data. These models can capture the dynamics of plant growth responses over time under prolonged stress conditions, such as drought, making them invaluable for predicting growth trajectories related to environmental variations (Das Choudhury et al. 2023). The capability of RNNs to manage sequences of data allows for tracking physiological changes in plants, thereby providing insights into recovery mechanisms or the gradual decline in plant health, which is critical for developing adaptive strategies in breeding programs aimed at enhancing stress tolerance (Islam and Yamane 2021). An integrated approach that combines CNNs and RNNs has shown promise for modeling abiotic stress in plants comprehensively. This fusion allows for spatial feature extraction through CNNs and temporal relationship modeling via RNNs, resulting in a robust framework for predicting plant stress responses under field conditions (Akram et al. 2023). Techniques that incorporate time-series image data further elucidate how persistently changing environmental conditions impact plant phenology, thereby aiding in the development of more precise models for sustainable agricultural practices (Yu et al. 2019). Moreover, advancements in drone technology and remote sensing significantly complement deep learning applications in abiotic stress modeling. These technologies facilitate real-time monitoring of crop conditions and health, providing critical data that enhance the efficiency of deep learning models when identifying and managing plant stresses (Gano et al. 2024). By integrating these technologies with ML approaches, researchers can better understand the interactions between plants and their environments, leading to improved crop yields even under stress conditions (Zheng et al. 2021).

6.6 Challenges and Future Directions

6.6.1 Data Availability and Quality

One of the major challenges in using ML for modeling plant abiotic stress is the availability of high-quality, labeled datasets. Collecting large datasets from field trials and HTP is expensive and labor-intensive, limiting the scope of ML applications. The application of ML in plant abiotic stress research holds promise for enhancing agricultural productivity, particularly in the face of increasing environmental challenges. One of the significant bottlenecks in this field is the scarcity of high-quality, labeled datasets required for effective ML model training and validation. The successful implementation of ML techniques relies heavily on the availability of comprehensive datasets, as they enable the models to learn complex patterns and relationships inherent in high-dimensional data (Chen et al. 2021; Mullainathan and Obermeyer 2017). Collecting large datasets from field trials and HTP presents numerous challenges. The processes are often expensive and labor-intensive, which can deter researchers from pursuing extensive data collection. This situation is further exacerbated by the underestimation of critical data preparation and curation activities, which are fundamental for ensuring data quality. Researchers often prioritize model development over rigorous data handling processes, potentially leading to suboptimal performance of ML systems. Particularly in the context of plant stress research, where environmental variability is significant, the production of high-fidelity datasets is paramount for any predictive modeling effort. Moreover, the nature of the datasets in plant science often involves imbalances and missing categories of responses. For effective ML applications, datasets must encapsulate a wide range of conditions and stress responses; instances where only specific forms of stress or outcomes are recorded can introduce biases into the models. Insufficient positive and negative samples is a common pitfall that impedes model robustness (Pillong et al. 2017). The curation of datasets that accurately reflect both the stress conditions and the variety of plant responses is essential for developing dependable models that support agricultural innovations. Furthermore, the democratization of data access has the potential to bridge these gaps. Emerging tools and platforms facilitate greater sharing and collaboration, allowing researchers to combine resources and create vast, high-quality datasets. These advancements can help alleviate the time and cost burdens associated with data acquisition in agriculture and plant science research (Tjaden and Tjaden 2023). Nevertheless, continuous efforts must be made to enhance data quality, ensuring that datasets are not only large but also comprehensive and well-labeled, allowing for effective ML applications in plant abiotic stress research.

6.6.2 Generalization

Models trained on specific datasets may not generalize well to other plant species or environments. Enhancing model robustness by incorporating data from diverse conditions and using transfer learning techniques may address this challenge. One crucial aspect of enhancing model generalization is the incorporation of data from multiple environmental conditions and plant species. As the physiological responses to abiotic stress factors, such as drought, salinity, and extreme temperatures, vary widely among species, models trained on a narrow dataset may misrepresent the tolerance mechanisms present in other species. For instance, studies illustrate that plants like *Arabidopsis thaliana* exhibit different responses to abiotic stresses when compared to crops like soybean (*Glycine max*), particularly in terms of hormonal signaling pathways, where ABA plays a significant role in mediating stress responses (Robles et al. 2015). Furthermore, variable expressions of auxin and other hormones among species have been documented, emphasizing the need for models that can account for this complexity (Rahman 2013). Transfer learning presents a promising solution to foster model robustness across different species and stressors. By leveraging existing models trained on well-studied species to enhance predictions for less-studied species, researchers can address the gap created by limited datasets. For instance, findings suggest that TFs like NAC and MYB, which have been characterized in model species, may also play pivotal roles in determining abiotic stress responses in a broader range of plant species (Xu et al. 2017). Such information could be utilized in training algorithms that adapt existing models to new datasets, thereby improving accuracy and increasing the predictive power of environmental responses (Suzuki et al. 2014). Moreover, diverse physiological mechanisms, such as ROS signaling and the role of osmolytes, underline the complexity of abiotic stress responses across different plant species. Compounds like trimethylamine *N*-oxide have been found to promote stress tolerance, which is a growing area of interest in developing generalized strategies for crop improvement (Catalá et al. 2021). At the same time, the responses can be significantly influenced by both genetic composition and environmental factors, revealing a need for models that can accommodate these variances (Musayev and Huseynova 2022).

6.6.3 Interpretability

While ML models, especially deep learning, provide high accuracy, they are often considered "black boxes," making it difficult to interpret the biological significance of the features they use for prediction. Developing interpretable ML methods that can offer biological insights is a growing area of research. The complexity of deep learning models raises concerns regarding their interpretability. This challenge is particularly pertinent when such models are applied to high-dimensional data prevalent in plant sciences. For instance, Singh et al. (2018) highlighted that while deep learning provides robust predictions for plant stress phenotyping, the lack of interpretability can obscure the biological implications of the model outputs. To bridge this gap, researchers advocate for interpretability techniques that can elucidate the workings of these black-box models. Methods like Local Interpretable Model-Agnostic Explanations (LIME) and Shapley values offer frameworks to understand feature importance and decision-making processes within ML models (Du et al. 2020; Henninger and Strobl 2023). These techniques are pivotal for reinforcing trust among end-users and ensuring that predictions can be translated into actionable agricultural practices. Recent developments in interpretable ML have led to a growing emphasis on integrating explainability into model design. The introduction of models with intrinsic interpretability aims to meet the desire for transparency in predictions (Sani et al. 2020). In plant research, this translates into selecting algorithms and representations that inherently provide meaningful insights while maintaining high predictive accuracy. Liu and Udell (2020) suggest that interpretability must coexist with model performance, underscoring the synergy between accuracy and understanding in deploying ML within scientific explorations. As stakeholders in the agricultural domain increasingly seek interpretable insights from ML applications, collaboration between ML specialists and biologists is essential to adapt these methods effectively to real-world problems (Marcinkevičs and Vogt 2020). Moreover, the field of plant abiotic stress research is witnessing the adaptation of ML methods that are designed to deliver insights alongside predictions. By addressing the limitations of traditional ML approaches, the integration of interpretability strategies fosters improved communication among interdisciplinary teams, which is fundamental for driving forward agricultural innovation in response to climate-induced stressors (Ruan et al. 2023).

6.7 Conclusion

ML provides a promising approach to understanding and predicting plant responses to abiotic stress. By leveraging vast datasets from HTP, omics technologies, and environmental monitoring, ML models can enhance our ability to breed stress-tolerant crops and optimize agricultural practices. However, challenges related to data quality, generalization, and interpretability remain and require further advancements in both computational techniques and experimental design.

The integration of ML into plant science has the potential to significantly accelerate progress in the fight against climate change and food insecurity. The integration of ML techniques into the study of plant abiotic stress responses represents a significant advancement in plant biology and agriculture. Through the analysis of large datasets encompassing various environmental conditions, genotypic variations, and physiological responses, researchers have begun to unravel the complex interactions between plants and their abiotic stressors, such as drought, salinity, temperature fluctuations, and nutrient deficiencies. ML models offer multiple advantages, including their ability to identify intricate patterns within high-dimensional data that are often overlooked by traditional statistical methods. The application of ML facilitates not only the prediction of plant responses to specific stress conditions but also the identification of key phenotypic traits and genomic factors that contribute to stress resilience. This predictive capacity is especially valuable in the face of climate change and increasing agricultural demands, where timely interventions can enhance crop yield and quality. Additionally, the adaptability of ML algorithms to various datasets, including imaging data, genomic data, and phenotypic measurements, allows for a more comprehensive and holistic approach to understanding plant biology. ML can aid in the development of phenotype prediction models that integrate environmental data, thus supporting plant breeding programs aimed at enhancing abiotic stress tolerance. Moreover, the synergy between ML and precision agriculture paves the way for innovations in real-time monitoring and intervention strategies. By employing ML algorithms to analyze data collected from sensors and drones, farmers can make informed decisions about irrigation, fertilization, and crop management tailored to specific environmental conditions and stress responses.

References

Abade, A., Ferreira, P.A., and de Barros Vidal, F. (2021). Plant diseases recognition on images using convolutional neural networks: A systematic review. *Computers and Electronics in Agriculture* 185: 106–125. https://doi.org/10.1016/J.COMPAG.2021.106125

Ahmed, B., Haque, M.A., Iquebal, M.A. et al. (2023). DeepAProt: Deep learning based abiotic stress protein sequence classification and identification tool in cereals. *Frontiers in Plant Science* 13: 1008756. https://doi.org/10.3389/FPLS.2022.1008756

Akram, N.A., Bashir, R., Ashraf, G. et al. (2023). Exogenous α-tocopherol regulates the growth and metabolism of eggplant (*Solanum melongena* L.) under drought stress. *Plants* 12 (2): 237. https://doi.org/10.3390/PLANTS12020237

Alhnaity, B., Pearson, S., Leontidis, G. et al. (2020). Using deep learning to predict plant growth and yield in greenhouse environments. *Acta Horticulturae* 1296 (1296): 425–431. https://doi.org/10.17660/ACTAHORTIC.2020.1296.55

Aloo, B.N., Dessureault-Rompré, J., Tripathi, V. et al. (2023). Signaling and crosstalk of rhizobacterial and plant hormones that mediate abiotic stress tolerance in plants. *Frontiers in Microbiology* 14: 1171104. https://doi.org/10.3389/FMICB.2023.1171104

Al-Tamimi, N., Langan, P., Bernád, V. et al. (2022). Capturing crop adaptation to abiotic stress using image-based technologies. *Open Biology* 12 (6): 210353. https://doi.org/10.1098/RSOB.210353

Amin, A.B., Rathnayake, K.N., Yim, W.C. et al. (2019). Crassulacean acid metabolism abiotic stress-responsive transcription factors: A potential genetic engineering approach for improving crop tolerance to abiotic stress. *Frontiers in Plant Science* 10: 129. https://doi.org/10.3389/FPLS.2019.00129

An, J.P., Li, R., Qu, F.J. et al. (2016). Apple F-box protein MdMAX2 regulates plant photomorphogenesis and stress response. *Frontiers in Plant Science* 7: 1685 (NOVEMBER 2016). https://doi.org/10.3389/FPLS.2016.01685

Aslam, N., Li, Q., Bashir, S. et al. (2024). Integrated review of transcriptomic and proteomic studies to understand molecular mechanisms of rice's response to environmental stresses. *Biology* 13 (9): 659. https://doi.org/10.3390/BIOLOGY13090659

Azad, M., Tohidfar, M., Ghanbari Moheb Seraj, R. et al. (2024). Identification of responsive genes to multiple abiotic stresses in rice (Oryza sativa): A meta-analysis of transcriptomics data. *Scientific Reports* 14 (1): 1–13. https://doi.org/10.1038/s41598-024-54623-7

Baddour, A.G. (2024). Response of soybean to gypsum and compost as diluents of salt stress under using cobalt and molybdenum. *Egyptian Journal of Soil Science* 64 (1): 0–0. https://doi.org/10.21608/EJSS.2023.248769.1687

Bailey-Serres, J., and Voesenek, L.A.C.J. (2008). Flooding stress: Acclimations and genetic diversity. *Annual Review of Plant Biology* 59: 313–339. https://doi.org/10.1146/ANNUREV.ARPLANT.59.032607.092752

Brahim, A.H., Jlidi, M., Daoud, L. et al. (2019). *Seed-Biopriming of Durum Wheat with Diazotrophic Plant Growth Promoting Bacteria (PGPB) Enhanced Tolerance to Fusarium Head Blight (FHB) and Salinity Stress.* https://doi.org/10.21203/RS.2.16636/V2

Catalá, R., López-Cobollo, R., Álvaro Berbís, M. et al. (2021). Trimethylamine *N* -oxide is a new plant molecule that promotes abiotic stress tolerance. *Science Advances* 7 (21): eabd9296. https://doi.org/10.1126/SCIADV.ABD9296

Chang, L., Jin, X., Rao, Y. et al. (2024). Predicting abiotic stress-responsive miRNA in plants based on multi-source features fusion and graph neural network. *Plant Methods* 20 (1): 33. https://doi.org/10.1186/S13007-024-01158-7

Chaudhary, K., Poirion, O.B., Lu, L. et al. (2018). Deep learning–based multi-omics integration robustly predicts survival in liver cancer. *Clinical Cancer Research* 24 (6): 1248–1259. https://doi.org/10.1158/1078-0432.CCR-17-0853

Chauhan, H., Khurana, N., Agarwal, P. et al. (2013). A seed preferential heat shock transcription factor from wheat provides abiotic stress tolerance and yield enhancement in transgenic Arabidopsis under heat stress environment. *PLoS ONE* 8 (11): e79577. https://doi.org/10.1371/JOURNAL.PONE.0079577

Chen, H., Tran, N., Thumati, A.S. et al. (2021). *Data Curation and Quality Assurance for Machine Learning-based Cyber Intrusion Detection*. https://doi.org/10.48550/ARXIV.2105.10041

Chen, S., Zhang, N., Zhou, G. et al. (2021). Knockout of the entire family of AITR genes in Arabidopsis leads to enhanced drought and salinity tolerance without fitness costs. *BMC Plant Biology* 21 (1): 137. https://doi.org/10.1186/S12870-021-02907-9

Cheng, C.Y., Li, Y., Varala, K. et al. (2021). Evolutionarily informed machine learning enhances the power of predictive gene-to-phenotype relationships. *Nature Communications* 12 (1): 5627. https://doi.org/10.1038/S41467-021-25893-W

Cheng, Y., Ma, Y., Zhang, N. et al. (2022). The R2R3 MYB transcription factor MYB71 regulates abscisic acid response in Arabidopsis. *Plants* 11 (10): 1369. https://doi.org/10.3390/PLANTS11101369

Chinnusamy, V., Zhu, J., and Zhu, J.K. (2007). Cold stress regulation of gene expression in plants. *Trends in Plant Science* 12 (10): 444–451. https://doi.org/10.1016/J.TPLANTS.2007.07.002

Chowdhury, R.H., Eti, F.S., Ahmed, R. et al. (2023). Drought-responsive genes in tomato: Meta-analysis of gene expression using machine learning. *Scientific Reports* 13 (1): 1–11. https://doi.org/10.1038/s41598-023-45942-2

Clemens, S. (2006). Toxic metal accumulation, responses to exposure and mechanisms of tolerance in plants. *Biochimie* 88 (11): 1707–1719. https://doi.org/10.1016/J.BIOCHI.2006.07.003

Danilevicz, M.F., Gill, M., Anderson, R. et al. (2022). Plant genotype to phenotype prediction using machine learning. *Frontiers in Genetics* 13: 822173. https://doi.org/10.3389/FGENE.2022.822173/PDF

Das Choudhury, S., Saha, S., Samal, A. et al. (2023). Drought stress prediction and propagation using time series modeling on multimodal plant image sequences. *Frontiers in Plant Science* 14: 1003150. https://doi.org/10.3389/FPLS.2023.1003150

Devireddy, A.R., Zandalinas, S.I., Fichman, Y. et al. (2021). Integration of reactive oxygen species and hormone signaling during abiotic stress. *The Plant Journal* 105 (2): 459–476. https://doi.org/10.1111/TPJ.15010

Dey, S., and Raichaudhuri, A. (2022). Abiotic stress in plants. *Advances in Plant Defense Mechanisms*. https://doi.org/10.5772/INTECHOPEN.105944

Dominiak-Świgoń, M., Olejniczak, P., Nowak, M. et al. (2019). Hyperspectral imaging in assessing the condition of plants: Strengths and weaknesses. *Biodiversity Research and Conservation* 55 (1): 25–30. https://doi.org/10.2478/BIORC-2019-0011

dos Santos, T.B., Ribas, A.F., de Souza, S.G.H. et al. (2022). Physiological responses to drought, salinity, and heat stress in plants: A review. *Stresses* 2 (1): 113–135. https://doi.org/10.3390/STRESSES2010009

Dossa, K., Mmadi, M.A., Zhou, R. et al. (2019). Depicting the core transcriptome modulating multiple abiotic stresses responses in sesame (Sesamum indicum L.). *International Journal of Molecular Sciences* 20 (16): 3930. https://doi.org/10.3390/IJMS20163930

Dresselhaus, T., and Hückelhoven, R. (2018). Biotic and abiotic stress responses in crop plants. *Agronomy* 8 (11): 267. https://doi.org/10.3390/AGRONOMY8110267

Du, M., Liu, N., and Hu, X. (2020). Techniques for interpretable machine learning. *Communications of the ACM* 63 (1): 68–77. https://doi.org/10.1145/3359786

Dutta, S., Islam, Z., Das, S. et al. (2025). Harmonizing plant resilience: Unveiling the symphony of membrane lipid dynamics in response to abiotic stresses: A review. *Discover Plants* 2 (1): 61. https://doi.org/10.1007/s44372-025-00152-0

Dutta, S., Zunjare, R.U., Sil, A. et al. (2024). Prediction of matrilineal specific patatin-like protein governing in-vivo maternal haploid induction in maize using support vector machine and di-peptide composition. *Amino Acids* 56 (1): 20. https://doi.org/10.1007/s00726-023-03368-0

Dwivedi, A.K., Singh, V., Anwar, K. et al. (2023). Integrated transcriptome, proteome and metabolome analyses revealed secondary metabolites and auxiliary carbohydrate metabolism augmenting drought tolerance in rice. *Plant Physiology and Biochemistry: PPB* 201: 107849. https://doi.org/10.1016/J.PLAPHY.2023.107849

Farooq, M.A., Gao, S., Hassan, M.A. et al. (2024). Artificial intelligence in plant breeding. *Trends in Genetics* 40 (10): 891–908. https://doi.org/10.1016/J.TIG.2024.07.001/ASSET/D5D2EA4C-5116-417E-9F70-A718A59BCD97/MAIN.ASSETS/GR2.JPG

Farooq, M., Wahid, A., Kobayashi, N. et al. (2009). Plant drought stress: Effects, mechanisms and management. *Agronomy for Sustainable Development* 29 (1): 185–212. https://doi.org/10.1051/AGRO:2008021

Feng, X., Zhan, Y., Wang, Q. et al. (2020). Hyperspectral imaging combined with machine learning as a tool to obtain high-throughput plant salt-stress phenotyping. *The Plant Journal* 101 (6): 1448–1461. https://doi.org/10.1111/TPJ.14597

Galieni, A., D'Ascenzo, N., Stagnari, F. et al. (2021). Past and future of plant stress detection: An overview from remote sensing to positron emission tomography. *Frontiers in Plant Science* 11: 609155. https://doi.org/10.3389/FPLS.2020.609155

Gano, B., Bhadra, S., Vilbig, J.M. et al. (2024). Drone-based imaging sensors, techniques, and applications in plant phenotyping for crop breeding: A comprehensive review. *The Plant Phenome Journal* 7 (1): e20100 https://doi.org/10.1002/PPJ2.20100

Ghosal, S., Blystone, D., Singh, A.K. et al. (2018). An explainable deep machine vision framework for plant stress phenotyping. *Proceedings of the National Academy of Sciences of the United States of America* 115 (18): 4613–4618. https://doi.org/10.1073/PNAS.1716999115/SUPPL_FILE/PNAS.1716999115.SAPP.PDF

Ghosh, D., and Xu, J. (2014). Abiotic stress responses in plant roots: A proteomics perspective. *Frontiers in Plant Science* 5 (JAN). https://doi.o)g/10.3389/FPLS.2014.00006

Gill, T., Gill, S.K., Saini, D.K. et al. (2022). A comprehensive review of high throughput phenotyping and machine learning for plant stress phenotyping. *Phenomics* 2 (3): 156. https://doi.org/10.1007/S43657-022-00048-Z

Guo, M., Liu, J.H., Ma, X. et al. (2016). The plant heat stress transcription factors (HSFs): Structure, regulation, and function in response to abiotic stresses. *Frontiers in Plant Science* 7 (FEB2016). https://doi.org/10.3389/FPLS.2016.00114

Gupta, C., Ramegowda, V., Basu, S. et al. (2021). Using network-based machine learning to predict transcription factors involved in drought resistance. *Frontiers in Genetics* 12: 652189. https://doi.org/10.3389/FGENE.2021.652189

Handayani, D.P., Adityansah, R., Sulastri. et al. (2024). The effect of endophytic halotolerant bacteria inoculation in the development of soybean seedling under salinity stress. *IOP Conference Series: Earth and Environmental Science* 1359 (1): 012122. https://doi.org/10.1088/1755-1315/1359/1/012122

Henninger, M., and Strobl, C. (2023). *Interpreting machine learning predictions with LIME and Shapley values: Theoretical insights, challenges, and meaningful interpretations*. https://doi.org/10.31234/OSF.IO/3SVB2

Herr, A.W., Adak, A., Carroll, M.E. et al. (2023). Unoccupied aerial systems imagery for phenotyping in cotton, maize, soybean, and wheat breeding. *Crop Science* 63 (4): 1722–1749. https://doi.org/10.1002/CSC2.21028

Hu, W., Ding, Z., Tie, W. et al. (2017). Comparative physiological and transcriptomic analyses provide integrated insight into osmotic, cold, and salt stress tolerance mechanisms in banana. *Scientific Reports* 7 (1): 43007. https://doi.org/10.1038/SREP43007

Hu, W., Ren, Q., Chen, Y. et al. (2021). Genome-wide identification and analysis of WRKY gene family in maize provide insights into regulatory network in response to abiotic stresses. *BMC Plant Biology* 21 (1): 427. https://doi.org/10.1186/S12870-021-03206-Z

Hu, Y., Zhi, L., Li, P. et al. (2022). The role of salicylic acid signal in plant growth, development and abiotic stress. *Phyton* 91 (12): 2591–2605. https://doi.org/10.32604/PHYTON.2022.023733

Huang, H., Ullah, F., Zhou, D.X. et al. (2019). Mechanisms of ROS regulation of plant development and stress responses. *Frontiers in Plant Science* 10: 800. https://doi.org/10.3389/FPLS.2019.00800

Huang, W., Hu, X., Ren, Y. et al. (2023). IPOP: An integrative plant multi-omics platform for cross-species comparison and evolutionary study. *Molecular Biology and Evolution* 40 (12): msad248. https://doi.org/10.1093/MOLBEV/MSAD248

Ibrahim, S.M., Ibrahim, M.S.A., Khaled, R. et al. (2023). Unlocking the genetic basis of abiotic stress tolerance in wheat: Insights from differential expression analysis and machine learning. *Highlights in BioScience*. https://doi.org/10.36462/H.BIOSCI.202307

Imran, M., Aaqil Khan, M., Shahzad, R. et al. (2021). Melatonin ameliorates thermotolerance in soybean seedling through balancing redox homeostasis and modulating antioxidant defense, phytohormones and polyamines biosynthesis. *Molecules* 26 (17): 5116. https://doi.org/10.3390/MOLECULES26175116

Iqbal, T., Elahi, A., Wijns, W. et al. (2022). Exploring unsupervised machine learning classification methods for physiological stress detection. *Frontiers in Medical Technology* 4: 782756. https://doi.org/10.3389/FMEDT.2022.782756

Islam, M.P., and Yamane, T. (2021). HortNet417v1—A deep-learning architecture for the automatic detection of pot-cultivated peach plant water stress. *Sensors* 21 (23): 7924. https://doi.org/10.3390/S21237924

Jenkins, G.I. (2009). Signal transduction in responses to UV-B radiation. *Annual Review of Plant Biology* 60: 407–431. https://doi.org/10.1146/ANNUREV.ARPLANT.59.032607.092953

Jeyasri, R., Muthuramalingam, P., Satish, L. et al. (2021). An overview of abiotic stress in cereal crops: Negative impacts, regulation, biotechnology and integrated omics. *Plants* 10 (7): 1472. https://doi.org/10.3390/PLANTS10071472

Jha, R.K., and Mishra, A. (2022). Introgression of SbERD4 gene encodes an early-responsive dehydration-stress protein that confers tolerance against different types of abiotic stresses in transgenic tobacco. *Cells* 11 (1). https://doi.org/10.3390/CELLS11010062

Kamarudin, M.H., and Ismail, Z.H. (2022). Lightweight deep CNN models for identifying drought stressed plant. In *IOP Conference Series: Earth and Environmental Science* 1091 (1): 012043. https://doi.org/10.1088/1755-1315/1091/1/012043

Kamran, M., Parveen, A., Ahmar, S. et al. (2020). An overview of hazardous impacts of soil salinity in crops, tolerance mechanisms, and amelioration through selenium supplementation. *International Journal of Molecular Sciences* 21 (1): 148. https://doi.org/10.3390/IJMS21010148

Kaniganti, S., Khisti, M., Kishor, P.B.K. et al. (2024). Genome-wide identification and characterization of the strigolactone (SL) pathway and associated genes in sorghum. *Current Trends in Biotechnology and Pharmacy* 18 (1): 1518–1530. https://doi.org/10.5530/CTBP.2024.1.1

Károly, A.I., Fullér, R., and Galambos, P. (2018). Unsupervised clustering for deep learning: A tutorial survey. *Acta Polytechnica Hungarica* 15 (8): 29–53. https://doi.org/10.12700/APH.15.8.2018.8.2

Kimotho, R.N., Baillo, E.H. and Zhang, Z. (2019). Transcription factors involved in abiotic stress responses in Maize (*Zea mays*L.) and their roles in enhanced productivity in the post genomics era. *PeerJ* 7 (7): e7211. https://doi.org/10.7717/PEERJ.7211

Kirongo, C., Omieno, K.K., Mutua, M. et al. (2019). Plant stress detection accuracy using deep convolution neural networks. *International Journal of Scientific Research in Computer Science, Engineering and Information Technology*: 263–270. https://doi.org/10.32628/CSEIT195447

Kopecká, R., Kameniarová, M., Černý, M. et al. (2023). Abiotic stress in crop production. *International Journal of Molecular Sciences* 24 (7): 6603. https://doi.org/10.3390/IJMS24076603/S1

Kotak, S., Larkindale, J., Lee, U. et al. (2007). Complexity of the heat stress response in plants. *Current Opinion in Plant Biology* 10 (3): 310–316. https://doi.org/10.1016/J.PBI.2007.04.011

Kumar, P., Paul, D., Jhajhriya, S. et al. (2024). Understanding heat-shock proteins' abundance and pivotal function under multiple abiotic stresses. *Journal of Plant Biochemistry and Biotechnology*: 1–22. https://doi.org/10.1007/s13562-024-00932-x

Lee, D.K., Jung, H., Jang, G. et al. (2016). Overexpression of the *OsERF71* transcription factor alters rice root structure and drought resistance. *Plant Physiology* 172 (1): 575–588. https://doi.org/10.1104/PP.16.00379

Li, J., Sun, S., Jiang, H. et al. (2022). Image recognition and empirical application of desert plant species based on convolutional neural network. *Journal of Arid Land* 14 (12): 1440–1455. https://doi.org/10.1007/S40333-022-0086-9

Liu, B., and Udell, M. (2020). *Impact of accuracy on model interpretations*. https://doi.org/10.48550/ARXIV.2011.09903

Liu, C., Xu, Y., Feng, Y. et al. (2019). Ectopic expression of mulberry G-Proteins alters drought and salt stress tolerance in tobacco. *International Journal of Molecular Sciences* 20 (1): 89. https://doi.org/10.3390/IJMS20010089

Liu, J., Gong, M., and Miao, Q. (2017). Modeling Hebb learning rule for unsupervised learning. *Proceedings of the Twenty-Sixth International Joint Conference on Artificial Intelligence* 0: 2315–2321. https://doi.org/10.24963/IJCAI.2017/322

López-Arredondo, D.L., Leyva-González, M.A., González-Morales, S.I. et al. (2014). Phosphate nutrition: Improving low-phosphate tolerance in crops. *Annual Review of Plant Biology* 65: 95–123. https://doi.org/10.1146/ANNUREV-ARPLANT-050213-035949

Lowe, A., Harrison, N., and French, A.P. (2017). Hyperspectral image analysis techniques for the detection and classification of the early onset of plant disease and stress. *Plant Methods* 13 (1): 80. https://doi.org/10.1186/S13007-017-0233-Z

Lu, M., Gao, P., Hu, J. et al. (2023). A classification method of stress in plants using unsupervised learning algorithm and chlorophyll fluorescence technology. *Frontiers in Plant Science* 14: 1202092. https://doi.org/10.3389/FPLS.2023.1202092

Ma, C., Xin, M., Feldmann, K.A. et al. (2014). Machine learning–based differential network analysis: A study of stress-responsive transcriptomes in *Arabidopsis*. *The Plant Cell* 26 (2): 520–537. https://doi.org/10.1105/TPC.113.121913

Malwattage, N.R., Wone, B., and Wone, B.W.M. (2024). A CAM-related NF-YB transcription factor enhances multiple abiotic stress tolerance in Arabidopsis. *International Journal of Molecular Sciences* 25 (13): 7107. https://doi.org/10.3390/IJMS25137107

Manzoor, H., Athar, H.U.R., Rasul, S. et al. (2016). Avenues for improving drought tolerance in crops by ABA regulation. *Water Stress and Crop Plants* 1–2: 177–193. https://doi.org/10.1002/9781119054450.CH13

Marcinkevičs, R. and Vogt, J.E. (2020). *Interpretability and Explainability: A Machine Learning Zoo Mini-tour*. https://doi.org/10.48550/ARXIV.2012.01805

Marschner, P. (2011). Marschner's mineral nutrition of higher plants: Third edition. *Marschner's Mineral Nutrition of Higher Plants: Third Edition*, 1–651. https://doi.org/ 10.1016/C2009-0-63043-9

Meher, P.K., Begam, S., Sahu, T.K. et al. (2022). ASRmiRNA: Abiotic stress-responsive miRNA prediction in plants by using machine learning algorithms with pseudo K-tuple nucleotide compositional features. *International Journal of Molecular Sciences* 23 (3): 1612. https://doi.org/10.3390/IJMS23031612

Meher, P.K., Sahu, T.K., Gupta, A. et al. (2024). ASRpro: A machine-learning computational model for identifying proteins associated with multiple abiotic stress in plants. *The Plant Genome* 17 (1): e20259. https://doi.org/10.1002/TPG2.20259

Meng, X., Liang, Z., Dai, X. et al. (2021). Predicting transcriptional responses to cold stress across plant species. *Proceedings of the National Academy of Sciences* 118 (10): e2026330118. https://doi.org/10.1073/PNAS.2026330118

Minerva, M.F., Thelma D.P., and Marvin D.M. (2024). Improving abiotic stress mitigation via predictive modeling of water quality parameters in recirculated aquaculture systems. *Journal of Electrical Systems* 20 (3s): 1382–1389. https://doi.org/ 10.52783/JES.1515

Mou, L., Bruzzone, L., and Zhu, X.X. (2019). Learning spectral-spatial-temporal features via a recurrent convolutional neural network for change detection in multispectral imagery. *IEEE Transactions on Geoscience and Remote Sensing* 57 (2): 924–935. https://doi.org/10.1109/TGRS.2018.2863224

Mullainathan, S., and Obermeyer, Z. (2017). Does machine learning automate moral hazard and error? *American Economic Review* 107 (5): 476–480. https://doi.org/10.1257/AER.P20171084

Munns, R., and Tester, M. (2008). Mechanisms of salinity tolerance. *Annual Review of Plant Biology* 59: 651–681. https://doi.org/ 10.1146/ANNUREV.ARPLANT.59.032607.092911/CITE/REFWORKS

Musayev, M., and Huseynova, T. (2022). Assessment of grapevine genetic resources based on physiological data. *International Journal of Minor Fruits, Medicinal and Aromatic Plants* 8 (2): 80–85. https://doi.org/10.53552/IJMFMAP.8.2.2022.80-85

Naik, B., Kumar, V., Rizwanuddin, S. et al. (2023). Genomics, proteomics, and metabolomics approaches to improve abiotic stress tolerance in tomato plant. *International Journal of Molecular Sciences* 24: 3025. https://doi.org/10.3390/IJMS24033025

Negrão, S., Schmöckel, S.M., and Tester, M. (2017). Evaluating physiological responses of plants to salinity stress. *Annals of Botany* 119 (1): 1–11. https://doi.org/10.1093/AOB/MCW191

Okhovatian-Ardakani, A.R., Mehrabanian, M., Dehghani, F. et al. (2010). Salt tolerance evaluation and relative comparison in cuttings of different pomegranate cultivar. *Plant, Soil and Environment* 56 (4): 176–185. https://doi.org/10.17221/158/ 2009-PSE

Panahi, B., Hassani, M., and Hosseinzaeh Gharajeh, N. (2025). Integrative analysis of RNA-Seq data and machine learning approaches to identify biomarkers for Rhizoctonia solani resistance in sugar beet. *Biochemistry and Biophysics Reports* 41: 101920. https://doi.org/10.1016/J.BBREP.2025.101920

Paul, N., Sunil, G.C., Horvath, D. et al. (2025). Deep learning for plant stress detection: A comprehensive review of technologies, challenges, and future directions. *Computers and Electronics in Agriculture* 229: 109734. https://doi.org/10.1016/J.COMPAG. 2024.109734

Pervaiz, T., Amjid, M.W., El-Kereamy, A. et al. (2022). MicroRNA and cDNA-microarray as potential targets against abiotic stress response in plants: Advances and prospects. *Agronomy* 12 (1). https://doi.org/10.3390/AGRONOMY12010011

Pillong, M., Marx, C., Piechon, P. et al. (2017). A publicly available crystallisation data set and its application in machine learning. *CrystEngComm* 19 (27): 3737–3745. https://doi.org/10.1039/C7CE00738H

Pradhan, U.K., Mahapatra, A., Naha, S. et al. (2024). ASPTF: A computational tool to predict abiotic stress-responsive transcription factors in plants by employing machine learning algorithms. *Biochimica et Biophysica Acta. General Subjects* 1868 (6): 130597. https://doi.org/10.1016/J.BBAGEN.2024.130597

Radwan, O. (2015). Genomics improvement of agronomic crops to abiotic stress. *Journal of Investigative Genomics* 2 (3). https:// doi.org/10.15406/JIG.2015.02.00026

Rahman, A. (2013). Auxin: A regulator of cold stress response. *Physiologia Plantarum* 147 (1): 28–35. https://doi.org/10.1111/ J.1399-3054.2012.01617.X

Rane, J., Singh, A.K., Kumar, M. et al. (2021). The adaptation and tolerance of major cereals and legumes to important abiotic stresses. *International Journal of Molecular Sciences* 22 (23): 12970. https://doi.org/10.3390/IJMS222312970

Rico-Chávez, A.K., Franco, J.A., Fernandez-Jaramillo, A.A. et al. (2022). Machine learning for plant stress modeling: A perspective towards hormesis management. *Plants* 11 (7): 970. https://doi.org/10.3390/PLANTS11070970

Robles, P., Micol, J.L., and Quesada, V. (2015). Mutations in the plant-conserved MTERF9 alter chloroplast gene expression, development and tolerance to abiotic stress in *Arabidopsis thaliana*. *Physiologia Plantarum* 154 (2): 297–313. https://doi.org/ 10.1111/PPL.12307

Roy, A., Dutta, S., Das, S. et al. (2024). Next-generation sequencing in the development of climate-resilient and stress-responsive crops: A review. *The Open Biotechnology Journal* 18 (1). https://doi.org/10.2174/0118740707301657240517063244

Ruan, S., Cang, H., Chen, H. et al. (2023). Hyperspectral classification of frost damage stress in tomato plants based on few-shot learning. *Agronomy* 13 (9): 2348. https://doi.org/10.3390/AGRONOMY13092348

Sah, S.K., Reddy, K.R., and Li, J. (2016). Abscisic acid and abiotic stress tolerance in crop plants. *Frontiers in Plant Science* 7: 571 (2016). https://doi.org/10.3389/FPLS.2016.00571

Saharan, B.S., Brar, B., Duhan, J.S. et al. (2022). Molecular and physiological mechanisms to mitigate abiotic stress conditions in plants. *Life* 12 (10): 1634. https://doi.org/10.3390/LIFE12101634

Sani, N., Lee, J., Nabi, R. et al. (2020). *A Semiparametric Approach to Interpretable Machine Learning*. https://doi.org/10.48550/ARXIV.2006.04732

Sarkar, S., Chidambaranathan, P., Jeughale, K. et al. (2022). *Common genes regulating multiple abiotic stress response in rice: An analysis using supervised and unsupervised machine learning models*. https://doi.org/10.21203/RS.3.RS-1861354/V1

Schmidt, G., Vågen, I.M., Hagen, S.F. et al. (2024). Health- and sensory-related phytochemicals and agronomic characteristics of 26 head cabbage cultivars. *ACS Food Science & Technology* 4 (10): 2341–2354. https://doi.org/10.1021/ACSFOODSCITECH.4C00370

Seo, D.H., Seomun, S., Choi, Y. Do. et al. (2020). Root development and stress tolerance in rice: The key to improving stress tolerance without yield penalties. *International Journal of Molecular Sciences* 21 (5): 1807. https://doi.org/10.3390/IJMS21051807

Shaik, R., and Ramakrishna, W. (2014). Machine learning approaches distinguish multiple stress conditions using stress-responsive genes and identify candidate genes for broad resistance in rice. *Plant Physiology* 164 (1): 481–495. https://doi.org/10.1104/PP.113.225862

Sharma, S.S., and Dietz, K.J. (2009). The relationship between metal toxicity and cellular redox imbalance. *Trends in Plant Science* 14 (1): 43–50. https://doi.org/10.1016/J.TPLANTS.2008.10.007

Sheikh, M., Iqra, F., Ambreen, H. et al. (2024). Integrating artificial intelligence and high-throughput phenotyping for crop improvement. *Journal of Integrative Agriculture* 23 (6): 1787–1802. https://doi.org/10.1016/J.JIA.2023.10.019

Shen, H., Hou, Y., and Wang, X. et al. (2024). Genome-wide identification, expression analysis under abiotic stress and co-expression analysis of MATE gene family in Torreya grandis. *International Journal of Molecular Sciences* 25 (7): 3859. https://doi.org/10.3390/IJMS25073859

Shinozaki, K., and Yamaguchi-Shinozaki, K. (2007). Gene networks involved in drought stress response and tolerance. *Journal of Experimental Botany* 58 (2): 221–227. https://doi.org/10.1093/JXB/ERL164

Singh, A.K., Ganapathysubramanian, B., Sarkar, S. et al. (2018). Deep learning for plant stress phenotyping: Trends and future perspectives. *Trends in Plant Science* 23 (10): 883–898. https://doi.org/10.1016/J.TPLANTS.2018.07.004

Singh, A., Ganapathysubramanian, B., Singh, A.K. et al. (2016). Machine learning for high-throughput stress phenotyping in plants. *Trends in Plant Science* 21 (2): 110–124. https://doi.org/10.1016/J.TPLANTS.2015.10.015

Smet, D., Opdebeeck, H., and Vandepoele, K. (2023). Predicting transcriptional responses to heat and drought stress from genomic features using a machine learning approach in rice. *Frontiers in Plant Science* 14: 1212073. https://doi.org/10.3389/FPLS.2023.1212073

Sun, Y., Ma, R., Yang, X. et al. (2024). A metabolomic analysis of tomato fruits in response to salt stress. *Horticulturae* 10 (12): 1303. https://doi.org/10.3390/HORTICULTURAE10121303

Suzuki, N., Rivero, R.M., Shulaev, V. et al. (2014). Abiotic and biotic stress combinations. *New Phytologist* 203 (1): 32–43. https://doi.org/10.1111/NPH.12797

Tang, Y., Cao, Z., Guo, N. et al. (2023). *Siam-Swin-Unet: A Siamese Swin-Unet for remote sensing image change detection*. https://doi.org/10.21203/RS.3.RS-3236851/V1

Tjaden, J., and Tjaden, B. (2023). MLpronto: A tool for democratizing machine learning. *Plos One* 18 (11): e0294924. https://doi.org/10.1371/JOURNAL.PONE.0294924

Tuğaç, M.G. (2021). GIS-based land suitability classification for wheat cultivation using fuzzy set model. *International Journal of Agriculture Environment and Food Sciences* 5 (4): 524–536. https://doi.org/10.31015/JAEFS.2021.4.12

Ullah, M.A., Hyder, S.I., Mahmood, I.A. et al. (2017). Inducing salt tolerance in wheat through inoculation with rhizobacteria containing 1-aminocyclopropane-1-carboxylate deaminase activity. *Journal of Agricultural Science and Practice* 2 (4): 54–57. https://doi.org/10.31248/JASP2017.055

Vaidya, S.N. (2023). Automated plant species detection: A CNN approach for image classification. *International Journal of Scientific Research in Engineering and Management* 07 (07). https://doi.org/10.55041/IJSREM25018

Verma, V., Ravindran, P., and Kumar, P.P. (2016). Plant hormone-mediated regulation of stress responses. *BMC Plant Biology* 16 (1): 86. https://doi.org/10.1186/S12870-016-0771-Y

Wang, P., Su, L., Gao, H. et al. (2018). Genome-wide characterization of bHLH genes in grape and analysis of their potential relevance to abiotic stress tolerance and secondary metabolite biosynthesis. *Frontiers in Plant Science* 9: 64. https://doi.org/10.3389/FPLS.2018.00064

Wang, Y., and Frei, M. (2011). Stressed food: The impact of abiotic environmental stresses on crop quality. *Agriculture, Ecosystems & Environment* 141 (3–4): 271–286. https://doi.org/10.1016/J.AGEE.2011.03.017

Waseem, M., Aslam, M.M., and Sahu, S.K. (2024). Understanding the mechanistic basis of plant adaptation to salinity and drought. *Functional Plant Biology* 51 (3). https://doi.org/10.1071/FP23216

Wei, Y., Hu, W., and Xia, F. et al. (2016). Heat shock transcription factors in banana: Genome-wide characterization and expression profile analysis during development and stress response. *Scientific Reports* 6 (1): 36864. https://doi.org/10.1038/SREP36864

Xiao, M., Li, Z., Zhu, L. et al. (2021). The multiple roles of ascorbate in the abiotic stress response of plants: Antioxidant, cofactor, and regulator. *Frontiers in Plant Science* 12: 598173. https://doi.org/10.3389/FPLS.2021.598173

Xu, D., Leister, D., and Kleine, T. (2017). Arabidopsis thaliana mTERF10 and mTERF11, but Not mTERF12, are involved in the response to salt stress. *Frontiers in Plant Science* 8: 1213. https://doi.org/10.3389/FPLS.2017.01213

Xu, L., Du, H., and Zhang, X. (2019). Spatial distribution characteristics of soil salinity and moisture and its influence on agricultural irrigation in the Ili River Valley, China. *Sustainability* 11 (24): 7142. https://doi.org/10.3390/SU11247142

Xu, X., Fonseca de Lima, C.F., Vu, L.D. et al. (2023). When drought meets heat: A plant omics perspective. *Frontiers in Plant Science* 14: 1250878. https://doi.org/10.3389/FPLS.2023.1250878/FULL

Yan, J., and Wang, X. (2022). Unsupervised and semi-supervised learning: The next frontier in machine learning for plant systems biology. *The Plant Journal* 111 (6): 1527–1538. https://doi.org/10.1111/TPJ.15905

Yang, J.F., Chen, W.J., Zhou, L.M. et al. (2022). Real-time fluorescence imaging of the abscisic acid receptor allows nondestructive visualization of plant stress. *ACS Applied Materials & Interfaces* 14 (25): 28489–28500. https://doi.org/10.1021/ACSAMI.2C02156

Yoon, Y., Seo, D.H., Shin, H. et al. (2020). The role of stress-responsive transcription factors in modulating abiotic stress tolerance in plants. *Agronomy* 10 (6): 788. https://doi.org/10.3390/AGRONOMY10060788

Yu, J., Schumann, A.W., Cao, Z. et al. (2019). Weed detection in perennial ryegrass with deep learning convolutional neural network. *Frontiers in Plant Science* 10: 1422. https://doi.org/10.3389/FPLS.2019.01422

Zandalinas, S.I., Fichman, Y., Devireddy, A.R. et al. (2020). Systemic signaling during abiotic stress combination in plants. *Proceedings of the National Academy of Sciences* 117 (24): 13810–13820. https://doi.org/10.1073/PNAS.2005077117

Zhang, X., Ibrahim, Z., Khaskheli, M.B. et al. (2024). Integrative approaches to abiotic stress management in crops: Combining bioinformatics educational tools and artificial intelligence applications. *Sustainability* 16 (17): 7651. https://doi.org/10.3390/SU16177651

Zhang, Y., Xu, J., Li, R. et al. (2023). Plants' response to abiotic stress: Mechanisms and strategies. *International Journal of Molecular Sciences* 24 (13): 10915. https://doi.org/10.3390/IJMS241310915

Zheng, C., Abd-elrahman, A., and Whitaker, V. (2021). Remote sensing and machine learning in crop phenotyping and management, with an emphasis on applications in strawberry farming. *Remote Sensing* 13 (3): 1–29. https://doi.org/10.3390/RS13030531

Zhu, J.K. (2016). Abiotic stress signaling and responses in plants. *Cell* 167 (2): 313–324. https://doi.org/10.1016/J.CELL.2016.08.029

Zhuang, J., Jin, X., Chen, Y. et al. (2023). Drought stress impact on the performance of deep convolutional neural networks for weed detection in Bahiagrass. *Grass and Forage Science* 78 (1): 214–223. https://doi.org/10.1111/GFS.12583

7

Machine Learning for Modeling Plant–Pathogen Interactions

Sourish Pramanik[1,2], *Dibyendu Seth*[3], *Sandip Debnath*[4*]

[1] *Department of Plant Pathology, Punjab Agricultural University, Ludhiana, Punjab*
[2] *Ethophilia Research Foundation, Santiniketan, West Bengal*
[3] *Department of Genetics and Plant Breeding, Chaudhary Charan Singh Haryana Agriculture University (CCSHAU), Hisar, Haryana*
[4] *Department of Genetics and Plant Breeding, Institute of Agriculture, Visva-Bharati, Sriniketan, Birbhum, West Bengal*

7.1 Introduction

Plants come into contact with a variety of pathogens, including bacteria, fungi, viruses, mycoplasma, and nematodes during their lifetime (Agrios 2005). These phytopathogens are responsible for the spread of infections and epidemics that have a major effect on agricultural output and the security of the world's food supply (Fang and Ramasamy 2015). Protecting crops from plant diseases is essential to guaranteeing the quantity and quality of agricultural produce. It is estimated that up to 40% of the world's agricultural production may be lost directly as a result of weeds, animals, and pathogens (Savary et al. 2012). Biotic stresses brought on by pathogens alone still account for 15% of worldwide agricultural output losses, despite improvements in crop protection techniques (Murmu et al. 2025). Therefore, to reduce losses in agricultural output and preserve food security, efficient plant protection measures are crucial. Researchers are working to improve their comprehension of plant diseases, especially the relationship between pathogenic microorganisms and hosts, to tackle and avoid future epidemics (Montesinos et al. 2002). The factors that contribute to the disease are very diverse in terms of biochemical structure and genetic makeup, and they come from multiple sources and only sometimes surpass their threshold of severity due to slight shifts in dynamic environmental equilibrium. Climate variability and alterations from ecological harmony, along with other as-yet-undiscovered parameters, contribute to the development of new diseases (Lahlali et al. 2024). In agricultural research, understanding plant–pathogen interactions (PPI) is essential for optimizing crop yield and sustainability. Plant health and disease susceptibility are determined by these interactions, which are controlled by molecular communication involving plants and pathogens. Although PPIs are essential for understanding infection processes, experimental methods including immune-precipitation, yeast two-hybrid, and X-ray crystallography are laborious, costly, and boast a restricted scope (Zhu et al. 2001). Although techniques like protein chips and tandem affinity purification (TAP)-tagging offer useful PPI data, they are time-consuming and prone to misleading errors, which makes computational approaches more necessary (Uetz et al. 2000; Aumentado-Armstrong et al. 2015). Computational techniques are moreover necessary to enable large-scale PPI identification (Shoemaker and Panchenko 2007; Lam and Chan 2010). Numerous tactics have been put forth, utilizing a range of data sources, including 3D structural data (Smith and Sternberg 2002), Annotations of Gene Ontology (GO) (Lee et al. 2006), gene fusion events, phylogenetic profiles (Huang and Zheng 2006), and patterns of co-evolution (Jothi et al. 2006). These approaches, however, frequently depend on preexisting information, which restricts their capacity to recognize new relationships. In order to overcome these obstacles, data-driven models with predictive powers are essential.

A thorough and methodical comprehension of the hereditary, cellular, physiological, biochemical, and phenotypic data of both static and dynamic biological functions of an organism as a whole comprises a systems biology approach (Barabási and Oltvai 2004; Garbutt et al. 2014). This all-encompassing method can reveal a biological system's evolving characteristics. In order to achieve this, systems biology requires a top-down approach that integrates various -omics methods related to systems analyses, where multiple elements (genes, RNA, proteins, metabolites, ions, and so forth) are

Machine Learning for Plant Biology, First Edition. Edited by Jen-Tsung Chen.

evaluated qualitatively and quantitatively, and their interactions are determined (Garbutt et al. 2014). Additionally, practical mathematical and computational models are recreated, which serve as comprehensive narratives of the elements to show and calculate the relationships between the various components of the system and explain its local and global characteristics (McCormack et al. 2016; Pritchard and Birch 2014). The accuracy of developed models may then be further increased by empirically verifying these models to generate new hypotheses. Lastly, computational methods ought to be time efficient for rapidly predicting the dynamics of complex systems utilizing the available data and the models created. Thus, this improves the in-depth comprehension of a biological system to find and recreate new biological characteristics (Barabási and Oltvai 2004; Pritchard and Birch 2014). Parallel to this, machine learning (ML) is a branch of artificial intelligence that frequently makes use of computational and statistical methods to enable scripts to "learn" from data without the need for programming. For modeling predictions, it makes use of various data patterns made up of "features" and "labels" (Camacho et al. 2018). The model can forecast any variety of outcomes after acquiring a collection of patterns. Therefore, for effective predictions, the kind of living system, data structure, and its sufficiency are crucial. A genetic sequence, gene expression patterns, metabolite concentrations, protein–protein interactions and other information can all be considered characteristics in biological applications. Unsupervised and supervised learning are the two main categories into which ML techniques are divided (McMurray and Hollich 2009).

As stated earlier, unsupervised and supervised approaches are the two main categories of ML, which is the use of statistical techniques to find patterns in data (Witten et al. 2005). Exploratory analysis (for example, K-means clustering) and dimensionality reduction (such as principal component analysis) are two examples of unsupervised ML techniques that use unlabeled training data. Usually used for classification issues, supervised learning makes use of partially labeled data (semi-supervised learning) or labeled training data. The goal of supervised ML is to use training data to develop a function that can categorize unknown data. A separate collection of tests for confirmation, choice of features for model training, training data with both positive and negative scenarios, and the right model are the main components of categorization. Although deep learning requires large quantities of data, which are seldom accessible in biological domains, it is a member of a family of models which lack explicit feature selection. Cluster learning, which combines many ML models that have learned given identical data and have different projections, can significantly increase prediction accuracy (Dietterich 2000). An ML model can be trained somewhat easily, but proper feature engineering and assessment take time, and there are frequent errors to steer clear of (Sperschneider 2020). Particularly in the post-genome age, ML has been proven to be vital for the discipline of bioinformatics as well (Larranaga et al. 2006). As high-throughput technologies have grown, sequencing data are coming from a variety of fields, including natural sciences, ecology, and pharmacology (Baldi and Brunak 2001). Large data sets make it difficult to organize, handle, interpret, analyze, and visualize, necessitating the use of gadgets, statistics, and algorithms (Ma et al. 2014). Certain machine-learning techniques have demonstrated efficacy in addressing particular molecular-level bioinformatics issues, such as binding sites (Krivák and Hoksza 2018), protein alignment (Mrozek et al. 2009), transcription start sites (Won et al. 2008), subcellular localization prediction (Garg et al. 2020), and splice sites (Pashaei et al. 2017), as well as secondary structure (Yoo et al. 2008) and protein-coding region identification (Salzberg 1995). To clarify the intricate interactions in biological systems, ML is used through the systems biology level to combine many -omics, including proteomics, metabolomics, interactomics, transcriptomics, and genomics (Camacho et al. 2018). Additionally, from uses of ML in plant disease and stress identification concerning disease or stress monitoring employing remote sensing data and drone-based imaging at the field scale to the prediction of pathogen effectors and plant resistance proteins at the genome scale have been studied in detail and complied (Singh et al. 2016; Sperschneider et al. 2020). In order to effectively evaluate high-throughput data and comprehend intricate biological systems, ML technologies are needed for PPI, which entail incredibly complex defense and counter-defense processes (Han 2019; Wang et al. 2019). This chapter outlines ML research on PPI, as well as their potential uses and applications in networks and sequence feature categorization.

7.2 Basics of PPI

Plants attempt to respond to various combinations of abiotic and biotic stress in order to thrive in periodically fairly harsh circumstances (Becklin et al. 2016). Bacteria, viruses, oomycetes, fungi, nematodes, and mycoplasma are among the most researched pathogenic microorganisms that change host physiological state as a result of pathogenesis under biotic stressors (Afzal et al. 2016; Kamoun et al. 2015; Lopez et al. 2015). Furthermore, a variety of molecular complexes that possess the ability to alter the prospective result of the chemical reactions have been gained by pathogens and host plants through competitive evolution (Abdullah et al. 2017). Moreover, crops' reactions to pathogens are also influenced by several abiotic environmental conditions, such as light and temperature, which in turn modulate the PPI (Huot et al. 2017). Hence, interactions between plants and pathogens are complex, which produces complex networks that include molecular mediators from both the hosts and the pathogens (Mukhtar et al. 2016; Abdullah et al. 2017). Pathogens use a variety of tactics to get past the plant defense system in order to effectively spread through plants. Known as a pathogen (microbe)-associated molecular patterns (PAMPs or MAMPs), some conserved, identifiable molecular fingerprints of pathogens, including bacterial flagellin, lipopolysaccharide, and peptidoglycan, can function as elicitors to initiate initial basal resistance (Grennan 2006). Secreting effector proteins inside host cells to impede plant defense responses is another efficient virulence tactic used by pathogens (Petre and Kamoun 2014). In order to identify pathogens and shield themselves against more infections, plants have developed two levels of protective mechanisms. Pattern-triggered immunity (PTI) is the initial tier of the plant immune response. To identify PAMPs or MAMPs, such as flagellin and EF-Tu, plant peripheral pattern recognition receptors (PRRs), which include receptor-like kinases (RLKs) and receptor-like proteins (RLPs), include a variety of ligand-binding ectodomains (Boller and Felix 2009). Known as effector-triggered immunity (ETI), the second line of defense is started by plant disease resistance (R) proteins that are used to identify effector molecules that enter the cytoplasm of plant cells during infection. The most typical class of R proteins is the nucleotide-binding site–leucine-rich repeat (NBS-LRR) class (Marone et al. 2013). Numerous intracellular host defense mechanisms, including ubiquitination systems, vesicle trafficking systems, hormone signaling, and MAPK signaling, may be hampered by the elicitation of MAMPs and effectors (Asai and Shirasu 2015). Numerous cellular reactions, including ion flow, the generation of reactive oxygen species, post-translational changes, and transcriptional reprogramming, can be triggered by an experience of invading at the cell surface (Mrozek et al. 2009). Together, these processes regulate a wide range of chemical and physical defenses that help plants defend themselves. To start a plant defensive reaction, intracellular NLR domain—containing proteins—detect non-self and modified-self ligands (Jones 2006; Jubic et al. 2019). Although unique immunogenic ligands from various sources are detected by plant immune receptors, it is unclear how plants combine several signals to produce particular immune responses (Tsuda and Somssich 2015). The degree how which immune responses are adapted to particular ligands that provide a customized defense response against an intruder is an unresolved subject (Tsuda et al. 2009). There exists both overlapping and heterogeneity to distinct immunogenic signals, according to the study of plant transcriptional reactions to various pathogens and ligands. The knowledge of the ultimate progression of disease consequences is quite restricted (Zhang et al. 2017), which may be thought of as several highways going to an identical destination, despite the various transcriptional reactions to the genetically heterogeneous pathogen isolates. Interconnected immune subnetworks may mediate such an immune reaction to a variety of inputs, giving a variety of microbial interactors resilience (Kim et al. 2014; Hilmer et al. 2017). Immune communication may direct replies from a variety of inputs into outputs that essentially coincide (Tsuda et al. 2009; Bjornson et al. 2021); however, it is evident that the triggering of distinct PRRs does not produce similar transcriptional outputs (Li et al. 2016). According to the newly reviewed emergent systems biology perspective of plant immunity, it is provided by a highly linked network of hormone cross-talk, intracellular signaling, receptor detection, and cellular outputs (Tsuda and Somssich 2015; Katagiri 2018; Delplace et al. 2022).

Therefore, the pathogen-activated plant defense signaling mechanism is a highly linked network, and in order to better understand the behavior of this intricate signaling network, ML algorithms and multi-omics data must be integrated. Interestingly, current transcriptome investigations of pathogens and plants show that the transcriptional regulatory networks of pathogens are shaped by host vulnerability or immune-mediated results (Lovelace et al. 2018; Wolf et al. 2018). According to Nobori et al. (2018), a conceptual framework is being set out to achieve this goal, emphasizing the cellular connections of networks between plants and pathogens. Thus, ML and systems biology are very relevant to multilayer network modeling, data analytics, and multi-omics data integration related to plant–pathogen interplay in order to forecast the result of a certain situation.

7.3 Basics of ML and Its Integration into Biological Systems

Researchers have contributed to the fast collaborative development of sequencing technology and sequence analysis tools since the first entire human genome was published in the early 21st century (Gauthier et al. 2019). The primary shift began in 2008, marking the start of the "Big Data" era when data generated by sophisticated sequencing technologies began to grow rapidly and manufacturing prices began to drop (Gauthier et al. 2019; Pal et al. 2020). Additionally, the computational analysis needed to be improved at this time. Despite the fact that bioinformatics has been around for over 50 years, we may contribute to its growth by virtue of the advent of next-generation sequencing (Gauthier et al. 2019). However, the computational analysis needed to be improved at this time. In order to handle sequence compilation and assessment, computational biology and bioinformatics are interdisciplinary fields that incorporate concepts from mathematics, statistics, and physics into specialized tools and automated analysis pipelines (Eddy 2008; Robinson et al. 2011; Albanese et al. 2015; Nielsen 2017; Kolmogorov et al. 2019; Kolmogorov et al. 2020; Patel et al. 2021). All digital techniques used to examine DNA, RNA, or peptide sequences and learn about their characteristics are referred to as biological sequence analysis (Prjibelski et al. 2018). A strong and widespread bioinformatic tool for characterizing biological sequences, alignment of biological sequences has contributed to various bioinformatics fields, including genome compilation and comparison, gene prediction and annotation, phylogenetic relationships, the study of conserved protein motifs, or protein structure prediction by sequence homology and species classification (Patel et al. 2021). In addition to sequence analysis, bioinformatics is used to identify gene expression and potential biological entity functions, supporting the analysis and categorization of biological pathways meant to address unresolved issues such as the formation of host defense responses, host–pathogen interactions, and species' modification to harsh environments. In fact, researchers are helping to bring about an upsurge in the fields of biology and bioinformatics, where ML and deep learning models have become increasingly popular. These are artificial intelligence domains that can carry out cognitive activities on their own (Choi et al. 2020).

ML is based on the idea that an algorithm or computer system may learn from experience to tackle a new issue (Jordan and Mitchell 2015; LeCun et al. 2015). The ability to interpret a collection of attributes that provide an algorithm-understandable description of the data is the foundation of problem-solving abilities. There are still a lot of issues to be resolved even if bioinformatics and ML have proven to be effective in studying vast amounts of biological data. Each study requires a different type of prior processing phase that helps minimize the data size for faster computing analysis and appropriately handles data extraction and comparison in order to produce informative results for the community. Raw data is unusable immediately as they are collected (Bhaya 2017; Han and Kamber 2006; Arnold et al. 2009). In actuality, noise, missing numbers, and inconsistent results from several studies are major threats to raw data (Bhaya 2017). The latter is particularly affecting experimental designs that integrate several "omics" in order to achieve knowledge mining. Bulk impacts and experiment-dependent technical differences frequently impact omics and multi-omics data, which can have a detrimental impact on the uniformity of the data processing process (Han and Li 2022). The intricate nature of biological data which corresponds to a high multidimensional space of depiction (curse of dimensionality), storage-intensive usage (curse of modularity), and generalization issues caused by the algorithm's inability to acquire variation in data and prejudices for projected future are some of the most prevalent obstacles that ML applications especially encounter in addition to data filtering and variability (L'Heureux et al. 2017).

An algorithm that can learn from data by gaining experience is called an ML model. "A computer program is said to learn from experience E with respect to some class of tasks T and performance measure P, if its performance at tasks in T, as measured by P, improves with experience E," Mitchell (1997) stated in his excellent explanation of the learning model. Four main categories of models can be used: (i) supervised learning, which is typically used for classification tasks; (ii) unsupervised learning, which is primarily used to find trends within the information in which known classes are not available; (iii) semi-supervised learning, which is utilized when only a portion of the data is suitable for supervised learning; and (iv) reinforcement learning, which is employed when the task can be completed in multiple accurate ways, the pursuit of which provides the algorithm with a "reward" (Choi et al. 2020).

7.3.1 Supervised Learning

The ML approach is instructed to link a collection of input data, referred to as the training set (70–80% of the total data), to the identified equivalent output. The system is then applied to unidentified information, known as the test set (~30–20%), with the goal of predicting or categorizing them according to the process of learning (Cunningham et al. 2008). In the training set, $Y = f(x)$, the algorithm must derive a function, f, that may describe the relationship between the input and its output label (Cunningham et al. 2008; James et al. 2023a; Choi et al. 2020). When dealing with issues like regression, categorization, or ranking, this method is most frequently employed (Mohri et al. 2018).

7.3.2 Unsupervised Learning

The understanding of categories for training raw data is currently not a prerequisite for unsupervised learning algorithms, in contrast to supervised learning. The algorithm may naturally discover the associations in the data through unsupervised learning (James et al. 2023b; Alloghani et al. 2020). Instead of generating estimations, the primary goal in this situation is to find commonalities among the assumed characteristics that describe the data (Hastie et al. 2001; James et al. 2023c). Because of this, this type of learning is typically employed for picking features, reducing dimensionality, or clustering (Mahesh 2020); the last one is occasionally utilized prior to the employment of supervised learning. Depending on the type of input data, several distance, similarity, and dissimilarity metrics can be employed to achieve these goals (Xu and Tian 2015). However, by appropriately translating the high-dimensional data into a more manageable and compact domain, reduced dimensionality is employed to make the learning work simpler. By emphasizing the retention of distinct characteristics that still effectively characterize the data while avoiding feature redundancies, unsupervised learning may also be utilized to save the computing time required for feature selection (Mitra et al. 2002).

7.3.3 Random Forests

For a more reliable result, the supervised learning method Random Forest incorporates decision tree predictors (Breiman 2001). A decision tree is a linked acyclic network with a single root beginning node and a single path for every pair of nodes. With each node representing a feature and each branch (or edge) branching from the node representing a value that the feature can take, decision trees actually generate judgments according to an array of precomputed attributes defining the data being studied. Therefore, the choice is formed by observing the tree's splitting at the root node until it reaches the path's last node, which is the decision tree's leaf (Saraswat 2021).

7.3.4 Naive Bayes

The Bayes rule and the presumption of conditional autonomy across the characteristics under consideration serve as the foundation for learning in naive Bayes (NB) classifiers. Even in cases when this independence assumption is broken, NBs can nevertheless achieve comparable accuracy and strong results in a classification setting (Zhang 2004). Given the prior understanding of circumstances that may be associated with the event (the ensemble of features that characterize the input data), the Bayes rule is used to predict the subsequent likelihood of the occurrence (the corresponding class of input data). Given their low variability, computational effectiveness, and resilience to distortion and data shortages, NB classifiers are often used in practice (Sammut and Webb 2011). On the basis of the total feature standards, Gaussian Naive Bayes is a straightforward classifier that optimizes the probability that a sample (such as a protein) belongs to one class or the other (Bishop and Nasrabadi 2006). Multinomial Naive Bayes is typically used for text classification, but it may also be employed to anticipate, given the designated feature collection and the assumption of feature independence, the conditional likelihood that a sample (such as a protein) would qualify for a particular class (Sammut and Webb 2011).

7.3.5 Neural Networks

Deep learning algorithms called neural networks are meant to mimic the human brain's neural network. Synapses between neurons are how the network develops, and the information is reinforced by giving different interactions throughout the network with varying weights (Kriegeskorte and Golan 2019). Neural networks typically consist of multiple successive layers with varying numbers of nodes. Each node is controlled by an activation function that modifies the way knowledge spreads throughout the network and affects the connection weights (Kriegeskorte and Golan 2019; Choi et al. 2020). Depending on the neural network's objective, activation functions are usually linear, sigmoid, or similar basic functions (Choi et al. 2020). The origin nodes in the input layer of multi-layer feed-forward neural networks are depicted by a vector of input data. Each subsequent layer, also known as a hidden layer, is set up to receive the signal from the preceding layer, analyze it, and then forward it to the subsequent layer until it reaches the output layer. In biology, neural networks are frequently employed for clustering tasks such as gene expression research, biomedical picture pattern recognition, or identifying the optimal abiotic component combination for plant development (Samborska et al. 2014; Karim et al. 2021). Furthermore, autoencoders (AE) can reduce complexity using multi-layer neural networks (Hinton and Salakhutdinov 2006).

7.3.6 Variational Autoencoders

An encoder, a decoder made up of mirroring multi-layer neural networks, and a feature reduction space in the center known as the latent space comprise autoencoders, which are deep learning algorithms (Pinaya et al. 2020; Bank et al. 2023).

The original data is mapped using the information space to the latent space by the encoder, and the initial information is reconstructed from the latent space by the decoder. One specific kind of autoencoder that uses a changed encoder and latent space for an improved data space recovery is designated as a Variational AutoEncoder (VAE) (Kingma and Welling 2022). For each data point, the encoder generates a sample size across the range of potential latent feature values, from which they may be further created (Kingma and Welling 2022; Bank et al. 2023). After the estimate of the lower limit of the greatest probability of the input data has been properly established and reparametrized, a subsection of the latent space is selected based on the latent distribution (Kingma and Welling 2022).

The assignment that an individual has to perform has a significant impact on how well ML models are evaluated (Choi et al. 2020). While the most common performance metric for regression tasks is the generalization error, which serves as a gauge of the model's suitability for applying to unknown information, in supervised learning, we can typically ascertain the amount that the model has gained knowledge and how effectively it relates the trained task to the test set. For instance, in classification tasks, the model is assessed based on the examples that are correctly or incorrectly classified (Suthaharan 2016). A confusion matrix serves as the foundation for performance evaluation in categorization (Somogyi 2021). Confusion matrix is a straightforward table that lists the algorithm's accurate and inaccurate output predictions together with the actual class memberships of each sample in the input dataset. As a result, the elements of each of the outcome measures—True Positive, False Positive, True Negative, and False Negative example quantities—can be easily calculated using this table. Precision is an extremely common and generic performance metric that can be obtained from the confusion matrix. Simply expressed, reliability is the number of right categorizations the model makes over all of the input data (Choi et al. 2020; Somogyi 2021). In contrast, the Mean Squared Error (MSE) and R^2 are the two straightforward metrics most frequently employed in regression to assess the model. MSE calculates the averaged squared difference between the latest estimated value and the initial value in the input data (Choi et al. 2020; Goodfellow et al. 2016), whereas R^2 (coefficient of determination) calculates the variance of the input data that the regression model can account for (James et al. 2023d).

The model assessment provides quantitative data on our model's capacity for abstraction, but it doesn't reveal how the model makes judgments, particularly which attributes are the most important. Stated differently, we are unable to derive qualitative insights into the trained model's decision-making processes from model results. This knowledge gap is caused by the model's clarity, which indicates how well a human can understand the processes that contribute to the model's conclusions (Chazette et al. 2021). Both unsupervised and supervised learning models can benefit from model comprehensibility but unsupervised models—where input data is entered and outputs are retrieved without a clear understanding of the intermediate process—need it most (Choi et al. 2020). Moreover, Researchers may frequently run into the following typical mistakes and misnomers when they start working with ML (Sperschneider 2020).

- Training a model to correctly identify unknown data is the main aim of ML. Considering a straightforward image categorization problem, such as determining whether a plant leaf is unhealthy from pictures. The classifier will make use of a collection of validation images that ought to be representative of the population, positive examples (pictures of sick leaves), negative examples (pictures of healthy leaves), and particular characteristics (such as the presence of chlorosis indicated by the ratio of yellow to green pixels). Positive examples are rare in biology and are frequently required for full instruction. The k-fold cross-validation is a popular method of performance evaluation in the absence of a distinct validation set (Wong 2015). The training data are divided into k-equal-sized sets for k-fold cross-validation. The one holdout set is used to test the classifier after it has been trained on k-1 datasets. After k repetitions of this process, an overall performance is provided.
- Training a model to only recall the training instances is the most frequent mistake; this is known as "overfitting" (Domingos 2012). In essence, an overfitted model is making projections at random for data that has not been observed, but having exceptional precision on the training set. It learns every nuance and noise in the training data, fitting the data excessively well. Overfitting can happen for a number of reasons, such as fitting an overly complicated function or retraining a model for an extended period of time. Without an independent validation set, overfitting is a major problem that is difficult to identify. If the precision of a prognosis looks too excellent to be true, one should always exercise caution.
- Since one will probably spend just 5% of their time training a model and 95% of their time identifying relevant characteristics and creating representations of them, it is imperative that they have an in-depth knowledge of the subject matter in which they are supposed to execute. In this context, Sperschneider (2020) has examined a suitable case. Considering the challenge of determining if a pathogen protein is a released effector that alters the host cell to promote

infection, for instance. Since the N-terminal sequence's amino acid composition will provide the highest signal for discriminating, one might utilize secreted effectors as positive testing data and non-secreted proteins as negative training data. However, a classifier will only detect the signal peptide composition at that point since signal peptides are also present in non-effectors that are released. This shall only become clear with a strong confirmational set if one is not familiar with basic biology. Additionally, while training models, a lack of knowledge of ML might cause serious issues. For instance, certain classifiers, like support vector machines, require appropriate feature scaling, while others, like naive Bayes, are invariant to it (Hsu et al. 2003). Inequitable training sets pose a further problem. When evaluating efficiency on unbalanced datasets, measures like recall, precision, kappa, F-score, or Receiver Operating Characteristic (ROC) curves should be used instead of accuracy (Chawla 2010).

- Furthermore, a large number of omics data are created in pertinent websites and databases, but in practice, most of the information is limited and imperfect; unworthy data would obscure the detection of biological features and compromise the precision of the model, particularly when incorporating omics data to define behavior across the entire system (Joyce and Palsson 2006).

7.4 ML in PPI

Numerous types of network elements that are supportive of cellular functions can produce every type of cell response for an ideal operational result. This encompasses metabolite reaction networks, gene co-expression, noncoding RNA (ncRNA)–DNA interactions, protein–DNA interactions, ncRNA–protein interactions, and protein–protein interactions. The operational modules integrated throughout these networks transform the transmission of data under diverse environmental and cellular settings (Garbutt et al. 2014). Conversely, pathogenic proteins interact with these elements and reorganize the distribution of data for their advantage (Mukhtar et al. 2011; Pajerowska-Mukhtar et al. 2013). Network biology–based investigations are essential for offering thorough and organized knowledge regarding such cross-species relationships (Pritchard and Birch 2014). By examining the many levels within biological systems, such network studies can uncover genotype–phenotype correlations (Ahmed et al. 2018). Similar to any sort of physical, electrical, social, or communication network, biological systems are made up of interacting parts called "nodes." In these systems, nodes are linked together according to their operational or tangible links, which are known as edges or linkages. Network topology can be defined by the kind of connection (e.g., one-to-one, one-to-many, many-to-one, and many-to-many linkages), which determines whether such relationships obey supervised or unsupervised depiction (McCormack et al. 2016). A network's architectural centralities may be categorized into three types according to the topological properties that are employed: (i) neighborhood-based centralities, including LocalRank, ClusterRank, degree H-index and coreness assess the impact and actions of its immediate neighbors. (ii) path-based centralities, including shortest route length, Katz centrality, eccentricity, betweenness centrality, information centrality and closeness centrality determine the gap among both nodes (Gräßler et al. 2012); and (iii) iterative refinement centralities, like a cumulative nomination, LeaderRank, PageRank, stochastic approach for link structure analysis (SALSA), and eigenvector centrality determine the complementary improvement contribution of a given node by figuring out how many neighbors it shares and how much of an impact they have on the network (Lü et al. 2016). The key nodes in plant–microbe interactions have been identified using a variety of iterative refinement centralities and structural centralities (Barabási and Oltvai 2004; Lü et al. 2016).

Furthermore, plant diseases alter the way that solar radiation reacts with the canopy, leaves, and energy balance of the plant; these modifications may be measured by photography and in-situ spectroscopy. These instruments, sometimes referred to as "hyperspectral imaging," may identify biotic stress in natural ecosystems swiftly, and non-destructively, and are scalable (Mahlein 2016; Fallon et al. 2020). Given the connection underlying light absorption at particular wavelengths and the biochemical, physical, structural, and architectural characteristics of plants, spectral information can be used to make conclusions regarding the condition of vegetation. The definition of "foliar active traits" has become an integrating framework in agricultural remote sensing in order to comprehend both natural heterogeneity in plant activity and fluctuations in stress responses (DuBois et al. 2018). Image spectroscopy can precisely measure and quantify a number of characteristics that have been demonstrated to be highly correlated with alterations in the activity of plants (Townsend et al. 2003; Wright et al. 2004). This method has been successfully applied to evolutionary biology (Cavender-Bares et al. 2016; Sapes et al. 2022) and biology of conservation (Asner et al. 2016), including plant metabolic functionality (Couture et al. 2016). The emerging field of plant disease sensing is the result of plant pathologists' recent efforts to capitalize on the "spectranomics" track established by ecologists (Sapes et al. 2021). Numerous detectable plant characteristics are impacted by both parasitic (Mahlein et al. 2019) and advantageous (Sousa et al. 2021) plant–microbe associations. Plant pathogens injure, impede, and/or modify plant function, therefore changing foliar structure, by such processes as generation of pathogen infection structures or secondary metabolites or effector proteins (Agrios 2009). Since the 1980s, late-stage plant disease has been detected using broadband and multispectral techniques that depend on near-infrared (NIR) reflectance indices and visible (Vis), such as the normalized difference vegetation index (NDVI) (Nagarajan et al. 1984). Given that their responsiveness to a variety of foliar properties (Curran 1989), such as water (Gao 1996), nutrient content (Gillon et al. 1999; Wang et al. 2020), physiology (Serbin et al. 2019), photosynthetic capacity (Oren et al. 1986), secondary metabolites (Couture et al. 2013), and phenolics (Kokaly and Skidmore 2015), all of which have been affected by early-stage disease, modifications to persistent, shortwave infrared (SWIR) wavelengths have proven useful for plant disease detecting.

Proximal and distant spectroscopy (Mahlein 2016) may be used to quantify the sum of both indirect and direct alterations that microbes confer on plant health through a statistical technique that includes deep learning hyperspectral data processing and nonparametric statistics (Jacquemoud et al. 1995). The ability to recognize plant diseases in the beginning stages may greatly assist sustainability. Satellite imagery powered plant disease detecting provides tracking at earlier unattainable dimensions, completing discrepancies in time and space among laborious field inspections whereas cutting the risk associated with subsequent evaluations and organizational assessments. The majority of detection techniques, from molecular assays to smartphone apps, demand engaged sampling since a human need to first check outdoors to track disease conditions. Furthermore, diseases in general may propagate for weeks or months before symptoms grow apparent, thereby by the point that disease is identified, it has probably accomplished substantial harm and it is frequently far too late to carry out rehabilitation. Disease monitoring can assist in identifying the molecular mechanisms behind disease patterns and trends at various sizes. By focusing on leaf-level functions, in-situ and proximal examination spectroscopy expands the field of agricultural systems biology research and helps assess theories regarding the functioning of plants in reaction to biotic stress. It also pushes the limits of what remote sensing may eventually be able to do. For instance, it is now well accepted that spectral systems may be used to detect diseases both before and after symptoms appear (Rumpf et al. 2010; Conard et al. 2020; Gold et al. 2020). Nevertheless, it is unclear how widespread or distinct these bio-spectral traits may be throughout disease kinds and periods, in addition to the fundamental pathosystem mechanisms that produce different spectroscopic patterns across young and sick plants. Around the globe, upcoming satellite networks like NASA's Surface Biology and Geology (SBG) (Cawse-Nicholson et al. 2021), ESA's Copernicus Hyperspectral Imaging Mission for the Environment (CHIME) (Nieke and Rast 2018), and Planet Lab's CarbonMapper (Anonymous 2021) will transform the accessibility of optical spectroscopic data. These devices will offer superior clarity (30 m) full-spectrum (400–2500 nm) hyperspectral imaging worldwide. When combined, these tools will deliver information at weekly or greater intervals and will have, perhaps the initial time, make such potent informational assets accessible to both high- and low-income nations (Gold 2021).

7.4.1 Integration of ML in PPI: Molecular Level

Large volumes of genomic, transcriptomic, and other omics data have been gathered since the development of high-throughput technologies, enabling computational simulation to explore biological processes throughout the metabolic and regulatory scales. Proteins are comprehensive elements that engage in multiple biological functions and affect immune systems in plants, even though plant immunity comprises metabolic, genetic, and regulatory processes in the cell. Owing mostly to the abundance of data from high-throughput sequencing, ML is becoming more and more important in the field of genomics knowledge discovery. ML has been effectively used to the characterization and projection of gene expression, as well as to the identification of promoters, enhancers, splicing, transcription factors, or RNA-binding proteins in regulatory genomics, epigenetics, proteomics, structural genomics, metabolomics, and genome annotation (Sperschneider 2020). Plant or pathogen genomics has seen relatively few uses of ML thus far, with the majority of these advancements concentrating on medical genomics (Ma et al. 2014). The scarcity of authentic data is a prevalent problem that results in a class disparity and a large number of poorly labeled samples. Additionally, high-dimensional genomic data with little information might exist. Such as in the case of transcriptome sequencing data, which usually contain knowledge about several thousand of genes despite an extremely limited data set. Precise selection of features is necessary to minimize the dimensions of the feature space since small sample numbers inherently increase the danger of overfitting.

Up till now, the primary application areas of ML in PPI consisted of the identification of pathogen effector proteins, the forecasting of gene regulatory networks, and genome-wide characterization for disease resistance. Gene regulatory networks may be deduced and genes implicated in PPI can be identified when comprehensive transcriptome datasets are accessible (Mochida et al. 2018). Genomic selection models are employed to pick candidates for disease resistance by predicting untested traits using marker and phenotypic data from experimental samples (Poland and Rutkoski 2016; Crossa et al. 2017). Once more, overfitting is a problem with marker data, as there are usually few phenotypic observations. The combined impact of genes for disease resistance, varied characteristics, epistasis, and the numerical categories commonly used to score disease severity make ML models ideal for genomic selection for disease resistance (Gonzalez-Camacho et al. 2018). Hyperspectral imaging and genomic selection are two examples of the extreme approaches that ML may integrate (Crossa et al. 2017). Furthermore, bacterial and fungal effectors with relevant nucleotide similarities or motifs have been predicted using ML. A defined section of the protein, including the N-terminal amino acids or the global characteristic of the amino acid sequence for bacterial and fungal effectors, correspondingly, can be used to represent an effector in a feature vector. Its purpose is to address overfitting error. To be more specific, "effectors" might be used as the positive set and "toxins" as the negative set while training the data. Because these enzymes are often distinct from effectors, this will result in a simple classifier with remarkable performance statistics. As a result, the training data need to ideally represent the population's actual distribution. The low base rate of effectors in a particular genome is another problem.

The majority of fungal pathogen effectors lack unified sequence motifs and conventional invasion systems, in contrast to the sophisticated secretion apparatus seen in bacterial pathogens. In order to anticipate fungal effectors from secretomes, Sperschneider et al. (2016) developed a prototype ML tool, EffectorP 1.0, built around NB, after evaluating many machines learning classifiers, including logistic regression (LR), NB, RF, and C4.5 decision tree. EffectorP1.0 attained a specificity and sensitivity of more than 80% and was based on a number of characteristics seen in known effectors, including cysteine concentration and protein length. Sperschneider et al. (2018a) increased the pool of training data and developed EffectorP 2.0 utilizing an EL technique in order to increase prediction accuracy. With an accuracy of 89%, EffectorP 2.0 outperformed comparable small-size algorithms with an accuracy of 59.8% and EffectorP 1.0 with an accuracy of 82%. Furthermore, a lot of fungal effectors seem to be later transferred inside the host cytoplasm after being transported via the apoplast (De Jonge et al. 2011). While cytoplasmic effectors penetrate plant cells to target various host cellular compartments, apoplastic effectors carry out their tasks in the plant's apoplastic space (Rafiqi et al. 2012). Apoplastic and intracellular localization predictions are pioneered by ApoplastP and LOCALIZER, correspondingly, using RF and SVM approaches (Sperschneider et al. 2017; Sperschneider et al. 2018b).

CC-NBS-LRR (CNLs), TIR-NBS-LRR (TNLs), NBS-LRR (NL), TIRNBS (TN), RLP, and RLK are among the common protein architectures found in R proteins (Sanseverino and Ercolano 2012). Most R genes form NBS-LRR proteins that have a leucine-rich repeat (LRR) part and a nucleotide-binding site (NBS). These proteins are categorized into two subclasses according to their domains: CC-NBS-LRR (CNL) are proteins with a coiled coil (CC) domain and TIR-NBS-LRR (TNL) are proteins with an N-terminal domain that is homologous to the Toll/interleukin-1 receptor (TIR) domain (Takken et al. 2006). To increase crop resistance to plant diseases, it is essential to discover the genes that provide resistance to these diseases. Two ML-based pipelines for SVM-based NBS-LRR protein prediction constructed are NBSPred and DRPPP (Kushwaha et al. 2016; Pal et al. 2016).

Different kinds of secretion systems (SS), including type I (T1SS), type III (T3SS), type IV (T4SS), and type VI (T6SS), regulate the mechanisms that move effector proteins through the cytoplasm into adjacent living cells in Gram-negative bacteria (Barret et al. 2013). T3SS have been linked to bacterial pathogenicity (Hueck 1998) and are implicated in flagellar biosynthesis (Dieplod and Armitage 2015). Certain ML techniques have been effectively used to predict T3SS effectors (T3SEs) based on their particular characteristics. SVM has been used extensively to detect T3SEs employing the effective technique for regression analysis and categorization. SVM may create a hyperplane with the maximum interval separation to the closest training data set of any category if it is assumed that a collection of training instances can be divided into two groups, such as "T3SEs" and "non-T3SEs." The classifier's generalization error decreases with increasing distance (Rumpf et al. 2010). In a 5-fold cross-validation examination, the BPBAac SVM classifier, which uses N-terminal sequence characteristics to identify T3SEs, obtained average sensitivity, selectivity, accuracy, AUC, and MCC values of 90.97%, 97.42%, 95.27%, 98.88%, and 0.89 (Wang et al. 2011). Using SVM, BEAN 2.0 is an unified web resource that can predict T3SE with 86.05% sensitivity, 100% specificity, and 95.35% accuracy (Dong et al. 2015). A novel predictor for ANN-based T3SE identification was introduced by Löwer and Schneider (2009). An MCC value of 0.63 was obtained from the mean cross-validation efficiency (Wang 2003). In order to identify T3SEs, Arnold et al. (2009) created the classifier EffectiveT3, which relies on the NB method. The Bayes theorem, on which the NB program is based, determines the likelihood of an effector protein according to the previous probability of circumstances (Sen et al. 2019). EffectiveT3 could detect T3SEs having a specificity of 71% and an accuracy of 85% after identifying the most significant characteristics from the N-terminal sequence's secondary structure characteristics, physicochemical characteristics, and amino acid content. Li et al. (2020)

and Wang et al. (2019) created EP3 and Bastion3 using the ensemble learning (EL) model in order to enhance precision. To improve predictive performance, EL, a meta-algorithm, integrates many ML approaches into a prediction model (Polikar 2012; Wan and Yang 2013). Additional bacterial conjugation machinery that facilitates the transfer of protein or DNA substrates to target cells include T6SS and T4SS (Christie et al. 2005). According to reports, a number of ML tools, including PredT4SE-Stack (Xiong et al. 2018), T4SEpre (Wang et al. 2014), PyPredT6 (Sen et al. 2019), and Bastion6 (Wang et al. 2018), may be utilized for the estimation of T4SEs and T6SEs. By including the PSSM structure feature into their stacked ensemble model, Xiong et al. (2018) created PredT4SE-Stack, which produced precision rates of 94.5% and 94.4% for SVM-based and LR-based metaclassifiers, respectively. Utilizing sequential, developmental, and physical parameters, Bastion6 is a two-layer SVM-based ensemble prediction model that has demonstrated outstanding results in *F*-value (0.946), overall accuracy (ACC: 0.943), area under the curve (AUC: 0.976), and Matthews correlation coefficient (MCC: 0.892) (Wang et al. 2018).

Furthermore, deep learning requires more training data than conventional methods, and it permits a limitless number of computational layers to extract characteristics from input data. Greater layers encode more complicated information, whereas lower levels abstract simpler ones (LeCun et al. 2015). As a result, when the model's complexity rises, so will the computing complexity. Nevertheless, deep learning offers outstanding effector prediction capability and has surmounted additional algorithmic constraints (Min et al. 2017; Lan et al. 2018). DeepT3 uses deep convolutional neural networks (CNNs) to classify protein sequences into T3SEs or non-T3SEs. Its mean precision over the specimen set was 92.6% (Xue et al. 2019). The other instance is ACNNT3, which predicts bacterial T3SEs by combining one-hot encoding with PSSM of 100 residues in the N-terminal domain as the characteristic input. The method outperforms the earlier techniques, such as Effective T3, DeepT3, BEAN2, and BPBAac with precision and MCC values of 96.7% and 0.92, respectively, employing a standard independent dataset (Li et al. 2020). Table 7.1 enlists the ML classifiers for R and effector proteins.

7.4.2 Integration of ML in PPI: Field Level

Novel technologies have been described as novel phenotyping techniques that have the ability to get around this bottleneck, such as optical and noninvasive sensors. Numerous optical imaging techniques that employ sensors such as 3D (Paulus et al. 2014), RGB (Fiorani and Schurr 2013), fluorescence (Chaerle et al. 2007), HSI (Ustin and Gamon 2010) and thermography

Table 7.1 Machine learning classifiers for R and effector proteins.

Classifier Model	Classifier	Application	References
NB	EffectiveT3	Bacterial effector predictor	Arnold et al. (2009)
	EffectorP V1.0	Fungal effector predictor	Sperschneider et al. (2016)
ANN	Lower and Schneider	Bacterial effector predictor	Löwer and Schneider (2009)
SVM	BPBAac	Bacterial effector predictor	Wang et al. (2011)
	BEAN 2.0	Bacterial effector predictor	Dong et al. (2015)
	LOCALIZER	Localization predictor	Sperschneider et al. (2018a)
	NBSPred	R protein predictor	Kushwaha et al. (2016)
	DRPPP	R protein predictor	Pal et al. (2016)
	T4SEpre	Bacterial effector predictor	Wang et al. (2014)
RF	ApoplastP	Localization predictor	Sperschneider et al. (2018b)
	T3SPs	Bacterial effector predictor	Yang et al. (2013)
EL	EffectorP V2.0	Fungal effector predictor	Sperschneider et al. (2018a)
	Bastion3	Bacterial effector predictor	Wang et al. (2019)
	EP3	Bacterial effector predictor	Li et al. (2020)
	Bastion6		Wang et al. (2018)
	PredT4SE-Stack		Xiong et al. (2018)
	PyPredT6		Sen et al. (2019)

(Oerke et al. 2006) may assess multiple plant characteristics and may be used in programmed, high-throughput phenotyping workflows. Hyperspectral imaging (HSI) represents one among many effective methods across those for evaluating active plant characteristics (Ustin and Gamon 2010). The spectral properties of plants may be noninvasively seen on various scales and across time using HSI. High spectral and spatial accuracy are made possible by sensors' sensitivity, and narrow wavelengths may be used to analyze reflectance per pixel. The primary determinants of the spectrum in the visual range (VIS, 400–700 nm) are compounds found in leaves, such as carotenoids and chlorophylls (Curran 1989). The primary factors influencing the near infrared spectrum (NIR, 700–1000 nm) include scattering mechanisms in the spongy mesophyll and absorptions from fatty acids, proteins, starch, and water each of which has a distinct absorption frequency (Kuska et al. 2015).

At the present time, every person is equipped with a digital sophisticated and modern imaging sensor with the smartphones and video cameras, which are the primary sources to acquire the digital imagery data of different plant organs from top to bottom, which if used with RGB sensors is advantageous for monitoring plants in seasons of disease breakouts. RGB-color images have been used to detect the biotic stress upon the diseased plants through red, green, and blue channels (Bock et al. 2008), addition to LAB, YCBCR, and HSV, which investigates lightness, color compression scheme, and hue saturation value, respectively. Several software packages, such as ASSESS 2.0, "Leaf Doctor," Scion Image software, and custom-made modules are available (Bock et al. 2010; Pethybridge and Nelson 2015; Tucker and Chakraborty 1997; Wijekoon et al. 2008), in which disease severity is extracted as diseased pixels or as a percentage after the background is masked from the object of interest. Sharpness, uniform focus, and illumination are vital for high precision and dependable outcomes from automated image analysis. Additionally, under natural conditions, the leaf orientation (imaging angle) and pixel size (distance between the object and the sensor) are added influences on the image quality. Heterogeneous circumstances and poor image quality frequently lead to difficulty in detection and low levels of accuracy.

Complex backgrounds with various real-world conditions, computational time complexity, infection level of disease, detection of multiple infections in the full sample of single or multiple leaves, segmentation sensitivity towards region of interest, life cycle of plant on the basis of infection, and a dearth of images collected and labeled from real-world scenarios in currently available datasets are the main challenges in the field of ML-based diseases image analysis. Spectral sensors are characterized based on their spectral resolution, which includes the number and width of assessed wavebands; based on either they are detected by a nonimaging or an imaging detector; also, on the variation of their spatial scale. With precise precision, deep learning technology can identify the presence of diseases and pests in fields. Deep learning's primary challenge is the requirement for enormous amounts of data. When setting the hyperparameter, there is a danger of overfitting if there is not enough data and there are too many classes, which will cause it to be set to local minima. But as digital technology has developed, a vast quantity of data can now be easily gathered, and utilizing a variety of deep learning algorithms, the best judgements may be made in terms of identifying various plant ailments. The special scale variably depends on the distance between the object and the sensor, which has essential effect on detection of plant–microbe interactions (West et al. 2003). Thus, airborne or far range sensors acquire low spatial ranges, which are suitable to detect patches with soil borne pathogens (Hillnhütter et al. 2011), or crop stand at the end state of diseases (Mahlein et al. 2012; Mewes et al. 2011; Steddom et al. 2005). Multispectral sensors are the first spectral sensors which evaluated spectral information of objects in apparently bigger wavebands. The optical properties of leaves are characterized by light reflected from leaf structure of internal leaf structure; or light transmitted through leaves or absorbed by chemical components, i.e., lignin, water, amino acids, and sugar, which results in complex phenomenon dependent on multiple biophysical and biochemical interactions. Changes in the chemical makeup of the tissue during pathogenesis, such as the succession of chlorotic and necrotic tissue, or the appearance of typical infection structures, such as powdery mildew hyphae and conidia or rust uredospore, can explain changes in reflectance brought on by plant pathogens and plant diseases. Disease pathogens, such as *Cercospora* causing leaf spots, induces degradation of plant leaf tissues due to toxins and enzymes, which induces the necrotic spots (Daub and Ehrenshaft 2000). However, the biotrophic pathogens, such as rusts and powdery mildew, form pustules and layer of spores on plant parts especially on the leaves, which has an impact on the PPI's optical characteristics (Gudkov et al. 2023). The changes in transpiration due to early infections can be assessed by infrared thermography (IRT) (Oerke et al. 2006), by means of change in temperature and plant water and the microclimate in the population (Lenthe et al. 2007). IRT can be useful at different temporal and spatial scales from airborne to small-scale applications, although it is affected by the variable climatic conditions. The mean difference in temperature of plants, single leaf, and the crop stands happens to be a useful indicator for selection of infection. The changes in transpiration rate can be variable in the plants from pathogens being leaf attacking, such as foliar pathogens of leaf spots and rusts; systemic infection of *Fusarium sp.*; or root pathogens *Rhizoctonia solani* or *Phytium spp* (Ray 2024). Table 7.2 enlists some of the plant disease and disorders detected by different researchers.

Table 7.2 List of diseases and disorders detected by different imaging sensors.

S. No.	Detected By	Pathogen/Disorder	References
1	Spectral properties	Foliar pathogens of sugar beet	Mahlein et al. (2010)
		Cercospora leaf spot, rust and powdery mildew	Rumpf et al. (2010)
		Fusarium graminearum in wheat	Bauriegel et al. (2011)
		Venturia inaequalis in apple	Delalieux et al. (2007)
		Phytophthora infestans in tomato	Wang et al. (2008)
		Yellow rust in wheat	Bravo et al. (2003)
		Heterodera schachtii in sugar beet	Hillnhütter et al. (2011)
		Rhizoctonia solani in sugar beet	Hillnhütter et al. (2011)
2	Hyperspectral imaging	Mycotoxin-producing pathogens in maize	Del Fiore et al. (2010)
		Orange rust in sugarcane	Apan et al. (2004)
		Yellow rust in wheat	Huang et al. (2007)
		Canker lesions of citrus fruits	Qin et al. (2009)
		Apple surface defects	Mehl et al. (2004)
		Rot of strawberries	ElMasry et al. (2007)

Thus, light reflectance difference among undamaged compared to that of a stressed leaf, remote sensors can identify plant diseases in the natural environment in real time, even prior visible systems are apparent. This makes it possible for farmers to employ focused, prompt safeguarding techniques. Sensors such as thermography and chlorophyll fluorescence track variations in the temperature of the leaf surface along with the amount of chlorophyll present, respectively. Although such sensors are effective in detecting initial stressful reactions, they are unable to discriminate among various diseases. It is to be taken special care that prior to categorization, the data must be adjusted geometrically and radiometrically (Landgrebe 2003). Major improvements have been made in the sector as a result of declining technological prices and the ability to analyze and store enormous volumes of data with supercomputing. Enormous numbers of photos or more are being employed for the development and evaluation of deep learning models, compared to just handful in earlier research (Ferentinos 2018). Though overfitting to specific lighting settings and chosen stressors in monitored conditions is a significant obstacle to moving identifying plant diseases from the glasshouse to the field, ML, and deep learning in particular, are essential for evaluating images with huge sample numbers (Mahlein 2015). In the wild, where many biotic and abiotic stressors exist concurrently and lighting is considerably more inconsistent, the classifiers need to operate effectively (Barbedo 2018). ML techniques for plant disease tracking shall be strengthened over time by the incorporation of various types of data (sensor data, satellite pictures, weather measurements) and a rise in labeled training data encompassing a variety of ailments.

7.5 Conclusion

Enormous amount of information is being generated and used more often in the field of PPI due to technical advancements and declining costs. This includes the utilization of robots, sensors, and drones is helping to digitize agricultural decision-making. These days, high-throughput sequencing is a standard procedure. Big data collection and the availability of supercomputing resources are driving a greater use of ML to enable scientific estimations at all scales, spanning the field to the molecular level. Thus far, the study of PPI have only just begun to explore this trend. To create reliable models that facilitate discovering information and have real-world applications, nevertheless, one must be knowledgeable about both the biological implementation area and ML theory. For ML-driven advancement in the discipline, specialists with backgrounds in both computer science and biology will be essential. In summary, ML offers a number of methods for identifying plant resistance proteins and pathogen effectors associated with PPI. It may also be utilized to investigate the interactome of plants and pathogens using a network model. The precise understanding of the plant defenses is still lacking, despite the fact that computational ecosystems have revealed several significant roles in pathogen invasion and plant defense mechanisms. ML techniques have a vast amount of room to develop in order to combine multi-omics data and determine the relationships between different signaling molecules. ML theory will advance in bioinformatics by incorporating more high-quality interactome records, and it may operate as a supportive tool for more in-depth analysis of PPI knowledge.

References

Abdullah, A.S., Moffat, C.S., Lopez-Ruiz, F.J. et al. (2017). Host–multi-pathogen warfare: Pathogen interactions in co-infected plants. *Frontiers in Plant Science* 8: 1806.

Afzal, Z., Howton, T.C., Sun, Y. et al. (2016). The roles of aquaporins in plant stress responses. *Journal of Developmental Biology* 4 (1): 9.

Agrios, G.N. (2009). Plant pathogens and disease: General introduction. In: Schaechter, M. (ed.) *Encyclopedia of Microbiology*, 3rd ed. Academic Press: Oxford, UK, 2009; pp. 613–646.

Agrios, G.N. (2005). *Plant Pathology*. Elsevier.

Ahmed, H., Howton, T.C., Sun, Y. et al. (2018). Network biology discovers pathogen contact points in host protein-protein interactomes. *Nature Communications* 9 (1): 2312.

Albanese, D., Fontana, P., De Filippo, C. et al. (2015). MICCA: a complete and accurate software for taxonomic profiling of metagenomic data. *Scientific Reports* 5 (1): 9743.

Alloghani, M., Al-Jumeily, D., Mustafina, J. et al. (2020). A systematic review on supervised and unsupervised machine learning algorithms for data science. In: Berry, M., Mohamed, A., Yap, B. (eds.) *Supervised and Unsupervised Learning for Data Science*, 3–21. Springer, Cham.

Anonymous. (2021). Carbon Mapper launches satellite program to pinpoint methane and carbon dioxide super emitters. https://www.prnewswire.co.uk/news-releases/carbon-mapper-launches-satellite-program-to-pinpoint-methane-and-carbon-dioxide-super-emitters-827980651.html (Accessed 29 April 2025).

Apan, A., Held, A., Phinn, S. et al. (2004). Detecting sugarcane "orange rust" disease using EO-1 hyperion hyperspectral imagery. *International Journal of Remote Sensing* 25 (2): 489–498.

Arnold, R., Brandmaier, S., Kleine, F. et al. (2009). Sequence-based prediction of type III secreted proteins. *PLoS Pathogens* 5 (4): e1000376.

Asai, S., and Shirasu, K. (2015). Plant cells under siege: Plant immune system versus pathogen effectors. *Current Opinion in Plant Biology* 28: 1–8.

Aumentado-Armstrong, T.T., Istrate, B., and Murgita, R.A. (2015). Algorithmic approaches to protein-protein interaction site prediction. *Algorithms for Molecular Biology* 10: 1–21.

Baldi, P., and Brunak, S. (2001). *Bioinformatics: The Machine Learning Approach*. MIT press.

Bank, D., Koenigstein, N., and Giryes, R. (2023). Autoencoders. In: Rokach, L., Maimon, O., Shmueli, E. (eds) *Machine Learning for Data Science Handbook: Data Mining and Knowledge Discovery Handbook*, 353–374. Springer, Cham.

Barabasi, A.L., and Oltvai, Z.N. (2004). Network biology: understanding the cell's functional organization. *Nature Reviews Genetics* 5 (2): 101–113.

Barbedo, J.G.A. (2018). Impact of dataset size and variety on the effectiveness of deep learning and transfer learning for plant disease classification. *Computers and Electronics in Agriculture* 153: 46–53.

Barret, M., Egan, F., and O'Gara, F. (2013). Distribution and diversity of bacterial secretion systems across metagenomic datasets. *Environmental Microbiology Reports* 5 (1): 117–126.

Bauriegel, E., Giebel, A., Geyer, M. et al. (2011). Early detection of Fusarium infection in wheat using hyper-spectral imaging. *Computers and Electronics in Agriculture* 75 (2): 304–312.

Becklin, K.M., Anderson, J.T., Gerhart, L.M. et al. (2016). Examining plant physiological responses to climate change through an evolutionary lens. *Plant Physiology* 172 (2): 635–649.

Bishop, C.M., and Nasrabadi, N.M. (2006). *Pattern Recognition and Machine Learning* (Vol. 4, No. 4, p. 738). Springer.

Bjornson, M., Pimprikar, P., Nürnberger, T. et al. (2021). The transcriptional landscape of Arabidopsis thaliana pattern-triggered immunity. *Nature Plants* 7 (5): 579–586.

Bock, C.H., Parker, P.E., Cook, A.Z. et al. (2008). Visual rating and the use of image analysis for assessing different symptoms of citrus canker on grapefruit leaves. *Plant Disease* 92 (4): 530–541.

Bock, C.H., Poole, G.H., Parker, P.E. et al. (2010). Plant disease severity estimated visually, by digital photography and image analysis, and by hyperspectral imaging. *Critical Reviews in Plant Sciences* 29 (2): 59–107.

Boller, T., and Felix, G. (2009). A renaissance of elicitors: Perception of microbe-associated molecular patterns and danger signals by pattern-recognition receptors. *Annual Review of Plant Biology* 60 (1): 379–406.

Bravo, C., Moshou, D., West, J. et al. (2003). Early disease detection in wheat fields using spectral reflectance. *Biosystems Engineering* 84 (2): 137–145.

Breiman, L. (2001). Random forests. *Machine Learning* 45: 5–32.

Camacho, D.M., Collins, K.M., Powers, R.K. et al. (2018). Next-generation machine learning for biological networks. *Cell* 173 (7): 1581–1592.

Cavender-Bares, J., Meireles, J.E., Couture, J.J. et al. (2016). Associations of leaf spectra with genetic and phylogenetic variation in oaks: Prospects for remote detection of biodiversity. *Remote Sensing* 8 (3): 221.

Cawse-Nicholson, K., Townsend, P.A., Schimel, D. et al. (2021). NASA's surface biology and geology designated observable: A perspective on surface imaging algorithms. *Remote Sensing of Environment* 257: 112349.

Chaerle, L., Leinonen, I., Jones, H.G. et al. (2007). Monitoring and screening plant populations with combined thermal and chlorophyll fluorescence imaging. *Journal of Experimental Botany* 58 (4): 773–784.

Chawla, N.V. (2010). Data mining for imbalanced datasets: An overview. In: Maimon, O., Rokach, L. (eds) *Data Mining and Knowledge Discovery Handbook*, 875–886. Springer, Boston, MA.

Chazette, L., Brunotte, W., and Speith, T. (2021). Exploring explainability: A definition, a model, and a knowledge catalogue. In *2021 IEEE 29th International Requirements Engineering Conference (RE)*, September 20–24, 2021, Norte Dam, USA, (pp. 197–208). IEEE.

Choi, R.Y., Coyner, A.S., Kalpathy-Cramer, J. et al. (2020). Introduction to machine learning, neural networks, and deep learning. *Translational Vision Science & Technology* 9 (2): 14–14.

Christie, P.J., Atmakuri, K., Krishnamoorthy, V. et al. (2005). Biogenesis, architecture, and function of bacterial type IV secretion systems. *Annual Review of Microbiology* 59 (1): 451–485.

Conrad, A.O., Li, W., Lee, D.Y. et al. (2020). Machine learning-based presymptomatic detection of rice sheath blight using spectral profiles. *Plant Phenomics*, 2020: 1–10.

Couture, J.J., Serbin, S.P., and Townsend, P.A. (2013). Spectroscopic sensitivity of real-time, rapidly induced phytochemical change in response to damage. *New Phytologist* 198 (1): 311–319.

Couture, J.J., Singh, A., Rubert-Nason, K.F. et al. (2016). Spectroscopic determination of ecologically relevant plant secondary metabolites. *Methods in Ecology and Evolution* 7 (11): 1402–1412.

Crossa, J., Pérez-Rodríguez, P., Cuevas, J. et al. (2017). Genomic selection in plant breeding: Methods, models, and perspectives. *Trends in Plant Science* 22 (11): 961–975.

Cunningham, P., Cord, M., and Delany, S.J. (2008). Supervised learning. In: Cord, M., Cunningham, P. (eds) *Machine Learning Techniques for Multimedia: Case Studies on Organization and Retrieval*. 21–49. Springer: Berlin, Heidelberg.

Curran, P.J. (1989). Remote sensing of foliar chemistry. *Remote Sensing of Environment* 30 (3): 271–278.

Daub, M.E., and Ehrenshaft, M. (2000). The photoactivated Cercospora toxin cercosporin: Contributions to plant disease and fundamental biology. *Annual Review of Phytopathology* 38 (1): 461–490.

De Jonge, R., Bolton, M.D., and Thomma, B.P. (2011). How filamentous pathogens co-opt plants: The ins and outs of fungal effectors. *Current Opinion in Plant Biology* 14 (4): 400–406.

Del Fiore, A., Reverberi, M., Ricelli, A. et al. (2010). Early detection of toxigenic fungi on maize by hyperspectral imaging analysis. *International Journal of Food Microbiology* 144 (1): 64–71.

Delalieux, S., Van Aardt, J.A.N., Keulemans, W. et al. (2007). Detection of biotic stress (Venturia inaequalis) in apple trees using hyperspectral data: Non-parametric statistical approaches and physiological implications. *European Journal of Agronomy* 27 (1): 130–143.

Delplace, F., Huard-Chauveau, C., Berthomé, R. et al. (2022). Network organization of the plant immune system: From pathogen perception to robust defense induction. *The Plant Journal* 109 (2): 447–470.

Diepold, A., and Armitage, J.P. (2015). Type III secretion systems: The bacterial flagellum and the injectisome. *Philosophical Transactions of the Royal Society B: Biological Sciences* 370 (1679): 20150020.

Dietterich, T.G. (2000). Ensemble methods in machine learning. *International Workshop on Multiple Classifier Systems*, 1–15. Springer: Berlin, Heidelberg.

Domingos, P. (2012). A few useful things to know about machine learning. *Communications of the ACM* 55 (10): 78–87.

Dong, X., Lu, X., and Zhang, Z. (2015). BEAN 2.0: An integrated web resource for the identification and functional analysis of type III secreted effectors. *Database 2015*: bav064.

DuBois, S., Desai, A.R., Singh, A. et al. (2018). Using imaging spectroscopy to detect variation in terrestrial ecosystem productivity across a water-stressed landscape. *Ecological Applications* 28 (5): 1313–1324.

Eddy, S.R. (2008). A probabilistic model of local sequence alignment that simplifies statistical significance estimation. *PLoS Computational Biology* 4 (5): e1000069.

ElMasry, G., Wang, N., ElSayed, A. et al. (2007). Hyperspectral imaging for nondestructive determination of some quality attributes for strawberry. *Journal of Food Engineering* 81 (1): 98–107.

Fallon, B., Yang, A., Lapadat, C. et al. (2020). Spectral differentiation of oak wilt from foliar fungal disease and drought is correlated with physiological changes. *Tree Physiology* 40 (3): 377–390.

Fang, Y., and Ramasamy, R.P. (2015). Current and prospective methods for plant disease detection. *Biosensors* 5 (3): 537–561.

Ferentinos, K.P. (2018). Deep learning models for plant disease detection and diagnosis. *Computers and Electronics in Agriculture* 145: 311–318.

Fiorani, F., and Schurr, U. (2013). Future scenarios for plant phenotyping. *Annual Review of Plant Biology* 64 (1): 267–291.

Gao, B.C. (1996). NDWI: A normalized difference water index for remote sensing of vegetation liquid water from space. *Remote Sensing of Environment* 58 (3): 257–266.

Garbutt, C.C., Bangalore, P.V., Kannar, P. et al. (2014). Getting to the edge: Protein dynamical networks as a new frontier in plant–microbe interactions. *Frontiers in Plant Science* 5: 312.

Garg, A., Singhal, N., Kumar, R. et al. (2020). mRNALoc: A novel machine-learning based in-silico tool to predict mRNA subcellular localization. *Nucleic Acids Research* 48 (W1): W239–W243.

Gauthier, J., Vincent, A.T., Charette, S.J. et al. (2019). A brief history of bioinformatics. *Briefings in Bioinformatics* 20 (6): 1981–1996.

Gillon, D., Houssard, C. and Joffre, R. (1999). Using near-infrared reflectance spectroscopy to predict carbon, nitrogen and phosphorus content in heterogeneous plant material. *Oecologia* 118: 173–182.

Gold, K.M. (2021). Plant disease sensing: Studying plant-pathogen interactions at scale. *Msystems* 6 (6): e01228–21.

Gold, K.M., Townsend, P.A., Chlus, A. et al. (2020). Hyperspectral measurements enable pre-symptomatic detection and differentiation of contrasting physiological effects of late blight and early blight in potato. *Remote Sensing* 12 (2): 286.

González-Camacho, J.M., Ornella, L., Pérez-Rodríguez, P. et al. (2018). Applications of machine learning methods to genomic selection in breeding wheat for rust resistance. *The Plant Genome* 11 (2): 170104.

Goodfellow, I., Bengio, Y., Courville, A. et al. (2016). *Deep Learning* (Vol. 1, No. 2). MIT Press: Cambridge.

Gräßler, J., Koschützki, D. and Schreiber, F. (2012). CentiLib: Comprehensive analysis and exploration of network centralities. *Bioinformatics* 28 (8): 1178–1179.

Grennan, A.K. (2006). Plant response to bacterial pathogens. Overlap between innate and gene-for-gene defense response. *Plant Physiology* 142 (3): 809–811.

Gudkov, S.V., Matveeva, T.A., Sarimov, R.M. et al. (2023). Optical methods for the detection of plant pathogens and diseases. *AgriEngineering* 5 (4): 1789–1812.

Han, G.Z. (2019). Origin and evolution of the plant immune system. *New Phytologist* 222 (1): 70–83.

Han, J., Kamber, M., and Mining, D. (2006). Concepts and techniques. *Morgan Kaufmann* 340 (1): 94104–103205.

Han, W., and Li, L. (2022). Evaluating and minimizing batch effects in metabolomics. *Mass Spectrometry Reviews* 41 (3): 421–442.

Hastie, T., Tibshirani, R., Friedman, J. et al. (2009). Unsupervised learning. In: Hastie, T., Tibshirani, R., Friedman, J. (eds.) *The Elements of Statistical Learning: Data Mining, Inference, and Prediction*, 485–585.

Hillmer, R.A., Tsuda, K., Rallapalli, G. et al. (2017). The highly buffered Arabidopsis immune signaling network conceals the functions of its components. *PLoS Genetics* 13 (5): e1006639.

Hillnhütter, C., Mahlein, A.K., Sikora, R.A. et al. (2011). Remote sensing to detect plant stress induced by Heterodera schachtii and Rhizoctonia solani in sugar beet fields. *Field Crops Research* 122 (1): 70–77.

Hinton, G.E., and Salakhutdinov, R.R. (2006). Reducing the dimensionality of data with neural networks. *Science* 313 (5786): 504–507.

Hsu, C.W., Chang, C.C., and Lin, C.J. (2003). A practical guide to support vector classification; Technical Report; Department of Computer Science and Information Engineering, University of National Taiwan: Taipei, Taiwan, pp. 1396–1400.

Huang, D.S., and Zheng, C.H. (2006). Independent component analysis-based penalized discriminant method for tumor classification using gene expression data. *Bioinformatics* 22 (15): 1855–1862.

Huang, W., Lamb, D.W., Niu, Z. et al. (2007). Identification of yellow rust in wheat using in-situ spectral reflectance measurements and airborne hyperspectral imaging. *Precision Agriculture* 8 (4): 187–197.

Hueck, C.J. (1998). Type III protein secretion systems in bacterial pathogens of animals and plants. *Microbiology and Molecular Biology Reviews* 62 (2): 379–433.

Huot, B., Castroverde, C.D.M., Velásquez, A.C. et al (2017). Dual impact of elevated temperature on plant defence and bacterial virulence in Arabidopsis. *Nature Communications* 8 (1): 1808.

Jacquemoud, S., Verdebout, J., Schmuck, G. et al. (1995). Investigation of leaf biochemistry by statistics. *Remote Sensing of Environment* 54 (3): 180–188.

James, G., Witten, D., Hastie, T. et al. (2023a). Linear regression. In: *An Introduction to Statistical Learning: With Applications in Python*, 69–134. Springer International Publishing: Cham.

James, G., Witten, D., Hastie, T. et al. (2023b). Statistical learning. In: *An Introduction to Statistical Learning: With Applications in Python*, 15–67. Springer International Publishing: Cham.

James, G., Witten, D., Hastie, T. et al. (2023c). Unsupervised learning. In: *An Introduction to Statistical Learning: With Applications in Python*, 503–556. Springer International Publishing: Cham.

Jones, J.D., and Dangl, J.L. (2006). The plant immune system. *Nature* 444 (7117): 323–329.

Jordan, M.I., and Mitchell, T.M. (2015). Machine learning: Trends, perspectives, and prospects. *Science* 349 (6245): 255–260.

Jothi, R., Cherukuri, P.F., and Tasneem, A. (2006). Co-evolutionary analysis of domains in interacting proteins reveals insights into domain–domain interactions mediating protein–protein interactions. *Journal of Molecular Biology* 362 (4): 861–875.

Joyce, A.R., and Palsson, B.Ø. (2006). The model organism as a system: Integrating "omics" data sets. *Nature Reviews Molecular Cell Biology* 7 (3): 198–210.

Jubic, L.M., Saile, S., Furzer, O.J. et al. (2019). Help wanted: Helper NLRs and plant immune responses. *Current Opinion in Plant Biology* 50: 82–94.

Kamoun, S., Furzer, O., Jones, J.D. et al. (2015). The top 10 oomycete pathogens in molecular plant pathology. *Molecular Plant Pathology* 16 (4): 413–434.

Karim, M.R., Beyan, O., Zappa, A. et al. (2021). Deep learning-based clustering approaches for bioinformatics. *Briefings in Bioinformatics* 22 (1): 393–415.

Katagiri, F. (2018). Plant immune signaling from a network perspective. *Plant Science* 276: 14–21.

Kim, Y., Tsuda, K., Igarashi, D. et al. (2014). Mechanisms underlying robustness and tunability in a plant immune signaling network. *Cell Host & Microbe* 15 (1): 84–94.

Kingma, D.P., and Welling, M. (2013). *Auto-encoding Variational Bayes. arXiv preprint arXiv:1312.6114*.

Kokaly, R.F., and Skidmore, A.K. (2015). Plant phenolics and absorption features in vegetation reflectance spectra near 1.66 μm. *International Journal of Applied Earth Observation and Geoinformation* 43: 55–83.

Kolmogorov, M., Bickhart, D.M., Behsaz, B. et al. (2020). metaFlye: Scalable long-read metagenome assembly using repeat graphs. *Nature Methods* 17 (11): 1103–1110.

Kolmogorov, M., Yuan, J., Lin, Y. et al. (2019). Assembly of long, error-prone reads using repeat graphs. *Nature Biotechnology* 37 (5): 540–546.

Kriegeskorte, N., and Golan, T. (2019). Neural network models and deep learning. *Current Biology* 29 (7): R231–R236.

Krivák, R., and Hoksza, D. (2018). P2Rank: Machine learning based tool for rapid and accurate prediction of ligand binding sites from protein structure. *Journal of Cheminformatics* 10: 1–12.

Kushwaha, S.K., Chauhan, P., Hedlund, K. et al. (2016). NBSPred: A support vector machine-based high-throughput pipeline for plant resistance protein NBSLRR prediction. *Bioinformatics* 32 (8): 1223–1225.

Kuska, M., Wahabzada, M., Leucker, M. et al. (2015). Hyperspectral phenotyping on the microscopic scale: towards automated characterization of plant-pathogen interactions. *Plant Methods* 11: 1–15.

L'heureux, A., Grolinger, K., Elyamany, H.F. et al. (2017). Machine learning with big data: Challenges and approaches. *IEEE Access* 5: 7776–7797.

Lahlali, R., Taoussi, M., Laasli, S.E. et al. (2024). Effects of climate change on plant pathogens and host-pathogen interactions. *Crop and Environment* 3 (3): 159–170.

Lam, W.W., and Chan, K.C. (2010). Discovering functional interdependence relationship in PPI networks for protein complex identification. *IEEE Transactions on Biomedical Engineering* 59 (4): 899–908.

Lan, K., Wang, D.T., Fong, S. et al. (2018). A survey of data mining and deep learning in bioinformatics. *Journal of Medical Systems* 42: 1–20.

Landgrebe, D.A. (2003). *Signal Theory Methods in Multispectral Remote Sensing*. John Wiley & Sons.

Larranaga, P., Calvo, B., Santana, R. et al. (2006). Machine learning in bioinformatics. *Briefings in Bioinformatics* 7 (1): 86–112.

LeCun, Y., Bengio, Y., and Hinton, G. (2015). Deep learning. *Nature* 521 (7553): 436–444.

Lee, H., Deng, M., Sun, F. et al. (2006). An integrated approach to the prediction of domain-domain interactions. *BMC Bioinformatics* 7: 1–15.

Lenthe, J.H., Oerke, E.C., and Dehne, H.W. (2007). Digital infrared thermography for monitoring canopy health of wheat. *Precision Agriculture* 8 (1): 15–26.

Li, B., Meng, X., Shan, L. et al. (2016). Transcriptional regulation of pattern-triggered immunity in plants. *Cell Host & Microbe* 19 (5): 641–650.

Li, J., Li, Z., Luo, J. et al. (2020). ACNNT3: Attention-CNN framework for prediction of sequence-based bacterial type III secreted effectors. *Computational and Mathematical Methods in Medicine* 2020 (1): 3974598.

Lopez, J.A., Sun, Y., Blair, P.B. et al. (2015). TCP three-way handshake: Linking developmental processes with plant immunity. *Trends in Plant Science* 20 (4): 238–245.

Lovelace, A.H., Smith, A., and Kvitko, B.H. (2018). Pattern-triggered immunity alters the transcriptional regulation of virulence-associated genes and induces the sulfur starvation response in Pseudomonas syringae pv. tomato DC3000. *Molecular Plant-Microbe Interactions* 31 (7): 750–765.

Löwer, M., and Schneider, G. (2009). Prediction of type III secretion signals in genomes of gram-negative bacteria. *PloS One* 4 (6): e5917.

Lü, L., Chen, D., Ren, X.L. et al. (2016). Vital nodes identification in complex networks. *Physics Reports* 650: 1–63.

Ma, C., Zhang, H.H., and Wang, X. (2014). Machine learning for big data analytics in plants. *Trends in Plant Science* 19 (12): 798–808.

Mahesh, B. (2020). Machine learning algorithms: A review. *International Journal of Science and Research (IJSR)[Internet]* 9 (1): 381–386.

Mahlein, A.K. (2016). Plant disease detection by imaging sensors: Parallels and specific demands for precision agriculture and plant phenotyping. *Plant Disease* 100 (2): 241–251.

Mahlein, A.K., Kuska, M.T., Thomas, S. et al. (2019). Quantitative and qualitative phenotyping of disease resistance of crops by hyperspectral sensors: Seamless interlocking of phytopathology, sensors, and machine learning is needed!. *Current Opinion in Plant Biology* 50: 156–162.

Mahlein, A.K., Oerke, E.C., Steiner, U. et al. (2012). Recent advances in sensing plant diseases for precision crop protection. *European Journal of Plant Pathology* 133 (1): 197–209

Mahlein, A.K., Steiner, U., Dehne, H.W. et al. (2010). Spectral signatures of sugar beet leaves for the detection and differentiation of diseases. *Precision Agriculture* 11 (4): 413–431.

Marone, D., Russo, M.A., Laidò, G. et al. (2013). Plant nucleotide binding site–leucine-rich repeat (NBS-LRR) genes: Active guardians in host defense responses. *International Journal of Molecular Sciences* 14 (4): 7302–7326.

McCormack, M.E., Lopez, J.A., Crocker, T.H. et al. (2016). Making the right connections: Network biology and plant immune system dynamics. *Current Plant Biology* 5: 2–12.

McMurray, B., and Hollich, G. (2009). Core computational principles of language acquisition: can statistical learning do the job? Introduction to special section. *Developmental Science* 12 (3): 365–368.

Mehl, P.M., Chen, Y.R., Kim, M.S. et al. (2004). Development of hyperspectral imaging technique for the detection of apple surface defects and contaminations. *Journal of Food Engineering* 61 (1): 67–81.

Min, S., Lee, B., and Yoon, S. (2017). Deep learning in bioinformatics. *Briefings in Bioinformatics* 18 (5): 851–869.

Mitchell, T.M., and Mitchell, T.M. (1997). *Machine Learning* (Vol. 1, No. 9). McGraw Hill.

Mitra, P., Murthy, C.A., and Pal, S.K. (2002). Unsupervised feature selection using feature similarity. *IEEE Transactions on Pattern Analysis and Machine Intelligence* 24 (3): 301–312.

Mochida, K., Koda, S., Inoue, K. et al. (2018). Statistical and machine learning approaches to predict gene regulatory networks from transcriptome datasets. *Frontiers in Plant Science* 9: 1770.

Mohri, M., Rostamizadeh, A. and Talwalkar, A. (2018). Foundations of machine learning. Adaptive computation and machine learning. Second edition; The MIT Press: Cambridge, MA.

Montesinos, E., Bonaterra, A., Badosa, E. et al. (2002). Plant-microbe interactions and the new biotechnological methods of plant disease control. *International Microbiology* 5: 169–175.

Mrozek, D., Malysiak-Mrozek, B. and Kozielski, S. (2009). Alignment of protein structure energy patterns represented as sequences of Fuzzy Numbers. *NAFIPS 2009-2009 Annual Meeting of the North American Fuzzy Information Processing Society*, 1–6. IEEE.

Mukhtar, M.S., Carvunis, A.R., Dreze, M. et al. (2011). Independently evolved virulence effectors converge onto hubs in a plant immune system network. *Science* 333 (6042): 596–601.

Mukhtar, M.S., McCormack, M.E., Argueso, C.T. et al. (2016). Pathogen tactics to manipulate plant cell death. *Current Biology* 26 (13): R608–R619.

Murmu, S., Chaurasia, H., Rao, A.R. et al. (2025). PlantPathoPPI: An ensemble-based machine learning architecture for prediction of protein-protein interactions between plants and pathogens. *Journal of Molecular Biology* 437: 169093.

Nagarajan, S., Seibold, G., Kranza, J. et al. (1984). Monitoring wheat rust epidemics with the Landsat-2 satellite. *Phytopathology* 74 (5): 585–587.

Nieke, J., and Rast, M. (2018). Towards the Copernicus hyperspectral imaging mission for the environment (CHIME). *IGARSS 2018: 2018 IEEE International Geoscience and Remote Sensing Symposium*, 157–159. IEEE.

Nielsen, H. (2017). Predicting secretory proteins with SignalP. *Protein Function Prediction: Methods and Potocols*: 59–73.

Nobori, T., Vel'asquez, A.C., and Wu, J. (2018). Transcriptome landscape of a bacterial pathogen under plant immunity. *Proceedings of the National Academy of Sciences of the United States of America* 115: E3055–E3064.

Oerke, E.C., Steiner, U., Dehne, H.W. et al. (2006). Thermal imaging of cucumber leaves affected by downy mildew and environmental conditions. *Journal of Experimental Botany* 57 (9): 2121–2132.

Oren, R., Schulze, E.D., Matyssek, R. et al. (1986). Estimating photosynthetic rate and annual carbon gain in conifers from specific leaf weight and leaf biomass. *Oecologia* 70: 187–193.

Pajerowska-Mukhtar, K.M., Emerine, D.K., and Mukhtar, M.S. (2013). Tell me more: Roles of NPRs in plant immunity. *Trends in Plant Science* 18 (7): 402–411.

Pal, S., Mondal, S., Das, G. et al. (2020). Big data in biology: The hope and present-day challenges in it. *Gene Reports* 21: 100869.

Pal, T., Jaiswal, V., and Chauhan, R.S. (2016). DRPPP: A machine learning based tool for prediction of disease resistance proteins in plants. *Computers in Biology and Medicine* 78: 42–48.

Pashaei, E., Ozen, M., and Aydin, N. (2017). Splice site identification in human genome using random forest. *Health and Technology* 7: 141–152.

Patel, H., Bhatt, D., Shakhreliya, S. et al. (2021). An introduction and applications of bioinformatics. In: Singh, V., Kumar, A. (eds) Advances in Bioinformatics. Springer, Singapore.

Paulus, S., Behmann, J., Mahlein, A.K. et al. (2014). Low-cost 3D systems: Suitable tools for plant phenotyping. *Sensors* 14 (2): 3001–3018.

Pethybridge, S.J., and Nelson, S.C. (2015). Leaf doctor: A new portable application for quantifying plant disease severity. *Plant Disease* 99 (10): 1310–1316.

Petre, B., and Kamoun, S. (2014). How do filamentous pathogens deliver effector proteins into plant cells?. *PLoS Biology* 12 (2): e1001801.

Pinaya, W.H.L., Vieira, S., Garcia-Dias, R. et al. (2020). Autoencoders. In: *Machine Learning*, 193–208. Academic Press.

Poland, J., and Rutkoski, J. (2016). Advances and challenges in genomic selection for disease resistance. *Annual Review of Phytopathology 54* (1): 79–98.

Polikar, R. (2012). Ensemble learning. In: Zhang, C., Ma, Y. (eds) *Ensemble Machine Learning: Methods and Applications*, 1–34.

Pritchard, L., and Birch, P.R. (2014). The zigzag model of plant–microbe interactions: Is it time to move on? *Molecular Plant Pathology* 15 (9): 865.

Prjibelski, A.D., Korobeynikov, A.I., and Lapidus, A.L. (2018). Sequence analysis. In: *Encyclopedia of Bioinformatics and Computational Biology: ABC of Bioinformatics*, 292–322.

Qin, J., Burks, T.F., Ritenour, M.A. et al. (2009). Detection of citrus canker using hyperspectral reflectance imaging with spectral information divergence. *Journal of Food Engineering* 93 (2): 183–191.

Rafiqi, M., Ellis, J.G., Ludowici, V.A. et al. (2012). Challenges and progress towards understanding the role of effectors in plant–fungal interactions. *Current Opinion in Plant Biology* 15 (4): 477–482.

Ray, R.V. (2024). Effects of pathogens and disease on plant physiology. In: *Agrios' Plant Pathology*, 63–92. Academic Press.

Robinson, J.T., Thorvaldsdóttir, H., Winckler, W. et al. (2011). Integrative genomics viewer. *Nature Biotechnology* 29 (1): 24–26.

Rumpf, T., Mahlein, A.K., Steiner, U. et al. (2010). Early detection and classification of plant diseases with support vector machines based on hyperspectral reflectance. *Computers and Electronics in Agriculture* 74 (1): 91–99.

Salzberg, S. (1995). Locating protein coding regions in human DNA using a decision tree algorithm. *Journal of Computational Biology* 2 (3): 473–485.

Samborska, I.A., Alexandrov, V., Sieczko, L. et al. (2014). Artificial neural networks and their application in biological and agricultural research. *Journal of NanoPhotoBioSciences* 2 (14): 2347–7342.

Sammut, C., and Webb, G.I. (eds.). (2011). *Encyclopedia of Machine Learning*. Springer Science & Business Media.

Sanseverino, W., and Ercolano, M.R. (2012). In silico approach to predict candidate R proteins and to define their domain architecture. *BMC Research Notes* 5: 1–11.

Saraswat, P. (2021). Supervised machine learning algorithm: A review of classification techniques. *Integrated Emerging Methods of Artificial Intelligence & Cloud Computing*: 477–482.

Savary, S., Ficke, A., Aubertot, J.N. et al. (2012). Crop losses due to diseases and their implications for global food production losses and food security. *Food Security* 4 (4): 519–537.

Sen, R., Nayak, L. and De, R.K. (2019). PyPredT6: A python-based prediction tool for identification of Type VI effector proteins. *Journal of Bioinformatics and Computational Biology* 17 (03): 1950019.

Serbin, S.P., Wu, J., Ely, K.S. et al. (2019). From the Arctic to the tropics: Multibiome prediction of leaf mass per area using leaf reflectance. *New Phytologist* 224 (4): 1557–1568.

Shoemaker, B.A., and Panchenko, A.R. (2007). Deciphering protein–protein interactions. Part II. Computational methods to predict protein and domain interaction partners. *PLoS Computational Biology* 3 (4): e43.

Singh, A., Ganapathysubramanian, B., Singh, A.K. et al. (2016). Machine learning for high-throughput stress phenotyping in plants. *Trends in Plant Science* 21 (2): 110–124.

Smith, G.R., and Sternberg, M.J. (2002). Prediction of protein–protein interactions by docking methods. *Current Opinion in Structural Biology* 12 (1): 28–35.

Somogyi, Z. (2021). Performance evaluation of machine learning models. In: *The Application of Artificial Intelligence: Step-by-Step Guide from Beginner to Expert*, 87–112. Springer International Publishing: Cham.

Sousa, D., Fisher, J.B., Galvan, F.R. et al. (2021). Tree canopies reflect mycorrhizal composition. *Geophysical Research Letters* 48 (10): e2021GL092764.

Sperschneider, J. (2020). Machine learning in plant–pathogen interactions: empowering biological predictions from field scale to genome scale. *New Phytologist* 228 (1): 35–41.

Sperschneider, J., Catanzariti, A.M., DeBoer, K. et al. (2017). LOCALIZER: Subcellular localization prediction of both plant and effector proteins in the plant cell. *Scientific Reports* 7 (1): 44598.

Sperschneider, J., Dodds, P.N., Gardiner, D.M. et al. (2018a). Improved prediction of fungal effector proteins from secretomes with EffectorP 2.0. *Molecular Plant Pathology* 19 (9): 2094–2110.

Sperschneider, J., Dodds, P.N., Singh, K.B. et al. (2018b). ApoplastP: Prediction of effectors and plant proteins in the apoplast using machine learning. *New Phytologist* 217 (4): 1764–1778.

Sperschneider, J., Gardiner, D.M., Dodds, P.N. et al. (2016). EffectorP: Predicting fungal effector proteins from secretomes using machine learning. *New Phytologist* 210 (2): 743–761.

Suthaharan, S. (2016). Supervised Learning Algorithms. In: Machine Learning Models and Algorithms for Big Data Classification. Integrated Series in Information Systems, vol 36. Springer, Boston, MA.

Takken, F.L., Albrecht, M., and Tameling, W.I. (2006). Resistance proteins: Molecular switches of plant defence. *Current Opinion in Plant Biology* 9 (4): 383–390.

Townsend, P.A., Foster, J.R., Chastain, R.A. et al. (2003). Application of imaging spectroscopy to mapping canopy nitrogen in the forests of the central Appalachian Mountains using Hyperion and AVIRIS. *IEEE Transactions on Geoscience and Remote Sensing* 41 (6): 1347–1354.

Tsuda, K., and Somssich, I.E. (2015). Transcriptional networks in plant immunity. *New Phytologist* 206 (3): 932–947.

Tsuda, K., Sato, M., Stoddard, T. et al. (2009). Network properties of robust immunity in plants. *PLoS Genetics* 5 (12): e1000772.

Tucker, C.C., and Chakraborty, S. (1997). Quantitative assessment of lesion characteristics and disease severity using digital image processing. *Journal of Phytopathology* 145 (7): 273–278.

Uetz, P., Giot, L., Cagney, G. et al. (2000). A comprehensive analysis of protein–protein interactions in Saccharomyces cerevisiae. *Nature* 403 (6770): 623–627.

Ustin, S.L., and Gamon, J.A. (2010). Remote sensing of plant functional types. *New Phytologist* 186 (4): 795–816.

Wan, S., and Yang, H. (2013). Comparison among methods of ensemble learning. *2013 International Symposium on Biometrics and Security Technologies*, 2–5 July 2013; Chengdu, Sichuan, China, 286–290. IEEE.

Wang, J., Li, J., Yang, B. et al. (2019). Bastion3: A two-layer ensemble predictor of type III secreted effectors. *Bioinformatics* 35 (12): 2017–2028.

Wang, J., Yang, B., Leier, A. et al. (2018). Bastion6: A bioinformatics approach for accurate prediction of type VI secreted effectors. *Bioinformatics* 34 (15): 2546–2555.

Wang, S.C. (2003). Artificial neural network. In: *Interdisciplinary Computing in Java Programming*, 81–100. Springer US: Boston, MA.

Wang, X., Zhang, M., Zhu, J. et al. (2008). Spectral prediction of Phytophthora infestans infection on tomatoes using artificial neural network (ANN). *International Journal of Remote Sensing* 29 (6): 1693–1706.

Wang, Y., Tyler, B.M. and Wang, Y. (2019). Defense and counterdefense during plant-pathogenic oomycete infection. *Annual Review of Microbiology* 73 (1): 667–696.

Wang, Y., Wei, X., Bao, H. det al. (2014). Prediction of bacterial type IV secreted effectors by C-terminal features. *BMC Genomics* 15: 1–14.

Wang, Y., Zhang, Q., Sun, M.A. et al. (2011). High-accuracy prediction of bacterial type III secreted effectors based on position-specific amino acid composition profiles. *Bioinformatics* 27 (6): 777–784.

Wang, Z., Chlus, A., Geygan, R. et al. (2020). Foliar functional traits from imaging spectroscopy across biomes in eastern North America. *New Phytologist* 228 (2): 494–511.

West, J.S., Bravo, C. and Oberti, R. (2003). The potential of optical canopy measurement for targeted control of field crop diseases. *Annual Review of Phytopathology* 41 (1): 593–614.

Wijekoon, C.P., Goodwin, P.H., and Hsiang, T. (2008). Quantifying fungal infection of plant leaves by digital image analysis using Scion Image software. *Journal of Microbiological Methods* 74 (2–3): 94–101.

Witten, I.H., Frank, E., Hall, M.A. et al. (2005). Practical machine learning tools and techniques. In: *Data Mining* (Vol. 2, No. 4, pp. 403–413). Elsevier: Amsterdam, The Netherlands.

Wolf, T., Kämmer, P., Brunke, S. et al. (2018). Two's company: Studying interspecies relationships with dual RNA-seq. *Current Opinion in Microbiology* 42: 7–12.

Won, H.H., Kim, M.J., Kim, S. et al. (2008). EnsemPro: An ensemble approach to predicting transcription start sites in human genomic DNA sequences. *Genomics* 91 (3): 259–266.

Wong, T.T. (2015). Performance evaluation of classification algorithms by k-fold and leave-one-out cross validation. *Pattern Recognition* 48 (9): 2839–2846.

Wright, I.J., Reich, P.B., Westoby, M. et al. (2004). The worldwide leaf economics spectrum. *Nature* 428 (6985): 821–827.

Xiong, Y., Wang, Q., Yang, J. et al. (2018). PredT4SE-stack: Prediction of bacterial type IV secreted effectors from protein sequences using a stacked ensemble method. *Frontiers in Microbiology* 9: 2571.

Xu, D. and Tian, Y. (2015). A comprehensive survey of clustering algorithms. *Annals of Data Science* 2 (2): 165–193.

Xue, L., Tang, B., Chen, W. et al. (2019). DeepT3: Deep convolutional neural networks accurately identify Gram-negative bacterial type III secreted effectors using the N-terminal sequence. *Bioinformatics* 35 (12): 2051–2057.

Yang, X., Guo, Y., Luo, J. et al. (2013). Effective identification of Gram-negative bacterial type III secreted effectors using position-specific residue conservation profiles. *PloS one* 8 (12): e84439.

Yoo, P.D., Zhou, B.B. and Zomaya, A.Y. (2008). Machine learning techniques for protein secondary structure prediction: An overview and evaluation. *Current Bioinformatics* 3 (2): 74–86.

Zhang, H. (2004). The optimality of naive Bayes. *Aa* 1 (2): 3.

Zhang, W., Corwin, J.A., Copeland, D. et al. (2017). Plastic transcriptomes stabilize immunity to pathogen diversity: the jasmonic acid and salicylic acid networks within the Arabidopsis/Botrytis pathosystem. *The Plant Cell* 29 (11): 2727–2752.

Zhu, H., Bilgin, M., Bangham, R. et al. (2001). Global analysis of protein activities using proteome chips. *Science* 293 (5537): 2101–2105.

8

Machine Learning-Enhanced Plant Disease Detection and Management

Lellapalli Rithesh[1*], *Sucharita Mohapatra*[2], *Shimi Jose*[3], *Juel Debnath*[1], *Gyanisha Nayak*[4], *Mehjebin Rahman*[5], *Soumya Shephalika Dash*[6], *Anwesha Sharma*[5], *Sneha Mohan*[1]

[1] *College of Agriculture, Kerala Agricultural University, Thiruvananthapuram, Kerala, India*
[2] *College of Agriculture, Odisha University of Agriculture and Technology, Bhubaneswar, Odisha, India*
[3] *Hawkesbury Institute for the Environment, Western Sydney University, Richmond, Australia*
[4] *Crop Improvement Division, ICAR-National Rice Research Institute, Cuttack, Odisha, India*
[5] *College of Agriculture, Assam Agricultural University, Assam, India*
[6] *Institute of Agriculture, Visva-Bharati, Sriniketan, West Bengal, India*

*Corresponding author: rithesh132@gmail.com

8.1 Introduction

Plant diseases are a major threat to global food security, causing 20–40% annual crop losses, as estimated by the Food and Agriculture Organization (FAO). Pathogens such as fungi, bacteria, viruses, and nematodes weaken plant growth, leading to economic losses and reduced agricultural productivity. The impact is particularly severe in developing countries, where food shortages and financial instability are exacerbated by crop failures. Early and accurate disease detection is crucial for effective management, helping farmers implement timely interventions, prevent disease spread, and reduce reliance on chemical pesticides. However, detecting plant diseases remains a challenge due to the diverse and often subtle symptoms exhibited by different pathogens. The complexity of plant–pathogen interactions further complicates diagnosis, making traditional detection methods inadequate for large-scale agricultural applications.

Traditional disease detection relies primarily on visual inspections, which, despite being commonly practiced, are subjective and prone to human error. More advanced techniques, such as biochemical tests, culture-based pathogen identification, and molecular methods like polymerase chain reaction (PCR) and enzyme-linked immunosorbent assay (ELISA), offer high accuracy but require specialized laboratories, trained personnel, and extensive processing time. These methods are not feasible for real-time field applications due to their high costs and logistical constraints. As a result, there is an increasing demand for innovative, efficient, and scalable plant disease detection technologies that can provide rapid, high-throughput, and precise results. The integration of technology-driven solutions is essential for improving agricultural disease management and ensuring sustainable crop production.

Machine learning (ML) has revolutionized plant disease detection by offering automated, accurate, and scalable solutions (Table 8.1). ML models analyze vast amounts of data, including plant images, spectral signatures, and environmental parameters, to identify disease patterns with high precision. Advanced techniques such as convolutional neural networks (CNNs), deep learning, and support vector machines (SVMs) have proven effective in classifying plant diseases based on leaf symptoms (Aliyu et al. 2020). Additionally, predictive ML models assess climate, soil, and epidemiological factors to forecast disease outbreaks, enabling farmers to take proactive control measures (Delfani et al. 2024). The integration of ML with IoT devices, remote sensing technologies, and smartphone-based diagnostic applications has further enhanced real-time disease monitoring. This chapter explores the transformative role of ML in plant pathology, highlighting its potential to overcome the limitations of conventional methods and revolutionize sustainable crop protection strategies. By harnessing AI-driven solutions, agriculture can shift towards a more resilient, data-driven approach to plant health management.

Machine Learning for Plant Biology, First Edition. Edited by Jen-Tsung Chen.

Table 8.1 Machine learning–enhanced plant disease detection.

Crop	Diseases	Method of Detection	References
	Tungro (*Rice tungro bacilliform virus, Rice tungro spherical virus*), bacterial blight (*Xanthomonas oryzae* pv. *oryzae*), blast (*Magnaporthe oryzae*), brown spot (*Bipolaris oryzae*)	Random Forest + Naive Bayes + SVM	Sudhesh et al. (2023)
	Sheath blight (*Rhizoctonia solani*), brown spots (*B. oryzae*), bacterial leaf blight (*X. oryzae pv. oryzae*), leaf blasts (*M. oryzae*), leaf smut (*Entyloma oryzae*)	Random Forest Classifier and R-CNN with VGG 16	Rajpoot et al. (2023)
	Leaf blast (*M. oryzae*), brown spot (*B. oryzae*)	YOLO v5 (Deep Neural Network-Based Model)	Mishra et al. (2023)
	Brown spot (*B. oryzae*), leaf blast (*M. oryzae*)	ResNet34 with self-attention layer	Stephen et al. (2023)
	Leaf smut (*E. oryzae*), Brown spot (*B. oryzae*), Bacterial leaf blight (*X. oryzae pv. oryzae*)	CNN with 15-layer architecture	Parasa et al. (2024)
	Sheath spot (*various fungi*), grassy stunt virus (*Rice grassy stunt virus*), brown spot (*B. oryzae*), sheath rot (*Sarocladium oryzae*), false smut (*Ustilaginoidea virens*), yellowing motel (*abiotic factors*), leaf scald (*Microdochium oryzae*), eyespot (*R. oryzae*), blast (*M. oryzae*), tungro virus (*Rice tungro bacilliform virus, Rice tungro spherical virus*), kernel smut (*Tilletia barclayana*), foot rot (*Fusarium spp.*), leaf smut (*E. oryzae*), sheath blight (*R. solani*), ragged stunt (*Rice ragged stunt virus*), bacterial leaf streak (*X. oryzae pv. oryzicola*), narrow brown leaf spot (*Cercospora janseana*), crown sheath rot (*S. oryzae*), flag leaf sheath (*S. oryzae*), powdery mildew (*Erysiphe graminis*), bacterial blight (*X. oryzae pv. oryzae*)	VGG16 + GoogleNet CNN	Yakkundimath et al. (2022)
	Leaf smut (*E. oryzae*), bacterial leaf blight (*X. oryzae pv. oryzae*), brown spot (*B. oryzae*)	DenseNet with multilayer perceptron (MLP)	Narmadha et al. (2022)
Oryza sativa	Neck blast (*M. oryzae*), tungro (*Rice tungro bacilliform virus, Rice tungro spherical virus*), bacterial blight (*X. oryzae pv. oryzae*), false smut (*U. virens*), blast (*M. oryzae*), brown spot (*B. oryzae*)	RDD_CNN (Automated Rice Disease Diagnosis System - RDDS)	Vasantha et al. (2022)
	Rice blast (*M. oryzae*), brown spot (*B. oryzae*), bacterial leaf blight (*X. oryzae pv. oryzae*)	SVM with GLCM & Intensity-Based Color Feature Extraction	Archana et al. (2022)
	Leaf blast (*M. oryzae*), brown spot (*B. oryzae*)	Deep learning-based object detection (E-crop doctor app)	Chaudhari et al. (2023)
	Red blight (*various fungi*), blast (*M. oryzae*), leaf smut (*E. oryzae*), bacterial leaf streak (*X. oryzae pv. oryzicola*), bacterial leaf blight (*X. oryzae pv. oryzae*), brown spot (*B. oryzae*), sheath blight (*R. solani*)	GoogLeNet (E-Inception Module)	Yang et al. (2023)
	Tungro (*Rice tungro bacilliform virus, Rice tungro spherical virus*), brown spot (*B. oryzae*), bacterial blight (*X. oryzae pv. oryzae*)	Vulture-Based Auto-Metric Graph Neural Network (VAGNN)	Rajasekhar et al. (2023)
	Fungal diseases like powdery mildew (*E. graminis*), rust (*Puccinia spp.*), septoria (*Septoria oryzae*)	Marker & Mask-Based Membership Filtering (M3FCM)	Bukar et al. (2023)

	Fusarium head blight (*F. graminearum*)	Logistic Regression, SVM, Artificial Neural Network	Almoujahed et al. (2022)
	Blast (*M. oryzae*), false smut (*U. virens*), brown spot (*B. oryzae*), Bakanae (*F. fujikuroi*), sheath blight (*R. solani*), sheath rot (*S. oryzae*), bacterial blight (*X. oryzae pv. oryzae*), bacterial sheath rot (*Pseudomonas fuscovaginae*), seedling blight (*various fungi*), bacterial wilt (*Ralstonia solanacearum*), neck blast (*M. oryzae*), stack burn (*Alternaria padwickii*), leaf smut (*E. oryzae*), leaf scald (*M. oryzae*), stem rot (*M. salvinii*), stripe blight (*Xanthomonas spp.*), bacterial leaf streak (*X. oryzae pv. oryzicola*)	CNNs, MobileNet-V2 with attention mechanism	Lu et al. (2017); Rahman et al. (2020)
	Northern maize leaf blight (*Setosphaeria turcica*)	Deep Neural Network (CEMLB-YOLO)	Leng et al. (2023)
	Common rust (*P. sorghi*), blight (*P. syringae*), gray leaf spot (*C. zeae-maydis*)	3D-DCNN with Ebola Optimization Search Algorithm	Sowmyalakshmi et al. (2021)
	Gray leaf spot (*C. zeae-maydis*), blight (*P. syringae*), common rust (*P. sorghi*)	AlexNet-Inception	Balasubramanian (2024)
Zea mays	Gibberella stalk rot (*F. graminearum*), anthracnose leaf blight (*Colletotrichum graminicola*), northern corn leaf spot (*Cochliobolus carbonum*), eyespot (*Kabatiella zeae*), anthracnose stalk rot (*C. graminicola*), southern rust (*P. polysora*)	Deep Learning (NPNet 19)	Nagaraju and Chawla (2023)
	Fungal foliar disease (*various fungi*), northern leaf blight (*Setosphaeria turcica*)	Attention U-Net Segmentation Model	Rai and Pahuja (2024)
	Cercospora leaf spot (*C. zeae-maydis*), grey leaf spot, common rust, northern and southern leaf blight, banded leaf and sheath blight (*R. solani*), rust (*P.* spp.)	CNN, VGG16, multiple instance learning, generic algorithm, deep CNN, Vision Transformer	Goyal et al. (2021); Borhani et al. (2022)
Pennisetum glaucum	Blast (*Pyricularia grisea*), rust (*P. substriata*)	Automatic & intelligent data collector & classifier	Kundu et al. (2021)
Phaseolus vulgaris	Angular leaf spot (*Pseudocercospora griseola*), bean rust (*Uromyces appendiculatus*), white mold (*Sclerotinia sclerotiorum*)	U-Net, VGG16, AlexNet, MobileNet-V2, DenseNet201, GoogLeNet, ELM, SSA, SVM, RF, ANN	Kursun et al. (2023); Shahoveisi et al. (2022)
Vigna radiata	Cercospora leaf spot (*C. canescens*), yellow mosaic (*Mungbean yellow mosaic virus - MYMV*), halo blight (*P. syringae pv. phaseolicola*), charcoal rot (*Macrophomina phaseolina*)	MobileNet-V2, lightweight Android smartphone-based deep learning model	Mallick et al. (2023)
Vigna mungo	Yellow mosaic disease (*Mungbean yellow mosaic virus - MYMV*), anthracnose (*Colletotrichum spp.*), leaf crinkle (*Urdbean leaf crinkle virus—ULCV*), powdery mildew (*E. polygoni*)	CNN VirLeafNet, SIDMO, deep ensemble	Hajare and Rajawat (2023); Joshi et al. (2021)
Glycine max	Brown leaf spot (*Septoria glycines*), frog eye leaf spot (*C. sojina*), phyllosticta leaf spot (*Phyllosticta sojicola*), soybean rust (*Phakopsora pachyrhizi*)	Residual attention network, ConvNeXt with CBAM	Wu et al. (2023)

(Continued)

Table 8.1 (Continued)

Crop	Diseases	Method of Detection	References
Helianthus annuus	Rust (*P. helianthi*), leaf downy mildew (*Plasmopara halstedii*), grey mold (*Botrytis cinerea*), flower downy mildew (*Plasmopara halstedii*)	Transfer learning, CNN (Xception, ResNet50, EfficientNetB4, MobileNet-V2)	Ghosh et al. (2023); Shahoveisi et al. (2023)
Arachis hypogaea	Bacterial wilt (*Ralstonia solanacearum*), scab (*Sphaceloma arachidis*), grey spot (*C. personata*), web blotch (*Phoma arachidicola*), anthracnose (*Colletotrichum spp.*), scorch (*Leptosphaerulina crassiasca*), leaf spot (*C. spp., Phaeoisariopsis personata*), rust (*P. arachidis*)	MLP-BP neural network, Xception network, attention-augmented branches, stack ensemble	Qi et al. (2021); Xu et al. (2023)
Gossypium spp.	Root rot (*Macrophomina phaseolina*), leaf curl (*Cotton leaf curl virus—CLCuV*), sooty mold (*Capnodium spp.*)	Semi-supervised classifier (K-means & SVM), RBF SVM, YOLOX, Spatial Pyramid Pooling	Noon et al. (2022); Wang et al. (2020)
Saccharum officinarum	White leaf phytoplasma (*Phytoplasma spp.*), smut (*Sporisorium scitamineum*), mosaic (*Sugarcane mosaic virus—SCMV*), red rot (*C. falcatum*), rust (*P. kuehnii*), yellow leaf disease (*Sugarcane yellow leaf virus—SCYLV*)	XGBoost, RF, DT, KNN, dual self-attention block, VGG19, ResNet50, XceptionNet, MobileNetV2, EfficientNet-B7	Daphal and Koli (2023); Narmilan et al. (2022)
Coffea spp.	Leaf rust (*Hemileia vastatrix*), phoma blight (*Phoma spp.*), cercospora leaf spot (*C. coffeicola*)	SVM, Ensemble-based	Novtahaning et al. (2022); Soares et al. (2022)
Camellia sinensis	Anthracnose (*Colletotrichum spp.*), brown blight (*Glomerella cingulata*), red leaf spot (*Cephaleuros virescens*), blight (*various fungi*), tea algae leaf spot (*Cephaleuros spp.*), black rot (*Corticium theae*), rust (*P. theae*)	Faster R-CNN, RGReLU, Res4net-CBAM, TSBA-YOLO, shuffle attention, LC3Net	Bhuyan et al. (2023); Jayapal and Poruran (2023)
Allium cepa	Anthracnose-twister (*C. gloeosporioides*), downy mildew (*Peronospora destructor*), purple blotch (*Alternaria porri*)	SVM, CNN, InceptionV3	Alberto et al. (2020); Kim et al. (2020)
Abelmoschus esculentus	Cercospora leaf spot (*C. malayensis*)	t-SNE, Class Activated Mapping	Rangarajan et al. (2022)
Raphanus sativus	Fusarium wilt (*F. oxysporum f. sp. raphani*)	CNN model (RadRGB)	Dang et al. (2020)
Manihot esculenta	Brown leaf spot (*C. henningsii*), red/green mite damage (*Oligonychus gossypii*), brown streak disease (*Cassava brown streak virus*), mosaic disease (*Cassava mosaic virus*)	CNN	Ramcharan et al. (2017)
Citrullus lanatus	Downy mildew (*Pseudoperonospora cubensis*)	MLP, DT	Abdulridha et al. (2022)
Brassica oleracea	Bacterial spot rot (*X. campestris*), black rot (*X. campestris pv. campestris*), downy mildew (*Hyaloperonospora brassicae*)	DenseNet201	Kanna et al. (2023)

Prunus persica	Brown rot (*Monilinia fructicola*), anthracnose (*Colletotrichum spp.*), scab (*Cladosporium carpophilum*), bacterial shot hole disease (*X. arboricola pv. pruni*), gummosis (*Botryosphaeria spp.*), powdery mildew (*Podosphaera pannosa*), bacterial spots (*X. campestris*), leaf curl (*Taphrina deformans*)	Xception Network ensemble with L2M Loss, AlexNet, VGGNet, YOLO-V3, LWNet model based on VGG19, Focal Loss Mask R-CNN and Mask Scoring R-CNN, EfficientNet	Akbar et al. (2022); Farman et al. (2022)
Fragaria ananassa	Grey mould (*Botrytis cinerea*), *Gnomonia fructicola* fall blight (*Gnomonia fructicola*), pestalotiopsis leaf spot (*Pestalotiopsis spp.*), common leaf spot (*Mycosphaerella fragariae*), anthracnose (*Colletotrichum spp.*)	Spatial convolutional self-attention-based transformer (SCSA-Transformer)	Li et al. (2023)
Pyrus communis	Foliar diseases (*various pathogens*), *Septoria piricola, A. alternata, Gymnosporangium haracannum*, powdery mildew (*Podosphaera leucotricha*), dry rot (*Neofabraea spp.*), fire blight (*Erwinia amylovora*), leaf spot (*various fungi*), leaf curl (*Taphrina deformans*)	CBAM attention mechanism + CNNs, ResNet50, ResNet101, DL CNN, Faster R-CNN, ensemble CNN	Alirezazadeh et al. (2023); Linker and Dafny-Yalin (2024)
Mangifera indica	Powdery mildew (*Oidium mangiferae*), anthracnose (*C. gloeosporioides*), dieback (*Lasiodiplodia theobromae*), Phoma blight (*Phoma spp.*), bacterial canker (*X. campestris pv. mangiferaeindicae*), red rust (*Cephaleuros virescens*), sooty mould (*Capnodium spp.*)	Ensembled stack deep neural network (ESDNN)	Gautam et al. (2023)
Citrus spp.	Black spot (*Guignardia citricarpa*), bacterial canker (*X. axonopodis pv. citri*), blight (*Phytophthora spp.*), scab (*Elsinoë spp.*), greening (HLB) (*Candidatus Liberibacter spp.*), melanose (*Diaporthe citri*), anthracnose (*Colletotrichum spp.*), sandpaper rust (*various fungi*), sunscald (*abiotic factor*)	CNN, MobileNet-V2, DenseNet201, Whale Optimization Algorithm (WOA), ResNet50, YOLO-V4, EfficientNet, Faster R-CNN, Enhanced CNN	Dou et al. (2023); Shastri et al. (2023)
Malus domestica	Scab (*Venturia inaequalis*), black rot (*Botryosphaeria obtusa*), cedar apple rust (*Gymnosporangium juniperi-virginianae*), leaf black disease (*various fungi*)	SVM, logistic regression, multilayer perceptron, CNN, Inception-V3, ResNet152, MobileNet, Faster R-CNN, CapsNet, DenseNet201, GoogleNet, UNet, Mask R-CNN	Haque et al. (2022); Karthickmanoj and Sasilatha (2024)
Punica granatum	*A. alternata*, anthracnose (*Colletotrichum* spp.), bacterial blight (*X. axonopodis pv. punicae*), cercospora leaf spot (*C. punicae*)	ResNet152, MobileNet, InceptionNet, Stacking Ensemble Learning	Karthickmanoj and Sasilatha (2024); Nirmal et al. (2023)

8.2 Fundamentals of ML

8.2.1 Fundamental Principles of ML

The objective of ML, a subset of artificial intelligence (AI), is to develop algorithms that enable computers to learn from data and make decisions or predictions autonomously, without explicit programming. This sector's capacity to process substantial data volumes and identify complex patterns, which traditional computing methods struggle to do, has garnered significant interest in recent years. ML is typically categorized into three fundamental types: supervised learning, unsupervised learning, and reinforcement learning (RL). Each category has a distinct purpose and is utilized across various domains, including banking, healthcare, agriculture, and autonomous systems.

8.2.1.1 Supervised Learning

In supervised learning, an algorithm is trained with labeled data, indicating that each training sample consists of input-output pairs. The model reduces the divergence between its predictions and the actual values to learn how to convert inputs into appropriate outputs. Once trained, the model can extend its learning to generate accurate predictions on new, untested data. Regression and classification problems represent two categories of supervised learning tasks. The focus of classification tasks is predicting discrete labels, such as determining whether an email is spam, while regression tasks concentrate on predicting continuous values, such as estimating property prices based on various variables. Decision trees, neural networks, and SVMs are exemplars of widely utilized supervised learning algorithms (Murphy 2012).

8.2.1.2 Unsupervised Learning

Unsupervised learning operates on unlabeled data, as opposed to supervised learning. The program aims to identify inherent links, structures, or patterns within the data autonomously. This method is highly effective for clustering, anomaly detection, and exploratory data analysis applications. In unsupervised learning, two fundamental techniques are clustering and dimensionality reduction. Clustering techniques such as K-means and hierarchical clustering significantly enhance customer segmentation and biological data analysis by grouping related data points based on their attributes. Dimensionality reduction approaches, such as principal component analysis (PCA), simplify complex datasets by eliminating extraneous features while retaining essential information (Hastie et al. 2009).

8.2.1.3 Reinforcement Learning

In RL, an agent acquires optimal decision-making skills by learning from errors and refining its actions based on feedback to maximize rewards. RL is being investigated in agriculture for adaptive crop management, enhancing irrigation, fertilization, and insect control. RL is generally represented as a Markov Decision Process (MDP), wherein it monitors environmental variables (e.g., soil moisture, plant development), executes actions (e.g., watering, fertilizing), and obtains rewards (e.g., enhanced yield, resource efficiency) to formulate optimal long-term plans. Prominent RL techniques encompass Deep Q-Network (DQN) for discrete decision-making, such as irrigation scheduling; Proximal Policy Optimization (PPO) for the optimization of fertilization and irrigation; and Deep Deterministic Policy Gradient (DDPG) for continuous tasks, including accurate nutrient delivery. RL provides advantages such as adaptive decision-making, optimized resource utilization, and automation, enhancing the viability of precision agriculture. Nonetheless, obstacles including data scarcity, intricate reward function formulation, and issues in generalization across many contexts impede its extensive use. To address these challenges, offline RL, which utilizes pre-collected datasets for model training, and large language models (LLMs), which improve RL techniques by translating agricultural data into comprehensible recommendations, are being investigated to promote smart farming.

8.2.2 Categories of ML Models Employed in Plant Disease Detection

Recently, there has been significant interest in employing ML for plant disease diagnostics due to its ability to evaluate large data sets and identify complex patterns. A variety of ML models have been employed in this domain, encompassing advanced methodologies such as deep learning and CNNs, as well as traditional algorithms like decision trees and SVMs (Kamilaris and Prenafeta-Boldú 2018).

8.2.2.1 Decision Trees

Decision trees are extensively employed for classification problems in plant disease detection owing to their clarity and simplicity. These models generate a hierarchical structure wherein each node signifies a decision criterion and each leaf denotes a classification outcome, achieved by recursively partitioning the dataset based on feature values. In agricultural contexts, decision trees effectively diagnose plant diseases by examining variables such as leaf color, texture, and form (Revathi and Hemalatha 2014). They have been efficiently employed in the detection of fungal infections, bacterial blight, and viral diseases by recognizing critical indicators that differentiate infected plants from healthy ones. Pruning and ensemble methods are two tactics that can mitigate the overfitting inherent in decision trees.

8.2.2.2 Support Vector Machines

Robust supervised learning models, such as SVMs, are particularly effective for binary classification tasks in the diagnosis of plant diseases. SVMs function by identifying the optimal hyperplane in a high-dimensional space that maximizes the margin between distinct classes. This characteristic enables SVMs to distinguish between healthy and diseased plants utilizing spectral data and leaf imagery (Revathi and Hemalatha 2014). Studies indicate that SVMs are effective in diagnosing diseases such as late blight, rust, and powdery mildew. The key advantages of SVMs are their robustness against overfitting, particularly when employing kernel functions, and their ability to manage high-dimensional data. Nevertheless, the processing difficulty of vast agricultural information may present a disadvantage.

8.2.2.3 Random Forests

Random forests (RFs) enhance predictive performance by amalgamating various decision tree–based ensemble learning methods, thereby mitigating overfitting and augmenting accuracy. A random subset of the training data and features is utilized to construct each tree in the forest, ensuring diversity in decision-making. RFs have been effectively utilized in plant disease detection for classifying leaf pictures, forecasting disease dissemination, and recognizing environmental elements that contribute to disease outbreaks (Kamilaris and Prenafeta-Boldú 2018). They are suitable for agricultural applications where data quality may be inconsistent because of their robustness and ability to handle missing data. Moreover, RFs offer feature importance rankings, aiding agronomists in comprehending the relevance of various plant health indicators in disease prediction.

8.2.2.4 Deep Learning

Deep learning has revolutionized plant disease diagnostics by employing multi-layered artificial neural networks to create hierarchical representations of agricultural data. Deep learning algorithms, unlike traditional ML models, may autonomously extract relevant features from raw data, eliminating the necessity for feature engineering. This capability has demonstrated particular utility when assessing complex datasets, including multispectral and hyperspectral imaging data (Mohanty et al. 2016). Deep learning algorithms can accurately diagnose diseases such as bacterial wilt, mosaic virus, and leaf spot. Deep learning techniques may be challenging to implement broadly in agricultural settings due to their substantial processing costs and extensive training data needs. A CNN transfer deep learning–based model for plant disease detection has been illustrated in Figure 8.1.

8.2.2.5 Convolutional Neural Networks

CNNs are a particular type of deep learning model utilized for image processing. Convolutional layers effectively extract spatial features from plant images, enabling CNNs to proficiently recognize anomalies in stem patterns, fruit texture, and

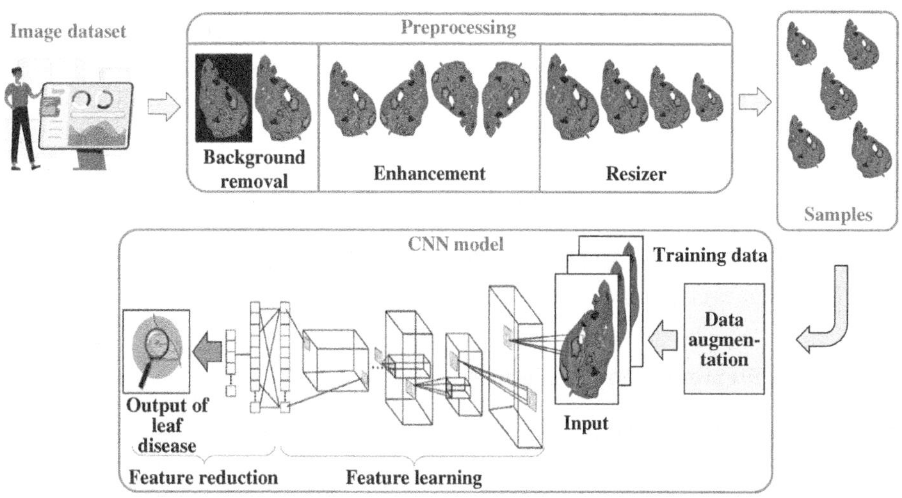

Figure 8.1 This is a CNN transfer deep learning-based model for plant disease detection. It begins with image preprocessing, including background removal, resizing, and enhancement. Data augmentation is applied to expand the dataset by introducing slight variations to the original images. The model then processes the images through multiple convolutional layers, pooling layers, and fully connected layers. These layers extract spatial-temporal features and optimize the model using backpropagation, ultimately generating disease classification output.

leaf structure. These networks have been extensively utilized in agricultural imaging tasks, such as pest identification, crop health assessment, and early disease detection (Sladojevic et al. 2016). CNN architectures like as ResNet, VGGNet, and U-Net have been customized for particular agricultural applications, exhibiting exceptional accuracy in plant disease identification (Mohanty et al. 2016). CNNs necessitate substantial computational resources and extensive labeled datasets for training, which may be a limitation in agricultural environments with constrained resources.

8.3 Data Collection and Preprocessing

ML models for plant disease detection and management require high-quality datasets. The accuracy and efficiency of these models heavily depend on how data is collected, processed, and fed into the learning algorithms (Sujatha et al. 2021).

8.3.1 Data Types

Data collection is the foundational step in building an ML model. The quality, diversity, and volume of the data significantly influence model performance. The major types of data used for plant disease detection and management, with their main challenges, include the following.

8.3.1.1 Image Data

High-resolution images of plant parts are captured using various devices, such as smartphones, drones, and specialized cameras. These images play a vital role in visually detecting plant diseases by identifying distinct symptoms (Li et al. 2021). However, challenges such as lighting inconsistencies, occlusions (e.g., overlapping leaves), background noise, and variations in resolution can affect the quality of the dataset.

8.3.1.2 Spectral Data

Hyperspectral and multispectral imaging technologies capture data across different wavelengths, enabling the detection of subtle physiological changes in plants that may not be visible to the human eye. While these techniques provide valuable insights, they require specialized sensors, involve complex preprocessing, and can be expensive to implement.

8.3.1.3 Sensor Data

Environmental parameters like soil moisture, temperature, and humidity, which influence disease development, are monitored through IoT-based field sensors (Ouhami et al. 2021). These real-time measurements are crucial for predictive modeling (Ahmed and Yadav 2023). However, maintaining sensor accuracy can be challenging due to factors such as calibration issues, inconsistent data, and occasional sensor malfunctions.

8.3.1.4 Genomic and Molecular Data

Advanced molecular techniques, such as DNA sequencing and PCR, help analyze pathogen genetics and plant–pathogen interactions (Wang et al. 2021). The genome-wide association studies and genomic selection (GS) have been used to identify important markers and select varieties with desirable disease-resistant traits in plants (Upadhyaya et al. 2024). However, handling such data comes with challenges, including high dimensionality, sequencing errors, and complex computational requirements.

8.3.1.5 Tabular Data

This includes structured datasets containing field observations, records of crop yields, and historical disease incidence. Such data provides valuable insights into disease trends and supports decision-making in plant disease management. However, common issues include missing values, data standardization difficulties, and potential errors from manual data entry (Domingues et al. 2022).

8.3.2 Data Preprocessing Techniques

The preprocessing methods applied to datasets vary based on the type of data being used. However, the fundamental steps remain consistent across different data types. Proper data preprocessing is essential for improving data quality, eliminating

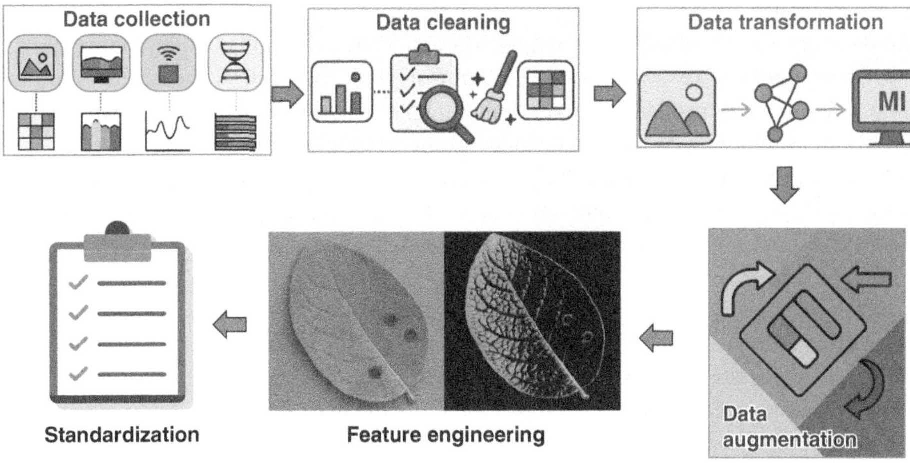

Figure 8.2 Workflow of general data preprocessing.

inconsistencies, and optimizing the performance of ML models in plant disease detection. By refining raw data through various preprocessing techniques, the reliability and accuracy of the models can be significantly enhanced, ensuring robust predictions and meaningful insights. A simple workflow of data preprocessing is illustrated in Figure 8.2.

8.3.2.1 Image Data Preprocessing

Preprocessing image data is a crucial step in ensuring consistency and improving the accuracy of ML models in plant disease detection. Image resizing and normalization help standardize image dimensions and pixel intensity values, making the dataset uniform and ensuring smooth model training (Kulkarni et al. 2021). Data augmentation techniques, including rotation, flipping, and contrast adjustments, artificially expand the dataset, which enhances model generalization and prevents overfitting. Noise reduction methods, such as Gaussian blur filtering, improve image clarity by minimizing background disturbances. Image segmentation isolates diseased regions from the background using techniques like thresholding and deep learning–based segmentation, allowing the model to focus on relevant features. Feature extraction is another essential preprocessing step, where techniques like CNNs, the Gray Level Co-occurrence Matrix (GLCM) (Nugroho et al. 2024), and handcrafted feature extraction methods are employed to capture critical characteristics such as color, texture, and shape, thereby improving classification accuracy.

8.3.2.2 Spectral Data Preprocessing

Spectral data preprocessing plays a significant role in leveraging hyperspectral and multispectral imaging for early disease detection. Band selection helps identify spectral bands that differentiate between healthy and diseased plants, improving model sensitivity (Sun et al. 2021). Noise filtering methods, including PCA, assist in reducing unnecessary spectral noise, preserving only relevant information. Normalization ensures that spectral data is standardized into reflectance values, maintaining consistency across samples and making it easier for models to compare different observations. Dimensionality reduction techniques, such as PCA and Linear Discriminant Analysis (LDA), simplify complex spectral datasets by reducing the number of variables without losing significant information, making computational processes more efficient (Cheshkova 2022).

8.3.2.3 Sensor Data Preprocessing

The data collected from environmental sensors, such as temperature, humidity, and soil moisture sensors, requires preprocessing to maintain its reliability for disease prediction models (Patel and Patel 2022). Data cleaning is essential to handle missing values and inconsistencies arising from sensor errors or environmental factors, ensuring dataset completeness. Anomaly detection techniques help identify and correct irregular sensor readings, preventing misleading inputs in the ML model. Feature engineering is another important step in which derived features, such as rolling averages or trend-based indicators, are generated from raw sensor data to enhance the predictive power of the model. Additionally, normalization and standardization techniques are applied to scale sensor measurements, ensuring that all variables have a uniform range and improving model performance (Gupta et al. 2021).

8.3.2.4 Genomic and Molecular Data Preprocessing

Preprocessing genomic and molecular data is essential for analyzing pathogen genetic sequences and plant resistance traits. Sequence alignment aligns DNA or RNA sequences to reference genomes (Johnson et al. 2023), allowing for the accurate identification of pathogens and genetic variations. Feature selection techniques help identify specific genes associated with disease resistance, refining model inputs and improving predictive accuracy. Dimensionality reduction methods, such as t-distributed stochastic neighbor embedding (t-SNE) (Chen et al. 2023), facilitate the visualization of high-dimensional genetic data, revealing underlying biological patterns. One-hot encoding is used to convert nucleotide sequences into a numerical format compatible with ML algorithms, enabling efficient computational processing (Shovon et al. 2023).

8.3.2.5 Tabular Data Preprocessing

Structured tabular data, including field observations and disease incidence records, requires careful preprocessing before being used in ML models. Handling missing values is a key step, where imputation techniques such as mean substitution or regression-based imputation methods help maintain dataset completeness. Encoding categorical data is necessary to transform qualitative variables, such as disease severity levels, into numerical representations using techniques like one-hot encoding, making them interpretable by ML algorithms (Wani et al. 2022). Feature scaling techniques, including normalization and standardization, are applied to ensure that variables are measured on comparable scales, preventing bias during model training. Data balancing methods, such as the synthetic minority over-sampling technique (SMOTE), help address class imbalance issues, ensuring that the model performs equally well across all disease categories (Sun et al. 2024). By implementing these preprocessing techniques across different types of data, ML models can operate more effectively, leading to more accurate and reliable plant disease detection and management systems.

8.4 Image-based Pathology Identification

ML, especially deep learning, has transformed image-based disease detection by facilitating accurate and swift diagnosis. CNNs examine extensive databases of annotated images to detect patterns and anomalies, rendering them exceptionally beneficial for medical and agricultural purposes. Smartphone-assisted systems in crop disease detection employ trained models to evaluate plant health, identify diseases, and provide remedies. These models enhance over time by assimilating fresh data, hence augmenting accuracy. Utilizing publicly accessible statistics, and scalable AI systems can enable farmers globally, mitigating agricultural losses and enhancing food security via early detection and intervention methods.

8.4.1 The Role of Computer Vision in Plant Diseases Diagnosis

ML methodologies are progressively employed to autonomously detect patterns or anomalies that signify the existence of crop diseases (Neupane and Baysal-Gurel 2021). Deep neural networks have lately been utilized in many domains through end-to-end learning, correlating inputs such as photos of damaged plants to outputs like disease categorization via interconnected layers executing mathematical operations. The difficulty involves constructing a deep network in which both the architecture and the functions of the nodes, along with the edge weights, efficiently convert inputs into precise outputs. Diverse deep learning architectures, such as CNNs, recurrent neural networks (RNNs), and autoencoders, have exhibited proficiency in identifying disease patterns within photos. Researchers have demonstrated considerable interest in deep learning methodologies due to their capacity to attain high accuracy across many applications. CNNs, a sophisticated branch of artificial intelligence, are renowned for their capacity to effectively analyze extensive and varied datasets, facilitating precise predictions in intricate situations and improving performance in numerous applications that demand high accuracy and profound feature extraction (Goodfellow et al. 2016).

The inaugural CNN design, LeNet-5, was presented in 1998. Research indicates that the principal factors enhancing CNN performance are essential changes and the creation of novel network blocks (Shahoveisi et al. 2023). CNNs are the most prevalent category of deep neural networks, drawing inspiration from the connection patterns observed in the human brain. They employ a hierarchical layer architecture to independently learn and extract features from input photos, adeptly capturing spatial relationships and augmenting pattern recognition for superior image analysis and processing. Each subsequent layer enhances the outputs of preceding layers, facilitating the identification of progressively intricate patterns (O'Shea and Nash 2015). Cutting-edge CNN architectures are engineered to address classification, segmentation, and object identification tasks. CNN models utilize convolutional and subsampling layers to encode input images for classification tasks, subsequently employing a classifier to ascertain the likelihood of each class. Similar designs can be employed for segmentation and object detection tasks, leveraging convolutional layers to extract pertinent information (Alom et al. 2019). Detecting plant leaf diseases is a distinct challenge compared to classification and segmentation tasks because of the varied manifestations of these diseases.

The primary objective of the approach is to link target categories with their respective locations. It tackles two fundamental inquiries: What constitutes the classification problem? What is the location of the object or item within the photograph (a regression issue)? Alshammari et al. (2022) suggested an adaptive evolutionary algorithm to optimize parameter selection. Numerous CNN designs have been developed throughout the years to address diverse needs. The most commonly utilized models comprise the Visual Geometry Group Network (VGGNet), Residual Network (ResNet), AlexNet, Densely Connected Convolutional Network (DenseNet), and MobileNet. VGGNet employs 3×3 convolutional filters and is recognized for its straightforward, sequential construction, whereas ResNet facilitates the training of exceptionally deep networks with up to 152 layers. Although VGGNet performs well on smaller datasets, it requires substantial processing resources and memory (Simonyan and Zisserman 2014).

CNN architectures are proficient in classification, segmentation, and object detection tasks; nonetheless, they necessitate substantial training data owing to their deep layers and multiple model parameters that improve generalization capabilities (Arivazhagan and Ligi 2018). Rice, a fundamental crop vital for global food security, has been the subject of research on disease and pest detection. Rahman et al. (2020) devised a methodology for detecting diseases and pests in rice plant photos with a two-stage CNN architecture tailored for mobile platforms. The research contrasted this model with memory-efficient designs including MobileNet, NasNet Mobile, and SqueezeNet. The findings indicated that the proposed architecture attained 93.3% accuracy while markedly decreasing model size to approximately 99% smaller than VGG16.

Chen et al. (2020) presented a network architecture named Mobile-DANet for the identification of maize crop diseases. This model, derived from DenseNet, incorporated depth-wise separable convolution within dense blocks and an embedded attention module to evaluate interchannel relationships and spatial characteristics. The model enhanced accuracy through transfer learning while preserving computational resources relative to deep CNNs.

Haque et al. (2022) performed a study utilizing 5939 photographs obtained from experimental fields to examine three maize diseases: southern corn leaf blight, northern corn leaf blight, and banded leaf and sheath blight. The collection included photos from three disease categories as well as a healthy category. To mitigate class imbalance, supplementary photos were produced using data augmentation methods, including rotation and brightness modification. Three models utilizing the Inception-v3 framework were trained via a baseline training methodology, with the top-performing model attaining a classification accuracy of 95.99% and an average recall of 95.96% on an independent test dataset. The comparative examination with advanced pretrained models validated the superiority of the suggested model.

Elfatimi et al. (2024) examined rust and angular leaf spot infections in bean crops with the MobileNet architecture. The research assessed three datasets of bean leaf images characterized by differing degrees of complexity. MobileNet was selected for its efficiency, providing superior performance with reduced parameters and expedited execution times. The model consistently attained over 92% accuracy across all datasets. Comparison analysis and the use of the GradCAM technique yielded insights into the model's decision-making process.

Ghosh et al. (2023) investigated sunflower disease identification with a hybrid deep learning methodology. Their approach utilized transfer learning in conjunction with a basic CNN, capitalizing on a limited dataset. Of the eight models evaluated across four disease categories (downy mildew, grey mold, leaf scars, and fresh leaf), the VGG19 + CNN hybrid model demonstrated exceptional performance across multiple metrics, including precision, recall, F1 score, accuracy, Hamming loss, Matthews's coefficient, Jaccard score, and Cohen's kappa.

Grapes, being a significant fruit crop, are particularly vulnerable to diseases. Xie et al. (2020) created a real-time detector for prevalent grape leaf diseases, such as black rot, black measles, leaf blight, and mites, with an enhanced deep CNN model. Shastri et al. (2023) developed an advanced convolutional neural network (E-CNN) for the identification of citrus diseases. Their model exhibited substantial enhancements compared to prior methodologies, attaining an average F1 score of 92.06%, precision of 95.14%, recall of 96.67%, recognition accuracy of 98%, and classification accuracy of 99%. These results exceeded prior methodologies by more than 6%, underscoring the capability of this strategy to improve disease management in the citrus sector.

8.4.2 Key Algorithms in Image Processing

You Only Look Once (YOLO) is an object detection framework that utilizes a grid-based methodology, analyzing a whole image in a single traversal of the network, resulting in remarkable speed (Redmon et al. 2016). Recent breakthroughs in the YOLO series indicate that YOLOX has achieved notable enhancements in object detection. YOLOX, developed on the YOLOv3-Darknet architecture, integrates a Spatial Pyramid Pooling (SPP) layer to improve feature extraction. This model has been utilized for the diagnosis of kiwi leaf disease, isolating leaves from intricate backgrounds, using semantic segmentation executed by DeepLabV3+ and UNet (Yao et al. 2022).

Omer et al. (2024) assembled a novel dataset comprising 3057 photos of cucumber leaves to investigate pest and disease detection. Utilizing this information, they constructed an improved model for detecting cucumber leaf diseases and pests, founded on YOLOv5l. Noon et al. (2022) presented an enhanced YOLOX model that incorporates a revised spatial pyramid pooling layer to optimize feature extraction at various sizes. By incorporating skip connections and an alpha Intersection over Union (αIoU)-based regression loss function, their model proficiently tackled issues associated with multi-disease occurrences and analogous symptom patterns. Upon evaluation with a dataset of 1112 cotton plant photos, the improved YOLOX-s model attained a mean average precision (mAP) of 73.13%, exceeding the original YOLOX by 3.27%. Wang et al. (2021) utilized a YOLO-based methodology that integrated a DenseNet block into the YOLO framework to enhance feature extraction for tomato disease diagnosis in intricate background scenarios.

8.5 Sensor-based Disease Monitoring

Recent years have seen improvements in sensor technology that have enabled healthier plants to be monitored much more accurately and crop disease to be detected earlier than before. For example, traditional techniques for disease identification, such as visual inspection and laboratory testing, can be time-consuming, labor-intensive, and lack real-time information on disease initiation and progress. In contrast, sensor-based monitoring may provide a more targeted and quantitative methodology to track the environmental conditions and physiological state of plants that drive disease development.

8.5.1 Role of Sensor-based Monitoring in Plant Disease Detection

Environmental conditions that promote the growth of disease-causing pathogens can cause crop diseases can be analyzed to predict disease risk. Sensors are essential to tracking these conditions and provide real-time data that can be analyzed to predict disease risk. The early detection enabled by sensors helps in minimizing crop losses, reduce the excessive use of chemical pesticides, and implement targeted disease management strategies.

Several types of sensors are used for plant disease monitoring:

- **Optical Sensors:** These sensors use visible, near-infrared, and thermal imaging to detect changes in leaf reflectance, chlorophyll fluorescence, and canopy temperature. They help identify symptoms like leaf yellowing, necrotic spots, or fungal infections before they become visible to the human eye (Bagha et al. 2021).
- **Electrochemical Sensors:** These sensors detect specific chemical compounds related to plant stress, such as volatile organic compounds (VOCs) released during pathogen infection or nutrient deficiencies in plants (Cellini et al. 2016).
- **Microsensors for Soil Health Monitoring:** Sensors embedded in the soil measures parameters like pH, electrical conductivity, and microbial activity, providing insights into conditions favorable for pathogen growth and disease spread (Joly et al. 2024).
- **Gas Sensors:** These sensors detect gases such as ethylene, a plant hormone associated with stress responses, and VOCs released by infected plants, indicating early-stage disease development (Dung et al. 2023).

The integration of these sensors into precision agriculture provides farmers with valuable insights to take timely action against potential disease outbreaks. The continuous flow of sensor data is processed and analyzed using advanced computational techniques, allowing the implementation of predictive analytics and automated decision-making in disease control.

8.5.2 Role of IoT in Disease Detection and Management

Plant disease monitoring has been transformed by the use of IoT devices in agriculture, which allow for real-time data collecting and analysis (Khanna et al. 2019). IoT-based monitoring systems use a network of smart sensors to measure important environmental variables that have a big impact on disease outbreaks and pathogen proliferation, like temperature, humidity, soil moisture, and leaf wetness. In order to enable prompt responses, these sensors are placed in agricultural fields to continually monitor plant health and identify early indicators of diseases.

IoT-based disease detection relies on wireless sensor networks (WSNs), which transmit collected data to cloud-based platforms or edge computing devices for analysis. For instance, sensors embedded in the soil can measure moisture levels, helping to identify conditions conducive to fungal infections like root rot. Similarly, temperature and humidity sensors mounted in the canopy can assess environmental conditions that favor bacterial or viral disease outbreaks (Foughali et al. 2018). These real-time data streams enable rapid responses, preventing the escalation of infections and reducing economic losses. Additionally, remote sensing technologies, such as drones and satellite imagery equipped with multispectral and hyperspectral sensors, have enhanced plant disease detection by capturing high-resolution images of crop fields. These technologies enable the early identification of stress patterns and disease symptoms, often before visible signs appear. Advanced image processing techniques, including ML models like CNNs, can analyze the spectral signatures of plant leaves to detect subtle changes in chlorophyll content, water stress, or disease lesions, allowing for early intervention (Khalid et al. 2023). Moreover, IoT-based monitoring facilitates automated alerts and predictive analytics by integrating with AI models. By processing sensor data in real time, these systems can identify patterns associated with disease outbreaks and send automated warnings to farmers via mobile applications (Khattab et al. 2019). This integration ensures that corrective actions such as targeted fungicide applications, irrigation adjustments, or pest control measures can be implemented immediately, minimizing crop damage and economic losses.

8.5.3 Disease Progression Using Environmental Sensors

Environmental sensors play a critical role in understanding and predicting disease progression in crops. The relationship between environmental conditions and pathogen development is well established, and real-time monitoring of these factors allows for early detection and intervention. The following environmental parameters are crucial for plants.

- **Temperature:** Temperature fluctuations influence the growth and survival of many plant pathogens. For example, fungal pathogens such as *Phytophthora* thrive in warm and humid conditions, while bacterial diseases like *Xanthomonas* proliferate under high temperatures. Continuous temperature monitoring helps in predicting when conditions become favorable for disease outbreaks, enabling farmers to take preventive actions (Trilles et al. 2019).
- **Humidity:** High relative humidity can accelerate the spread of many plant diseases, especially fungal infections such as powdery mildew and late blight. Sensors detecting prolonged periods of high humidity can signal an increased risk of pathogen sporulation and transmission, prompting timely application of control measures such as ventilation adjustments or fungicide applications.
- **Soil Moisture:** Soil moisture levels are crucial for root health and can influence the development of soilborne diseases like *Rhizoctonia* and *Fusarium* wilt. Sensors measuring soil moisture provide early warnings about excessive or insufficient water availability, helping to optimize irrigation strategies and prevent conditions conducive to disease proliferation.
- **Leaf Wetness:** Prolonged leaf wetness creates an ideal environment for pathogens that rely on water films for infection, such as *Pseudomonas syringae* and *Botrytis cinerea*. Leaf wetness sensors track moisture retention on plant surfaces, offering insights into disease risk and allowing farmers to adjust irrigation schedules or deploy protective measures like row covers or chemical treatments (Rowlandson et al. 2015).

By analyzing real-time data from these sensors, predictive models can estimate the likelihood of disease outbreaks and suggest targeted interventions. ML algorithms can integrate historical disease patterns with current sensor readings to generate early warning systems, reducing reliance on reactive disease management strategies and promoting proactive control measures.

8.5.4 Integration with ML

ML plays a crucial role in enhancing IoT-based disease monitoring by analyzing large volumes of sensor data to detect patterns and predict potential disease outbreaks. ML algorithms, including supervised and unsupervised learning techniques, process the data collected from IoT sensors to classify disease presence, assess severity, and recommend optimal interventions (Buja et al. 2021). One of the key applications of ML in plant disease management is anomaly detection, where ML models analyze environmental sensor readings to identify deviations from normal conditions that may indicate disease onset. For instance, decision trees, RFs, and deep learning models can be trained on historical sensor data to recognize conditions that typically precede outbreaks of fungal or bacterial infections. By continuously learning from new data, these models improve their predictive accuracy over time.

ML algorithms also facilitate disease classification using image-based analysis in conjunction with IoT sensors. When combined with remote sensing technologies such as drones, ML-powered image recognition models can identify visual symptoms of diseases, such as leaf discoloration, necrosis, or lesions. CNNs are particularly effective in analyzing multispectral and hyperspectral images to distinguish between different plant diseases with high precision. Furthermore, by suggesting specific disease management techniques based on environmental factors, ML-driven predictive models aid precision agriculture. For instance, in order to predict the probability of particular disease outbreaks, decision support systems employ ML to evaluate complicated information, such as historical climatic data, real-time sensor measurements, and disease incidence reports. Before the illness spreads widely, these forecasts assist farmers in taking preventive measures like modifying the frequency of irrigation, using pesticides as efficiently as possible, or putting biological control techniques into practice (Gao et al. 2020) Furthermore, blockchain technology and ML models can be combined to guarantee data confidentiality and transparency in plant health monitoring. Stakeholders in the agricultural supply chain can access and validate plant health records by safely storing sensor data and ML-generated predictions on decentralized ledgers. This integration facilitates early detection, reaction, traceability and mitigation of plant diseases.

The combination of IoT and ML fosters a data-driven approach to plant disease management, ensuring that farmers receive accurate and timely insights into crop health. By leveraging automated data collection and advanced analytics, this integrated system enhances early disease detection, optimizes resource utilization, and contributes to sustainable agricultural practices. The continuous improvement of sensor technologies and ML algorithms will further enhance the precision and efficiency of plant disease monitoring and management in the future.

8.6 Genomic Approaches for Disease Prediction

ML can improve the performance and interpretation of trait association (Hu et al. 2020). The ability to use larger and more diverse datasets for trait association has been made possible by several technical developments. By evaluating genetic data, ML can identify QTLs, genes, and alleles that contribute to different traits, such as disease resistance. It can identify the complex relationships between pathogens and plants and estimate the efficiency of genes in plants that are important for defense against them (Sperschneider 2020).

The increased affordability of plant genotyping has led to the public availability of thousands of genotypes of key crop plants, such as rice, maize, and wheat (Wang et al. 2018; Sun et al. 2018; Thomas et al. 2023). Genome-wide association studies (GWAS) are a potent method for examining complex traits in plants, which effectively map traits based on phenotypes and genotypes (Cortes et al. 2021). In both rice and wheat, GWAS has discovered more than hundreds of marker trait associations (MTA). For example, in rice, 12 bacterial blight MTA (Lu et al. 2021), 27 sheath blight MTA (Zhang et al. 2019), 11 bacterial leaf streak MTA (Sattayachiti et al. 2020), more than 200 MTA blast MTA, and 15 rice false smut MTA (Long et al. 2020) have been discovered through GWAS. According to a recent research, in addition to dominant R genes, these loci also include susceptibility genes (Xu et al. 2023). However, it is challenging to select and utilize these MTAs in plant breeding. Different models of GS have been used in breeding of plants within the last two decades (Xu et al. 2021). The GEBV (genomic estimated breeding values) of genotyped individuals can be estimated using different GS models, which can improve the selection accuracy and decrease the breeding time (Crossa et al. 2021). GWAS-based GS tools have been developed using genome-wide marker data. To find the ideal amount of markers and choose the best predictive model, for instance, the GMStool employs suitable statistical and ML-based models (Jeong et al. 2020).

For GS in plants, numerous nonparametric ML methods have been used. These methods include neural networks (NNs), reproducing kernel Hilbert spaces (RKHSs), SVMs, and RFs (Rutkoski et al. 2012; Perez-Rodriguez et al. 2012; Xu et al. 2018). Additionally, it has been observed that ML for GS can increase resistance to plant diseases such as wheat rust (Ornella et al. 2017), wheat fusarium head blight (Arruda et al. 2015), and maize leaf blight (Technow et al. 2013). In particular, the RKHS technique achieved a maximum prediction accuracy of 57.50% for fusarium head blight resistance in a research examining the usefulness of ML in GS for disease resistance in wheat (Rutkoski et al. 2012). Multi-layer perceptron (MLP) was utilized in estimation of different disease resistance; grey leaf spot (Montesinos-López et al. 2020), and Septoria leaf spot (Pérez-Rodríguez et al. 2020) in maize, and fusarium head blight in wheat (Perez-Rodríguez et al. 2020). With predictive capacities varying from 15% to 72% across strains, the genomic best linear unbiased prediction (GBLUP) and genomic Bayesian was applied to rice for prediction of blast disease resistance (Huang et al. 2019).

8.6.1 ML for Identification of Disease Resistance Traits Using Genomic Data

Liu et al. (2024) created three new ML models: RF classification plus kinship: RFC_K, light gradient boosting machine plus kinship: lightGBM_K, and support vector classification plus kinship: SVC_K, by combining phenotypes, population kinship (K), and GWAS results of disease resistance data. When they compared these three models with five other ML techniques, they discovered that these three models were able to predict a wide range of diseases with great accuracy: up to 95% for blast resistance and 85% for both sheath blight resistance and black-streaked dwarf virus in rice. Additionally, the models demonstrated up to 93% and 90% accuracy for wheat stripe rust resistance and wheat blast, respectively. They used a different rice population to test the generalizability of these three models. When compared to real spray inoculation results, these models achieved a high prediction accuracy of 90% for rice blast resistance. In addition to providing a reliable method for predicting disease resistance, strategy developed by Liu et al. (2024) facilitates genome-based crop breeding by utilizing ML.

Puccinia melanocephala is the source of brown rust in sugarcane, which reduces yield and results in significant financial losses. To eradicate the disease, genotypes with innate resistance can be used. Sugarcane cultivars have distinct sets of chromosomes (numbers ranging from 80 to 130), a polyploid genome (6× to 14×), a high percentage of repetitive regions, an estimated whole-genome size of 10 Gb, and repeated existence of aneuploidy at the locus level depending on the number of homologous chromosomes in hybrid varieties. Because of this, basic statistical models struggle to capture the dataset's nonlinear characteristics. So, Aono et al. (2020) employed ML methods to find resistance genes using eight ML algorithms; decision tree (DT), adaptive boosting (AB), Gaussian process (GP), Gaussian naive Bayes (GNB), K-nearest neighbor (KNN), multilayer perceptron (MLP) neural network, RF, and support vector machine (SVM) were assessed in a study involving 180 offspring and their parents. The marker dataset was reduced by determining feature importance using five distinct Feature Selection techniques: RF, gradient tree boosting (GTB), L1-based feature selection using a linear support vector classification system, highly randomized trees, and univariate feature selection using ANOVA. The final dataset had a 95% accuracy rate and contained 131 single-nucleotide polymorphisms (SNPs). Numerous identified regions co-localized with previously identified QTLs.

Sugarcane yellow leaf (SCYL), a major disease of sugarcane, is caused by *sugarcane yellow leaf virus* (SCYLV). SCYL is largely unidentified in the plant, and breeding plant genotypes resistant to it is an effective approach to mitigate damage and losses. Pimenta et al. (2021) predicted genotypes linked to SCYLV resistance using mixed linear model (MLM) and FarmCPU models of GWAS and ML techniques in conjunction with feature selection. The genotype dataset included amplified fragment length polymorphisms (AFLP), simple-sequence repeats (SSR), insertions and deletions (indels), and single nucleotide polymorphisms (SNPs). When tested with the entire marker set, eight different ML methods; AB, DT, GNB, GP, KNN, MLP, RF, and SVM showed low accuracy; however, feature selection plus multilayer perceptron neural network (MLP) increased accuracy to 95%. In agreement with Aono et al. (2020), the study identified that a smaller marker set consisting of 120–190 SNPs produced better accuracy than a larger dataset. Another study was conducted to identify QTLs governing resistance to the sugarcane mosaic virus (SCMV) using GWAS, ML, and feature selection. To predict resistant and susceptible genotypes, GWAS was carried out using MLM approach, eight ML models, and three feature selection techniques on a heterogeneous panel of 97 genotypes. A smaller dataset of 73 SNPs produced an average accuracy of 90.2% for several ML methods, with MLP achieving the best accuracy of 99.7%, in line with earlier research (Pimenta et al. 2023). In order to discover the molecular pathways behind resistance, they extended this study by confirming the markers derived from ML and annotating those using RNA-Sequencing data.

Coffee leaf rust, caused by *Hemileia vastatrix*, is the primary fungal disease that affects coffee cultivation worldwide. A total of 137 markers (1 unique primer, 4 RAPD, 58 SSR, and 74 AFLP) were used to genotype 245 Arabica coffee plants (Sousa et al. 2021). To estimate resistance to coffee leaf rust, the study used DT models with refinements, such as boosting, bagging, and RF. The outcomes were compared with those from ANN, Bayesian Generalized Linear Regression (BGLR), and Bayesian Generalized Linear Regression (GBLASSO). The accuracy ranged from 30.6% to 60.7% on average, which was better than the BGLR approach. Results from the literature were compared with the top 10% of the most significant markers found by each method. DT with boosting and GBLASSO approaches shared a total of 11 markers in common. A comparison with the literature showed an average of 9.3% of predicted markers were present in regions of QTLs linked with leaf rust resistance. In comparison to the GBLASSO technique, the study found that the ML models had higher overall accuracy and that the authors had found more markers linked to the trait (Sousa et al. 2021).

Gonzalez-Camacho et al. (2016) used phenotypic and genomic data as input variables to evaluate two ML models: probabilistic neural network (PNN) and multi-layer perceptron (MLP) to predict the possibility of membership of one individual in a phenotypic class of interest. A total of 33 data sets [17 wheat and 16 maize including maize grey leaf spot (GLS) resistance] with different trait–environment combinations were examined. PNN consistently outperformed MLP in classifying all 33 datasets including GLS resistance phenotype, with scores 0.05 higher than MLP, according to the AUROC score, an evaluation metric used to determine how well a model can distinguish across classes. In GS, PNN with Gaussian radial basis functions appears promising to find the best individuals.

For the genetic prediction of disease resistance, a number of studies combined statistical and ML models. For instance, González-Camacho et al. (2012) used two nonlinear ML techniques; reproducing kernel Hilbert spaces and radial basis function neural networks, along with linear Bayesian least absolute shrinkage and selection operator regression (LASSO), to analyze maize genotyping data that included 55,000 markers. Along with other agronomic characteristics, they evaluated 300 maize lines for resistance to diseases such GLS and northern corn leaf blight. The study found that ML models demonstrated a greater correlation for resistance prediction, signifying that they could identify epistatic patterns in the marker set. Another study conducted by Ornella et al. (2014) favored Support Vector Classification with the linear kernel (SVC-lin) among three classifiers and six other regressor models that were assessed for estimating disease resistance for stem rust and GLS across 16 wheat and 14 maize datasets (Ornella et al. 2014). A study by Ornella et al. (2012) demonstrated that statistical methods performed well with high prediction accuracy for genomic prediction of resistance to yellow and stem rust. Five CIMMYT wheat populations were used to test Bayesian LASSO, support vector regression, and ridge regression, with linear or radial basis function kernel models. Because of the additive effects of the two traits, linear statistical models Bayesian LASSO and ridge regression showed higher prediction accuracy than support vector regression (Ornella et al. 2012).

8.7 Advances in ML Techniques for Disease Management

8.7.1 AI-driven Disease Forecasting Models

Accurate disease forecasting is essential for proactive disease management. ML models leverage vast datasets, including historical disease occurrences, weather patterns, soil conditions, and remote sensing data to predict potential outbreaks before visible symptoms appear. Advanced models, such as Long Short-Term Memory (LSTM) networks, RF regression, and Gradient Boosting Machines (GBM), are employed to analyze complex interactions among environmental variables and pathogen life cycles. Hybrid AI approaches, which integrate deep learning with geospatial analysis, have significantly improved the precision of disease forecasting. For instance, CNNs applied to satellite imagery and drone-based hyperspectral data, enhance the early detection of stress symptoms in crops, allowing farmers to intervene before substantial yield losses occur. Additionally, cloud-based AI platforms aggregate real-time data from global agricultural networks, enabling large-scale disease surveillance and early warning systems for major crops, such as wheat, rice, and maize.

Several studies have demonstrated the effectiveness of ML for disease forecasting. Yuan et al. (2025) employed LSTM networks to predict wheat rust outbreaks based on climate and pathogen data and achieved high accuracy. Similarly, Sahoo et al. (2025) utilized RF algorithms to estimate the risk of rice blast disease by leveraging climate datasets. Yadav and Yadav (2025) developed a ML-driven early warning system for grape powdery mildew, integrating weather conditions and disease history to predict outbreaks in vineyards. Shahi et al. (2023) successfully used CNN models applied to UAV-based hyperspectral imaging to detect early symptoms of maize diseases, demonstrating the potential of AI in early disease detection and management.

8.7.2 Predictive Modeling for Crop Management

Predictive modeling has revolutionized disease management by optimizing decision-making processes related to crop protection strategies. ML algorithms, such as SVMs, Bayesian networks, and Decision Trees, analyze past disease incidences, weather fluctuations, and soil health metrics to recommend precise management interventions. These models can guide farmers in selecting the most effective crop rotation strategies, fungicide application schedules, and nutrient management plans tailored to specific regional conditions. RL techniques further enhance decision-support systems by continuously refining recommendations based on real-time field data. For instance, adaptive models trained on pathogen resistance patterns can suggest targeted interventions that minimize chemical use, while maximizing disease suppression. Furthermore, integrating ML-driven insights with precision agriculture tools, such as smart irrigation systems and automated sprayers, ensures optimized resource allocation and reduces input costs and environmental impacts.

The application of predictive modeling in agriculture has been well documented. Gammanpila et al. (2024) developed an AI-based model that predicts apple scab infections and recommended optimized fungicide application schedules to reduce unnecessary chemical use. Deininger et al. (2020) demonstrated how ML models can be used to suggest ideal crop rotation patterns for soybean farming to minimize fungal infections. A Bayesian network-based decision support system by Radwan et al. (2024) was utilized in potato farming to optimize fungicide applications based on real-time weather data. Hadri (2024) employed RL to suggest optimal disease control strategies for wheat stripe rust, showcasing the dynamic potential of AI-driven decision-making in disease management.

8.7.3 Transfer Learning and Model Generalization

One of the primary challenges in applying ML to plant disease management is the need for extensive labeled datasets for each crop and disease. Transfer learning addresses this limitation by enabling the adaptation of pretrained models to new agricultural contexts with minimal retraining. This technique significantly reduces computational costs and accelerates model deployment, making AI-driven disease management accessible to a broader range of crops and farming systems. Deep-learning architectures, such as CNNs and transformer models, can be fine-tuned using small-scale, domain-specific datasets while retaining knowledge from previous training on large, diverse datasets. For example, a CNN trained in wheat rust detection can be adapted to identify late blight in potatoes, with minor modifications to its final layers. Few-shot learning and domain adaptation techniques further enhance the model's robustness, ensuring consistent performance across different climatic zones and farming conditions.

Öztürk et al. (2023) applied CNN-based transfer learning to diagnose multiple crop diseases using a single model trained on diverse datasets, significantly reducing the need for extensive retraining. Kunduracioglu and Pacal (2024) demonstrated how transfer learning can be utilized to adapt disease detection models originally trained on wheat to identify powdery mildew in grapevines. Rezaei et al. (2024) explored few-shot learning techniques to enhance citrus disease detection, effectively improving classification accuracy with minimal training data. Lu et al. (2022) employed Generative Adversarial Networks (GANs) to generate synthetic plant disease images, which improved the robustness of ML models trained on small datasets, enabling more accurate predictions across different crops and regions.

8.8 Challenges and Perspectives

Despite considerable progress in ML for plant disease identification and control, various difficulties impede its broad adoption, requiring enhancements in efficacy, accessibility, and dependability. Deep learning models frequently operate as opaque systems, constraining interpretability for farmers and agronomists. Explainable AI (XAI) can improve transparency, hence cultivating trust and usability. The lack of high-quality annotated datasets continues to impede progress due to crop diversity, geographical differences, and environmental influences. Techniques such as synthetic data generation, augmentation, and collaborative data-sharing platforms can enhance model resilience. ML models also encounter difficulties in generalizing across various plant species and areas; transfer learning and domain adaption strategies may alleviate this issue. Insufficient computational resources in developing nations further hinder ML implementation; nevertheless, the incorporation of edge computing and lightweight ML algorithms can facilitate real-time illness diagnosis on mobile devices and low-power hardware. Integrating ML with conventional diagnostic techniques, including biochemical testing and expert knowledge, can improve precision and reliability. Hybrid approaches that integrate AI-driven and manual methods should be investigated to optimize effectiveness. Ethical problems, such as data ownership and privacy, must be handled through secure, farmer-centric data-sharing frameworks that incorporate equitable pay mechanisms. Definitive legislative frameworks are crucial for guaranteeing data security, precision, and the ethical implementation of AI in agriculture. Moreover, numerous farmers lack the technical proficiency to effectively employ ML techniques; creating user-friendly interfaces, mobile applications, and training initiatives can address this deficiency, facilitating wider adoption. Through interdisciplinary collaboration, ethical policy, and sustainable technological integration, ML can transform plant disease control, enhancing global food security and agricultural resilience.

8.9 Conclusion

Agriculture is fundamental to economic stability in emerging nations but is increasingly jeopardized by plant diseases. Traditional diagnostic techniques are deficient in precision and scalability, requiring sophisticated solutions for the early detection of diseases. Automated plant disease identification, utilizing ML and image-based technologies, provides a revolutionary method that improves accuracy, hastens detection, and facilitates real-time monitoring of crop health. These breakthroughs combine convolutional CNNs with phytopathological knowledge to accurately diagnose and classify plant diseases. Furthermore, they provide real-time pathogen monitoring, autonomous disease control, and cooperative disease tracking. Nonetheless, the deployment of these technologies in low-income areas is impeded by hardware expenses, infrastructural deficiencies, and the requirement for specific technical proficiency. Notwithstanding these obstacles, ML has transformed plant disease management by enhancing diagnostic precision, efficiency, and prediction powers. Advancements in transfer learning, edge computing, and explainable AI will augment model reliability and application. The integration of ML with sustainable agriculture practices, conventional diagnostic techniques, and novel artificial intelligence advances offers a promising trajectory. Effective utilization of ML-driven plant disease management necessitates interdisciplinary collaboration, investment in superior data collecting, and solutions centered on farmers. Future research must focus on enhancing model precision, scalability, and cost-efficiency across various agricultural systems. As technology advances, collaborative initiatives among researchers, politicians, and agricultural stakeholders will be essential for guaranteeing global food security and sustainable agricultural output.

References

Abdulridha, J., Ampatzidis, Y., Qureshi, J. et al. (2022). Identification and classification of downy mildew severity stages in watermelon utilizing aerial and ground remote sensing and machine learning. *Frontiers in Plant Science* 13: 791018.

Ahmed, I. and Yadav, P.K. (2023). Plant disease detection using machine learning approaches. *Expert Systems 40* (5): e13136.

Akbar, M., Ullah, M., Shah, B. et al. (2022). An effective deep learning approach for the classification of bacteriosis in peach leaves. *Frontiers in Plant Science* 13: 1064854.

Alberto, R.T., Rivera, J.C.E., Biagtan, A.R. et al. (2020). Extraction of onion fields infected by anthracnose- twister disease in selected municipalities of Nueva Ecija using UAV imageries. *Spatial Information Research* 28: 383–389.

Alirezazadeh, P., Schirrmann, M. and Stolzenburg, F. (2023). Improving deep learning- based plant disease classification with attention mechanism. *Gesunde Pflanz* 75: 49–59.

Aliyu, M.A., Mokji, M.M.M., and Sheikh, U.U.U. (2020). Machine learning for plant disease detection: An investigative comparison between support vector machine and deep learning. *IAES International Journal of Artificial Intelligence 9* (4): 670.

Almoujahed, M.B., Rangarajan, A.K., Whetton, R.L. et al. (2022). Detection of fusarium head blight in wheat under field conditions using a hyperspectral camera and machine learning. *Computers and Electronics in Agriculture 203*: 107456.

Alom, M.Z., Taha, T.M., Yakopcic, C. et al. (2019). A state-of-the-art survey on deep learning theory and architectures. *Electronics 8* (3): 292.

Alshammari, H., Gasmi, K., Krichen, M. et al. (2022). Optimal deep learning model for olive disease diagnosis based on an adaptive genetic algorithm. *Wireless Communications and Mobile Computing 2022* (1): 8531213.

Aono, A.H., Costa, E.A., Rody, H.V.S. et al. (2020). Machine learning approaches reveal genomic regions associated with sugarcane brown rust resistance. *Scientific Reports* 10: 20057.

Archana, K.S., Srinivasan, S., Bharathi, S.P. et al. (2022). A novel method to improve computational and classification performance of rice plant disease identification. *The Journal of Supercomputing 6*: 1–21.

Arivazhagan, S. and Ligi, S.V. (2018). Mango leaf diseases identification using convolutional neural network. *International Journal of Pure and Applied Mathematics 120* (6): 11067–11079.

Arruda, M.P, Brown, P.J, Lipka, A.E. et al. (2015). Genomic selection for predicting fusarium head blight resistance in a wheat breeding program. *Plant Genome* 8 (3): plantgenome2015.01.0003.

Bagha, H., Yavari, A., and Georgakopoulos, D. (2021). IoT-based plant health analysis using optical sensors in precision agriculture. *2021 Digital Image Computing: Techniques and Applications (DICTA)*, 01–08. IEEE.

Balasubramanian, K. (2024). Feature analysis and classification of maize crop diseases employing AlexNet-inception network. *Multimedia Tools and Applications 83* (9): 26971–26999.

Bhuyan, P., Singh, P.K. and Das, S.K. (2023). Res4net-CBAM: A deep CNN with convolution block attention module for tea leaf disease diagnosis. *Multimedia Tools and Applications* 83: 48925–48947.

Borhani, Y., Khoramdel, J. and Najafi, E. (2022). A deep learning based approach for automated plant disease classification using vision transformer. *Scientific Reports 12* (1): 11554.

Buja, I., Sabella, E., Monteduro, A.G. et al. (2021). Advances in plant disease detection and monitoring: From traditional assays to in-field diagnostics. *Sensors 21* (6): 2129.

Bukar, U.A., Sayeed, M.S., Razak, S.F.A. et al. (2023). A method for analyzing text using VOSviewer. *MethodsX 11*: 102339.
Cellini, A., Biondi, E., Blasioli, S. et al. (2016). Early detection of bacterial diseases in apple plants by analysis of volatile organic compounds profiles and use of electronic nose. *Annals of Applied Biology 168* (3): 409–420.
Chaudhari, D.J. and Malathi, K. (2023). Detection and prediction of rice leaf disease using a hybrid CNN-SVM model. *Optical Memory and Neural Networks 32* (1): 39–57.
Chen, T., Yang, W., Zhang, H. et al. (2020). Early detection of bacterial wilt in peanut plants through leaf-level hyperspectral and unmanned aerial vehicle data. *Computers and Electronics in Agriculture 177*: 105708.
Chen, Y., Pan, J., and Wu, Q. (2023). Apple leaf disease identification via improved CycleGAN and convolutional neural network. *Soft Computing 27* (14): 9773–9786.
Cheshkova, A.F. (2022). A review of hyperspectral image analysis techniques for plant disease detection and identification. *Vavilov Journal of Genetics and Breeding 26* (2): 202.
Cortes, L.T., Zhang, Z., and Yu, J. (2021). Status and prospects of genome-wide association studies in plants. *Plant Genome* 14 (1): e20077.
Crossa, J., Fritsche-Neto, R., Montesinos-Lopez, O.A. et al. (2021). The modern plant breeding triangle: Optimizing the use of genomics, phenomics, and Environics data. *Front Plant Sci* 12: 651480.
Daphal, S.D. and Koli, S.M. (2023). Enhancing sugarcane disease classification with ensemble deep learning: A comparative study with trans fer learning techniques. *Heliyon* 9: e18261.
Deininger, K., Ali, D.A., Kussul, N. et al. (2020). Using machine learning to assess yield impacts of crop rotation. *Policy Research Working Paper* (9306).
Delfani, P., Thuraga, V., Banerjee, B. et al. (2024). Integrative approaches in modern agriculture: IoT, ML and AI for disease forecasting amidst climate change. *Precision Agriculture 25* (5): 2589–2613.
Domingues, T., Brandão, T., and Ferreira, J.C. (2022). Machine learning for detection and prediction of crop diseases and pests: A comprehensive survey. *Agriculture 12* (9): 1350.
Dou, S., Wang, L., Fan, D. et al. (2023). Classification of citrus huanglongbing degree based on CBAM- MobileNetV2 and transfer learning. *Sensors* 23: 5587.
Dung, T.T., Lee, U.J., Oh, Y. et al. (2023). Gas sensor-based techniques for detecting ethylene (C_2H_4). *Applied Science and Convergence Technology 32* (4): 82–88.
Elfatimi, E., Eryiğit, R., and Shehu, H.A. (2024). Impact of datasets on the effectiveness of MobileNet for beans leaf disease detection. *Neural Computing and Applications 36* (4): 1773–1789.
Farman, H., Ahmad, J., Bilal, J. et al. (2022). EfficientNet-based robust recognition of peach plant diseases in field images. *Computers, Materials, and Continua* 71: 2073–2089.
Foughali, K., Fathallah, K., and Frihida, A. (2018). Using cloud IOT for disease prevention in precision agriculture. *Procedia Computer Science 130*: 575–582.
Gammanpila, H.W., Sashika, M.N., and Priyadarshani, S.V.G.N. (2024). Advancing horticultural crop loss reduction through robotic and AI technologies: Innovations, applications, and practical implications. *Advances in Agriculture 2024* (1): 2472111.
Gao, D., Sun, Q., Hu, B. et al. (2020). A framework for agricultural pest and disease monitoring based on internet-of-things and unmanned aerial vehicles. *Sensors 20* (5): 1487.
Gautam, V., Ranjan, R.K., Dahiya, P. et al. (2023). ESDNN: A novel ensembled stack deep neural network for mango leaf disease classification and detection. *Multimedia Tools and Applications* 83: 10989–11015.
Ghosh, P., Mondal, A.K., Chatterjee, S. et al. (2023). Recognition of sunflower diseases using hybrid deep learning and its explainability with AI. *Mathematics 11* (10): 2241.
González-Camacho, J.M., Crossa, J., Pérez-Rodríguez, P. et al. (2016). Genome-enabled prediction using probabilistic neural network classifiers. *BMC Genomics* 17: 208.
González-Camacho, J.M., de Los Campos, G., Pérez, P. et al. (2012). Genome-enabled prediction of genetic values using radial basis function neural networks. *Theoretical and Applied Genetics* 125: 759–771.
Goyal, L., Sharma, C.M., Singh, A. et al. (2021). Leaf and spike wheat disease detection and classification using an improved deep convolutional architecture. *Informatics in Medicine Unlocked 25*: 100642.
Gupta, D., Sharma, P., Choudhary, K. et al. (2021). Artificial plant optimization algorithm to detect infected leaves using machine learning. *Expert Systems 38* (6): e12501.
Hadri, M.I. (2024). Smart Deep System to Detect and Diagnose Wheat Disease. MSc thesis, University of Guelma, Algeria.
Hajare, N. and Rajawat, A.S. (2023). Black gram disease classification via deep ensemble model with optimal training. *International Journal of Image and Graphics 25* (04): 2550033. Available from https: //doi.org/10.1142/S0219467825500330.

Haque, M.A., Marwaha, S., Deb, C.K. et al. (2022). Deep learning-based approach for identification of diseases of maize crop. *Scientific Reports 12* (1): 6334.

Hastie, T., Tibshirani, R., and Friedman, J. (2009). *The Elements of Statistical Learning: Data Mining, Inference, and Prediction*. Springer New York, NY.

Goodfellow, I., Bengio, Y., Courville, A., & Bengio, Y. (2016). *Deep learning* (Vol. 1, No. 2). Cambridge: MIT press.

Hu, T., Darabos, C. and Urbanowicz, R. (2020). Editorial: Machine learning in genome-wide association studies. *Frontiers in Genetics* 11: 593958.

Huang, M., Balimponya, E.G., Mgonja, E.M. et al. (2019). Use of genomic selection in breeding rice (Oryza sativa L.) for resistance to rice blast (Magnaporthe oryzae). *Molecular Breeding* 39 (8): 114.

Jayapal, S.K. and Poruran, S. (2023). Enhanced disease identification model for tea plant using deep learning. *Intelligent Automation & Soft Computing* 35: 1261–1275.

Jeong, S., Kim, J.Y., and Kim, N. (2020). GMStool: GWAS-based marker selection tool for genomic prediction from genomic data. *Scientific Reports* 10 (1): 19653.

Johnson, M.A., Vinatzer, B.A., and Li, S. (2023). Reference-free plant disease detection using machine learning and long-read metagenomic sequencing. *Applied and Environmental Microbiology 89* (6): e00260–23.

Joly, M., Marlet, M., Barreau, D. et al. (2024). Monitoring of ammonium and nitrate ions in soil using ion-sensitive potentiometric microsensors. *Sensors (Basel, Switzerland) 24* (22): 7143.

Joshi, R.C., Kaushik, M., Dutta, M.K. et al. (2021). VirLeafNet: Automatic analysis and viral disease diagnosis using deep learning in Vigna mungo plant. *Ecological Informatics* 61: 101197.

Kamilaris, A. and Prenafeta-Boldú, F.X. (2018). Deep learning in agriculture: A survey. *Computers and Electronics in Agriculture 147*: 70–90. https://doi.org/10.1016/j.compag.2018.02.016

Kanna, G.P., Kumar, S.J.K.J., Kumar, Y. et al. (2023). Advanced deep learning techniques for early disease prediction in cauliflower plants. *Scientific Reports* 13: 18475.

Karthickmanoj, R. and Sasilatha, T. (2024). Development of plant disease detection for smart agriculture. *Multimedia Tools and Applications* 83: 54391–54410.

Khalid, M., Sarfraz, M.S., Iqbal, U. et al. (2023). Real-time plant health detection using deep convolutional neural networks. *Agriculture 13* (2): 510.

Khanna, A. and Kaur, S. (2019). Evolution of Internet of Things (IoT) and its significant impact in the field of precision agriculture. *Computers and Electronics in Agriculture 157*: 218–231.

Khattab, A., Habib, S.E., Ismail, H. et al. (2019). An IoT-based cognitive monitoring system for early plant disease forecast. *Computers and Electronics in Agriculture 166*: 105028.

Kim, W.S., Lee, D.H., and Kim, Y.J. (2020). Machine vision-based automatic disease symptom detection of onion downy mildew. *Computers and Electronics in Agriculture* 168: 105099.

Kulkarni, P., Karwande, A., Kolhe, T. et al. (2021). Plant disease detection using image processing and machine learning. *arXiv preprint arXiv:2106.10698*.

Kundu, N., Rani, G., Dhaka, V.S. et al. (2021). IoT and interpretable machine learning based framework for disease prediction in pearl millet. *Sensors 21* (16): 5386.

Kunduracioglu, I. and Pacal, I. (2024). Advancements in deep learning for accurate classification of grape leaves and diagnosis of grape diseases. *Journal of Plant Diseases and Protection 131* (3): 1061–1080.

Kursun, R., Bastas, K.K., and Koklu, M. (2023). Segmentation of dry bean (Phaseolus vulgaris L.) leaf disease images with U-net and classification using deep learning algorithms. *European Food Research and Technology* 249: 2543–2558.

Leng, S., Musha, Y., and Yang, Y. (2023). CEMLB-YOLO: Efficient detection model of maize leaf blight in complex field environments. *Applied Sciences 13* (16): 9285.

Li, G., Jiao, L., Chen, P. et al. (2023). Spatial convolutional self-attention-based transformer module for strawberry disease identification under complex background. *Computers and Electronics in Agriculture* 212: 108121.

Li, L., Zhang, S., and Wang, B. (2021). Plant disease detection and classification by deep learning—A review. *IEEE Access 9*: 56683–56698.

Linker, R. and Dafny-Yalin, M. (2024). Mapping fire blight cankers and autumn blooming in pear trees using faster R-CNN. *Precision Agriculture* 25: 396–411.

Liu, A., Zuo, S.M.B, Peng, S. et al. (2024). Development of machine learning methods for accurate prediction of plant disease resistance. *Engineering* 40 (2024): 100–110.

Long, W., Yuan, Z., Fan, F. et al. (2020). Genome-wide association analysis of resistance to rice false smut. *Molecular Breeding* 40 (5): 46.

Lu, J., Wang, C., Zeng, D. et al. (2021). Genome-wide association study dissects resistance loci against bacterial blight in a diverse rice panel from the 3000 rice genomes project. *Rice (NY)* 14 (1): 22.

Lu, Y., Chen, D., Olaniyi, E. et al. (2022). Generative adversarial networks (GANs) for image augmentation in agriculture: A systematic review. *Computers and Electronics in Agriculture 200*: 107208.

Mallick, M.D.T., Biswas, S., Das, A.K. et al. (2023). Deep learning based automated disease detection and pest classification in Indian mung bean. *Multimedia Tools and Applications* 82: 12017–12041.

Mishra, N., Seetha, J., Kumar, A.G.D. et al. (2023). Design an ant lion-based yolo-V5 model for prediction and classification of paddy leaf disease. *International Journal of Intelligent Systems and Applications in Engineering 11* (6s): 599–612.

Mohanty, S.P., Hughes, D.P., and Salathé, M. (2016). Using deep learning for image-based plant disease detection. *Frontiers in Plant Science* 7: 1419. https://doi.org/10.3389/fpls.2016.01419

Montesinos-López, O.A., Montesinos-López, J.C., Singh P. et al. (2020). A multivariate Poisson deep learning model for genomic prediction of count data. *G3:Genes, Genomes, Genetics* 10 (11): 4177–90.

Murphy, K.P. (2012). *Machine Learning: A Probabilistic Perspective. MIT Press.*

Nagaraju, M. and Chawla, P. (2023). Maize crop disease detection using NPNet-19 convolutional neural network. *Neural Computing and Applications 35* (4): 3075–3099.

Narmadha, R.P., Sengottaiyan, N., and Kavitha, R.J. (2022). Deep transfer learning based rice plant disease detection model. *Intelligent Automation & Soft Computing 31* (2).

Narmilan, A., Gonzalez, F., Salgadoe, A.S.A. et al. (2022). Detection of white leaf disease in sugarcane using machine learning techniques over UAV multispectral images. *Drones* 6: 230.

Neupane, K. and Baysal-Gurel, F. (2021). Automatic identification and monitoring of plant diseases using unmanned aerial vehicles: A review. *Remote Sensing 13* (19): 3841.

Nirmal, M.D., Jadhav, P.P., and Pawar, S. (2023). Pomegranate leaf disease detection using supervised and unsupervised algorithm techniques. *Cybernetics and Systems*: 1–12. https://doi.org/10.1080/01969722.2023.2166192

Noon, S.K., Amjad, M., Qureshi, M.A. et al. (2022). Handling severity levels of multiple co-occurring cotton plant diseases using improved YOLOX model. *IEEE Access 10*: 134811–134825.

Novtahaning, D., Hasnain Ali, S., and Jae-Mo, K. (2022). Deep learning ensemble-based automated and high-performing recognition of coffee leaf disease. *Agriculture* 12: 1909.

Nugroho, H., Pramudito, W.A., and Laksono, H.S. (2024). Gray level co-occurrence matrix (GLCM)-based feature extraction for rice leaf diseases classification. *Buletin Ilmiah Sarjana Teknik Elektro 6* (4): 392–400.

Omer, S.M., Ghafoor, K.Z., and Askar, S.K. (2024). Lightweight improved yolov5 model for cucumber leaf disease and pest detection based on deep learning. *Signal, Image and Video Processing 18* (2): 1329–1342.

Ornella, L., González-Camacho, J.M., Dreisigacker S. et al. (2017). Applications of genomic selection in breeding wheat for rust resistance. In: *Wheat rust diseases* (ed. Periyannan, S.), 173–182. Springer: Berlin.

Ornella, L., Pérez, P., Tapia, E. et al. (2014). Genomic-enabled prediction with classification algorithms. *Heredity* 112: 616–626.

Ornella, L., Singh, S., Perez, P. et al. (2012). Genomic prediction of genetic values for resistance to wheat rusts. *The Plant Genome* 5: 17.

O'shea, K. and Nash, R. (2015). An introduction to convolutional neural networks. *arXiv preprint arXiv:1511.08458.*

Ouhami, M., Hafiane, A., Es-Saady, Y. et al. (2021). Computer vision, IoT and data fusion for crop disease detection using machine learning: A survey and ongoing research. *Remote Sensing 13* (13): 2486.

Öztürk, C., Taşyürek, M., and Türkdamar, M.U. (2023). Transfer learning and fine-tuned transfer learning methods' effectiveness analyse in the CNN-based deep learning models. *Concurrency and Computation: Practice and Experience 35* (4): e7542.

Parasa, G., Arulselvi, M.M., and Razia, S. (2024). Identification of diseases in paddy crops using CNN. *11* (6s): 548–557.

Patel, K. and Patel, A. (2022). Plant disease diagnosis using image processing techniques—A review on machine and deep learning approaches. *Ecology, Environment and Conservation 28*: 351–362.

Pérez-Rodríguez, P., Flores-Galarza, S., Vaquera-Huerta, H. et al. (2020). Genome-based prediction of Bayesian linear and non-linear regression models for ordinal data. *Plant Genome* 13 (2): e20021.

Pérez-Rodríguez, P., Gianola, D., González-Camacho, J.M. et al. (2012). Comparison between linear and non-parametric regression models for genome-enabled prediction in wheat. *G3: Genes, Genomes, Genetics* 2 (12): 1595–605.

Pimenta, R.J.G., Aono, A.H., Burbano, R.C.V. et al. (2021). Genome-wide approaches for the identification of markers and genes associated with sugarcane yellow leaf virus resistance. *Scientific Reports* 11: 15730.

Pimenta, R.J.G., Aono, A.H., Burbano, R.C.V. et al. (2023). Multiomic investigation of sugarcane mosaic virus resistance in sugarcane. *The Crop Journal* 11: 1805–1815.

Qi, H., Yu, L., Ding, Q. et al. (2021). Automatic identification of peanut-leaf diseases based on stack ensemble. *Applied Sciences* 11: 1950.

Radwan, M., Alhussan, A.A., Ibrahim, A. et al. (2024). Potato leaf disease classification using optimized machine learning models and feature selection techniques. *Potato Research*: 68: 897–921 (2025). https://doi.org/10.1007/s11540-024-09763-8

Rahman, C.R., Arko, P.S., Ali, M.E. et al. (2020). Identification and recognition of rice diseases and pests using convolutional neural networks. *Biosystems Engineering 194*: 112–120.

Rai, C.K. and Pahuja, R. (2024). Northern maize leaf blight disease detection and segmentation using deep convolution neural networks. *Multimedia Tools and Applications 83* (7): 19415–19432.

Rajasekhar, V., Arulselvi, G. and Babu, K.S. (2023). Optimization-based auto-metric graph neural network framework for rice leaf disease classification. *International Journal of Intelligent Systems and Applications in Engineering 11* (6s): 563–575.

Rajpoot, V., Tiwari, A. and Jalal, A.S. (2023). Automatic early detection of rice leaf diseases using hybrid deep learning and machine learning methods. *Multimedia Tools and Applications 82* (23): 36091–36117.

Rangarajan, A.K., Balu, E.J., Boligala, M.S. et al. (2022). A low-cost UAV for detection of Cercospora leaf spot in okra using deep convolutional neural network. *Multimedia Tools and vApplications* 81: 21565–21589.

Redmon, J., Divvala, S., Girshick, R. et al. (2016). You only look once: Unified, real-time object detection. *Proceedings of the IEEE Conference on Computer Vision and Pattern Recognition*, IEEE, Las Vegas, NV, USA, 2016, pp. 779–788. https://doi.org/10.1109/CVPR.2016.91

Revathi, P. and Hemalatha, M. (2014). Identification of cotton diseases based on cross information gain deep forward neural network classifier with PSO feature selection. *International Journal of Engineering and Technology 5* (6): 4637–4642.

Rezaei, M., Diepeveen, D., Laga, H. et al. (2024). Plant disease recognition in a low data scenario using few-shot learning. *Computers and Electronics in Agriculture 219*: 108812.

Rowlandson, T., Gleason, M., Sentelhas, P. et al. (2015). Reconsidering leaf wetness duration determination for plant disease management. *Plant Disease 99* (3): 310–319.

Rutkoski, J., Benson, J., Jia, Y. et al. (2012). Evaluation of genomic prediction methods for fusarium head blight resistance in wheat. *Plant Genome* 5 (2): 51–61.

Sahoo, S., Singha, C., Govind, A. et al. (2025). Leveraging ML to predict climate change impact on rice crop disease in Eastern India. *Environmental Monitoring and Assessment 197* (4): 1–30.

Sattayachiti, W., Wanchana, S., Arikit, S. et al. (2020). Genome-wide association analysis identifies resistance loci for bacterial leaf streak resistance in rice (Oryza sativa L.). *Plants* 9 (12): 1673.

Shahi, T.B., Xu, C.Y., Neupane, A. et al. (2023). Recent advances in crop disease detection using UAV and deep learning techniques. *Remote Sensing 15* (9): 2450.

Shahoveisi, F., Riahi Manesh, M., and Del Río Mendoza, L.E. (2022). Modeling risk of Sclerotinia sclerotiorum-induced disease development on canola and dry bean using machine learning algorithms. *Scientific Reports* 12: 864.

Shahoveisi, F., Taheri Gorji, H., Shahabi, S. et al. (2023). Application of image processing and transfer learning for the detection of rust disease. *Scientific Reports 13* (1): 5133.

Shastri, R., Chaturvedi, A., Mouleswararao, B. et al. (2023). An automatic detection of citrus fruits and leaves diseases using enhanced convolutional neural network. *Remote Sensing in Earth Systems Sciences 6* (3): 123–134.

Shovon, M.S.H., Mozumder, S.J., Pal, O.K. et al. (2023). PlantDet: A robust multi-model ensemble method based on deep learning for plant disease detection. *IEEE Access 11*: 34846–34859.

Simonyan, K. and Zisserman, A. (2014). Very deep convolutional networks for large-scale image recognition. *arXiv preprint arXiv:1409.1556.*

Sladojevic, S., Arsenovic, M., Anderla, A. et al. (2016). Deep neural networks based recognition of plant diseases by leaf image classification. *Computational Intelligence and Neuroscience 2016*: 3289801. https://doi.org/10.1155/2016/3289801

Soares, A.D.S., Vieira, B.S., Bezerra, T.A. et al. (2022). Early detection of coffee leaf rust caused by Hemileia vastatrix using multispectral images. *Agronomy* 12: 2911.

Sousa, I.C.D., Nascimento, M., Silva, G.N. et al. (2021). Genomic prediction of leaf rust resistance to Arabica coffee using machine learning algorithms. *Scientia Agricola* 78: e20200021.

Sowmyalakshmi, R., Jayasankar, T., Pilllai, V.A. et al. 2021. An optimal classification model for rice plant disease detection. *Computers, Materials, & Continua 68*: 1751–1767.

Sperschneider, J. (2020). Machine learning in plant–pathogen interactions: Empowering biological predictions from field scale to genome scale. *New Phytologist* 228: 35–41.

Stephen, A., Punitha, A. and Chandrasekar, A. (2023). Designing self attention-based ResNet architecture for rice leaf disease classification. *Neural Computing and Applications 35* (9): 6737–6751.

Sudhesh, K.M., Sowmya, V., Kurian, S. et al. (2023). AI based rice leaf disease identification enhanced by Dynamic Mode Decomposition. *Engineering Applications of Artificial Intelligence 120*: 105836.

Sujatha, R., Chatterjee, J.M., Jhanjhi, N.Z. et al. (2021). Performance of deep learning vs machine learning in plant leaf disease detection. *Microprocessors and Microsystems 80*: 103615.

Sun, S., Zhou, Y., Chen, J. et al. (2018). Extensive intraspecific gene order and gene structural variations between Mo17 and other maize genomes. *Nature Genetics* 50 (9): 1289–1295.

Sun, H., Zhou, L., Shu, M. et al. (2024). Estimation of peanut southern blight severity in hyperspectral data using the synthetic minority oversampling technique and fractional-order differentiation. *Agriculture 14* (3): 476.

Sun, Y., Liu, Y., Zhou, H. et al. (2021). Plant diseases identification through a discount momentum optimizer in deep learning. *Applied Sciences 11* (20): 9468.

Technow, F., Bürger, A., and Melchinger, A.E. (2013). Genomic prediction of northern corn leaf blight resistance in maize with combined or separated training sets for heterotic groups. *G3: Genes, Genome, Genetics* 3 (2): 197–203.

Thomas, W.J.W., Zhang, Y., Amas, J.C. et al. (2023). Innovative advances in plant genotyping. In: *Plant Genotyping* (ed. Shavrukov, Y.), 451–465. Springer: Berlin.

Trilles Oliver, S., González-Pérez, A., and Huerta Guijarro, J. (2019). Adapting models to warn fungal diseases in vineyards using in-field internet of things (IoT) nodes. *Sustainability 11* (2): 416.

Upadhyaya, S.R., Danilevicz, M.F., Dolatabadian, A. et al. (2024). Genomics-based plant disease resistance prediction using machine learning. *Plant Pathology 73* (9): 2298–2309.

Vasantha, S.V., Samreen, S., and Aparna, Y.L. (2022). Rice disease diagnosis system (RDDS). *Computers, Materials & Continua 73* (1).

Wang, W., Mauleon, R., Hu, Z. et al. (2018). Genomic variation in 3,010 diverse accessions of Asian cultivated rice. *Nature* 557 (7703): 43–9.

Wang, T., Thomasson, J.A., Yang, C. et al. (2020). Automatic classification of cotton root rot disease based on UAV remote sensing. *Remote Sensing* 12: 1310.

Wang, X. and Liu, J. (2021). Tomato anomalies detection in greenhouse scenarios based on YOLO-Dense. *Frontiers in Plant Science 12*: 634103.

Wang, Y., Zhou, M., Zou, Q. et al. (2021). Machine learning for phytopathology: From the molecular scale towards the network scale. *Briefings in Bioinformatics 22* (5): bbab037.

Wani, J.A., Sharma, S., Muzamil, M. et al. (2022). Machine learning and deep learning based computational techniques in automatic agricultural diseases detection: Methodologies, applications, and challenges. *Archives of Computational methods in Engineering 29* (1): 641–677.

Wu, Q., Ma, X., Liu, H. et al. (2023). A classification method for soybean leaf diseases based on an improved ConvNeXt model. *Scientific Reports* 13: 19141.

Xie, X., Ma, Y., Liu, B. et al. (2020). A deep-learning-based real-time detector for grape leaf diseases using improved convolutional neural networks. *Frontiers in Plant Science 11*: 751.

Xu, Y., Bai, L., Liu, M. et al. (2023). Identification of two novel rice S genes through combination of association and transcription analyses with gene-editing technology. *Plant Biotechnology Journal* 21 (8): 1628–41.

Xu, Y., Ma, K., Zhao, Y. et al. (2021). Genomic selection: A breakthrough technology in rice breeding. *The Crop Journal* 9 (3): 669–77.

Xu, Y., Wang, X., Ding, X. et al. (2018). Genomic selection of agronomic traits in hybrid rice using an NCII population. *Rice* 11: 32.

Xu, L., Cao, B., Ning, S. et al. (2023). Peanut leaf disease identification with deep learning algorithms. *Molecular Breeding* 43: 25.

Yadav, A. and Yadav, K. (2025). Portable solutions for plant pathogen diagnostics: Development, usage, and future potential. *Frontiers in Microbiology 16*: 1516723.

Yakkundimath, R., Saunshi, G., Anami, B. et al. (2022). Classification of rice diseases using convolutional neural network models. *Journal of The Institution of Engineers (India): Series B 103* (4): 1047–1059.

Yang, L., Yu, X., Zhang, S. et al. (2023). Stacking-based and improved convolutional neural network: A new approach in rice leaf disease identification. *Frontiers in Plant Science 14*: 1165940.

Yao, J., Wang, Y., Xiang, Y. et al. (2022). Two-stage detection algorithm for kiwifruit leaf diseases based on deep learning. *Plants 11* (6): 768.

Yuan, Z., Zhang, Y., Bi, G. et al. (2025). Exploring the feasibility of deep learning models for long-term disease prediction: A case study for wheat yellow rust in England. *arXiv preprint arXiv:2501.15677.*

Zhang, F., Zeng, D., Zhang, C.S. et al. (2019). Genome-wide association analysis of the genetic basis for sheath blight resistance in rice. *Rice* 12 (1): 93.

9

Machine Learning for Analyzing and Integrating Multiple Omics

Sarfraz Ahmad[1], Saniya[2], Mohammad Ghani Raghib[3], Rubina Khan[4], Vikas Belwal[5], Pankaj Kumar[6]*

[1] *Department of Genetics and Plant Breeding, College of Agriculture Mandawa, Swami Keshwanand Rajasthan Agriculture University, Bikaner, Rajasthan, India*
[2] *Department of Horticulture, ICAR-National Research Centre for Grapes, Pune, Maharashtra, India*
[3] *Department of Horticulture, Faculty of Agriculture, Nangarhar University, Afghanistan*
[4] *Department of Genetics and Plant Breeding, Sri Karan Narendra College of Agriculture Jobner, Sri Karan Narendra Agriculture University, Jobner, Rajasthan, India*
[5] *Department of Tree Improvement, College of Forestry Ranichauri, VCSG Uttarakhand University of Horticulture and Forestry, Bharsar, Uttarakhand, India*
[6] *Department of Crop Improvement, College of Forestry Ranichauri, VCSG Uttarakhand University of Horticulture and Forestry, Bharsar, Uttarakhand, India*

*Corresponding author: sarfraz_firoz@rediffmail.com

9.1 Introduction

The rapid advancement of high-throughput omics technologies has enabled researchers to generate vast amounts of biological data. These technologies include genomics, transcriptomics, proteomics, metabolomics, and epigenomics collectively known as multi-omics. By capturing diverse molecular profiles, multi-omics approaches offer unprecedented insights into cellular mechanisms, gene regulation, and disease progression. However, the complexity, heterogeneity, and sheer volume of these datasets pose significant analytical challenges. Machine learning (ML) has emerged as a powerful tool for extracting meaningful insights from multi-omics data. By leveraging ML algorithms, researchers can efficiently analyze high-dimensional data, identify key biomarkers, and predict biological outcomes. Unlike traditional statistical approaches, ML models are well-suited for discovering complex patterns in nonlinear, multi-source datasets, making them ideal for multi-omics integration (Kang et al. 2022). Integrating multiple omics layers is essential for understanding biological systems comprehensively. Single-omics approaches often provide fragmented insights due to the complexity of molecular interactions. For instance, genomic data reveals an organism's genetic blueprint, but understanding gene expression patterns, protein interactions, and metabolic pathways requires integrating transcriptomic, proteomic, and metabolomic data (Ab Basit Ahangera et al. 2025). ML models enable the effective fusion of these datasets, facilitating holistic insights into complex biological phenomena.

The limitations of single-omics studies have driven the increasing adoption of multi-omics integration strategies. Biological processes are highly interconnected; for example, genetic mutations may alter transcriptional profiles, which in turn influence protein abundance and metabolic pathways. Relying solely on one omics layer often overlooks these interactions, resulting in incomplete or biased conclusions (Agamah et al. 2022). By integrating multi-omics data, researchers can achieve a systems-level understanding of complex traits and diseases. This integration has proven especially valuable in fields such as cancer research, where genomic, transcriptomic, and proteomic data are combined to identify potential biomarkers and therapeutic targets (Chakraborty et al. 2018). Similarly, in plant biology, integrating multi-omics data has enabled researchers to understand stress responses, metabolic pathways, and gene–environment interactions (Pinu et al. 2019).

Significant technological advancements have enabled comprehensive multi-omics studies. The following key innovations have contributed to the rapid growth of multi-omics research: Advances in next-generation sequencing (NGS) have enabled rapid and cost-effective sequencing of entire genomes. Techniques such as whole-genome sequencing (WGS) and whole-exome sequencing (WES) provide insights into genetic variation, mutations, and structural variants (Satam 2023). Genomic data is foundational in multi-omics research, providing a blueprint for understanding downstream

Machine Learning for Plant Biology, First Edition. Edited by Jen-Tsung Chen.

biological processes. RNA sequencing (RNA-seq) and microarrays have revolutionized transcriptomics research by enabling genome-wide quantification of gene expression. Single-cell RNA sequencing (scRNA-seq) has further enhanced the resolution of transcriptomic data, enabling the study of cellular heterogeneity in complex tissues (Shaw et al. 2021). Mass spectrometry (MS) and tandem MS (MS/MS) have significantly improved protein identification, quantification, and post-translational modification analysis (Neagu et al. 2022). Proteomic data is crucial for understanding molecular signaling pathways and protein interactions. Metabolomic profiling through techniques such as nuclear magnetic resonance (NMR) spectroscopy and gas chromatography-mass spectrometry (GC-MS) has enabled the comprehensive characterization of small molecules. Metabolomics data provides valuable insights into cellular metabolism and phenotypic changes (Muthubharathi et al. 2021). Techniques such as whole-genome bisulfite sequencing (WGBS) and chromatin immunoprecipitation sequencing (ChIP-seq) have advanced epigenomic research. Epigenomics data reveals key regulatory mechanisms that influence gene expression without altering DNA sequences (Chenarani et al. 2021).

Despite the potential of multi-omics integration, several challenges hinder effective analysis. Multi-omics datasets often differ in terms of data structure, measurement scale, and dynamic range. For example, genomic data is largely categorical (e.g., nucleotide sequences), while transcriptomic and proteomic data are continuous with varying signal intensities (Subramanian et al. 2020). Integrating these heterogeneous data types requires specialized algorithms capable of aligning and transforming diverse datasets into a unified format. Omics datasets often contain thousands of variables (features) relative to a limited number of samples. This "curse of dimensionality" poses a significant risk of overfitting and poor model generalization. Dimensionality reduction techniques such as PCA, t-SNE, and autoencoders are commonly employed to manage this challenge (Hira et al. 2021). Omics experiments frequently contain noisy measurements and missing values. Robust data imputation techniques such as KNN imputation, multiple imputation by chained equations (MICE), and matrix factorization are widely used to address these issues (Flores et al. 2023). Multi-omics data integration requires careful handling of feature dependencies, biological context, and temporal variability. Integrative ML models must effectively combine disparate datasets while preserving meaningful biological relationships.

ML algorithms have revolutionized multi-omics research by offering scalable, data-driven approaches for pattern recognition and predictive modeling. ML models excel at analyzing high-dimensional data, identifying complex interactions, and discovering novel biomarkers. Supervised ML models such as random forests, support vector machines (SVM), and neural networks have shown remarkable success in disease classification, phenotype prediction, and biomarker discovery (Galal et al. 2022). Unsupervised learning methods such as K-means clustering, hierarchical clustering, and t-SNE are widely employed to detect hidden patterns, patient subtypes, and molecular signatures without predefined labels (Trezza et al. 2024). Deep learning architectures such as convolutional neural networks (CNNs) and recurrent neural networks (RNNs) have significantly improved predictive accuracy in multi-omics integration. Emerging models like graph neural networks (GNNs) are increasingly used to capture complex molecular interactions in biological networks (Waqas et al. 2024). Recent advancements such as transfer learning and federated learning have further enhanced multi-omics integration. These models leverage pre-trained networks or distributed learning frameworks to improve model performance on complex omics datasets.

This chapter explores the pivotal role of ML in multi-omics integration. It highlights the characteristics of omics data, preprocessing strategies, core ML techniques, and key applications in fields such as precision medicine, plant biology, and biotechnology. This chapter also discusses challenges and future directions for enhancing ML-driven multi-omics integration.

9.2 Characteristics of Omics Data

Omics data, generated from large-scale biological studies, exhibit unique characteristics that pose both challenges and opportunities for ML-based analysis. These data types capture various biological layers, including genomics, transcriptomics, proteomics, metabolomics, and epigenomics, each contributing to a comprehensive understanding of cellular processes (Figure 9.1). Understanding the distinct characteristics of omics data is crucial for designing robust ML models capable of deriving meaningful insights. This section explores the key features of omics data, emphasizing data structure, dimensionality, heterogeneity, sparsity, noise, and dynamic nature.

9.2.1 High Dimensionality and Low Sample Size

One of the defining characteristics of omics data is its high dimensionality, where the number of features (e.g., genes, proteins, metabolites) far exceeds the number of samples. For example, a typical RNA-seq dataset may contain expression

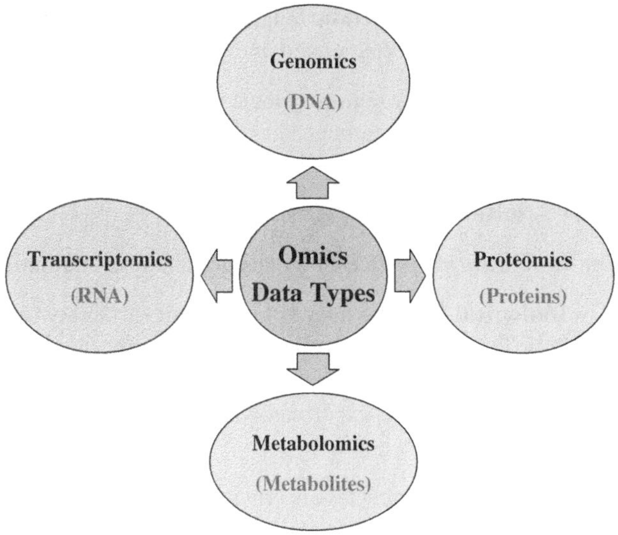

Figure 9.1 Multi-omics data types categorized based on structural and functional levels.

levels for 20,000+ genes, while the number of biological samples may be limited to only a few hundred or fewer (Wu and Zhang 2020).

- **Genomics Data:** DNA sequencing datasets often contain millions of single-nucleotide polymorphisms (SNPs), requiring dimensionality reduction or feature selection for effective analysis (Jain et al. 2023).
- **Proteomics and Metabolomics Data:** These datasets involve thousands of measured proteins or metabolites, often with a limited number of biological replicates due to experimental constraints.

The HDLSS nature of omics data presents a significant challenge for ML models, increasing the risk of overfitting and limiting model generalizability. Feature selection techniques such as LASSO regression, recursive feature elimination (RFE), and principal component analysis (PCA) are widely employed to mitigate this issue (Mahendran et al. 2020).

9.2.2 Data Heterogeneity

Omics data originate from multiple biological sources, including DNA, RNA, proteins, and metabolites, each governed by distinct biochemical properties and measurement techniques.

- **Genomics Data:** Encodes stable genetic information in the form of nucleotide sequences, SNPs, and structural variations.
- **Transcriptomics Data:** Captures dynamic gene expression profiles, influenced by cellular state, environmental conditions, and temporal changes (Oh and Li 2021).
- **Proteomics Data:** Reflects the complex interactions and post-translational modifications of proteins, offering insights into functional pathways (Pascovici et al. 2018).
- **Metabolomics Data:** Represents the end products of cellular processes, making it highly sensitive to environmental factors and biological conditions (Ahmad et al. 2023).

Integrating these heterogeneous data sources is essential to understanding complex biological systems, yet their diverse structures often require specialized ML techniques such as multi-view learning, graph-based models, or tensor factorization (Nguyen and Wang 2020).

9.2.3 Sparsity and Missing Values

Omics data are frequently sparse, with numerous missing values due to limitations in detection sensitivity, experimental variability, or technical issues.

- **Proteomics and Metabolomics Data:** Mass spectrometry–based platforms often generate sparse datasets where certain biomolecules fall below detection thresholds, resulting in zero-inflated data distributions (Gregorich et al. 2024).

- **Transcriptomics Data:** Single-cell RNA sequencing (scRNA-seq) often suffers from "dropout events," where some genes are undetected in certain cells despite being expressed (Lu et al. 2021).

Addressing sparsity requires robust imputation strategies such as k-nearest neighbors (KNN), MICE, and deep learning models like denoising autoencoders (DAEs) (Barrabés et al. 2024).

9.2.4 Noise and Technical Variability

Omics data are often affected by substantial noise originating from both biological and technical sources.

- **Biological Noise:** Arises from inherent variability in biological systems, including stochastic gene expression, metabolic fluctuations, and cellular heterogeneity.
- **Technical Noise:** Introduced during sample preparation, instrument calibration, or sequencing processes. For instance, RNA-seq experiments may suffer from polymerase chain reaction (PCR) amplification bias or batch effects (Li et al. 2015).

Noise reduction strategies are crucial in ML pipelines and often include data smoothing, noise-filtering algorithms, and batch effect correction methods such as ComBat and SVA (Albert 2014).

9.2.5 Dynamic and Temporal Variability

Omics data are dynamic in nature, reflecting temporal changes in gene expression, protein abundance, and metabolite concentrations.

- **Time-series Data:** Transcriptomics data often captures dynamic gene expression profiles over time, requiring specialized ML models such as RNNs and long short-term memory (LSTM) networks to track temporal dependencies (Duan et al. 2024).
- **Circadian Rhythms:** Metabolomics data may fluctuate across biological cycles, emphasizing the need for temporal feature engineering in predictive models.

Dynamic data integration remains a growing area of interest in systems biology, enabling researchers to model biological processes as evolving networks.

9.2.6 Multicollinearity and Feature Interdependence

Omics data often exhibit strong correlations between features, posing challenges for ML models.

- **Genomics Data:** Gene regulatory networks demonstrate tightly linked interactions where the expression of one gene may influence several others.
- **Metabolomics Data:** Metabolic pathways are interconnected, leading to high collinearity among related metabolites.

To address multicollinearity, ML techniques such as elastic net regression, partial least squares (PLS), and correlation-based feature selection (CFS) are commonly applied to extract informative features while minimizing redundancy (Zhang et al. 2022).

9.2.7 High Biological Variability

Biological systems are inherently diverse, influenced by genetic backgrounds, environmental conditions, and developmental stages.

- **Plant omics Data:** Crop species such as *Arabidopsis thaliana, Oryza sativa* (rice), and *Zea mays* (maize) often exhibit substantial genetic and phenotypic variability, complicating model generalization (Kaur et al. 2021).
- **Human Multi-omics Data:** Variations in age, ethnicity, and lifestyle factors introduce additional layers of complexity in clinical studies.

Ensuring model robustness requires advanced ML approaches like domain adaptation, transfer learning, and federated learning for improved generalization across diverse datasets (Huang et al. 2024).

9.2.8 Class Imbalance in Omics Data

Omics datasets frequently contain imbalanced class distributions, particularly in disease studies or rare phenotype predictions.

- **Cancer Genomics:** Tumor subtypes often occur in low frequency, leading to imbalanced datasets that hinder ML model performance.
- **Stress-responsive Genes in Plants:** Genes responsible for drought, salinity, or pathogen resistance may represent a minor proportion of the entire transcriptome.

Addressing class imbalance requires methods such as Synthetic Minority Over-sampling Technique (SMOTE), weighted loss functions, or ensemble strategies for improved classification performance (Maldonado et al. 2022).

9.2.9 Data Integration Complexity

Integrating diverse omics datasets is challenging due to differences in data types, scales, and feature distributions.

- **Genomics Data:** Discrete data such as SNPs and mutation counts.
- **Transcriptomics Data:** Continuous gene expression levels.
- **Proteomics and Metabolomics Data:** Semi-continuous data with zero-inflated distributions.

Successful integration methods include matrix factorization, canonical correlation analysis (CCA), and deep learning frameworks designed to unify complex biological layers (Moon et al. 2022).

Understanding the unique characteristics of omics data is essential for developing effective ML models for multi-omics integration. Challenges such as high dimensionality, noise, sparsity, and dynamic variability necessitate specialized pre-processing, feature selection, and integration strategies. As ML techniques continue to evolve, incorporating biological insights into model design will be critical for improving predictive accuracy and advancing discoveries in systems biology, precision medicine, and plant genomics.

9.3 Data Preprocessing for Multi-omics Integration

Data preprocessing is a crucial step in multi-omics integration, significantly influencing the quality and reliability of downstream ML models. Multi-omics data are inherently complex, heterogeneous, and often noisy, necessitating robust preprocessing methods to ensure accurate integration and analysis. Effective preprocessing mitigates data inconsistencies, reduces noise, and enhances feature compatibility across various omics platforms. This section explores key preprocessing steps, including data normalization, imputation, dimensionality reduction, feature selection, and data transformation, with insights into their role in improving multi-omics data integration.

9.3.1 Data Normalization and Standardization

Multi-omics datasets are generated using distinct platforms, resulting in differences in data scales, distributions, and measurement units. Normalization and standardization are essential to harmonize these data types for effective integration.

- **Normalization** scales data to a specific range (e.g., 0 to 1) to ensure comparability between features. Common methods include min-max scaling, z-score normalization, and log transformation. For instance, RNA-seq data are often transformed using log2(FPKM + 1) or TPM (Transcripts Per Million) to normalize gene expression values across samples (Zhao et al. 2021).
- **Standardization** involves rescaling data to have zero mean and unit variance. Techniques like z-score transformation are widely used in metabolomics to correct for batch effects and improve downstream clustering and classification tasks (Misra 2020).

Proper normalization ensures that features with different ranges (e.g., genomic variants, gene expression values, and metabolite concentrations) contribute equally to ML models.

9.3.2 Data Cleaning and Outlier Detection

Outliers are prevalent in multi-omics datasets due to experimental errors, biological variability, or technical noise. Identifying and addressing these anomalies is crucial for model stability.

- **Univariate Methods:** Techniques such as the IQR (interquartile range) and z-score methods are effective for detecting extreme values in individual features.
- **Multivariate Approaches:** PCA or Mahalanobis distance methods can capture multivariate outliers by identifying points that deviate from the expected correlation structure.

For instance, in metabolomics studies, batch effects and instrument drift are common sources of outliers, necessitating quality control steps such as median fold change normalization (Sanchez-Illana et al. 2018).

9.3.3 Data Imputation for Missing Values

Missing data are frequent in multi-omics experiments due to technical limitations, measurement errors, or data sparsity. Proper imputation strategies are crucial to maintain data integrity.

- **Statistical Imputation Methods:** Mean, median, or mode imputation methods are suitable for handling small amounts of missing data in low-dimensional datasets.
- **ML-based Imputation:** Advanced methods such as KNN, random forests, and MICE can effectively reconstruct missing values by leveraging relationships between features (Samad et al. 2022).
- **Deep Learning Imputation Models:** Autoencoders and variational autoencoders (VAEs) have shown promise in recovering missing data patterns in complex multi-omics datasets. For example, deep learning-based imputation methods have significantly improved accuracy in metabolomic profiling where metabolite peaks may be undetectable in some samples (Cuperlovic-Culf et al. 2022).

9.3.4 Batch Effect Correction

Batch effects arise from technical variability across different experimental batches, platforms, or instruments. Uncorrected batch effects can obscure biological signals and reduce the reliability of ML models.

- **Combat Method:** A Bayesian framework designed to adjust batch effects in microarray and RNA-seq data (Zhang et al. 2020).
- **SVA (Surrogate Variable Analysis):** Identifies hidden sources of variability and adjusts for them to improve signal detection.
- **Harmony Algorithm:** Widely used in single-cell multi-omics integration to align datasets across different batches (Cao et al. 2020).

Batch effect correction is critical for integrating multi-omics data from large-scale consortiums such as The Cancer Genome Atlas (TCGA) or the 1000 Genomes Project.

9.3.5 Dimensionality Reduction

Multi-omics data are typically high-dimensional, posing challenges for ML models due to the "curse of dimensionality." Dimensionality reduction methods improve computational efficiency, enhance model performance, and reduce noise.

- **PCA:** A linear method that projects data into lower-dimensional space while retaining maximum variance.
- **t-Distributed Stochastic Neighbor Embedding:** A nonlinear method that preserves local data structure for improved clustering in complex datasets.
- **Uniform Manifold Approximation and Projection:** An emerging dimensionality reduction technique that excels in visualizing complex biological patterns in high-dimensional omics data.

In practice, PCA is often applied in genomics for visualizing population structure, while UMAP has gained popularity for integrating single-cell transcriptomics and proteomics data (Yang et al. 2021).

9.3.6 Feature Selection

Feature selection is vital for reducing redundancy, improving model interpretability, and enhancing predictive accuracy. Methods for feature selection include:

- **Filter Methods:** Techniques like correlation-based filtering and mutual information analysis rank features based on statistical relevance.
- **Wrapper Methods:** Algorithms such as RFE iteratively select the most informative features for ML models.
- **Embedded Methods:** These approaches incorporate feature selection directly into ML models (e.g., LASSO regression, random forest feature importance).

For example, feature selection has been instrumental in identifying key genomic markers associated with cancer subtypes or stress-responsive genes in crops (Fan et al. 2025).

9.3.7 Data Integration Strategies

Data integration techniques enable multi-omics layers to be combined for comprehensive analysis. Integration strategies can be broadly classified as:

- **Early Integration (Data Fusion):** Combines raw omics data at the input level to construct a unified feature matrix.
- **Intermediate Integration (Feature-Level Integration):** Extracts relevant features from each omics layer before combining them for ML modeling.
- **Late Integration (Decision-Level Integration):** Combines individual model predictions from each omics layer to generate a final consensus prediction.

For instance, in cancer genomics, late integration models have demonstrated superior performance in predicting patient survival by aggregating individual predictions from genomic, transcriptomic, and proteomic models (Boehm et al. 2022).

9.3.8 Data Transformation and Encoding

Certain ML models require specific data formats for optimal performance. Transformation and encoding techniques address these requirements.

- **One-hot Encoding:** Frequently used for categorical genomic data such as SNP variants or mutation profiles.
- **Embedding Techniques:** Word2Vec or BioVec embeddings are employed to encode biological sequences for deep learning models.
- **Wavelet Transformation:** Applied in time-series metabolomics data to capture temporal patterns in metabolic pathways (Smilde et al. 2010).

Data transformation ensures compatibility between diverse omics data types and facilitates robust ML model development.

9.3.9 Data Augmentation Techniques

Data augmentation artificially increases dataset size by generating synthetic data points. This strategy is particularly useful when working with limited omics datasets.

- **SMOTE (Synthetic Minority Over-sampling Technique):** Widely applied in genomics to address class imbalances by synthesizing new samples from minority classes (Chawla et al. 2002).
- **GANs (Generative Adversarial Networks):** Used to create synthetic transcriptomic profiles for improved model generalization in cancer prediction (Gutta et al. 2023).

Augmentation methods enhance model robustness and prevent overfitting, especially in datasets with small sample sizes.

9.3.10 Data Quality Assessment

Assessing data quality is a final yet crucial step in preprocessing. Common strategies include:

- **Correlation Analysis:** Ensures consistency between omics layers.
- **Biological Validation:** Confirms data-driven insights align with known biological processes.
- **Cross-validation Techniques:** Evaluates model performance using partitioned data to ensure generalizability.

For instance, quality assessment has been vital in the Human Microbiome Project to ensure reproducibility across large-scale omics data collections (Moossavi et al. 2021).

9.4 ML Techniques for Multi-omics Analysis

ML techniques have emerged as powerful tools for integrating and analyzing multi-omics data, enabling researchers to uncover complex biological patterns, identify biomarkers, and predict phenotypic outcomes. Due to the inherent complexity and heterogeneity of omics data, specialized ML algorithms are employed to extract meaningful insights from diverse datasets. This section explores key ML techniques used in multi-omics analysis, including supervised learning, unsupervised learning, deep learning, and network-based approaches, with emphasis on their applications, strengths, and limitations in integrative biology.

9.4.1 Supervised Learning Techniques

Supervised learning models are widely applied in multi-omics data analysis for classification, regression, and predictive modeling. These models require labeled training data, where features (e.g., gene expression levels, metabolite concentrations) are mapped to known outcomes such as disease status, plant stress response, or crop yield.

9.4.1.1 Support Vector Machines

SVM is a robust algorithm particularly effective for high-dimensional omics data. By constructing a hyperplane that maximizes the margin between classes, SVM efficiently handles complex biological data with small sample sizes.

- **Application in Multi-omics Analysis:** SVM has been used to predict cancer subtypes by integrating transcriptomics, proteomics, and epigenomics data (Huang et al. 2018).
- **In Plant Sciences:** SVM models have been employed to classify drought-tolerant and susceptible rice cultivars by integrating gene expression and metabolite profiles (Thanmalagan et al. 2022).

9.4.1.2 Random Forest

RF is an ensemble method that constructs multiple decision trees, making it well-suited for multi-omics data integration. RF offers robustness against overfitting and effectively manages missing data.

- **Application in Disease Prediction:** RF has been utilized to identify biomarkers for breast cancer prognosis by integrating transcriptomics and methylation data (Mirza et al. 2023).
- **In Plant Genomics:** RF models have shown success in predicting plant stress resistance traits using combined genomics, transcriptomics, and metabolomics data (Zhang et al. 2024).

9.4.1.3 Elastic Net and LASSO Regression

Elastic Net and LASSO (Least Absolute Shrinkage and Selection Operator) are effective feature selection methods for high-dimensional omics data. These techniques apply L1 or L2 regularization to select informative features while reducing model complexity.

- **In Cancer Genomics:** LASSO regression has been applied to identify prognostic gene signatures by integrating multi-omics data (Liu et al. 2017).
- **In Plant Metabolomics:** Elastic Net has been employed to correlate metabolite profiles with drought-tolerance traits in *Arabidopsis thaliana* (Nazari et al. 2023).

9.4.2 Unsupervised Learning Techniques

Unsupervised learning techniques are widely used to explore hidden patterns, group samples, and reduce dimensionality in multi-omics datasets without relying on predefined labels.

9.4.2.1 Principal Component Analysis

PCA is a popular dimensionality reduction technique that projects multi-omics data into a lower-dimensional space, preserving maximum variance.

- **Application in Disease Research:** PCA has been employed to identify molecular subtypes of cancer by integrating genomic and epigenomic data.
- **In Plant Metabolomics:** PCA effectively reveals clustering patterns in metabolic profiles under different environmental conditions.

9.4.2.2 K-means Clustering

K-means clustering is a distance-based method that groups samples into distinct clusters based on feature similarity.

- **In Human Multi-omics Studies:** K-means has been employed to classify patients with distinct metabolic profiles linked to diabetes risk (Wang et al. 2023).
- **In Plant Research:** K-means clustering has been used to group rice genotypes based on multi-omics data linked to drought tolerance (Devarajan et al. 2022).

9.4.2.3 Hierarchical Clustering

Hierarchical clustering organizes data into a tree-like structure (dendrogram), useful for visualizing relationships between biological samples.

- **In Systems Biology:** Hierarchical clustering is frequently applied to identify co-expressed genes, regulatory modules, and metabolic pathways (Galán-Vásquez and Perez-Rueda 2019).

9.4.3 Deep Learning Techniques

Deep learning models have gained prominence in multi-omics analysis due to their ability to automatically extract complex patterns from large, high-dimensional datasets.

9.4.3.1 Convolutional Neural Networks

CNNs are particularly effective for extracting spatial patterns from structured data such as genomic sequences or protein structures.

- **In Genomics Research:** CNN models have been employed to predict DNA-binding protein sites and identify genetic mutations linked to diseases (Hu et al. 2019).
- **In Plant Biology:** CNNs have demonstrated success in predicting stress-responsive gene networks using multi-omics data.

9.4.3.2 Recurrent Neural Networks

RNNs are specialized for sequential data, making them suitable for temporal omics data such as gene expression time-series or metabolite flux profiles.

- **In Cancer Progression Models:** RNNs have been employed to model dynamic transcriptional changes in tumor progression.
- **In Plant Stress Response:** RNN-based models have improved predictions of gene regulatory changes in plants under drought conditions (Mochida et al. 2018).

9.4.3.3 Autoencoders

Autoencoders are powerful tools for dimensionality reduction and data integration in multi-omics research. These models encode input data into a compressed representation and reconstruct the original data.

- **In Disease Biomarker Discovery:** Autoencoders have identified latent features for breast cancer prediction by integrating transcriptomics and methylation data (Guo et al. 2019).
- **In Plant Genomics:** Autoencoders have successfully extracted key features for predicting crop yield under variable environmental conditions.

9.4.4 Network-based Approaches

Network-based ML techniques model biological interactions by constructing graphs representing gene regulatory networks, protein–protein interactions, or metabolic pathways.

9.4.4.1 Graph Neural Networks

GNNs are ideal for modeling complex biological relationships in multi-omics data. These models capture both node-specific features (e.g., gene expression) and edge-based relationships (e.g., protein–protein interactions).

- **In Human Genomics:** GNNs have been applied to identify disease-associated gene networks in Alzheimer's research (Hernández-Lorenzo et al. 2022).
- **In Plant Systems Biology:** GNN-based frameworks have predicted stress-responsive genes in crops like *Oryza sativa* by integrating transcriptomics and metabolomics data (Kaminuma 2022).

9.4.4.2 Bayesian Networks

Bayesian networks model probabilistic dependencies between biological variables, making them useful for multi-omics integration.

- **In Cancer Genomics:** Bayesian models have reconstructed signaling pathways to identify key regulatory genes driving tumor progression (Gendelman et al. 2017).
- **In Plant Genomics:** Bayesian approaches have successfully modeled gene–environment interactions in stress-responsive plants (Mohanta et al. 2017).

9.4.5 Hybrid and Ensemble Learning Approaches

Combining multiple ML models often improves predictive performance in multi-omics data analysis. Hybrid approaches integrate different learning algorithms, while ensemble models combine predictions from multiple models.

- **In Disease Prediction:** Hybrid approaches combining CNNs and random forests have improved the prediction of cancer outcomes by integrating genomic, proteomic, and clinical data (Sathyamoorthi et al. 2025).
- **In Plant Phenotyping:** Ensemble models combining SVM, RF, and k-NN have improved drought stress prediction in wheat using transcriptomics and metabolomics data (Gupta et al. 2024).

ML techniques play a pivotal role in multi-omics integration, enabling researchers to explore complex biological networks, predict phenotypic traits, and identify biomarkers with precision. While supervised and unsupervised methods remain widely used, deep learning and network-based models are increasingly driving advancements in multi-omics research. The integration of these ML approaches, combined with innovative data preprocessing strategies, is essential for unraveling complex biological insights and advancing precision agriculture, medicine, and systems biology.

9.5 Applications of ML in Multi-omics Research

ML has revolutionized multi-omics research by enabling the integration of diverse biological data types to extract meaningful insights, predict phenotypes, and identify key biomarkers. With the rapid growth of genomic, transcriptomic, proteomic, metabolomic, and epigenomic data, ML techniques offer robust solutions to manage data complexity and uncover biological mechanisms. This section explores key applications of ML in multi-omics research across various domains such as biomarker discovery, disease diagnosis, drug development, plant breeding, and environmental studies.

9.5.1 Biomarker Discovery in Human Health

Biomarker identification is crucial for early diagnosis, prognosis, and therapeutic decision-making in diseases. Integrating multi-omics data enhances biomarker discovery by capturing interactions between genes, proteins, and metabolites.

9.5.1.1 Cancer Biomarker Identification

Cancer is a complex disease driven by genetic, epigenetic, and metabolic alterations. ML models have successfully integrated multi-omics data to reveal predictive biomarkers for early detection and personalized treatment strategies.

- **Random Forest (RF)** models have been applied to integrate transcriptomic, proteomic, and methylation data to identify gene signatures predicting breast cancer survival (Mirza et al. 2023).
- **Support Vector Machines** has successfully combined RNA-seq data and DNA methylation profiles to identify biomarkers for colorectal cancer diagnosis (Zhang et al. 2021).
- **Deep Learning (DL)** models, particularly convolutional neural networks (CNNs), have been used to predict cancer subtypes based on integrated genomic, epigenomic, and transcriptomic data.

9.5.1.2 Cardiovascular and Metabolic Disorders

ML models are extensively used to identify biomarkers for cardiovascular diseases and metabolic disorders by combining transcriptomics, lipidomics, and metabolomics data.

- **Elastic Net Regression** successfully predicted lipid biomarkers associated with cardiovascular disease risk by integrating lipidomics and proteomics data (Nurmohamed et al. 2023).
- **K-means Clustering** has been employed to stratify individuals based on multi-omics profiles to assess diabetes risk (Babu and Snyder 2023).

9.5.2 Disease Diagnosis and Classification

Integrative ML approaches play a pivotal role in predicting disease progression, stratifying patients, and improving diagnostic accuracy.

9.5.2.1 Cancer Classification Models

ML models have improved the accuracy of cancer classification by leveraging multi-omics data.

- **Autoencoder-based Neural Networks** have extracted key features from transcriptomic and proteomic data to distinguish aggressive and nonaggressive prostate cancer.
- **Bayesian Networks** have been employed to identify gene-gene interactions associated with leukemia subtypes (Sun et al. 2023).

9.5.2.2 Neurological Disorders

Multi-omics integration has advanced the identification of biomarkers and pathways linked to neurodegenerative diseases such as Alzheimer's and Parkinson's.

- **Graph Neural Networks** have been applied to model protein–protein interaction networks and predict Alzheimer's associated gene modules (Thapa et al. 2025).
- **Random Forest Models** have identified metabolite biomarkers linked to cognitive decline using integrated metabolomic and genomic data.

9.5.3 Drug Discovery and Personalized Medicine

ML techniques are transforming drug discovery by predicting drug targets, optimizing drug combinations, and improving patient-specific treatment strategies.

9.5.3.1 Drug–Target Interaction Prediction

Predicting interactions between drugs and target molecules is crucial for drug development.

- **Deep Learning Models** have demonstrated exceptional performance in predicting drug-target interactions by combining transcriptomic, proteomic, and structural data.
- **Random Forest Algorithms** have been used to predict kinase inhibitors by integrating proteomics and molecular docking data (Cao et al. 2013).

9.5.3.2 Personalized Medicine

ML models are widely used to tailor treatment strategies based on patient-specific omics profiles.

- **Multi-task Learning Models** have integrated transcriptomics, clinical data, and genomic profiles to predict individual responses to immunotherapy (Li et al. 2024).
- **Reinforcement Learning** has been employed to optimize chemotherapy regimens by modeling dynamic molecular interactions in cancer cells.

9.5.4 Applications in Plant Sciences

ML-based multi-omics integration has gained significant traction in plant sciences for trait prediction, crop improvement, and stress resilience. Key applications of ML in multi-omics research in plant science are presented in Table 9.1.

9.5.4.1 Crop Trait Prediction

Predicting desirable crop traits using integrated genomic, transcriptomic, and metabolomic data has significantly enhanced plant breeding.

- **Random Forest Models** have been applied to predict drought tolerance in *Oryza sativa* by combining transcriptomics and metabolomics data (Lu et al. 2023).
- **Convolutional Neural Networks** have successfully predicted fruit yield in *Solanum lycopersicum* by integrating genomic and proteomic data (Naik et al. 2023).

9.5.4.2 Stress Response Prediction

Plant responses to environmental stress factors such as drought, salinity, and heat can be predicted using ML-driven multi-omics integration.

- **Support Vector Machines** has been employed to predict heat-tolerant wheat varieties by integrating gene expression and metabolite data (Zahravi et al. 2024).
- **Bayesian Networks** have been utilized to identify stress-responsive pathways in maize under drought conditions (Lie et al. 2021).

9.5.5 Environmental and Ecological Applications

ML techniques are increasingly employed to study environmental interactions, microbial ecosystems, and soil health through multi-omics integration.

9.5.5.1 Soil Microbiome Analysis

ML models have enabled the prediction of soil microbial diversity and its influence on crop productivity by integrating metagenomics and metabolomics data.

- **Random Forest Models** have been employed to predict soil microbial diversity in organic farmlands based on metagenomics data.
- **Hierarchical Clustering** has revealed key microbial communities contributing to soil fertility (Li et al. 2020).

9.5.5.2 Environmental Stress Monitoring

Multi-omics data integration with ML techniques has improved environmental monitoring and stress assessment.

Table 9.1 *Applications of machine learning in multi-omics research.*

Application Area	ML Technique	Omics Types	Crop	Remarks	Reference
Gene function prediction	Random Forest (RF), Support Vector Machine (SVM)	Genomics, Transcriptomics	*Arabidopsis thaliana*	Identified key genes regulating drought tolerance and root development	Meher et al. (2024)
Yield prediction and trait analysis	Convolutional Neural Networks (CNNs), Decision trees	Genomics, Phenomics	*Oryza sativa* (Rice)	Enhanced prediction of grain yield by integrating genotypic and environmental data	Niazian and Niedbała (2020)
Disease resistance analysis	Extreme Gradient Boosting (XGBoost), k-Nearest Neighbors (k-NN)	Genomics, Metabolomics	*Solanum lycopersicum* (Tomato)	Identified candidate genes linked to resistance against *Phytophthora infestans*	Gerakari et al. (2025)
Stress response analysis	Artificial Neural Networks (ANNs), Random Forest (RF)	Transcriptomics, Proteomics	*Glycine max* (Soybean)	Revealed gene networks regulating drought stress response pathways	Herrmann et al. (2024)
Nutrient uptake and transport	Support Vector Regression (SVR), Bayesian Networks	Transcriptomics, Metabolomics	*Zea mays* (Maize)	Identified key transporters involved in phosphate uptake under phosphorus stress	Wu et al. (2024)
Secondary metabolite biosynthesis	Gaussian Mixture Models (GMM), K-means Clustering	Metabolomics, Genomics	Plant biology system	Studied candidate genes involved in metabolite biosynthesis	Yan and Wang (2022)
Crop improvement and breeding	Gradient Boosting Machines (GBM), Deep Neural Networks (DNNs)	Genomics, Transcriptomics	*Triticum aestivum* (Wheat)	Enhanced selection of high-yielding genotypes using predictive modeling	Uribeetxebarria et al. (2023)
Metabolic pathway reconstruction	Deep Belief Networks (DBN), Multi-Layer Perceptron (MLP)	Metabolomics, Proteomics	*Brassica napus* (Rapeseed)	Identified novel metabolic pathways linked to lipid biosynthesis	Zhang et al. (2025)
Identification of regulatory networks	Elastic Net Regression, LASSO	Genomics, Epigenomics	*Capsicum annuum* (Pepper)	Revealed transcription factors involved in fruit ripening and color development	Gerakari et al. (2025)
Environmental adaptation analysis	Random Forest (RF), K-means Clustering	Genomics, Transcriptomics, Epigenomics	*Medicago truncatula*	Identified adaptive genes controlling nitrogen fixation under varying environmental conditions	Knaack et al. (2022)

- **K-means Clustering** has been applied to group plants based on stress-adaptive traits by integrating metabolomics and proteomics data (Pan et al. 2020).
- **Autoencoders** have been used to predict ecosystem changes resulting from environmental stressors (Singh et al. 2018).

9.5.6 Systems Biology and Pathway Analysis

ML-driven multi-omics integration is essential in reconstructing biological pathways, inferring gene regulatory networks, and modeling metabolic pathways.

- **Bayesian Networks** have modeled gene regulatory interactions in *Arabidopsis thaliana*, identifying key genes involved in secondary metabolite biosynthesis (Lv et al. 2014).
- **Graph Neural Networks** have reconstructed protein-protein interaction networks to uncover novel gene functions in rice under nutrient-deficient conditions (Sharma et al. 2022).

9.6 Challenges and Future Directions

The integration of ML techniques with multi-omics data has proven to be a powerful tool in deciphering complex biological systems. However, this emerging field still faces several technical, biological, and computational challenges that hinder its full potential. Addressing these limitations is crucial for advancing multi-omics research and improving practical applications in healthcare, agriculture, and environmental sciences. This section explores the major challenges and outlines future directions for enhancing ML-driven multi-omics integration.

9.6.1 Key Challenges in ML-driven Multi-omics Research

9.6.1.1 Data Heterogeneity and Dimensionality

Multi-omics data are highly heterogeneous, originating from diverse sources such as genomics, transcriptomics, proteomics, and metabolomics. These data types differ in scale, resolution, and data distribution, making their integration complex (Subramanian et al. 2020). For instance, while genomic data are often categorical or binary, metabolomic data are typically continuous with varying dynamic ranges. The significant differences in data structure require sophisticated data transformation and normalization techniques to achieve meaningful integration.

Moreover, the high dimensionality of omics data, with thousands of features and limited sample sizes, increases the risk of overfitting in ML models. Dimensionality reduction methods such as PCA, t-distributed Stochastic Neighbor Embedding (t-SNE), and Uniform Manifold Approximation and Projection (UMAP) can mitigate this issue but may result in information loss if not applied carefully (Ballard et al. 2024).

9.6.1.2 Data Quality and Noise

Multi-omics datasets often suffer from noise, batch effects, and missing values due to variability in experimental protocols, sample preparation, and instrument conditions. Such inconsistencies pose significant challenges for ML algorithms, which rely heavily on accurate and complete data (Flores et al. 2023). Although imputation techniques like k-nearest neighbors (KNN) and MICE can help address missing data, they may introduce biases in downstream analyses.

9.6.1.3 Model Interpretability and Biological Relevance

While deep learning models such as convolutional neural networks (CNNs) and autoencoders have achieved remarkable accuracy in multi-omics research, their black-box nature often limits interpretability (Hussein et al. 2024). The inability to explain model decisions hinders the identification of biologically meaningful insights, reducing confidence in the results. Techniques like SHapley Additive exPlanations (SHAP), Local Interpretable Model-agnostic Explanations (LIME), and integrated gradients are being developed to improve model interpretability in biological contexts (Parisineni and Pal 2024).

9.6.1.4 Integration of Multi-scale Data

Multi-omics data are collected at different biological scales from molecular and cellular levels to tissue and organism-wide systems. Aligning these datasets across scales while maintaining biological coherence remains a significant challenge.

Advanced graph-based approaches and tensor factorization models have shown promise in modeling complex interactions but require further refinement for broader applicability (Ding et al. 2024).

9.6.1.5 Computational and Resource Limitations

The complexity and volume of multi-omics data demand substantial computational resources, especially for deep learning frameworks. Training these models often requires high-performance computing clusters and extensive memory, posing accessibility barriers for researchers with limited resources (Mahmood et al. 2024). Furthermore, integrating multiple omics layers requires data harmonization strategies, which are still evolving.

9.6.2 Future Directions in ML-driven Multi-omics Research

9.6.2.1 Advancing Data Integration Techniques

Future research must focus on developing robust data integration frameworks that effectively combine heterogeneous omics data. Hybrid approaches that blend traditional statistical models with ML techniques are emerging as promising solutions. For example, Variational Autoencoders (VAEs) and multi-view learning models are being explored to improve feature extraction and integration efficiency (Ma and Zhang et al. 2019).

9.6.2.2 Improving Model Interpretability

Incorporating explainable artificial intelligence (XAI) methods into multi-omics analysis will enhance biological insights and clinical applicability. Combining GNNs with attention mechanisms is expected to improve the interpretability of complex biological pathways and networks (Hussein et al. 2024).

9.6.2.3 Enhancing Data Quality and Standardization

Developing standardized protocols for data acquisition, processing, and storage is crucial to improving data reliability and reproducibility. Open-source platforms such as Galaxy, OmicsDI, and MetaboLights have facilitated data sharing; however, further efforts are required to implement universal data curation standards (Yurekten et al. 2024).

9.6.2.4 Leveraging Transfer Learning and Few-shot Learning

Given the limited sample sizes common in multi-omics studies, transfer learning and few-shot learning approaches are gaining popularity. These methods allow ML models trained on large datasets to generalize effectively on smaller omics datasets, improving prediction performance in rare diseases and underrepresented conditions (Assimakopoulos Pan et al. 2024).

9.6.2.5 Advancing Cloud-based Solutions

The adoption of cloud computing platforms such as Google Cloud, Amazon Web Services (AWS), and Microsoft Azure has improved access to large-scale ML models and computational resources. Developing scalable and user-friendly cloud solutions can accelerate multi-omics research by lowering technical barriers for researchers.

9.6.2.6 Ethical and Regulatory Considerations

As ML-based multi-omics research advances toward clinical applications, addressing ethical concerns regarding data privacy, security, and bias will be crucial. Transparent data sharing policies, coupled with robust encryption strategies, will ensure data integrity and patient confidentiality (Wekesa and Kimwele 2023).

9.7 Conclusion

ML has emerged as a transformative tool in the integration and analysis of multi-omics data, offering powerful solutions for addressing the complexity and heterogeneity inherent in biological systems. By leveraging advanced algorithms such as deep learning, random forests, and SVM, researchers can uncover intricate relationships between genomic, transcriptomic, proteomic, and metabolomic data. These insights have significantly contributed to advancements in personalized medicine, crop improvement, and biomarker discovery. Despite the remarkable progress, challenges such as data inconsistency, computational demands, and model interpretability remain prominent obstacles. Future advancements will require improved data integration frameworks, enhanced explainable AI techniques, and scalable computational infrastructure. Moreover, collaboration between bioinformaticians, data scientists, and domain experts will be essential to develop effective ML pipelines tailored for multi-omics research. As ML techniques continue to evolve, they hold tremendous promise for unraveling complex biological networks, facilitating precision agriculture, improving disease diagnosis, and accelerating therapeutic development. By addressing current limitations and investing in interdisciplinary collaboration, ML-driven multi-omics research is poised to unlock groundbreaking insights that will shape the future of biological and biomedical sciences.

References

Ab Basit Ahangera, S.W.A., Ab Naffi Ahangera, S.K., Assadc, A. et al. (2025). Multi-omics data tools and integration approaches. In: *Multi-Omics Technology in Human Health and Diseases: Genomics, Epigenomics, Transcriptomics, Proteomics, Metabolomics, Radiomics, Multi-omics*, 13. Elsevier.

Agamah, F.E., Bayjanov, J.R., Niehues, A. et al. (2022). Computational approaches for network-based integrative multi-omics analysis. *Frontiers in Molecular Biosciences* 9: 967205.

Ahmad, S., Belwal, V., Punia, S.S. et al. (2023). Role of plant secondary metabolites and phytohormones in drought tolerance: a review. *Gesunde Pflanzen* 75 (4): 729–746.

Albert, F.F. (2014). *Machine Learning Methods for the Analysis of Liquid Chromatography-Mass Spectrometry Datasets in Metabolomics.* Doctoral dissertation, Universitat Politècnica de Catalunya.

Assimakopoulos, F., Vassilakis, C., Margaris, D. et al. (2024). Artificial intelligence tools for the agriculture value chain: Status and prospects. *Electronics*, 13 (22): 4362.

Babu, M. and Snyder, M. (2023). Multi-omics profiling for health. *Molecular & Cellular Proteomics* 22 (6): 100561.

Ballard, J.L., Wang, Z., Li, W. et al. (2024). Deep learning-based approaches for multi-omics data integration and analysis. *BioData Mining* 17 (1): 38.

Barrabés, M., Perera, M., Moriano, V.N. et al. (2024). Advances in biomedical missing data imputation: a survey. *IEEE Access.* Vol. 13, pp. 16918–16932.

Boehm, K.M., Khosravi, P., Vanguri, R. et al. (2022). Harnessing multimodal data integration to advance precision oncology. *Nature Reviews Cancer* 22 (2): 114–126.

Cao, D.S., Zhou, G.H., Liu, S. et al. (2013). Large-scale prediction of human kinase–inhibitor interactions using protein sequences and molecular topological structures. *Analytica Chimica Acta* 792: 10–18.

Cao, K., Bai, X., Hong, Y.et al. (2020). Unsupervised topological alignment for single-cell multi-omics integration. *Bioinformatics* 36 (Supplement_1): i48–i56.

Chakraborty, S., Hosen, M.I., Ahmed, M. et al. (2018). Onco-multi-OMICS approach: a new frontier in cancer research. *BioMed Research International* 2018 (1): 9836256.

Chawla, N.V., Bowyer, K.W., Hall, L.O. et al. (2002). SMOTE: Synthetic minority over-sampling technique. *Journal of Artificial Intelligence Research* 16: 321–357.

Chenarani, N., Emamjomeh, A., Allahverdi, A. et al. (2021). Bioinformatic tools for DNA methylation and histone modification: A survey. *Genomics* 113 (3): 1098–1113.

Cuperlovic-Culf, M., Nguyen-Tran, T., and Bennett, S.A. (2022). Machine learning and hybrid methods for metabolic pathway modeling. *Computational Biology and Machine Learning for Metabolic Engineering and Synthetic Biology* 2553: 417–439.

Devarajan, A.K., Truu, M., Gopalasubramaniam, S.K. et al. (2022). Application of data integration for rice bacterial strain selection by combining their osmotic stress response and plant growth-promoting traits. *Frontiers in Microbiology* 13: 1058772.

Ding, J., Liu, C., Zheng, Y. et al. (2024). Artificial intelligence for complex network: Potential, methodology and application. *arXiv preprint arXiv:2402.16887.*

Duan, J., Xiong, J., Li, Y. et al. (2024). Deep learning based multimodal biomedical data fusion: An overview and comparative review. *Information Fusion* 112: 102536.

Fan, B.L., Chen, L.H., Chen, L.L. et al. (2025). Integrative multi-omics approaches for identifying and characterizing biological elements in crop traits: Current progress and future prospects. *International Journal of Molecular Sciences* 26 (4): 1466.

Flores, J.E., Claborne, D.M., Weller, Z.D. et al. (2023). Missing data in multi-omics integration: Recent advances through artificial intelligence. *Frontiers in Artificial Intelligence* 6: 1098308.

Galal, A., Talal, M., and Moustafa, A. (2022). Applications of machine learning in metabolomics: Disease modeling and classification. *Frontiers in Genetics* 13: 1017340.

Galán-Vásquez, E., and Perez-Rueda, E. (2019). Identification of modules with similar gene regulation and metabolic functions based on co-expression data. *Frontiers in Molecular Biosciences* 6: 139.

Gendelman, R., Xing, H., Mirzoeva, O.K. et al. (2017). Bayesian network inference modeling identifies TRIB1 as a novel regulator of cell-cycle progression and survival in cancer cells. *Cancer Research* 77 (7): 1575–1585.

Gerakari, M., Katsileros, A., Kleftogianni, K. et al. (2025). Breeding of solanaceous crops using AI: Machine learning and deep learning approaches—a critical review. *Agronomy* 15 (3): 757.

Gregorich, M., Kammer, M., Mischak, H. et al. (2024). Prediction modelling with many correlated and zero-inflated predictors: assessing a nonnegative garrote approach. *arXiv preprint arXiv:2402.02064*.

Guo, Y., Shang, X., and Li, Z. (2019). Identification of cancer subtypes by integrating multiple types of transcriptomics data with deep learning in breast cancer. *Neurocomputing* 324: 20–30.

Gupta, A., Kaur, L., and Kaur, G. (2024). Development of a two-layer meta-classifier–based drought stress detection system for wheat crop sustainability. *Sustainable Development Using Geospatial Techniques* 1–40.

Gutta, C., Morhard, C., and Rehm, M. (2023). Applying a GAN-based classifier to improve transcriptome-based prognostication in breast cancer. *PLOS Computational Biology* 19 (4): e1011035.

Hernández-Lorenzo, L., Hoffmann, M., Scheibling, E. et al. (2022). On the limits of graph neural networks for the early diagnosis of Alzheimer's disease. *Scientific Reports* 12 (1): 17632.

Herrmann, P.S.D.P., Santos Luccas, M.D., Ferreira, E.J. et al. (2024). Application of electronic nose and machine learning used to detect soybean gases under water stress and variability throughout the daytime. *Frontiers in Plant Science* 15: 1323296.

Hira, M.T., Razzaque, M.A., Angione, C. et al. (2021). Integrated multi-omics analysis of ovarian cancer using variational autoencoders. *Scientific Reports* 11 (1): 6265.

Hu, S., Ma, R., and Wang, H. (2019). An improved deep learning method for predicting DNA-binding proteins based on contextual features in amino acid sequences. *PLoS One* 14 (11): e0225317.

Huang, S., Cai, N., Pacheco, P.P. et al. (2018). Applications of support vector machine (SVM) learning in cancer genomics. *Cancer Genomics & Proteomics* 15 (1): 41–51.

Huang, W., Ye, M., Shi, Z. et al. (2024). Federated learning for generalization, robustness, fairness: a survey and benchmark. *IEEE Transactions on Pattern Analysis and Machine Intelligence* 46 (12): 9387–9406.

Hussein, A., Prasad, M., and Braytee, A. (2024). *Explainable AI Methods for Multi-Omics Analysis: A Survey. arXiv preprint arXiv:2410.11910.*

Jain, S.K., Wettberg, E.J.V., Punia, S.S. et al. (2023). Genomic-mediated breeding strategies for global warming in chickpeas (Cicer arietinum L.). *Agriculture* 13 (9): 1721.

Kaminuma, E. (2022). Introduction of deep learning approaches in plant omics research. In: *Plant Omics: Advances in Big Data Biology*, 217–223. CABI: GB.

Kang, M., Ko, E., and Mersha, T.B. (2022). A roadmap for multi-omics data integration using deep learning. *Briefings in Bioinformatics* 23 (1): bbab454.

Kaur, B., Sandhu, K.S., Kamal, R. et al. (2021). Omics for the improvement of abiotic, biotic, and agronomic traits in major cereal crops: applications, challenges, and prospects. *Plants* 10 (10): 1989.

Knaack, S.A., Conde, D., Chakraborty, S. et al. (2022). Temporal change in chromatin accessibility predicts regulators of nodulation in Medicago truncatula. *BMC Biology* 20 (1): 252.

Li, R., Pang, Z., Zhou, Y. et al. (2020). Metagenomic analysis exploring taxonomic and functional diversity of soil microbial communities in sugarcane fields applied with organic fertilizer. *BioMed Research International* 2020 (1): 9381506.

Li, X., Nair, A., Wang, S. et al. (2015). Quality control of RNA-seq experiments. *RNA Bioinformatics*: 137–146. New York, NY: Springer New York.

Li, Y., Wu, X., Fang, D. et al. (2024). Informing immunotherapy with multi-omics driven machine learning. *npj Digital Medicine* 7 (1): 67.

Liu, C., Wang, X., Genchev, G.Z. et al. (2017). Multi-omics facilitated variable selection in cox-regression model for cancer prognosis prediction. *Methods* 124: 100–107.

Lu, S., Chen, Y., Wang, S. et al. (2023). Combined metabolomic and transcriptomic analysis reveals key components of OsCIPK17 overexpression improves drought tolerance in rice. *Frontiers in Plant Science* 13: 1043757.

Lu, T., Park, S., Zhu, J. et al. (2021). Overcoming expressional drop-outs in lineage reconstruction from single-cell RNA-sequencing data. *Cell Reports* 34 (1): 1–13.

Lv, Q., Cheng, R., and Shi, T. (2014). Regulatory network rewiring for secondary metabolism in Arabidopsis thaliana under various conditions. *BMC Plant Biology* 14: 1–12.

Ma, T. and Zhang, A. (2019). Integrate multi-omics data with biological interaction networks using Multi-view Factorization AutoEncoder (MAE). *BMC Genomics* 20 (Suppl 11): 944.

Mahendran, N., Durai Raj Vincent, P.M., Srinivasan, K. et al. (2020). Machine learning based computational gene selection models: a survey, performance evaluation, open issues, and future research directions. *Frontiers in Genetics* 11: 603808.

Mahmood, U., Li, X., Fan, Y. et al. (2022). Multi-omics revolution to promote plant breeding efficiency. *Frontiers in Plant Science* 13: 1062952.

Maldonado, S., Vairetti, C., Fernandez, A. et al. (2022). FW-SMOTE: A feature-weighted oversampling approach for imbalanced classification. *Pattern Recognition* 124: 108511.

Meher, P.K., Hati, S., Sahu, T.K. et al. (2024). SVM-root: Identification of root-associated proteins in plants by employing the support vector machine with sequence-derived features. *Current Bioinformatics* 19 (1): 91–102.

Mirza, Z., Ansari, M.S., Iqbal, M.S. et al. (2023). Identification of novel diagnostic and prognostic gene signature biomarkers for breast cancer using artificial intelligence and machine learning assisted transcriptomics analysis. *Cancers* 15 (12): 3237.

Misra, B.B. (2020). Data normalization strategies in metabolomics: current challenges, approaches, and tools. *European Journal of Mass Spectrometry* 26 (3): 165–174.

Mochida, K., Koda, S., Inoue, K. et al. (2018). Statistical and machine learning approaches to predict gene regulatory networks from transcriptome datasets. *Frontiers in Plant Science* 9: 1770.

Mohanta, T.K., Bashir, T., Hashem, A. et al. (2017). Systems biology approach in plant abiotic stresses. *Plant Physiology and Biochemistry* 121: 58–73.

Moon, S., Hwang, J., and Lee, H. (2022). SDGCCA: supervised deep generalized canonical correlation analysis for multi-omics integration. *Journal of Computational Biology* 29(8): 892–907.

Moossavi, S., Fehr, K., Khafipour, E. et al. (2021). Repeatability and reproducibility assessment in a large-scale population-based microbiota study: Case study on human milk microbiota. *Microbiome* 9: 1–10.

Muthubharathi, B.C., Gowripriya, T., and Balamurugan, K. (2021). Metabolomics: Small molecules that matter more. *Molecular Omics* 17 (2): 210–229.

Naik, B., Kumar, V., Rizwanuddin, S. et al. (2023). Genomics, proteomics, and metabolomics approaches to improve abiotic stress tolerance in tomato plant. *International Journal of Molecular Sciences* 24(3): 3025.

Nazari, L., Ghotbi, V., Nadimi, M. et al. (2023). A novel machine-learning approach to predict stress-responsive genes in Arabidopsis. *Algorithms* 16 (9): 407.

Neagu, A.N., Jayathirtha, M., Baxter, E. et al. (2022). Applications of tandem mass spectrometry (MS/MS) in protein analysis for biomedical research. *Molecules* 27 (8): 2411.

Nguyen, N.D. and Wang, D. (2020). Multiview learning for understanding functional multiomics. *PLoS Computational Biology* 16 (4): e1007677.

Niazian, M. and Niedbała, G. (2020). Machine learning for plant breeding and biotechnology. *Agriculture* 10 (10): 436.

Nurmohamed, N.S., Kraaijenhof, J.M., Mayr, M. et al. (2023). Proteomics and lipidomics in atherosclerotic cardiovascular disease risk prediction. *European Heart Journal* 44 (18): 1594–1607.

Oh, V.K.S. and Li, R.W. (2021). Temporal dynamic methods for bulk RNA-Seq time series data. *Genes* 12 (3): 352.

Pan, J., Li, Z., Dai, S. et al. (2020). Integrative analyses of transcriptomics and metabolomics upon seed germination of foxtail millet in response to salinity. *Scientific Reports* 10 (1): 13660.

Parisineni, S.R.A. and Pal, M. (2024). Enhancing trust and interpretability of complex machine learning models using local interpretable model agnostic Shap explanations. *International Journal of Data Science and Analytics* 18 (4): 457–466.

Pascovici, D., Wu, J.X., McKay, M.J. et al. (2018). Clinically relevant post-translational modification analyses—Maturing workflows and bioinformatics tools. *International Journal of Molecular Sciences* 20 (1): 16.

Pinu, F.R., Beale, D.J., Paten, A.M. et al. (2019). Systems biology and multi-omics integration: Viewpoints from the metabolomics research community. *Metabolites* 9 (4): 76.

Samad, M.D., Abrar, S., and Diawara, N. (2022). Missing value estimation using clustering and deep learning within multiple imputation framework. *Knowledge-Based Systems* 249: 108968.

Sanchez-Illana, Á., Pérez-Guaita, D., and Cuesta-García, D. et al. (2018). Model selection for within-batch effect correction in UPLC-MS metabolomics using quality control-support vector regression. *Analytica Chimica Acta* 1026: 62–68.

Satam, H., Joshi, K., Mangrolia, U. et al. (2023). Next-generation sequencing technology: Current trends and advancements. *Biology* 12 (7): 997.

Sathyamoorthi, K., Abishek, V.P., and Venkataramana, L.Y. (2025). Enhancing breast cancer survival prognosis through omic and non-omic data integration. *Clinical Breast Cancer* 25 (1): 27–37.

Sharma, M., Nath, K., and Sharma, R.K. et al. (2022). Ensemble averaging of transfer learning models for identification of nutritional deficiency in rice plant. *Electronics* 11 (1): 148.

Shaw, R., Tian, X., and Xu, J. (2021). Single-cell transcriptome analysis in plants: Advances and challenges. *Molecular Plant* 14(1): 115–126.

Singh, A.K., Ganapathysubramanian, B., Sarkar, S. et al. (2018). Deep learning for plant stress phenotyping: trends and future perspectives. *Trends in Plant Science* 23 (10): 883–898.

Smilde, A.K., Westerhuis, J.A., Hoefsloot, H.C.J. et al. (2010). Dynamic metabolomic data analysis: A tutorial review. *Metabolomics* 6: 3–17.

Subramanian, I., Verma, S., Kumar, S. et al. (2020). Multi-omics data integration, interpretation, and its application. *Bioinformatics and Biology Insights* 14: 1177932219899051.

Sun, N.A., Wang, Y.U., Chu, J. et al. (2023). Bayesian approaches in exploring gene-environment and gene-gene interactions: A comprehensive review. *Cancer Genomics & Proteomics* 20 (6suppl): 669–678.

Thanmalagan, R.R., Roy, A., Jayaprakash, A. et al. (2022). Comprehensive meta-analysis and machine learning approaches identified the role of novel drought specific genes in Oryza sativa. *Plant Gene* 32: 100382.

Thapa, K., Kinali, M., Pei, S. et al. (2025). Strategies to include prior knowledge in omics analysis with deep neural networks. *Patterns* 6 (3): 1–19.

Trezza, A., Visibelli, A., Roncaglia, B. et al. (2024). Unsupervised learning in precision medicine: unlocking personalized healthcare through AI. *Applied Sciences* 14 (20): 9305.

Uribeetxebarria, A., Castellón, A., and Aizpurua, A. (2023). Optimizing wheat yield prediction integrating data from Sentinel-1 and Sentinel-2 with CatBoost algorithm. *Remote Sensing* 15(6): 1640.

Wang, C., Li, Y., Wang, J. et al. (2023). Unsupervised cluster analysis of clinical and metabolite characteristics in patients with chronic complications of T2DM: An observational study of real data. *Frontiers in Endocrinology* 14: 1230921.

Waqas, A., Tripathi, A., Ramachandran, R.P. et al. (2024). Multimodal data integration for oncology in the era of deep neural networks: A review. *Frontiers in Artificial Intelligence* 7: 1408843.

Wekesa, J.S. and Kimwele, M. (2023). A review of multi-omics data integration through deep learning approaches for disease diagnosis, prognosis, and treatment. *Frontiers in Genetics* 14: 1199087.

Wu, C., Luo, J., and Xiao, Y. (2024). Multi-omics assists genomic prediction of maize yield with machine learning approaches. *Molecular Breeding* 44 (2): 14.

Wu, Y. and Zhang, K. (2020). Tools for the analysis of high-dimensional single-cell RNA sequencing data. *Nature Reviews Nephrology* 16 (7): 408–421.

Yan, J. and Wang, X. (2022). Unsupervised and semi-supervised learning: the next frontier in machine learning for plant systems biology. *The Plant Journal* 111 (6): 1527–1538.

Yang, Y., Sun, H., Zhang, Y. et al. (2021). Dimensionality reduction by UMAP reinforces sample heterogeneity analysis in bulk transcriptomic data. *Cell Reports* 36 (4): 1–15.

Yurekten, O., Payne, T., Tejera, N. et al. (2024). MetaboLights: Open data repository for metabolomics. *Nucleic Acids Research* 52 (D1): D640–D646.

Zahravi, M., Amirbakhtiar, N., Arhsad, Y. et al. (2024). Identifying key traits for heat stress tolerance in wheat using machine learning. *Iranian Journal of Genetics and Plant Breeding* 13 (1): 61–84.

Zhang, W., Kasun, L.C., Wang, Q.J. et al. (2022). A review of machine learning for near-infrared spectroscopy. *Sensors* 22 (24): 9764.

Zhang, X., Ibrahim, Z., Khaskheli, M.B. et al. (2024). Integrative approaches to abiotic stress management in crops: Combining bioinformatics educational tools and artificial intelligence applications. *Sustainability* 16 (17): 7651.

Zhang, Y., Liu, Y., Zong, Z. et al. (2025). Integrative omics analysis reveals the genetic basis of fatty acid composition in Brassica napus seeds. *Genome Biology* 26 (1): 83.

Zhang, Y., Parmigiani, G., and Johnson, W.E. (2020). ComBat-seq: Batch effect adjustment for RNA-seq count data. *NAR Genomics and Bioinformatics* 2 (3): lqaa078.

Zhang, Y., Wu, Y., Gong, Z.Y. et al. (2021). Distinguishing rectal cancer from colon cancer based on the support vector machine method and RNA-sequencing data. *Current Medical Science* 41 (2): 368–374.

Zhao, Y., Li, M.C., Konaté, M.M. et al. (2021). TPM, FPKM, or normalized counts? A comparative study of quantification measures for the analysis of RNA-seq data from the NCI patient-derived models repository. *Journal of Translational Medicine* 19 (1): 269.

10

Machine Learning for Plant Single-Cell RNA Sequencing

Mani Manoj[1], Ravichandran Sneha[1], Esakkimuthu Balaji[2], Vadivelu Bharathi[3], Thamaraiselvan Nandhini Devi[4], Ramasamy Manikandan[5], Jeyabal Philomenathan Antony Prabhu[1], Arumugam Vijaya Anand[1]**

[1] *Department of Human Genetics and Molecular Biology, Bharathiar University, Coimbatore, Tamil Nadu, India*
[2] *Sudha Fertility Hospital, Hyderabad, Telangana, India*
[3] *Department of Biochemistry, Thanthai Periyar Government College, Tiruchirappalli, Tamil Nadu, India*
[4] *Department of Biomedical Engineering, Selvam College of Technology, Namakkal, Tamil Nadu, India*
[5] *Department of Biochemistry, Shrimati Indira Gandhi College, Tiruchirappalli, Tamil Nadu, India*

*Corresponding authors: avahgmb@buc.edu.in; jtonybio@gmail.com

10.1 Introduction to Plant Single-cell RNA Sequencing

Single-cell RNA sequencing (scRNA-seq) has revolutionized our understanding of plants by enabling the comprehensive analysis of gene expression at the single cell level, thus uncovering cellular heterogeneity that is grossly underestimated by traditional methods. The novel technology enables the detection of common and rare cell types and their spatial and developmental associations, which are crucial in understanding plant biology and their response to environmental stresses (Bawa et al. 2022). scRNA-seq has been highly influential in model plants, crop plants, and trees, as it gives important insights into cell-type-specific expression profiles and aids in marker gene identification using databases designed specifically for this purpose (Sun et al. 2024). However, the rigid cell walls typical of plant cells require a protoplasting step, which introduces biases that make data interpretation complex. ScRNA-seq has proved to be instrumental in plant science in a number of species ranging from model species *Arabidopsis thaliana* to food crops *Oryza sativa* and shows great potential for maximizing plant biosystem design and synthetic biology. This includes genetic and epigenetic manipulations towards improving crops for improvement, yield, quality, and environmental sustainability (Islam et al. 2024). There is a need to overcome these biases to obtain accurate clustering and annotation of cell types, as supported by experiments in tobacco BY-2 cells (Yao et al. 2024). In addition, the combination of scRNA-seq with multi-omics approaches provides an integrated view of plant systems, although there are limitations in its full realization (Bian et al. 2024). The development of spatial transcriptomics (ST) in combination with scRNA-seq further enables the study of heterogeneity of gene expression and spatial organization of genes in tissues, thus offering a platform to complex cell classification and physiological investigation (Lv et al. 2024). As technologies advance to reduce prices and maximize methodologies, scRNA-seq is set to become a powerful tool for leading plant research, understanding environmental stress responses, and other critical biological processes (Bawa et al. 2022).

scRNA-seq data analysis is accompanied by many crucial challenges, mainly contributed by data high dimensionality, high noise, and inherent information loss. One of the main challenges is the efficient choice of informative genes required to cluster, a key requirement for the identification of cell types and tracking biological processes. scAIG addresses the challenge by using smart gene selection and a scale-invariant distance measure, which eventually enhances clustering accuracy and visualization quality, yielding the identification of intricate gene expression patterns (Zhang et al. 2023). The second main consideration, differential expression analysis, addresses the identification of the gene markers required to identify cell types but presents a challenge due to a high amount of data generated and the need for sound statistical methodology. Available methodology is currently classified into several categories, each presenting unique limitations requiring

Machine Learning for Plant Biology, First Edition. Edited by Jen-Tsung Chen.

breakthrough for optimal analytical achievement (Das et al. 2022). Central unsupervised clustering, which is crucial for discovering new cell types, is computationally demanding due to the nature of the data and, further, the necessity of accurate biological interpretation and annotation of the clusters (Nossier et al. 2023; Kiselev et al. 2019). Furthermore, the field faces bioinformatics and computational challenges as the nature of the scRNA-seq data requires new analytical paradigms for optimal utilization of the potential of the technology for disease study and the investigation of biological processes (Poirion et al. 2016).

Machine learning (ML) plays a pivotal role in the analysis of plant scRNA-seq data, particularly in identifying cell-type marker genes and annotating cell identities. Various ML methods, such as random forest and support vector machines (SVMs), have been employed to classify cell types in Arabidopsis root datasets, leading to the discovery of numerous new marker genes that outperform existing markers in terms of accuracy and sensitivity (Yan et al. 2020). The development of interpretable ML models, such as AtML and SPmarker, has further enhanced the ability to predict cell stages and identify novel biomarkers, providing insights into cellular heterogeneity and developmental trajectories in plant roots (Yu et al. 2024). These models not only improve the precision of cell-type annotations but also facilitate cross-species mapping by identifying orthologous marker genes in non-Arabidopsis species, thereby expanding the repertoire of known markers significantly (Yan et al. 2022). The integration of ML in scRNA-seq data analysis addresses major computational challenges, such as the high dimensionality of data, and advances the classification, clustering, and visualization of single-cell transcriptomics data (Vrahatis et al. 2020).

10.2 ML Techniques in scRNA-seq Data Analysis

10.2.1 Preprocessing and Quality Control

10.2.1.1 Data Filtering and Normalization

In the processing of plant scRNA-seq data, filtering and normalization are essential steps for conducting valid downstream analysis. Normalization is intended to correct technical and biological variation and render gene counts comparable between and across cells. Different methods have been developed, each addressing different sources of dispersion and making different assumptions about count data (Duran et al. 2024). For example, SCnorm employs a quantile-regression method to robustly normalize scRNA-seq data, correcting for artifacts introduced by conventional methods in single-cell conditions (Bacher et al. 2016). Partial least squares (PLS) regression with adaptive fuzzy weights is another method that allows for the heterogeneity of gene expression and removes unwanted biases, with performance shown across different datasets (Singh et al. 2024). scLENS also employs L2 normalization to avoid signal distortion during log normalization and employs random matrix theory for noise filtering, enabling the detection of biological signals without manual tuning (Kim et al. 2024). Differential gene expression and clustering are two downstream analyses that are greatly impacted by the normalization technique choice, and no single strategy is always better than another. Instead, it is recommended to assess normalizing techniques using performance criteria such as silhouette width and k-nearest neighbor batch-effect tests. In addition, the incorporation of normalization as part of other data processing steps, such as dimensionality reduction and clustering, can improve scRNA-seq data analysis for efficiency and accuracy, as shown by the comparison of different combinations of methods (Xu et al. 2022).

10.2.1.2 Batch Effect Correction

Batch effect correction in plant scRNA-seq data analysis is a crucial step to obtain sound interpretation of biological signals, as batch effects would hide true biological variability. Various solutions have been proposed to solve the challenge, each with particular methods and advantages. Fast-scBatch applies neural network-based distance matrix adjustment to correct nonlinear and complex batch effects to yield a corrected expression matrix that allows downstream analysis and outperforms current practice in clustering scores like Adjusted Rand Index (ARI) and Adjusted Mutual Information (AMI) (Cheng et al. 2023). The Biological-noise Decoupling Autoencoder and Central-cross Loss model provides a second option as it reconstructs raw data and uses a hierarchical clustering tree that preserves rare cell types and minimizes batch effects to a minimum (Di et al. 2024). The Joint and Individual Variation Explained method, initially proposed for bulk sequencing, has been applied in scRNA-seq to split datasets into joint structures that preserve biological variability and perform better in such measures as kBET and LISI (Hastings et al. 2024). Adversarial Information Factorization offers a good approach to correction based on preserving relative gene expression and performing well under low signal-to-noise ratios as well as batch-specific cell types (Monnier and Cournède 2024). Finally, the iterative supervised mutual nearest neighbor refinement method extends batch effect correction by iteratively refining mutual nearest neighbors, maximizing mixing of cells between batches and enabling easy identification of differently expressed genes (Yang et al. 2021).

10.2.1.3 Dimensionality Reduction Techniques

Dimensionality reduction is an essential step in the analysis of scRNA-seq data because of the high dimensionality and complexity of the data, with thousands of cells and tens of thousands of transcripts. Principal Component Analysis (PCA), t-distributed Stochastic Neighbor Embedding (t-SNE), and Uniform Manifold Approximation and Projection (UMAP) are the most widely used methods, each having different strengths. PCA is an easy method but might not be able to retain the complexity of heterogeneous cell types, while t-SNE is highly accurate but computationally expensive, and UMAP is highly stable and preserved in the original data structure (Linderman 2021). Nonlinear methods such as Pairwise Controlled Manifold Approximation (PaCMAP) and compactness preservation-PaCMAP (CP-PaCMAP) have been proposed to more effectively preserve local and global structures, improving visualization and clustering of scRNA-seq data (Yousuff et al. 2024). Exhaustive benchmarking experiments have compared more than 33 dimensionality reduction methods; Linear Discriminant Analysis (LDA) and Potential of Heat-diffusion for Affinity-based Trajectory Embedding (PHATE) have been found to be high-performance algorithms. These experiments employ internal validation measures (IVMs) to independently measure clustering performance regardless of downstream analysis, allowing for robust evaluation of dimensionality reduction methods (Koch et al. 2021). The method of choice is generally a function of the particular requirements of the analysis, e.g., the requirement for computational efficiency, accuracy, or preservation of data structure, and may involve hyperparameter tuning with caution (Xiang et al. 2021).

10.2.2 Feature Selection and Gene Expression Clustering

10.2.2.1 Highly Variable Gene Selection

In scRNA-seq data analysis, the identification of highly variable gene (HVG) is crucial to overcome high dimensionality and noise issues. Various methods have been proposed to enhance HVG selection. In a thorough comparison of 47 HVG selection techniques, Zhao et al. found that hybrid approaches that combined the advantages of several baseline techniques outperformed single-method approaches. Zhao et al. also created a R package to include the mixHVG strategy, which integrates the best characteristics from many approaches into data analysis workflows (Zhao et al. 2024). Penalized regression as a feature selection technique was investigated by Puliparambil et al., who pointed out that the Sparse Group Lasso (SGL) is a very useful technique for identifying genes that are differentially expressed (DE) in heterogeneous datasets, like plant species (Sen Puliparambil et al. 2022). Townes et al. suggested using multinomial models and deviance for HVG selection. Their approach performed better in clustering analysis than conventional techniques, demonstrating a productive substitute technique for reducing the dimensionality of scRNA-seq data (Townes et al. 2019). According to Islam et al., scRNA-seq has the potential to be used in synthetic biology and plant systems biology as a tool that can reveal cell-type heterogeneity and gene expression patterns. It also plays a crucial role in facilitating the advancement of plant biosystems design (Islam et al. 2024). In addition, the scAIG framework employs in another intelligent gene selection step that enhances clustering and visualization, again highlighting the efficiency of proper HVG selection for scRNA-seq analysis (Zhang et al. 2023).

10.2.2.2 Clustering Algorithms

Clustering algorithms like K-means, hierarchical clustering, and Density-Based Spatial Clustering of Applications with Noise (DBSCAN) are pivotal in the analysis of plant scRNA-seq data, and each has some pros and cons. K-means, a very used algorithm, is typically customized to mitigate scRNA-seq data's high dimensionality and noise. For example, the Dissimilarity-Density-Dynamic Radius K-means (D3K) algorithm uses a dynamic radius to compensate for the noise and yield improved clustering results, mitigating the conventional k-means shortcomings in processing the complexity of scRNA-seq data (Liu et al. 2022). Hierarchical clustering, another legacy algorithm, is used to mitigate scRNA-seq data's dropout of low-expression genes and technical biases, thereby making it suitable to unveil cell subtypes and lineages (Petegrosso et al. 2020). DBSCAN, a density-based algorithm, is less common in scRNA-seq application but is capable of processing clusters of different shapes without a specified number of clusters, which can be useful in detecting rare types of cells (Chowdhury 2021). Although useful, these algorithms have challenges such as inconsistent performance across different datasets with complex structures, which requires continuous research to develop stable and efficient processes (Li et al. 2020). Clustering is further enhanced by the use of statistical and ML techniques such as cell-specific normalization and imputation of dropouts, which enables more accurate identification of cell clusters and potential markers in plant scRNA-seq data. Generally, although these clustering algorithms are central to scRNA-seq analysis, continuous developments remain important to counteract their current shortcomings as well as to improve their application on various datasets (Peng et al. 2020).

10.2.2.3 Deep Learning for Feature Selection

Deep learning (DL) has emerged as a potent feature selector for scRNA-seq data, overcoming difficulties such as high-dimensionality, noise, and sparsity that are typical of such datasets. Research into different DL-based feature selections for scRNA-seq data analysis shows that DL offers an alternative mechanism by using neural networks to evaluate gene significance, whereas conventional methods rely on differential distribution models to identify significant genes (Huang et al. 2023). One such technique is the scFSNN approach, which improves classification accuracy in scRNA-seq data by including feature selection into neural network training. This allows for automated gene selection and control of the false discovery rate (Peng et al. 2024). The scGDC algorithm further augments auto-encoders using a self-representation layer to learn deep features and perceive the topological structure of cells, greatly enhancing clustering performance and the detection of rare cell types by projecting cells onto various subspaces (Wang and Ma 2022). An extensive review of DL in scRNA-seq analysis demonstrates the versatility of these methods, such as variational autoencoders (VAEs) and generative adversarial networks (GANs), which are specifically designed to accomplish specific data processing tasks, thereby offering a robust framework for addressing the computational demands posed by large-scale single-cell datasets (Flores et al. 2022).

10.2.3 Cell-type Identification and Annotation

10.2.3.1 Supervised vs. Unsupervised Learning Approaches

In plant scRNA-seq applications, supervised and unsupervised learning techniques are both characterized by peculiar advantages and limitations. Supervised learning techniques, as seen in the SuperCT framework, increase the accuracy and stability in defining single-cell transcriptomic profiles using labeled datasets to train classifiers capable of efficiently detecting known cell types (Xie et al. 2019). Such techniques are optimal where high-quality reference datasets are available, as they are likely to outperform unsupervised techniques in most applications, excluding the detection of new cell types. However, supervised techniques require large quantities of labeled data, which are likely to be time-consuming and costly to obtain, especially in plant systems where such data might be scarce (Yan and Wang 2022). On the other hand, unsupervised learning techniques, which are typically used in initial scRNA-seq data analysis, allow for the clustering of cells without pre-existing labels and are thus optimal in new cell-type discovery and processing of datasets with limited labeled data. While such techniques are comparatively less accurate in detecting known cell types, they are invaluable in cases where reference datasets are non-informative or skewed (Sun et al. 2022). The combination of self-supervised and unsupervised learning techniques, as seen in the scAnCluster framework, further enhances the ability to cluster and annotate cells by making use of available cell-type labels from reference data, essentially filling the gap between supervised and unsupervised techniques (Chen et al. 2020).

10.2.3.2 Decision Trees, Random Forest, and SVMs

In plant scRNA-seq studies, a variety of ML methods, including Decision Trees, Random Forests, and SVMs, have played a vital role in optimizing cell-type prediction and marker gene discovery. Especially, Random Forests and SVMs have performed better in identifying cell types from *Arabidopsis* root scRNA-seq datasets, resulting in the identification of 600 new marker genes, with Random Forests demonstrating outstanding improvement in the expression of marker genes in multiple cellular clusters (Yan et al. 2020). Similarly, the NS-Forest algorithm, inspired by Random Forest techniques, has been demonstrated to be efficient for marker gene detection with high accuracy and specificity of classification, as indicated by its use in analyzing human organ data (Liu et al. 2024). Decision Trees, due to their inherent interpretability to provide clear explanations for results, have been utilized in explaining gene regulation networks and making predictions regarding cell fate decisions and hence offering an interpretable framework for understanding trees of lineages in single cells (Veleslavov and Stumpf 2020). Additionally, ensemble learning techniques, like NRTPredictor, which combines multiple models, have shown high predictive accuracy in identifying the cell states of rice roots. This highlights the value of ensemble algorithms in tackling the challenge posed by the sparse and high-dimensional nature of scRNA-seq data (Wang et al. 2023). Moreover, novel ensemble models that integrate a number of different-depth regression trees have been formulated to improve efficiency and performance for decision tree–based methods (Mahani and Ziarati 2023).

10.2.3.3 DL-based Cell-type Classification

DL methods for cell-type classification in plant scRNA-seq have made giant leaps, especially using convolutional neural networks (CNNs) and recurrent neural networks (RNNs). The scCAEs approach combines convolutional autoencoders and soft K-means clustering to effectively combat high dropout rates and noise in scRNA-seq data and ensure successful clustering and consistency across datasets with heterogeneous batch effects (Hu et al. 2022). In contrast, scDeepInsight adopts a supervised approach by projecting scRNA-seq data onto images, which are then passed through CNNs like EfficientNet-b3, attaining a mean accuracy of 87.5% in cell-type annotation, surpassing other competitive approaches (Jia et al. 2023). MulCNN goes a step further in cell-type identification by using a multi-level CNN that achieves gene expression features while filtering out noise simultaneously, demonstrating better performance on various species datasets (Jiao et al. 2023). Also, single-cell RNA-seq data using residual graph CNNs with contrastive learning (scRGCL) utilizes residual graph convolutional networks and contrastive learning to learn high-order features and prominent cell-to-cell differential features, outperforming other methods at cell-type annotation on human and mouse datasets (Yuan et al. 2025).

10.2.4 Trajectory and Pseudotime Inference

10.2.4.1 Single-cell Developmental Trajectories

scRNA-seq has become a revolutionary technology in plant biology, allowing the in-depth investigation of developmental paths at the cellular level. In *Arabidopsis thaliana*, scRNA-seq has been used to examine root cells, identifying distinct subpopulations and developmental stages, such as quiescent center cells, rare cell types, and tracing the transition from meristematic to mature stages in root epidermal cells (Ryu et al. 2019). Likewise, scRNA-seq has revealed the differentiation path of female germline cells in *Arabidopsis*, identifying dynamic patterns of gene expression and key transcription factors for germline cell fate specification (Hou et al. 2021). In cotyledon development, scRNA-seq has been used to identify cell clusters and developmental paths, using methods like t-SNE and RNA velocity to track gene expression dynamics and identify new marker genes for vascular tissues (Verhage 2022). In developing leaf blade cells of the allotetraploid peanut, scRNA-seq has been used to chart the transcriptome landscape of leaf blade cells, identifying developmental paths and transcription factor networks leading to differentiation directions, such as palisade cells to spongy cells (Liu et al. 2021). These studies in sum prove the capacity of scRNA-seq in revealing the complexity of plant cell differentiation and development and providing insights into cellular heterogeneity and molecular mechanisms governing plant growth and development (Zheng et al. 2021).

10.2.4.2 Pseudotime Estimation

Monocle, Slingshot, and DL methods are relevant in pseudotime estimation of scRNA-seq data in plants, each with strengths and limitations. Monocle, especially its third version, is widely utilized for trajectory inference to allow scientists to sequence cells along a pseudotime trajectory that captures biological processes like differentiation and development. This aforementioned method has been useful in numerous studies, including the study of the mouse mammary gland, where it allowed the identification of pseudotime-related genes to understand cellular differentiation and potential biomarkers (Cheng et al. 2023). Slingshot, another deeply familiar method, is characterized by its ability to handle branching trajectories that are complicated and intrinsic in representing varied developmental pathways in plant cells. It provides a solid framework for trajectory inference through the utilization of multiple forms of data and adjustment of confounding variables, as evident from studies in *Arabidopsis* and other organisms (Reid and Wernisch 2016). Although DL methods are not mentioned by context, growing exploration is essential in the potential enhancement they bring to pseudotime estimation through the modeling of nonlinear interactions and the input of data of high dimensionality. The field of pseudotime analyses in scRNA-seq research in plants is advanced with high accuracy and credibility when combined with conventional techniques like TSCAN, which sequences cells, based on the dynamics of gene expression, and PseudotimeDE, which takes uncertainties in pseudotime inference into account (Ji and Ji 2019; Song and Li 2020).

10.2.4.3 Application of Graph Neural Networks

Graph Neural Networks (GNNs) have been increasingly applied in scRNA-seq data analysis for solving a range of analytical problems, ranging from data sparsity, noise contamination, and the intricacy of gene expression patterns. Predicting gene regulatory networks (GRNs), which are used to forecast the regulatory relationships between genes, is one of the main uses for GNNs. For example, the GNNLink framework shows higher performance on a variety of datasets by using known GRNs to predict putative regulatory linkages by framing the task as a graph link prediction problem (Mao et al. 2023). Similarly, the scRegNet unites GNNs with pre-trained single-cell foundation models for enhanced GRN inference, unifying extensive contextualized representations with structured knowledge, and outperforming state-of-the-art approaches in capturing intricate regulatory interactions among different cell types (Kommu et al. 2024). Besides GRN inference, GNNs are also applied in unsupervised analysis of scRNA-seq, as demonstrated by the CellVGAE model, which utilizes a variational graph autoencoder coupled with graph attention layers for dimensionality reduction as well as clustering, outperforming other approaches in extracting meaningful features from the dataset (Buterez et al. 2020). By simulating cell-to-cell and heterogeneous patterns of gene expression interactions, the scGNN framework further illustrates the usefulness of GNNs and enables effective gene imputation and cell clustering capabilities (Wang et al. 2021). Additionally, by building cell-to-cell graphs during training, the scE2EGAE expands on scRNA-seq data processing and enhances denoising and clustering performance using a graph autoencoder with integrated differentiable edge sampling (Wang et al. 2025).

10.2.5 Differential Gene Expression Analysis

10.2.5.1 ML Models for Identifying DE Genes

ML models have emerged as central to the identification of DE genes in plant scRNA-seq data, illuminating cellular heterogeneity and gene expression dynamics. The hurdle model with correlated random effects, as developed by Sekula et al., breaks the zero-inflated nature of scRNA-seq data, precisely modeling cell-to-cell variability and detecting DE genes that other approaches fail to detect (Sekula et al. 2019). Likewise, the Tweedie models, e.g., a zero-inflated model, provide a versatile solution to handle heavy tails and sparsity of scRNA-seq data, surpassing other DE methods in statistical power and regulation of false discovery rate across different experimental platforms (Mallick et al. 2022). Yan et al. demonstrated how ML technologies, such as random forest and SVMs, can be used to classify cell types and find new marker genes in plant scRNA-seq. This significantly improves the sensitivity and accuracy of cell-type assignment (Yan et al. 2020). Furthermore, by utilizing a Markov Random Field technique, the MARBLES model enhances the identification of DE genes by utilizing data from related cell types and taking sample-level variability into consideration. This makes it more effective than existing methods in identifying DE genes and pathways linked to disease (Zhu et al. 2022). Figure 10.1 conceptually illustrates how AI can be leveraged to process plant sc-RNA-seq data and identify gene expression differences across diverse cell types.

10.2.5.2 Bayesian Methods for Gene Expression Inference

Because of their capacity to handle the inherent noise and unpredictability that characterize plant scRNA-seq data sets, Bayesian approaches have emerged as powerful techniques for gene expression estimation. By addressing both biological variation and measurement noise, sanity, a Bayesian normalizing process, is a useful way to tackle the problem of quantifying gene expression states. It estimates changes in expression in terms of fold change rather than absolute values. Also, it is superior to existing normalization techniques in detecting genes with differential expression and classifying cells into distinct subtypes. It does this by directly estimating log expression levels from raw unique molecular identifier (UMI) counts without the need for tuning parameters (Breda et al. 2021). Similarly, bayNorm uses a binomial model of mRNA capture with empirical Bayes priors to scale and estimate scRNA-seq counts using a Bayesian technique. In comparison to alternative methods, it lowers batch effects, improves the sensitivity and accuracy of differential expression analysis, and correctly imputes missing values (Tang et al. 2020). By reconstructing hidden gene and cell matrices, Bfimpute, another Bayesian technique, attempts to recover dropout events in sparse gene expression matrices. It is better than other imputation methods with higher accuracy and has the ability to incorporate additional gene or cell-related data to further enhance its performance (Wen et al. 2022).

10.2.5.3 Application of Neural Networks in Predicting Gene Regulation

The use of neural networks to predict gene regulation from plant scRNA-seq data has been investigated through a number of novel methods, each addressing the sparsity and noise in the data in the problem. GNNs have been used to successfully predict GRNs by re-casting the problem as a task of graph link prediction. For example, the GNNLink framework uses a graph convolutional network to improve gene feature refinement and predict regulatory interdependencies, with robust performance across a variety of datasets (Mao et al. 2023). In a similar vein, GRNNLink uses graph RNNs to learn gene correlations, with high accuracy for GRN prediction (Bai and Wang 2024). Another method, scRegNet, uses GNNs with pre-trained single-cell foundation models to improve the inference of cell-type-specific GRNs, with enhanced performance over current approaches by combining rich contextual representations with structured knowledge from established networks (Kommu et al. 2024). For plant biology, the combination of scRNA-seq data with DAP-seq and ATAC-seq data has

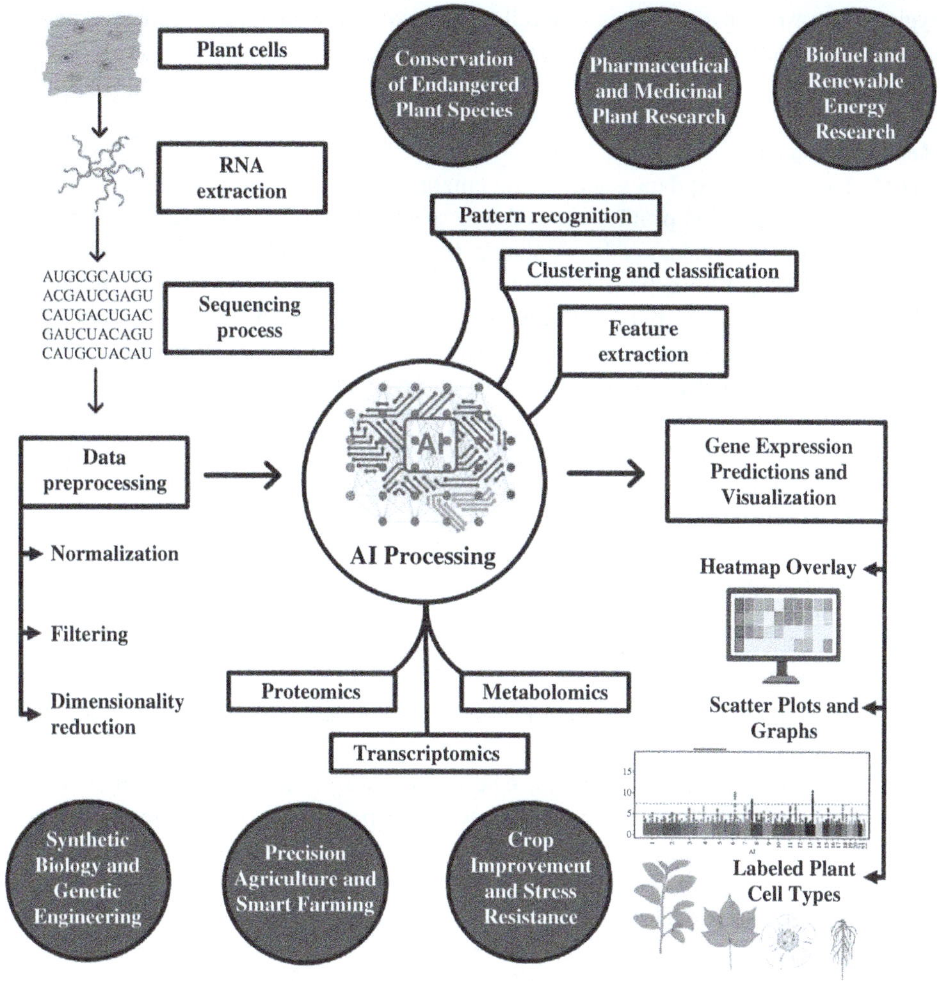

Figure 10.1 AI-assisted identification of differentially expressed genes in plant cells.

been used to predict regulatory networks, using ML methods for feature choice and stability selection to find key regulatory genes (Li et al. 2021). Also, CNNs have been used to extract gene expression information from 2D co-expression matrices, with better accuracy for gene interaction predictions by reducing the effect of noise and dropout in scRNA-seq data (Mao et al. 2023).

10.2.6 Cell–Cell Interaction and Network Analysis

10.2.6.1 Graph-based Approaches for Inferring Cell Communication

Graph-based approaches have become pivotal to inferring cell communication in plant scRNA-seq data analysis, leveraging the power of GNNs to model complex cellular interactions. One example is DeepTalk, which uses a graph attention network (GAT) to integrate scRNA-seq with ST data. This enables precise cell-type identification and spatially resolved cell–cell communication (CCC) inference at single-cell resolution, with improved performance across multiple datasets (Yang et al. 2024). Similarly, scE2EGAE enhances scRNA-seq data analysis by learning cell-to-cell graphs through a graph autoencoder, improving denoising and scRNA-seq data clustering, thereby allowing for the understanding of inter-cellular relationships (Wang et al. 2025). GTAD also uses a GAT to integrate scRNA-seq and ST-seq data, effectively capturing spatial relationships and topological structures in tissues, which improves cell-type identification and provides a robust understanding of tissue microenvironments (Zhang et al. 2024). By predicting CCC in scRNA-seq datasets using a graph-based DL approach that considers gene-gene networks, pathway, and protein complex impacts, GraphComm goes one step further and offers insights into cellular crosstalk and how it affects downstream pathways (So et al. 2023).

Lastly, scGPCL uses graph prototypical contrastive learning to encode cell representations, overcoming problems such as dropout phenomena in scRNA-seq data and improving clustering outputs by leveraging relational information in the data (Lee et al. 2023).

10.2.6.2 Co-expression Network Analysis Using ML

ML co-expression network analysis of plant scRNA-seq data is a new and emerging area of research that improves our knowledge of plant biology by outlining gene expression profiles and cell-type specific markers. Co-expression networks are a necessity in understanding large transcriptomic data, and by using this, researchers are able to pick out modules of co-expressed genes that could have common biological functions, therefore providing information regarding plant biology and allowing for integration of genomic data to obtain more biologically informative information (Rao and Dixon 2019). scRNA-seq data integration across plant species due to common gene family expansions is avoided using co-expression to pinpoint non-orthologous gene pairs with identical expression patterns, therefore enabling integration of cross-species data (Passalacqua and Gillis 2023). Random forest and SVMs are ML techniques that have been found useful in cell-type classification and novel marker gene discovery in *Arabidopsis* roots, with the markers having better performance than the existing markers (Yan et al. 2020). Additionally, single-cell gene co-expression network research has shown that Arabidopsis root growth and metabolism are regulated by gene expression programs (GEPs), with developmental regulators and metabolic processes unique to certain cell types being discovered (Han et al. 2024). Use of CNNs in co-expression analysis is a new method, with better accuracy than existing approaches and the potential to infer causal relationships from sequence and epigenetic data (Yuan and Bar-Joseph 2018).

10.2.6.3 Integration with ST

Traditional scRNA-seq provides useful information on transcriptional heterogeneity but does not locate these variations within the context of space within tissues, which is an essential feature in studying complex biological systems (Adema et al. 2024). To address this limitation, various integration techniques have been put forward, each with different strengths and limitations, thus allowing researchers to select the most appropriate method depending on their individual needs (Yan et al. 2024). For instance, technologies like Stereo-seq combines spatial information with single-cell transcriptomics, providing enhanced resolution that can transform plant transcriptome analysis and improve crop quality (Bawa et al. 2022). Additionally, the combination of scRNA-seq with ST can reveal intercellular communication pathways and tissue dynamics, which cannot be revealed using scRNA-seq (Longo et al. 2021). Metabolic RNA labeling-based time-resolved scRNA-seq is another example of this combination. When paired with ST, it creates a potent tool for spatiotemporal analysis that reveals intricate gene regulation processes in disease contexts (Chen et al. 2024).

10.3 Deep Learning for Plant scRNA-seq Analysis

10.3.1 Autoencoders for Dimensionality Reduction and Noise Removal

10.3.1.1 VAEs in scRNA-seq

By offering an integrated platform for dimensionality reduction, clustering, and the explanation of subtle gene expression patterns, VAEs have become a reliable technique for the analysis of scRNA-seq data. VAEs can work directly with raw count data, thus removing the need for rigorous preprocessing, and provide a latent representation that explains heterogeneity among cell populations, which is valuable in cell-type identification and the explanation of underlying biological processes. By learning a low-dimensional latent space that reflects significant biological information, models like scVAE are designed to address the high-dimensional nature of scRNA-seq data. In plant tissue, this provides efficient cell clustering, which is useful for recognizing different cell types (Grønbech et al. 2020). This is further improved by ScInfoVAE through the use of mutual information regularization, which increases the accuracy of clustering between datasets and makes the latent space easier to understand (Pan et al. 2023). By using biological ontologies and adversarial tactics to untangle covariate effects, models like OntoVAE and COBRA address the interpretability difficulty in VAEs and make the latent representations more physiologically relevant. Such models can predict context-dependent regulatory networks and pathways, providing important insights into the underlying biological processes taking place in plant cells (Doncevic et al. 2024). Although VAEs offer several benefits, they can have challenges in terms of interpretability, especially when latent variables are neglected due to overly expressive decoding distributions (Pan et al. 2023). Furthermore, while VAEs can facilitate correspondence between latent and experimental variables, improving correspondence can occasionally impair clustering performance, suggesting a balance between interpretability and clustering performance (Kimmel 2020).

10.3.1.2 GANs for Data Augmentation

GANs have been discovered to be a potent computational technique for dataset augmentation in plant-based data, effectively addressing the problem of data scarcity, class imbalance, as well as generation of high-quality synthetic images. GANs provide strong benefits in agricultural applications, particularly where the acquisition of large-scale datasets is frequently hampered by environmental and biological limitations. Through the generation of high-quality images, GANs enhances the accuracy of plant disease diagnosis, crop classification, and a variety of other agronomic applications. Importantly, GANs are employed to generate artificial images of diseased plants, thereby expanding training datasets to enhance model accuracy. For instance, in paddy disease diagnosis conditions, the application of data generated by GANs has been proven to enhance the performance of classification algorithms such as SVMs, with an accuracy of 96.23% (Lamba et al. 2022). GANs also overcome the limitation of dataset imbalance in crop classification by generating additional samples for under-represented classes. This process has led to improved mapping accuracy and intersection over union (IoU) measures, thereby enhancing dataset balance and overall model effectiveness (Shumilo et al. 2023). GANs are also employed in the fields of plant classification and weed detection, effectively overcoming issues of occlusion and dataset imbalance. The creation of high-fidelity synthetic samples using GANs greatly enhances the predictive ability and resilience of models in these particular applications (ur Rahman et al. 2024). In contrast to conventional augmentation techniques, GANs can create completely new images that are not present in the real world, therefore enhancing the diversity of datasets as well as their realism (Lamba et al. 2022). Empirical studies have demonstrated that models trained on datasets augmented using GANs have better performance than models that make use of conventional augmentation techniques, as indicated by better accuracy measures in a range of applications (Shumilo et al. 2023). In spite of the significant benefits provided by GANs for data augmentation in plant applications, new technologies such as diffusion models are proving to be more effective in the creation of high-fidelity synthetic images. For example, models such as RePaint have shown better performance in the identification of plant diseases than GANs, which indicates a possible revolution in the techniques used for synthetic data generation (Muhammad et al. 2023).

10.3.2 Convolutional and RNNs

10.3.2.1 CNNs for ST in Plants

The use of CNNs in plant ST is an innovative computational method of demystifying sophisticated gene expression patterns without compromising their spatial organization. This method is most useful in plant science, given that ST addresses the challenges involved in tissue sectioning and multi-dimensional data analysis. More importantly, CNNs with hexagonal convolutional architectures enhance the spatial domain detection accuracy and computational complexity in plant tissue. Hexagonal Convolutional Neural Networks (HexCNNs) are tailored to overcome the limitations involved in using traditional unsupervised methods in ST, which fall short of addressing data noise and dropout events. By applying hexagonal convolutional processes, the networks maximize spatial continuity preservation and effectively reduce the contribution of missing gene expression signals by aggregating information from nearby nodes (Gao et al. 2022). The hexagonal architecture maximizes the modeling of spatial relations in plant tissues by providing a more geometrically consistent and biologically relevant representation of gene expression patterns. The structural advantage is most crucial to the accurate detection of spatial domains, given that it provides signal integration from surrounding cells with minimal distortion while enabling better signal integration from neighbor cells compared to the standard grid-based method. By demystifying intricate spatial interdependencies, the hexagonal organization maximizes the detection of unique transcriptional landscapes, therefore the resolution and interpretability of spatial transcriptomic analysis in plant science (Gao et al. 2023). The combination of ST and CNNs provides the detection of individual cell types, cell fate trajectory reconstruction, and the analysis of intercellular interactions in plant tissues. This approach is of great significance as the plant biology science seeks to bridge the gap with advances in single-cell technologies that have been highly optimized for animal and human biological systems. Through the utilization of CNN-driven spatial transcriptomic analysis, scientists can have a more realistic understanding of the dynamics of plant cellular, developmental processes, and tissue organization at a higher resolution (Yin et al. 2023). The combination of ST with CNNs enables the discovery of previously unknown genes and clarification of complex gene regulation networks. This achievement provides useful information that can be utilized in crop improvement strategies, with emphasis on the optimization of yields, enhancement of quality factors, and resistance development against environmental stressors. Through the preservation of spatial context in measurements of gene expression, this approach offers a more realistic understanding of plant molecular processes, with a final aim of cultivating more productive and resilient crop varieties (Hu et al. 2024). Although CNN, particularly hexagonal CNN variants, are of very high potential in the application of ST in plant systems, there are still great challenges in sample preparation and data analysis steps. Anticipated advances in both technological equipment and methodological approaches are expected to enhance the resolution and efficiency of ST, thus enabling deeper, integrative knowledge of plant molecular biology as well as improved accuracy of spatial domain demarcation (Yin et al. 2023).

10.3.2.2 RNNs for Temporal Gene Expression Analysis

RNNs have been found to be extremely useful in computational analysis of temporal patterns of gene expression in plants, hence enabling explanation of GRNs and gene interactions over time. These architectures of neural networks are best suited to sequential biological data and in modeling temporal dependencies, which are essential to explain the dynamic regulation of gene expression. Of significance, attention-augmented versions of RNNs have been found to have better predictive accuracy in modeling gene expression trajectories and in distinguishing different GRN topologies. This predictive accuracy is further enhanced by the use of graph theoretical methods in interpreting attention distributions, enabling hierarchical description of GRN architectures. RNNs, especially those employing dual-attention mechanisms, have been used successfully in accurate prediction of gene expression time series. These models enable classification of GRNs based on the interpretability of RNN parameters, which is essential to explain the complex dynamics of gene interactions (Monti et al. 2022). RNNs are utilized in reverse engineering GRNs with temporal gene expression datasets. By systematically perturbing trained neural architectures, scientists can determine key regulatory genes and build potential GRN models to be further validated and studied (Krishna et al. 2005; Zhao et al. 2019). Regularized neural networks, which are a specialized type of RNNs, show robust performance in handling high-dimensional, noise-perturbed time-series gene expression data. These models perform better than conventional classification algorithms by capturing the dynamic features of gene expression patterns accurately despite stochastic fluctuations and complex regulatory interactions (Liang and Kelemen 2005). RNNs can model temporal gene expression dynamics in order to make predictions on potential transcription factor targets. Through gene knockout simulation, the models make predictions of expression trajectories and predict GRNs with high accuracy, surpassing most traditional methods in regulatory interaction reconstruction (Zhao et al. 2019). While RNNs are of immense value to the investigation of temporal dynamics of gene expression, there are still a number of challenges that persist, such as that large datasets are needed to enable optimal model training and that their use is computationally costly.

10.3.3 Transfer Learning in Plant scRNA-seq

10.3.3.1 Pretrained Models for Plant Gene Expression Prediction

Pretrained computational models for plant gene expression prediction have shown significant value in understanding the mechanisms of controlling gene regulation in plant systems. These models utilize sophisticated DL and ML algorithms to predict gene expression from a wide array of genomic and epigenomic features. Their application in plant genomics has enabled the identification of key regulatory elements and prediction of gene expression patterns in various species. Specifically, DL architectures have been designed to predict gene expression profiles derived from flanking regions in plant species like *Arabidopsis thaliana, Solanum lycopersicum, Sorghum bicolor*, and *Zea mays* with predictive accuracies higher than 80%. These models highlight the importance of untranslated regions (UTRs) and show high predictive capability across species. The computational models have the ability to identify evolutionarily conserved regulatory motifs as well as species-specific features. This ability is of utmost importance for understanding the mechanisms of genetic variation and dynamic regulation of gene expression. By separating conserved elements which reflect basic regulatory functions across taxa from unique, species-specific features that can influence adaptive responses, these models give significant insights into the molecular basis of transcriptional regulation and phenotypic diversity (Peleke et al. 2024). Additionally, ML models including chromatin modifications like H3K4me3 and H3K36me3 have proven very effective in predicting gene expression in *Arabidopsis* and rice. These results suggest that a small set of histone modifications can be sufficient in reaching high predictive accuracy (Wu et al. 2019). The incorporation of transcription factor binding data significantly enhances the predictive power of such models, with the implication that a limited set of transcription factors have an outsize regulatory

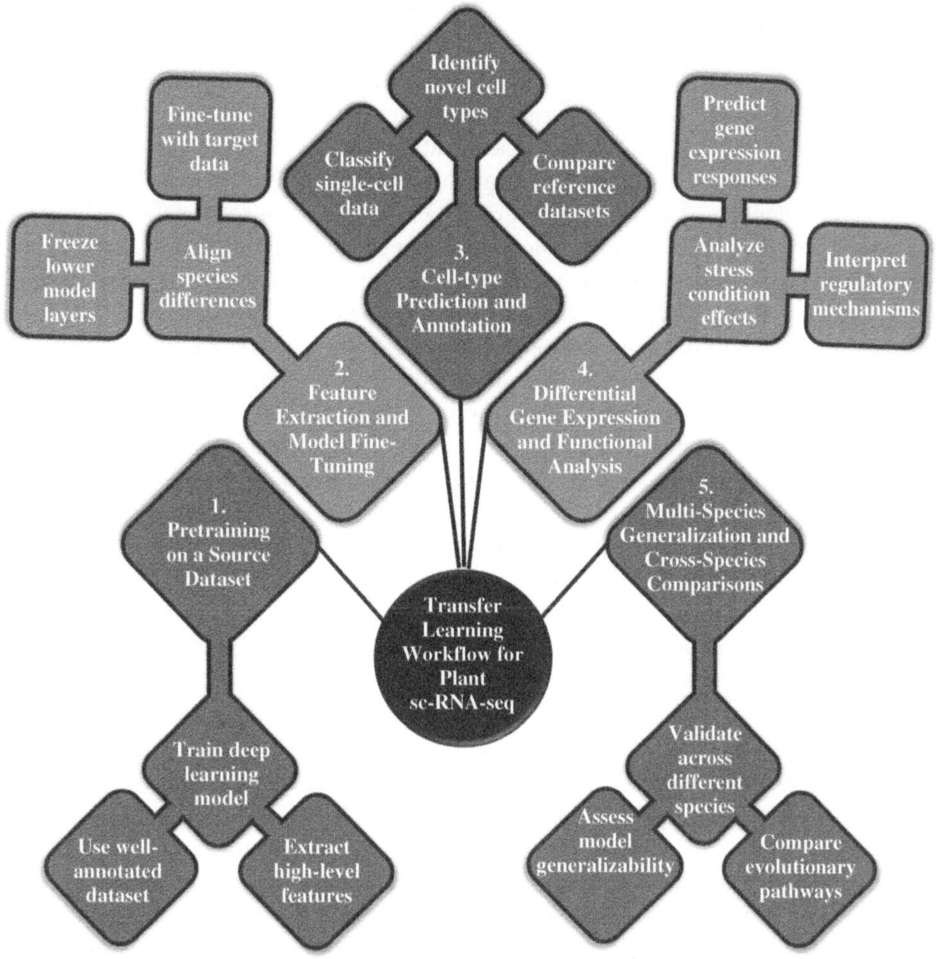

Figure 10.2 Transfer learning workflow for plant sc-RNA-seq.

effect on gene expression levels. Such an integrated model not only enhances model precision but also highlights the central role played by transcription factor dynamics in transcriptional regulation processes in various biological systems (Wu et al. 2019). AgroNT model, a state-of-the-art foundational language model, has been trained on genomic sequences for 48 plant species. The model has superior predictive power in regulatory annotation, promoter and terminator efficiency, and explaining tissue-specific gene expression and thus forms a strong reference framework for plant genomics research (Mendoza-Revilla et al. 2024). TeaTFactor, a state-of-the-art DL model based on the BERT framework, is designed for the identification and classification of transcription factors with high precision in *Camellia sinensis*. Leveraging contextualized embeddings and self-attention, it surpasses the predictive accuracy and generalization performance of traditional models. Its extreme high capacity for sequence motif detection, binding sites, and regulatory interactions makes it a valuable resource for the identification of transcriptional regulatory networks in the genomics of tea plants (Tang et al. 2024). While pretrained models demonstrate encouraging results, their application across plant species and across environmental contexts continues to pose strong challenges. The complexity of plant genomes and the central role played by environmental factors in regulating gene expression means that these models need to be constantly optimized and fine-tuned. Figure 10.2 illustrates the workflow of transfer learning for plant sc-RNA-seq analysis.

10.3.3.2 Domain Adaptation Techniques for Cross-species Analysis

Cross-species transcriptomic analysis in plants requires domain adaptation techniques for improving the transferability of ML models trained in one species to another. These methods resolve domain shifts caused by interspecies variations in gene expression patterns, regulatory network topologies, and underlying statistical distributions, thereby enhancing model robustness and predictive accuracy across heterogeneous plant species. Multi-Representation Subdomain Adaptation Network with Uncertainty Regularization (MSUN) is a cutting-edge framework that employs multi-representation modules to learn fine-grained feature embeddings and utilizes subdomain adaptation to tackle interclass similarity and intraclass variability. This method is extremely effective in reducing interdomain differences and weakening uncertainty in domain transfer, thereby achieving superior accuracy in plant disease classification across heterogeneous datasets (Wu et al. 2023). A deep transfer learning model with mixed subdomain adaptation is introduced for cross-species plant disease diagnosis. This method produces hybrid domain images by combining feature representations of source and target domains, thereby enhancing cross-domain correlation. Employing subdomain alignment across multiple levels of features, the method ensures effective knowledge transfer, particularly where the source and target domains are of low statistical similarity. This method significantly improves the model's ability to learn fine-grained pathological features, thereby outperforming existing methods in plant disease classification accuracy (Yan et al. 2021). Domain-adaptive neural networks are employed to enhance cross-species prediction of transcription factor binding by resolving domain shifts caused by species-specific repetitive sequences. These models employ adaptive feature regularization to constrain learning of species-specific sequence patterns, thereby enhancing generalization across species. This method significantly improves predictive accuracy by rectifying error in repetitive elements, thereby enhancing cross-species transcription factor binding inference (Cochran et al. 2022). The Discriminative Heterogeneous Maximum Mean Discrepancy (DMMD) model seeks to solve feature representation differences that are experienced during classification of cross-species data. By optimizing the domain alignment within a shared latent subspace and maintaining important discriminative information in the process, DMMD minimizes probabilistic divergence among the various domains. This approach significantly minimizes interspecies feature disparity, thus enhancing classification accuracy and model generalization across various biological datasets (Li and Cai 2021).

10.4 Integrating Multi-omics Data with ML

The integration of scRNA-seq and single-cell ATAC-seq (scATAC-seq) techniques in plant studies offers a comprehensive approach to understanding cellular identity and GRNs. This integration allows researchers to infer gene-regulatory relationships by combining gene expression data with chromatin accessibility profiles, which is crucial for elucidating the complex regulatory mechanisms in plants (Chen et al. 2019). In plant studies, such as those involving Arachis hypogaea, simultaneous single-nucleus RNA sequencing (snRNA-seq) and snATAC-seq have been employed to explore transcriptome and chromatin accessibility landscapes, revealing insights into cell differentiation and gene expression pathways (Liu et al. 2022). Tools like ScReNI have been developed to integrate scRNA-seq and scATAC-seq data, enabling the inference of cell-specific regulatory networks and identification of cell-enriched regulators, which are essential for understanding plant cell heterogeneity and function (Xu et al. 2024). Additionally, platforms like HyDrop have been introduced to enhance the scalability and cost-effectiveness of these techniques, providing a flexible and open-source solution for high-throughput single-cell sequencing, which can be adapted for plant studies (De Rop et al. 2022). Methods such as scJoint further facilitate the integration of large-scale single-cell datasets by employing transfer learning to achieve high cell-type label accuracy and meaningful joint visualizations, thus overcoming the challenges posed by the heterogeneity of data modalities (Lin et al. 2022). Recent advances have enabled the use of multi-omics approaches to provide a holistic view of plant physiology, allowing researchers to correlate different data sets and gain deeper insights into cellular and organismal functions (Yan et al. 2022). For instance, in the context of abiotic stress signaling, integrating proteomics and metabolomics data helps in understanding the cascading mechanisms from stress perception to physiological adjustments, as demonstrated in studies using the model plant *Arabidopsis thaliana* (Duruflé and Déjean 2023). Moreover, systematic multi-omics integration approaches have been developed to address the challenges posed by large, poorly annotated plant genomes and diverse secondary metabolites, offering methodological guidelines for successful data assimilation and modeling (Jamil et al. 2020). A practical application of this integration is seen in the study of sugarcane varieties under drought conditions, where network-based approaches combining proteomics and metabolomics data revealed distinct metabolic adjustments between tolerant and susceptible varieties, highlighting the potential of regularized canonical correlation analysis (rCCA) in inferring protein–metabolite associations (Budzinski et al. 2019). These advancements underscore the importance of integrating metabolomics and proteomics data to advance plant systems biology, offering new insights into plant resilience and adaptation mechanisms.

10.5 Challenges and Future Directions

Integrating ML with plant sc-RNA-seq presents promising future directions for enhancing crop resilience and yield. The application of ML, particularly DL models like CNNs, can significantly improve the interpretation of complex genetic and epigenetic data, which is crucial for understanding plant responses to environmental stresses and improving crop traits (Küçükrecep et al. 2024). The use of ML in genomic selection has already shown potential in predicting crop performance under varying environmental conditions, which is essential for developing climate-resilient crops (Saimon et al. 2023). Furthermore, AI-powered genomic research can accelerate breeding cycles and enhance predictive accuracy by analyzing vast datasets, thereby supporting the development of high-yielding crop varieties. The integration of AI with high-throughput phenotyping and multi-omics data can optimize agricultural practices, as AI systems can identify complex patterns and relationships within these datasets (Negus et al. 2024). However, challenges such as fragmented data sources and variability in phenotyping protocols need to be addressed through standardized data integration frameworks and advanced analytical tools (Wójcik-Gront et al. 2024). Additionally, the application of ML in plant breeding and biotechnology, including the use of neural networks and SVMs, has demonstrated efficiency in handling large-scale, complex data, which is crucial for interpreting genotype–environment interactions and optimizing breeding strategies (Niazian and Niedbała 2020).

The application of ML algorithms to plant scRNA-seq data presents several primary challenges, due to the inherent complexity and high dimensionality of the data. One significant challenge is the ultra-high dimensionality of scRNA-seq data, which complicates classification, clustering, and visualization tasks, as these require sophisticated computational methods to manage the vast amount of information generated by sequencing technologies (Vrahatis et al. 2020). Additionally, the data is often sparse and noisy, which obscures the underlying biological signals and complicates the extraction of meaningful patterns. The integration of multiple data modalities, such as gene expression and epigenetic modifications, is crucial for enhancing prediction accuracy and biological interpretability, yet remains a complex task due to the diverse nature of the data. Furthermore, the rapid expansion of single-cell sequencing to millions of cells exacerbates data handling challenges, necessitating the development of new bioinformatics algorithms and tools for effective analysis (Adil et al. 2021). Another challenge is the limited availability of high-quality annotated datasets, which hinders the training and validation of ML models. Moreover, the complex correlations within biological tissues make it difficult to accurately reconstruct cellular states and spatial contexts, posing additional hurdles for traditional feature engineering-based methods. Despite these challenges, advancements in DL offer promising solutions by automatically identifying meaningful patterns in high-dimensional data, although further development is needed to fully address these issues (Ge et al. 2024). Figure 10.3 summarizes both the persistent challenges and the promising avenues for future development in the application of ML to plant sc-RNA-seq data.

10.6 Conclusion

By using potent approaches for the analysis of vast and complicated data sets, ML has completely changed the interpretation of plant scRNA-seq data. ML approaches enhance the precision and effectiveness of scRNA-seq data processing, from feature selection and initial data cleaning to cell-type categorization and developmental trajectory prediction. Furthermore, sophisticated DL models, including autoencoders and CNNs, have significantly improved our capacity to derive meaningful biological information from plant single-cell data. Furthermore, a deeper understanding of gene regulation and cellular processes is made possible by the integration of multi-omics data using ML approaches. While it has its advantages, computational scalability, variability of data, and ethical concerns must be addressed to allow more widespread applications. Future advancements in AI-based techniques will boost scRNA-seq analysis accuracy and speed even more. Plant scientists may discover new knowledge about plant genetics, development, and stress responses using ML, which can lead to crop improvement and agricultural innovation.

Acknowledgement

All the authors are thankful to their respective universities and institutions for their valuable support.

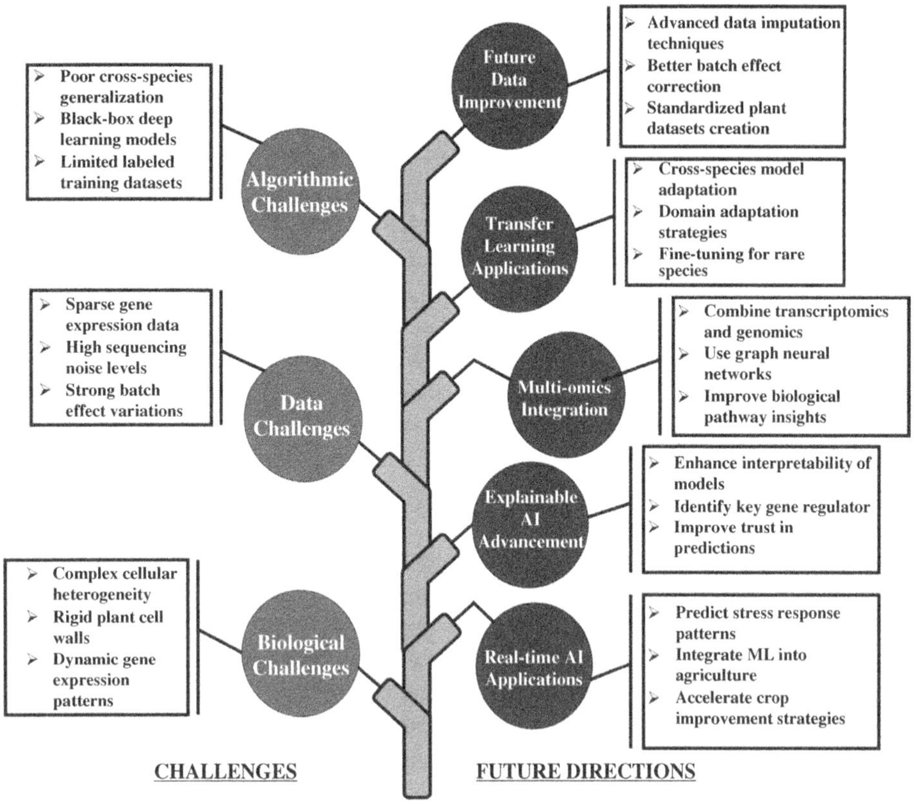

Figure 10.3 Challenges and future directions in ML for plant sc-RNA-seq.

Data Availability

The authors confirm that the data supporting the findings of this study are available within the book/chapter.

References

Adema, K., Schon, M.A., Nodine, M.D. et al. (2024). Lost in space: What single-cell RNA sequencing cannot tell you. *Trends in Plant Science* 9 (9): 1018–1028.

Adil, A., Kumar, V., Jan, A.T. et al. (2021). Single-cell transcriptomics: Current methods and challenges in data acquisition and analysis. *Frontiers in Neuroscience* 15: 591122.

Bacher, R., Chu, L.F., Leng, N. et al. (2016). SCnorm: A quantile-regression based approach for robust normalization of single-cell RNA-seq data. *bioRxiv* 090167.

Bai, L., and Wang, S. (2024). GRNNLink: Predicting gene regulatory links from single-cell RNA-seq data using graph recurrent neural network. *2024 IEEE International Conference on Bioinformatics and Biomedicine (BIBM)*, Lisbon, Portugal (03–06 December 2024), 491–496. IEEE.

Bawa, G., Liu, Z., Yu, X. et al. (2022). Single-cell RNA sequencing for plant research: Insights and possible benefits. *International Journal of Molecular Sciences* 23 (9): 4497.

Bian, J., Zhuang, Z., Ji, X. et al. (2024). Research progress of single-cell transcriptome sequencing technology in plants. *Agronomy* 14 (11): 2530.

Breda, J., Zavolan, M., and van Nimwegen, E. (2021). Bayesian inference of gene expression states from single-cell RNA-seq data. *Nature Biotechnology* 39 (8): 1008–1016.

Budzinski, I.G.F., de Moraes, F.E., Cataldi, T.R. et al. (2019). Network analyses and data integration of proteomics and metabolomics from leaves of two contrasting varieties of sugarcane in response to drought. *Frontiers in Plant Science* 10: 1524.

Buterez, D., Bica, I., Tariq, I. et al. (2022). CellVGAE: An unsupervised scRNA-seq analysis workflow with graph attention networks. *Bioinformatics* 38 (5): 1277–1286.

Chen, F., Tian, L., Fei, T. et al. (2024). Fast-scBatch: Batch effect correction using neural network-driven distance matrix adjustment. *Proceedings of the 15th ACM International Conference on Bioinformatics, Computational Biology and Health Informatics*, Shenzhen, China (22–25 November 2024), 1–6.

Chen, J., Xiao, W., Zhang, E. et al. (2024). Benchmarking unpaired single-cell RNA and single-cell ATAC integration. *Cancer Research* 84 (6_Supplement): 4943–4943.

Chen, L., Zhai, Y., He, Q. et al. (2020). Integrating deep supervised, self-supervised and unsupervised learning for single-cell RNA-seq clustering and annotation. *Genes* 11 (7): 792.

Cheng, J., Smyth, G.K., and Chen, Y. (2023). Unraveling the timeline of gene expression: A pseudotemporal trajectory analysis of single-cell RNA sequencing data. *F1000Research* 12: 684.

Chowdhury, H.A. (2021). Effective clustering of scRNA-seq data to identify biomarkers without user input. *Proceedings of the AAAI Conference on Artificial Intelligence* 35 (18): 15710–15711.

Cochran, K., Srivastava, D., Shrikumar, A. et al. (2022). Domain-adaptive neural networks improve cross-species prediction of transcription factor binding. *Genome Research* 32 (3): 512–523.

Cuevas-Diaz Duran, R., Wei, H., and Wu, J. (2024). Data normalization for addressing the challenges in the analysis of single-cell transcriptomic datasets. *BMC Genomics* 25 (1): 444.

Das, S., Rai, A., and Rai, S.N. (2022). Differential expression analysis of single-cell RNA-Seq data: Current statistical approaches and outstanding challenges. *Entropy* 24 (7): 995.

De Rop, F.V., Ismail, J.N., González-Blas, C.B. et al. (2022). Hydrop enables droplet-based single-cell ATAC-seq and single-cell RNA-seq using dissolvable hydrogel beads. *Elife* 11: e73971.

Di, Z., Yang, B., Li, M. et al. (2024). Batch effects correction in scRNA-seq based on biological-noise decoupling autoencoder and central-cross loss. *Computational Biology and Chemistry* 113: 108261.

Doncevic, D., Ramirez Alvarez, C., Li, A. et al. (2024). Prediction of context-specific regulatory programs and pathways using interpretable deep learning. *bioRxiv* 2024–11.

Duruflé, H., and Déjean, S. (2023). Multi-omics data integration in the context of plant abiotic stress signaling. In: *Plant Abiotic Stress Signaling*, 295–318. Springer US: New York, NY.

Flores, M., Liu, Z., Zhang, T. et al. (2022). Deep learning tackles single-cell analysis—A survey of deep learning for scRNA-seq analysis. *Briefings in Bioinformatics* 23 (1): bbab531.

Gao, J., Hu, K., Zhang, F. et al. (2023). Hexagonal image segmentation on spatially resolved transcriptomics. *Methods* 220: 61–68.

Gao, J., Zhang, F., Hu, K. et al. (2022). Hexagonal convolutional neural network for spatial transcriptomics classification. *2022 IEEE International Conference on Bioinformatics and Biomedicine (BIBM)*, Las Vegas, NV, USA (06–08 December 2022), 200–205. IEEE.

Ge, S., Sun, S., Xu, H. et al. (2024). Deep learning in single-cell and spatial transcriptomics data analysis: Advances and challenges from a data science perspective. *arXiv preprint arXiv: 2412.03614*.

Grønbech, C.H., Vording, M.F., Timshel, P.N. et al. (2020). scVAE: variational auto-encoders for single-cell gene expression data. *Bioinformatics* 36 (16): 4415–4422.

Han, E., Geng, Z., Qin, Y. et al. (2024). Single-cell network analysis reveals gene expression programs for Arabidopsis root development and metabolism. *Plant Communications* 5 (8): 100978.

Hastings, J., Lee, D., and O'Connell, M.J. (2024). Batch-effect correction in single-cell RNA sequencing data using JIVE. *Bioinformatics Advances* 4 (1): vbae134.

Hou, Z., Liu, Y., Zhang, M. et al. (2021). High-throughput single-cell transcriptomics reveals the female germline differentiation trajectory in Arabidopsis thaliana. *Communications Biology* 4 (1): 1149.

Hu, H., Li, Z., Li, X. et al. (2022). ScCAEs: Deep clustering of single-cell RNA-seq via convolutional autoencoder embedding and soft K-means. *Briefings in Bioinformatics* 23 (1): bbab321.

Hu, Y., Dash, L., May, G. et al. (2024). Harnessing single-cell and spatial transcriptomics for crop improvement. *Plants* 13(24): 3476.

Huang, H., Liu, C., Wagle, M.M. et al. (2023). Evaluation of deep learning-based feature selection for single-cell RNA sequencing data analysis. *Genome Biology* 24 (1): 259.

Islam, M.T., Liu, Y., Hassan, M.M. et al. (2024). Advances in the application of single-cell transcriptomics in plant systems and synthetic biology. *BioDesign Research* 6: 0029.

Jamil, I.N., Remali, J., Azizan, K.A. et al. (2020). Systematic multi-omics integration (MOI) approach in plant systems biology. *Frontiers in Plant Science* 11: 944.

Ji, Z., and Ji, H. (2019). Pseudotime reconstruction using TSCAN. *Computational Methods for Single-Cell Data Analysis* 1935: 115–124.

Jia, S., Lysenko, A., Boroevich, K.A. et al. (2023). scDeepInsight: A supervised cell-type identification method for scRNA-seq data with deep learning. *Briefings in Bioinformatics* 24 (5): bbad266.

Jiao, L., Ren, Y., Wang, L. et al. (2023). MulCNN: An efficient and accurate deep learning method based on gene embedding for cell type identification in single-cell RNA-seq data. *Frontiers in Genetics* 14: 1179859.

Kim, H., Chang, W., Chae, S.J. et al. (2024). scLENS: Data-driven signal detection for unbiased scRNA-seq data analysis. *Nature Communications* 15 (1): 3575.

Kimmel, J.C. (2020). Disentangling latent representations of single cell RNA-seq experiments. *bioRxiv* 2020–03.

Kiselev, V.Y., Andrews, T.S., Hemberg, M. (2019). Challenges in unsupervised clustering of single-cell RNA-seq data. *Nature Reviews Genetics* 20 (5): 273–282.

Koch, F.C., Sutton, G.J., Voineagu, I. et al. (2021). Supervised application of internal validation measures to benchmark dimensionality reduction methods in scRNA-seq data. *Briefings in Bioinformatics* 22 (6): bbab304.

Kommu, S., Wang, Y., Wang, Y. et al. (2024). Gene regulatory network inference with joint representation from graph neural network and single-cell foundation model. *bioRxiv*.

Krishna, A., Narayanan, A., and Keedwell, E.C. (2005). Neural networks and temporal gene expression data. *Workshops on Applications of Evolutionary Computation*, Lausanne, Switzerland (30 March–1 April 2005), 64–73. Springer: Berlin, Heidelberg.

Küçükrecep, A., Yıldız, Ş., Tekdal, D. et al. (2024). Deep learning for genomics and epi-genomics in horticulture crops improvement. In: *CRISPRized Horticulture Crops*, 217–232. Academic Press.

Lamba, S., Baliyan, A., and Kukreja, V. (2022). Generative adversarial networks based data augmentation for paddy disease detection using support vector machine. *2022 10th International Conference on Reliability, Infocom Technologies and Optimization (Trends and Future Directions) (ICRITO)*, Noida, India, (13–14 October 2022), 1–5. IEEE.

Lee, J., Kim, S., Hyun, D. et al. (2023). Deep single-cell RNA-seq data clustering with graph prototypical contrastive learning. *Bioinformatics* 39 (6): btad342.

Li, L., and Cai, M. (2019). Cross-species data classification by domain adaptation via discriminative heterogeneous maximum mean discrepancy. *IEEE/ACM Transactions on Computational Biology and Bioinformatics* 18 (1): 312–324.

Li, R., Guan, J., and Zhou, S. (2020). Single-cell RNA-seq data clustering: A survey with performance comparison study. *Journal of Bioinformatics and Computational Biology* 18 (04): 2040005.

Li, S., Yan, H., and Lee, J. (2021). Identification of gene regulatory networks from single-cell expression data. *Modeling Transcriptional Regulation: Methods and Protocols* 2328: 153–170.

Liang, Y., and Kelemen, A. (2005). Temporal gene expression classification with regularised neural network. *International Journal of Bioinformatics Research and Applications* 1 (4): 399–413.

Lin, Y., Wu, T.Y., Wan, S. et al. (2022). scJoint integrates atlas-scale single-cell RNA-seq and ATAC-seq data with transfer learning. *Nature Biotechnology* 40 (5): 703–710.

Linderman, G.C. (2021). Dimensionality reduction of single-cell RNA-seq data. *RNA Bioinformatics* 331–342.

Liu, A., Peng, B., Pankajam, A.V. et al. (2024). Discovery of cell type classification marker genes from single cell RNA sequencing data using NS-Forest. *biorxiv: the Preprint Server for Biology*.

Liu, G., Li, M., Wang, H. et al. (2022). D3k: The dissimilarity-density-dynamic radius k-means clustering algorithm for scRNA-seq data. *Frontiers in Genetics* 13: 912711.

Liu, H., Deng, Q., Du, P. et al. (2022). Simultaneous establishing single-cell transcriptome atlas and chromatin accessibility landscapes in allotetraploid leguminous plant.

Liu, H., Hu, D., Du, P. et al. (2021). Single-cell RNA-seq describes the transcriptome landscape and identifies critical transcription factors in the leaf blade of the allotetraploid peanut (Arachis hypogaea L.). *Plant Biotechnology Journal* 19 (11): 2261–2276.

Longo, S.K., Guo, M.G., Ji, A.L. et al. (2021). Integrating single-cell and spatial transcriptomics to elucidate intercellular tissue dynamics. *Nature Reviews Genetics* 22 (10): 627–644.

Lv, Z., Jiang, S., Kong, S. et al. (2024). Advances in single-cell transcriptome sequencing and spatial transcriptome sequencing in plants. *Plants* 13 (12): 1679.

Mahani, E.A., and Ziarati, K. (2023). A consolidated tree structure combining multiple regression trees with varying depths, resulting in an efficient ensemble model. *Informatica* 47 (9).

Mallick, H., Chatterjee, S., Chowdhury, S. et al. (2022). Differential expression of single-cell RNA-seq data using Tweedie models. *Statistics in Medicine* 41 (18): 3492–3510.

Mao, G., Pang, Z., Zuo, K. et al. (2023). Gene regulatory network inference using convolutional neural networks from scRNA-seq data. *Journal of Computational Biology* 30 (5): 619–631.

Mao, G., Pang, Z., Zuo, K. et al. (2023). Predicting gene regulatory links from single-cell RNA-seq data using graph neural networks. *Briefings in Bioinformatics* 24 (6): bbad414.

Mendoza-Revilla, J., Trop, E., Gonzalez, L. et al. (2024). A foundational large language model for edible plant genomes. *Communications Biology* 7 (1): 835.

Monnier, L., and Cournède, P.H. (2024). A novel batch-effect correction method for scRNA-seq data based on Adversarial Information Factorization. *PLoS Computational Biology* 20 (2): e1011880.

Monti, M., Fiorentino, J., Milanetti, E. et al. (2022). Prediction of time series gene expression and structural analysis of gene regulatory networks using recurrent neural networks. *Entropy* 24 (2): 141.

Muhammad, A., Salman, Z., Lee, K. et al. (2023). Harnessing the power of diffusion models for plant disease image augmentation. *Frontiers in Plant Science* 14: 1280496.

Negus, K.L., Li, X., Welch, S.M. et al. (2024). The role of artificial intelligence in crop improvement. *Advances in Agronomy* 184: 1–66.

Niazian, M., and Niedbała, G. (2020). Machine learning for plant breeding and biotechnology. *Agriculture* 10 (10): 436.

Nossier, M., Moussa, S.M., and Badr, N. (2023). Single-cell RNA-Seq data clustering: Highlighting computational challenges and considerations. *2023 IEEE International Conference on Bioinformatics and Biomedicine (BIBM)*, Istanbul, Turkiye, (05–08 December 2023), 4228–4234. IEEE.

Pan, W., Long, F., and Pan, J. (2023). ScInfoVAE: Interpretable dimensional reduction of single cell transcription data with variational autoencoders and extended mutual information regularization. *Biodata Mining* 16 (1): 17.

Passalacqua, M.J., and Gillis, J. (2023). Coexpression enhances cross-species integration of scRNA-seq across diverse plant species. *bioRxiv* 2023–11.

Peleke, F.F., Zumkeller, S.M., Gültas, M. et al. (2024). Deep learning the cis-regulatory code for gene expression in selected model plants. *Nature Communications* 15 (1): 3488.

Peng, L., Tian, X., Tian, G. et al. (2020). Single-cell RNA-seq clustering: Datasets, models, and algorithms. *RNA Biology* 17 (6): 765–783.

Peng, M., Lin, B., Zhang, J. et al. (2024). scFSNN: A feature selection method based on neural network for single-cell RNA-seq data. *BMC Genomics* 25 (1): 264.

Petegrosso, R., Li, Z., and Kuang, R. (2020). Machine learning and statistical methods for clustering single-cell RNA-sequencing data. *Briefings in Bioinformatics* 21 (4): 1209–1223.

Poirion, O.B., Zhu, X., Ching, T. et al. (2016). Single-cell transcriptomics bioinformatics and computational challenges. *Frontiers in Genetics* 7: 163.

Rao, X., and Dixon, R.A. (2019). Co-expression networks for plant biology: Why and how. *Acta biochimica et biophysica Sinica* 51 (10): 981–988.

Reid, J.E., and Wernisch, L. (2016). Pseudotime estimation: Deconfounding single cell time series. *Bioinformatics* 32 (19): 2973–2980.

Ryu, K.H., Huang, L., Kang, H.M. et al. (2019). Single-cell RNA sequencing resolves molecular relationships among individual plant cells. *Plant Physiology* 179 (4): 1444–1456.

Saimon, A.S.M., Moniruzzaman, M., Islam, M.S. et al. (2023). Integrating genomic selection and machine learning: A data-driven approach to enhance corn yield resilience under climate change. *Journal of Environmental and Agricultural Studies* 4 (2): 20–27.

Sekula, M., Gaskins, J., and Datta, S. (2019). Detection of differentially expressed genes in discrete single-cell RNA sequencing data using a hurdle model with correlated random effects. *Biometrics* 75 (4): 1051–1062.

Sen Puliparambil, B., Tomal, J.H., and Yan, Y. (2022). A novel algorithm for feature selection using penalized regression with applications to single-cell RNA sequencing data. *Biology* 11 (10): 1495.

Shumilo, L., Okhrimenko, A., Kussul, N. et al. (2023). Generative adversarial network augmentation for solving the training data imbalance problem in crop classification. *Remote Sensing Letters* 14 (11): 1129–1138.

Singh, V., Kirtipal, N., Lim, S. et al. (2024). Normalization of single-cell RNA-seq data using partial least squares with adaptive fuzzy weight. *bioRxiv* 2024–08.

So, E., Hayat, S., Nair, S.K. et al. (2023). GraphComm: A graph-based deep learning method to predict cell-cell communication in single-cell RNAseq data. *bioRxiv* 2023–04.

Song, D., and Li, J.J. (2021). PseudotimeDE: Inference of differential gene expression along cell pseudotime with well-calibrated p-values from single-cell RNA sequencing data. *Genome Biology* 22 (1): 124.

Sun, X., Lin, X., Li, Z. et al. (2022). A comprehensive comparison of supervised and unsupervised methods for cell type identification in single-cell RNA-seq. *Briefings in Bioinformatics* 23 (2): bbab567.

Sun, Y., Sun, J., Lin, C. et al. (2024). Single-cell transcriptomics applied in plants. *Cells* 13 (18): 1561.

Tang, Q., Xiang, Y., Gao, W. et al. (2024). TeaTFactor: A prediction tool for tea plant transcription factors based on BERT. *IEEE/ACM Transactions on Computational Biology and Bioinformatics* 21 (6): 2123–2132.

Tang, W., Bertaux, F., Thomas, P. et al. (2020). bayNorm: Bayesian gene expression recovery, imputation and normalization for single-cell RNA-sequencing data. *Bioinformatics* 36 (4): 1174–1181.

Townes, F.W., Hicks, S.C., Aryee, M.J. et al. (2019). Feature selection and dimension reduction for single-cell RNA-seq based on a multinomial model. *Genome Biology* 20: 1–16.

ur Rahman, Z., Asaari, M.S.M., Ibrahim, H. et al. (2024). Generative adversarial networks (GANs) for image augmentation in farming: A review. *IEEE Access*. Vol. 21, no. 6, pp. 2123–2132.

Veleslavov, I.C., and Stumpf, M.P. (2020). Decision tree models and cell fate choice. *bioRxiv* 2020–12.

Verhage, L. (2022). Single but not alone: The transcriptomes of 14,000 single cells from developing cotyledon veins. *The Plant Journal* 110 (1): 5–6.

Vrahatis, A.G., Tasoulis, S.K., Maglogiannis, I. et al. (2020). Recent machine learning approaches for single-cell RNA-seq data analysis. *Advanced Computational Intelligence in Healthcare-7: Biomedical Informatics* 65–79.

Wang, H., and Ma, X. (2022). Learning deep features and topological structure of cells for clustering of scRNA-sequencing data. *Briefings in Bioinformatics* 23 (3): bbac068.

Wang, H., Lin, Y.N., Yan, S. et al. (2023). NRTPredictor: Identifying rice root cell state in single-cell RNA-seq via ensemble learning. *Plant Methods* 19 (1): 119.

Wang, J., Ma, A., Chang, Y. et al. (2021). scGNN is a novel graph neural network framework for single-cell RNA-seq analyses. *Nature Communications* 12 (1): 1882.

Wang, S., Liu, Y., Zhang, H. et al. (2025). scE2EGAE: Enhancing single-cell RNA-seq data analysis through an end-to-end cell-graph-learnable graph autoencoder with differentiable edge sampling. *Biology direct* 20 (1): 66.

Wen, Z.H., Langsam, J.L., Zhang, L. et al. (2022). A Bayesian factorization method to recover single-cell RNA sequencing data. *Cell Reports Methods* 2 (1).

Wójcik-Gront, E., Zieniuk, B., and Pawełkowicz, M. (2024). Harnessing AI-powered genomic research for sustainable crop improvement. *Agriculture* 14 (12): 2299.

Wu, X., Fan, X., Luo, P. et al. (2023). From laboratory to field: Unsupervised domain adaptation for plant disease recognition in the wild. *Plant Phenomics* 5: 0038.

Wu, Z., Tang, J., Zhuo, J. et al. (2019). Chromatin signature and transcription factor binding provide a predictive basis for understanding plant gene expression. *Plant and Cell Physiology* 60 (7): 1471–1486.

Xiang, R., Wang, W., Yang, L. et al. (2021). A comparison for dimensionality reduction methods of single-cell RNA-seq data. *Frontiers in Genetics* 12: 646936.

Xie, P., Gao, M., Wang, C. et al. (2019). SuperCT: A supervised-learning framework for enhanced characterization of single-cell transcriptomic profiles. *Nucleic Acids Research*, 47 (8): e48–e48.

Xu, L., Xue, T., Ding, W. et al. (2022). Comparison of scRNA-seq data analysis method combinations. *Briefings in Functional Genomics* 21 (6): 433–440.

Xu, X., Liang, Y., Tang, M. et al. (2024). ScReNI: Single-cell regulatory network inference through integrating scRNA-seq and scATAC-seq data. *bioRxiv* 2024–09.

Yan, C., Zhu, Y., Chen, M. et al. (2024). Integration tools for scRNA-seq data and spatial transcriptomics sequencing data. *Briefings in Functional Genomics* 23 (4): 295–302.

Yan, H., Lee, J., Song, Q. et al. (2022). Identification of new marker genes from plant single-cell RNA-seq data using interpretable machine learning methods. *New Phytologist* 234 (4): 1507–1520.

Yan, H., Song, Q., Lee, J. et al. (2020). Identification of cell-type marker genes from plant single-cell RNA-seq data using machine learning. *bioRxiv* 10 (2020.11): 22–393165.

Yan, J., and Wang, X. (2022). Unsupervised and semi-supervised learning: The next frontier in machine learning for plant systems biology. *The Plant Journal* 111 (6): 1527–1538.

Yan, K., Guo, X., Ji, Z. et al. (2021). Deep transfer learning for cross-species plant disease diagnosis adapting mixed subdomains. *IEEE/ACM Transactions on Computational Biology and Bioinformatics* 20 (4): 2555–2564.

Yan, S., Bhawal, R., Yin, Z. et al. (2022). Recent advances in proteomics and metabolomics in plants. *Molecular Horticulture* 2 (1): 17.

Yang, W., Wang, P., Xu, S. et al. (2024). Deciphering cell–cell communication at single-cell resolution for spatial transcriptomics with subgraph-based graph attention network. *Nature Communications* 15 (1): 7101.

Yang, Y., Li, G., Xie, Y. et al. (2021). iSMNN: Batch effect correction for single-cell RNA-seq data via iterative supervised mutual nearest neighbor refinement. *Briefings in Bioinformatics* 22 (5): bbab122.

Yao, J., Shang, N., Chen, H. et al. (2024). Sub-clustering, marker identification and multi-omics integration influenced by protoplasting in plant scRNA-seq data analysis. *bioRxiv*, 2024–12.

Yin, R., Xia, K., and Xu, X. (2023). Spatial transcriptomics drives a new era in plant research. *The Plant Journal* 116 (6): 1571–1581.

Yousuff, M., Babu, R., and Rathinam, A. (2024). Nonlinear dimensionality reduction based visualization of single-cell RNA sequencing data. *Journal of Analytical Science and Technology* 15 (1): 1.

Yu, S., Fu, X., Deng, X. et al. (2024). AtML: Accurate identification of Arabidopsis thaliana root cell identity via single-cell sequencing analysis using interpretable machine learning.

Yuan, L., Sun, S., Jiang, Y. et al. (2025). scRGCL: A cell type annotation method for single-cell RNA-seq data using residual graph convolutional neural network with contrastive learning. *Briefings in Bioinformatics* 26 (1): bbae662.

Yuan, Y., and Bar-Joseph, Z. (2018). CNNC: Convolutional neural networks for co-expression analysis. *bioRxiv* 365007.

Zhang, T., Zhang, Z., Li, L. et al. (2024). GTAD: A graph-based approach for cell spatial composition inference from integrated scRNA-seq and ST-seq data. *Briefings in Bioinformatics* 25 (1): bbad469.

Zhang, X., Jin, K., Ni, T. et al. (2023). Interpretable scRNA-seq analysis with intelligent gene selection.

Zhao, R., Lu, J., Zhou, W. et al. (2024). A systematic evaluation of highly variable gene selection methods for single-cell RNA-sequencing. *bioRxiv* 2024–08.

Zhao, Y., Joshi, P., and Shin, D.G. (2019). Recurrent neural network for gene regulation network construction on time series expression data. *2019 IEEE International Conference on Bioinformatics and Biomedicine (BIBM)*, 610–615. IEEE.

Zheng, H.X., Wu, F.H., Li, S.M. et al. (2021). Single-cell profiling lights different cell trajectories in plants. *Abiotech* 2: 64–78.

Zhu, B., Li, H., Zhang, L. et al. (2022). A Markov random field model-based approach for differentially expressed gene detection from single-cell RNA-seq data. *Briefings in Bioinformatics* 23 (5): bbac166.

11

Machine Learning for Plant Genomic Prediction

Natasha Charaya[1], *Sonika Kalia*[2*], *Indra Rautela*[2*], *Poorvi Yadav*[2], *Poornima Bhardwaj*[2], *Ritakshi Nautiyal*[2], *Monika Kalia*[3], *Vinay Sharma*[4]

[1] *Department of Biotechnology, Guru Jambheshwar University of Science & Technology (GJUS& T) Hisar, Haryana, India*
[2] *Department of Biotechnology, School of Applied and Life Sciences, Uttaranchal University, Dehradun, India*
[3] *Department of Entomology, Dr YS Parmar University of Horticulture & Forestry, COH & F, Neri, Hamirpur (HP), India*
[4] *Center of Excellence in Genomics and Systems Biology (CEGSB), and Center for Prebreeding Research (CPBR), International Crops Research Institute for the Semi-Arid Tropics (ICRISAT), Patancheru, India*

*Corresponding authors: sonikakalia.266@gmail.com; rautela.indra7@gmail.com

11.1 Introduction

In order to address the issues of the security of food for the world's expanding population, which is predicted to reach more than nine billion people by the 21st century, one of the main objectives of breeding of plants is to boost crop yield, bringing new varieties more quickly (He and Li 2020; Qaim 2020). However, severe weather, ever-changing conditions, plant diseases, insect pests, and limited land resources have made it difficult for crop breeders to achieve optimal performance. Breeders may be able to meet the anticipated rise in food demand by creating varieties with noticeably improved agronomic performance with the use of accurate crop performance forecasts (Bali and Singla 2022). Breeders can use genome-wide marker sets to assess genomics-based breeding values using genomic prediction (GP) (Meuwissen et al. 2001; Crossa et al. 2017). For plant breeding programs, genomic selection (GS) begins with GP (Heffner et al. 2009). Finding superior germplasm is essential to modern plant breeding, and machine learning (ML) offers the means to make genotype selections possible through drawn data. The growing amount of data produced by phenotyping and genotyping through high-throughput technology (Lin 2017; Perakakis et al. 2018) and supplemented by rich data from weather stations (Costa-Neto et al. 2021; Shahhosseini et al. 2021) is a major driver behind the use of ML in plant breeding. More data necessitates a shift in perspective because enormous datasets may not be well-suited for standard statistical techniques, and new algorithms may be able to reveal previously undiscovered, significant patterns in the data. Breeders can select the best machine-learning technique for a task and parameterize it using their knowledge by being aware of and comprehending the various kinds of machine-learning techniques (Zampieri et al. 2019).

11.1.1 Genomic Prediction

The process of predicting phenotypic values using genotypic data produced by high-throughput genotyping methods, such as genotype-by-sequencing, is known as genome prediction (Hayes et al. 2001). Single-nucleotide polymorphisms (SNPs) in relation to a reference genome are commonly used to describe this observed genotype. GP has the potential to improve crop genetic performance in addition to helping us comprehend the fundamental genetic structure of genomes. GP's capacity to directly reveal complex features like yield or disease/pest resistance from genotypes is having a significant impact on plant breeding (Bernardo 2008; Scheben et al. 2016; Xu et al. 2020). To determine how SNP combinations affect the phenotype being studied, GP models are trained on a genotyped and phenotyped population. After each organism has been genotyped, these models are used to predict the phenotypes of a population (test). The predicted values are then

Machine Learning for Plant Biology, First Edition. Edited by Jen-Tsung Chen.

compared to the actual values. Because just a portion of a population needs to be phenotyped and because GP speeds up the discovery of desirable genotypes, it can help breeders save time and money (Crossa et al. 2017; Mahood et al. 2020).

11.1.2 Genome Selection

By forecasting the breeding values of untested individuals, GS was initially suggested as a solution to the drawbacks of marker-assisted selection (Meuwissen et al. 2021). Regardless of their impact or importance, GS incorporates markers found throughout the genome for GP (Crossa et al. 2017). Over the past ten years, GS has gained significant appeal due to the significant decrease in genotyping costs, which has resulted in significant genetic gains in numerous plant breeding initiatives (Crossa et al. 2017; Heffner et al. 2009; Bernardo 2008). Conventional approaches, such as regression-based methods, have limitations when it comes to analyzing high-dimensional data and capturing complicated and multivariate interactions between response and predictor factors (Meuwissen et al. 2001). Without requiring any prior knowledge of prospective genetic models, ML techniques can capture nonlinear correlations within genotypes (González-Recio et al. 2014; Abdollahi-Arpanahi et al. 2020).

11.1.3 Machine Learning

In the words of Arthur Samuel, ML is the study that teaches computers to learn without explicit programming. The need for ML is growing as an outcome of the large amount of data that is present. Using a given data set, ML made algorithms that carry out tasks. It is a branch of artificial intelligence that finds extensive use in both industry and research. Supervised studies aim at identifying a product (which can be either a numeral value in the event of regression or, in the case of categorizing, a distinct label) for an object given a collection of input features that define it. The supervised learning method accomplishes this by modifying a predictive model's variables to function well on labeled training data, which consists of components and related known products. Predictions for fresh, untested test data can then be made using the generated models. Conversely, unsupervised learning looks for patterns in unlabeled data. It is more difficult to measure whether an unsupervised type of design execute well than a supervised one; because unsupervised methods do not take outputs into account, there is usually no thing as "training information" for them. But, by enabling us to concentrate on pertinent patterns, the patterns discovered by unsupervised algorithms can aid in the understanding of vast information and can be helpful in preparing data for more successful supervised type of learning. The diagram, which is depicted in Figure 11.1 "Choosing the right machine learning algorithm for a task using a simplified decision tree," offers a comprehensive visual

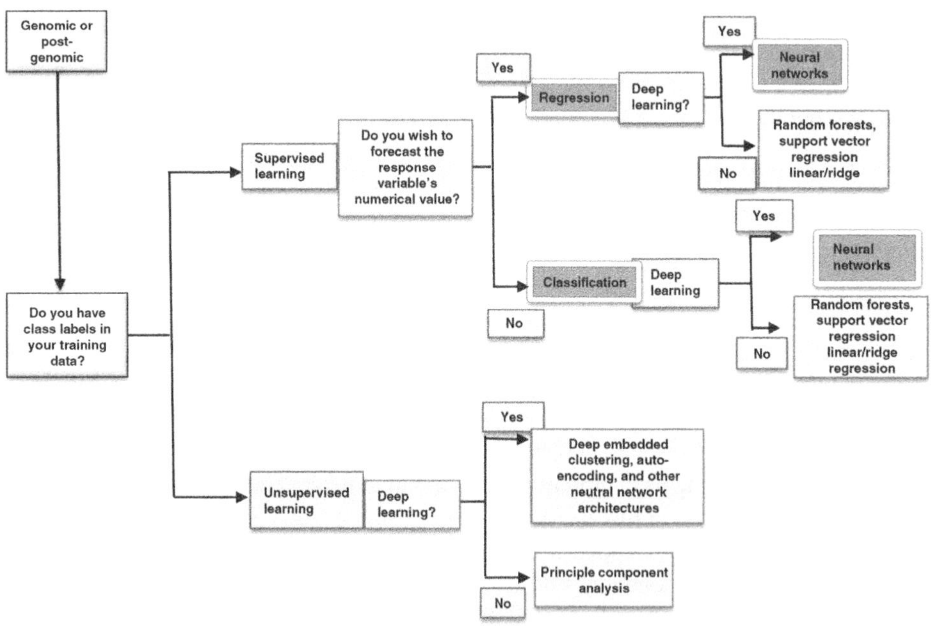

Figure 11.1 Choosing the right machine learning algorithm for a task using a simplified decision tree.

guide to assist users in choosing the best machine learning algorithm given the problem type (e.g., classification, regression, or clustering), dataset size, interpretability needs, and training time constraints. This streamlined decision tree is especially helpful for novices or in the early stages of model selection since it helps effectively reduce the number of algorithm options (Mahood et al. 2020; Van Dijk et al. 2021).

11.2 Types of Models Used in ML

For a thorough description of the model architecture, including the data input layer, processing steps, and output predictions, see Figure 11.2. This figure makes it easier to see how data moves through different learning components.

11.2.1 Supervised Models

In the supervised learning ML task, a function mapping a variable to a result is learnt using sample pairs of input-output. Strategies for supervised type of ML require outside help. Train and evaluation datasets are separated from the starting dataset. The final value of the training dataset must be predicted or grouped together. In order to predict or categorize the test dataset, each algorithm looks for some kind of arrangement in the training dataset as shown in Figure 11.3. Starting with labeled data, the supervised learning workflow goes through model training, assessment, and prediction to produce precise results.

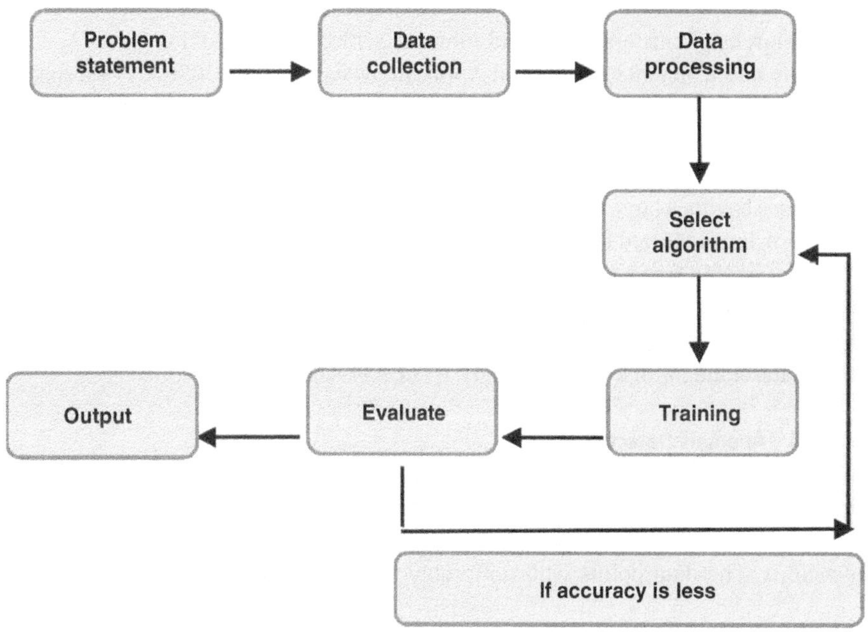

Figure 11.2 Machine learning model.

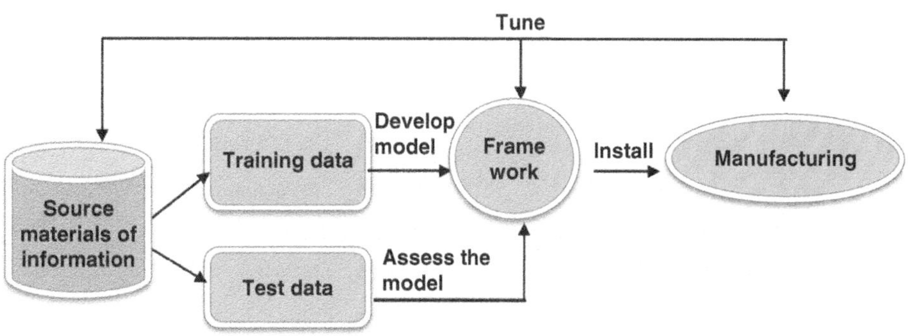

Figure 11.3 Supervised learning workflow.

11.2.1.1 Classification

The purpose of the classifier algorithms is to determine whether a new data point falls into one of the many predefined classes. Labeled examples from each category are used in the classification models to facilitate learning. They find relationships and correlations in the data during this process that aid in differentiating class one from the other classes. The model can then apply these class labels to data points that are not visible after it has learnt these patterns.

11.2.1.2 Regression

Regression algorithms use input features as the foundation for forecasting a continuous output variable. This value could be anything from forecasting sales to predicting stock market or real estate prices to predicting customer churn, or the likelihood that customers will stick around. In order to comprehend the relationship between the output variable and the continuous features, regression models employ features (Austin and Brunner 2004).

11.2.2 Unsupervised Models

Because there are no teachers or correct answers, unlike the supervised learning mentioned above these, are known as unsupervised learning. The algorithms are responsible for identifying and displaying the data's interesting structure. Its main uses are in feature reduction and clustering. Without requiring labeled training data, the algorithm finds patterns in the dataset to create clusters or groupings, as illustrated in Figure 11.4.

11.2.2.1 Clustering

Imagine being presented with a basket of unlabeled fruits. Clustering algorithms are used to group fruits based on their inherent similarities. On the other hand, hierarchical clustering facilitates the study of the system of groups by constructing a hierarchy of clusters (Kuhn and Johnson 2013; Maronna 2011).

Density Based Spatial Clustering of Applications with Noise (DBSCAN) is a spatial clustering algorithm that can identify clusters of high-density data points, even in areas with outliers or a dearth of data.

11.2.2.2 Dimensionality Reduction

Sometimes having a large feature space (dimensions) makes it challenging to both visualize and analyze the data. Reducing the dimensions required to preserve the essential features is the aim of dimensionality reduction techniques. Information is concentrated in some dimensions with the highest variations because principal component analysis (PCA) identifies the dimensions of greatest importance. This provides an opportunity for more effective visualization and accelerates model training. Although it is designed for classification tasks and focuses on dimensions that can distinguish the current classes in the dataset, linear discriminant analysis (LDA) is similar to PCA.

11.2.2.3 Anomaly Detection

Finding data points that significantly deviate from the majority can also be accomplished through unsupervised learning. These anomalies or outliers may be detected by the statistical model as indicators of fraud, mistakes, or even something out of the ordinary. The local outlier factor (LOF) compares the local density of a particular data point with those in its vicinity. The data points with noticeably lower densities are then identified as outliers or possible anomalies. One

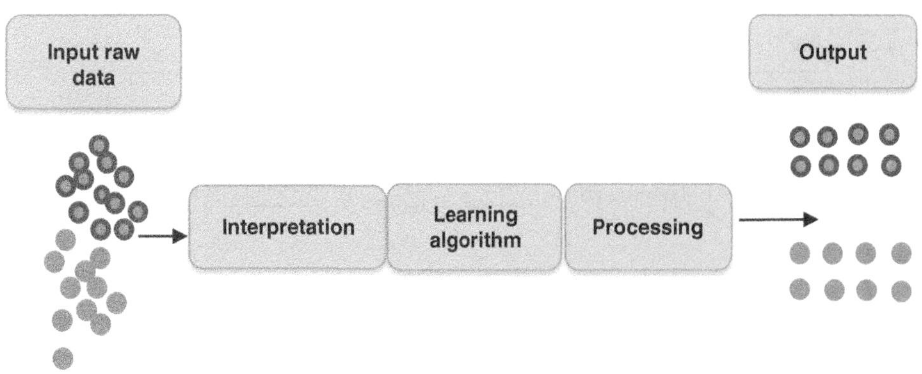

Figure 11.4 Unsupervised learning.

alternative method is Isolation Forest, which recursively separates data points based on their characteristics. Since anomalies frequently require fewer steps than an average normal point, they are typically easy to consider (Kuhn and Johnson 2013; Hastie et al. 2009).

11.2.3 Semi-supervised Model

Furthermore, unsupervised learning solves the problem when there is no labeled data, whereas supervised type of learning is a type of learning with labeled data. Finally, the gap between the two is filled by semi-supervised learning. Training with both labeled and unlabeled data sets reveals the advantages of both methods. This is particularly true when unlabeled data is clearly widely available and labeled data may be scarce or prohibitively expensive to obtain.

11.2.3.1 Transductive SVM

In semi-supervised type of learning, transductive support vector machines (TSVM) have been used extensively to handle partially labeled data. Due to a lack of knowledge regarding its generalization-based foundation, there has been mystery surrounding it. In order to maximize the margin between the labeled and unlabeled data, it is used to label the unlabeled data. It is an NPhard problem for TSVM to find an exact solution.

11.2.3.2 Generative Models

A generative model is one that has the ability to produce data. It models the class (i.e., the entire data) as well as the features. The distribution of the mixture can be verified with just one labeled example percomponent (Littell et al. 2006).

11.2.4 Deep Learning

To identify complex patterns, a subfield of ML known as "deep learning" uses multi-layered artificial neural networks (ANNs).

11.2.4.1 Neural Networks

A neural network is a group of algorithms designed to mimic the way the human brain works in order to find fundamental trends in a dataset. Neural networks in this context refer to artificial or natural neuronal systems. Because neural networks can adjust to changing input, the output criteria don't need to be redesigned in order for the network to produce the best possible result. Neural networks arc an idea that originated in Figure 11.5, "Unsupervised Neural Network," which shows the structure and operation of an unsupervised neural network.

1. **Artificial Neural Networks (ANNs)**
   This well-known model describes the composition and operation of the human brain. It is utilized for a variety of ML tasks and is made up of interconnected nodes based on different layers.
2. **Convolutional Neural Networks (CNNs)**
   A CNN is a type of deep learning model that uses input data to automatically determine the spatial hierarchies of features. This model is frequently applied to image classification and recognition.
3. **Recurrent Neural Networks (RNNs)**
   Sequential data processing is the focus of this model. It permits the memory input that neural network architectures are known for (Gazestani and Lewis 2019; Maronna 2011; Goodfellow et al. 2016).

Figure 11.5 Unsupervised neural network.

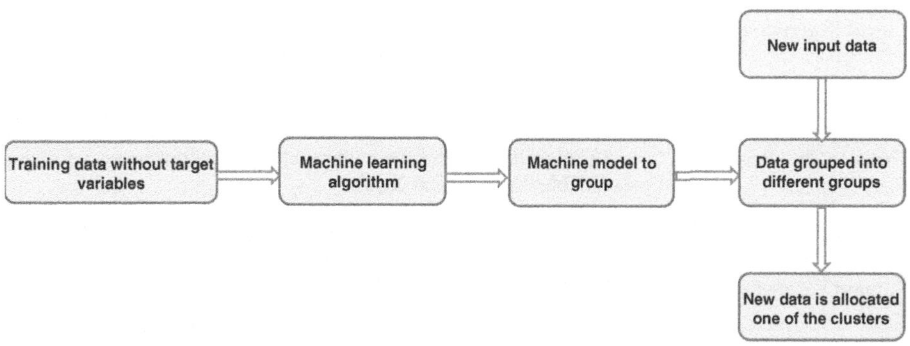

11.3 Methods

The terminology used in traditional statistics frequently deviates from the literature on ML. This section describes supervised ML techniques, which fall into four categories: neural networks, tree ensembles, kernel methods, and linear methods, using algebraic terminology that is familiar to plant breeders (Goodfellow et al. 2016; Deisenroth et al. 2020).

Selecting the appropriate algorithm is essential in machine learning for precise forecasts and insightful analysis. Regression and classification are two of the most popular forms of supervised learning. Regression algorithms are used to forecast continuous values, whereas classification algorithms are used to predict categorical outcomes. To assist practitioners in choosing the best tool for their particular task, the table below (Tables 11.1 and 11.2) lists some of the most popular algorithms in both categories.

11.3.1 Linear Methods

In the literature on ML, the variable that respond is represented as a linear combination of explanatory variables known as characteristics. Despite being referred to as "linear," these techniques may or may not have linear features. The formula for linear models is $y = Xb + e$. Scores at various marker loci serve as the features in GP, while marker effects serve as the coefficients. One efficient method for solving equation systems of linear models is coordinate descent, an iterative procedure that marginalizes the model to solve one univariate equation at a time.

11.3.2 Kernel Methods

In plant breeding, kernel methods—also referred to as kernels—are employed for pedigree-based selection, GP, and spatial adjustments. They produce the same result as linear approaches by computing a symmetric matrix K, or kernel, from features (X). Additionally, when linear methods are challenging, kernel approaches aid in the computation of interactions, which is occasionally required for GP. A linear system of equations is used to calculate predictions made with kernel approaches (Misztal and Legarra 2017; de Los Campos et al. 2010).

Table 11.1 Common classification algorithms.

Logistic Regression	When dealing with binary classification issues, like spam/not spam, this method works very well (Navya et al. 2025).
Support Vector Machine	This tool works well for classification tasks, especially when working with a lot of data features (Nayak et al. 2015).
Decision Tree	The procedure entails building class predictions using features after generating a decision tree with branches (Song and Lu 2015).
Random Forest	To improve accuracy and avoid overfitting, the model builds a decision tree ensemble, performing well on training data but poorly on unseen data (Salman et al. 2024).
K-Nearest Neighbors	For a given data point, the function labels the closest neighbors (Kramer 2013).

Table 11.2 Common regression algorithms.

Linear Regression	In order to comprehend the connection between features and continuous output, the model fits the depth of a line to the data. (Hope 2020).
Polynomial Regression	This approach uses complex polynomial functions, such as cubic and quadratic, to account for nonlinear data relationships, much like linear regression (Maulud and Abdulazeez 2020).
Decision Tree Regression	The algorithm makes several branching decisions using a decision tree to predict a continuous output variable (Song and Lu 2015).
Random Forest Regression	To guarantee reliable and error-free results, the system creates a regression prediction using several decision trees (Prasath et al. 2023).
Support Vector Regression	In order to find the hyperplane that best depicts continuous output data, the text adapts Support Vector Machine concepts for regression tasks (Kecman 2005).

11.3.3 Neural Networks

Neural networks generalize linear methods using latent spaces (H) by converting concealed spaces into a linear sequence of features (XB) using an activation function (α). Latent variables that correlate to a column of H are called nodes. Latent spaces can be used to predict response variables, or they can be developed into more intricate latent spaces with hidden layers, like $Hi + 1 = \alpha(Hi\ B)$. Latent spaces get more complicated and nonlinear as the total quantity of layers rises (Goodfellow et al. 2016).

A "test set" is created to determine convergence in order to assess the network's prediction power. The network is fitted and residuals are dispersed using basic activation functions. Tanh is a nonlinear function that uses linear feature combinations to translate input values into −1 and 1. It may be advantageous to normalize response variables because it is susceptible to input scale. Without altering positive values, ReLU sets the corresponding value to zero. The ReLU activation function provides a general reconstruction for any polynomial equation as the number of nodes rises. DNN assumptions based on ReLU activation can be enhanced by the dropout technique, which disregards a percentage of randomly chosen nodes in each iteration (Boris 2019).

11.3.4 Tree Ensembles

Regression tree-based models don't use explicit algebra; instead, they are algorithmic. Ensembles—the process of averaging several trees using sampling strategies like bagging and boosting—are beneficial for regression trees, which are poor learners. Three elements serve as the foundation for the hierarchical implementation of tree ensembles:

1. **Recursive Partitioning:** specifies how to divide a single feature (x) into two groups, or nodes, according to a metric that needs to be kept to a minimum. The most widely used metric for modeling continuous variables is the squared error. The Classification and Regression Tree Algorithm is the primary algorithm used to determine the partitioning threshold. When recursive partition is utilized to a plural SNP marker (x) with genotypes using numbers written as 012, there are only two feasible breaking locations in the genomic information prediction: "$x \leq 1$" or "$x < 1$," which advocate for (AA,Aa)/(aa) and (AA)/(Aa,aa), correspondingly. In the node with two classes, the same marker might offer additional data partitioning. The output of the recursive division function must contain the split point that reduces the square of the error and the quantity of difference explained. Trees will be constructed using both pieces of information.
2. **Tree Building:** Recursive splits are used to build trees sequentially. By using recursive partitioning to determine which feature offers the greatest reduction in squared error, a tree is constructed given the set of features (X). This will specify which feature and the location of the tree's initial split. To create the second split rule, the data is divided into two nodes, and each feature in X is reevaluated within each node. Until there is no more squared error reduction, the procedure is repeated within each succeeding node. By restricting the number of assessments per node, the tree depth, and establishing a minimum reduction of the squared deviation that justifies partitioning, the complex nature of the regression structures is managed.
3. **Ensemble Method:** There are two methods for assembling a group of trees: bagging and boosting. While boosting trees produce a model known as gradient boosting machine, or "GBM," with variants called AdaBoost and XGBoost (Leo 2001) boosting trees produce a model known as random forest (Chen and Carlos 2016).

11.4 Cross-validation

Since it is expensive to gather new data, prediction models are typically assessed using available data. Two subsets of data can be distinguished; one is used for testing and the other for training. We refer to this process as cross-validation (CV).

11.4.1 Strategies for CV

1. **Random K-fold:** In this program, a percentage of 1/k of the collection of data is sampled for examination, and the predicted metric takes an average across examples. The term "random k-fold validation" is frequently used for strategy evaluations because it gives an impression of the projection on fresh participants for a noted set of environments. The dataset's population structure may affect the precise interpretation. It offers a broad evaluation of the model's predictive ability in GP.

2. **Leave-one-out:** Some instances for information with known arrangement that can be assessed using this methodology include genotype-by-environment interactions and the estimation of undetectable populations (leave-population-out). Which set of observations is being omitted determines how the result should be interpreted. The term "random data fractures" fail to minimize structural data damage, but LOO can. In CV, evaluations that appear in both the validation and training sets are referred to as contamination (Zhan et al. 2019).
3. **Holdout:** The goal of this plan is to mimic the prediction model's actual use. Typically, the model is trained using data from previous months of growth to estimate environments and hiring applicants for the most recent period of growth. New developing reaches with linkage disequilibrium structures commonly seen in the training dataset are the source of these selection candidates (Pszczola and Calus 2016).

11.4.2 Validation Metrics

1. **Correlation:**
   It is quantified using the Pearson or Spearman parameters among the actual and predicted values. One measure that accounts for variations in magnitude and ranking is the Pearson correlation coefficient. By dividing Pearson correlations, which are also known as predictability in this context, by the squared root of heritability, one can rescale them into accuracy (Christina et al. 2013). The primary substitute for Pearson's coefficient is Spearman's coefficient, also known as rank-order correlation. Although it is a direct indicator of ranking, it does not capture magnitude. In GP, the Spearman correlation is used to gauge how well two selection techniques align overall (Heslot et al. 2012).
2. **Prediction Error:**
   Although it is not commonly used for being chosen in plant production, it performs well in prototypes of disease danger, event possibilities, and other assumptions where a drop in prediction values is detrimental. When cross-validating neural networks and tree-based machines, the mean squared prediction error (MSPE) is a frequently used metric. It can be mathematically calculated from the hat matrix without the use of CV techniques in linear frameworks and is nested within the training set.
3. **Success:**
   Success is for design whose choices adhere to precise guidelines that define success. Success is a Bernoulli process at the observation level, which relates to whether or not predetermined criteria were met. Since genotypes are either selected or not, One example of a success metric is the selection process itself. At the community level, victory is a linear process since it counts a set that are profitable attempts within a limited amount of tries. Accuracy is calculated as the percentage of correctly classified observations from a confusion matrix when there are multiple classes of success. It is possible to customize a CV to the particular application by selecting the right scheme and metric. By Daetwyler et al. (2013), standards for comparing GP techniques aimed at additive genetics were established (Crossa et al. 2017).

11.4.3 Information

Since confirmations are accomplished both from and within sub-populations, CV research also provides light on population differentiation. According to Werner et al. (2020), there are potential hazards and variations in predictability when validations are carried out within or across families (Salcedo-Sanz et al. 2020). CV allows for the study of predictability on several levels when population structure-defining factors, such as sub-population structure, are known (Xavier et al. 2020). The following succinctly describes the theory behind the data collected by molecular markers and the anticipated genetic data collected under different validation scenarios: Linkage is a single family. Linkage and LD within the family. Cross-family: Linkage, LD, and pedigree ties. Leave-family-out: LD and pedigree ties. The following is a summary of the three information sources: Linkage disequilibrium (LD) is a phrase used for defining the connections among indicators and genetic variation (QTL) that are determined across unconnected founder individuals; connection, also known as co-segregation, captures the inheritance of parental haplotypes; and pedigree relationship refers to population structure. Knowing the sources of genomic information has been shown to have some practical implications (Legarra et al. 2008). The discrepancy in precision reported between and among populations is referred to in the field of ML writing as the "Simpson paradox." The degree to which the different prediction models accurately represent the various information sources is probably going to vary. Furthermore, depending on the parameterization, the information's nature may alter. It is possible to learn about additive, dominance, and epistatic effects from molecular markers. On the other hand, semi-parametric techniques (such as DNN, RKHS, and SVR) might not be able to outperform more straightforward linear techniques in situations where the genetic architecture is primarily additive (Montesinos-López et al. 2021).

11.5 Applications of ML

Explaining a complex trait like yield in terms of the genomic, phenotypic, and environmental data that are currently available is a primary goal in both plant science and breeding. ML and other methods seek to achieve this goal by:

1. Identifying QTLs, or genomic regions linked to a particular trait. This is referred to as genome-wide association mapping (for diversity panels) or QTL mapping (for experimental populations).
2. Evaluating the genetic architecture by calculating the contributions of each locus and the percentage of trait variance that can be accounted for by all of the loci taken together. This is related to the interest in genetic correlations among traits, which measure the extent of genetic signal overlap.
3. Employing GP to forecast anticipated trait values for recently discovered genetic variants for which only indicator data are available.

Plant breeders are depending more and more on marker-assisted selection, which requires QTL mapping and GS (Crossa et al. 2017), which chooses material based on GPs rather than phenotypic values. The genes that underlie QTLs and genetic architecture are of primary interest to biologists.

Variable choice, calculation, and projection from a data-analytic standpoint are the aforementioned objectives 1–3. Prediction has been the primary focus of ML techniques, including support vector machines (SVM), random forests, gradient tree boosting (González-Recio et al. 2014), and others. A range of architectures for GP with DL have been proposed more recently; an overview is given in Azodi et al. (2020); Pérez-Enciso and Zingaretti (2019). Although some authors have investigated ML techniques for QTL mapping (primarily for pre-screening), their application is still restricted because practitioners frequently need p values or other confidence metrics.

Therefore, we will focus on GP. Environmental variables are necessary to capture most genotype-by-environment (G × E) interactions. Prediction for unobserved environments is one of two significant recent developments concerning GP in plants. These variables enable GP to be extrapolated to the desired surroundings and describe the various training environments (Montesinos-López et al. 2018). Precision for an intricate target trait, including yield or stress tolerance, can be increased by using supplementary characteristics, or traits, which are simple to measure and typically not of direct interest.

When GP is carried out in plants, the well-established random-effects or mixed-effects frameworks for statistical inheritance are usually still in use because they can perform GP and estimate effect sizes at the same time. Despite this, ML can offer a number of benefits over parametric random-effects models.

First, when secondary traits, like image value, are more intricate, the latter call for manually created features.

Secondly, when prevalent beliefs like Gaussianity cannot be true, as is the scenario, for example, for characteristics calculated on an ordinal scale, ML methods (DL in particular) may prove more flexible (Montesinos-López et al. 2021).

In nonadditive genetic variance, ML can increase accuracy, especially when strong dominance and epistatic effects are present. Real and simulated animal characteristics in sizable populations show this. Further research is necessary to compare DL to random-effects models and achieve comparable improvements in plant populations. In contrast to mixed models and ML approaches, ML can also improve plant breeding (Abdollahi-Arpanahi et al. 2020).

11.5.1 Combining Trials to Increase the Sample Size

The first challenge in using automated learning in this setting is the small number of samples, at least in comparison to human or animal genetics. Although characteristics one another can be highly dimensional, there are typically hundreds of genotypes, with a few exceptions. This issue might be resolved by some recent ML advancements (like transfer learning or synthetic data generation, which are covered below). Entrusting genetic variants to sets of tests and training as optimally as possible may enhance accuracy (Rincent et al. 2012).

Larger populations are necessary for successful ML utilization in the plant sciences, which study a wide variety of organisms, gene sets, and environments. This necessitates measuring distinct sets of genetic variants in independent experiments, which can be difficult because of inconsistent experiments. Non-ML-related issues like improved control over the surroundings, precise measurement expanding capacity, and sound research methodology will be critical to ML advancements in this field (Selby 2019).

11.5.2 An Explanation of the G × E Interaction Using Data Features

At the cultivation or plot level, an increasing number of ecological factors with high time conclusion are available. Then, the main obstacle is to predict genetic placing in each unusual surroundings based on ecological variables that explain G × E interactions, shown that this is possible with random-effects models. Montesinos-López et al. (2018) employed feedforward networks, although only in a few settings. Without mentioning within-environment accuracy, report good accuracy across environments using DL as well. More effort is needed to select the most relevant environmental variables and characterize meaningful based on data environmental features (Khaki and Lizhi 2019).

11.5.3 Exploiting Information from Secondary Traits

The accessibility of these attributes raises the question of how and when they can improve GP for the desired trait. This is the case when there is a sufficient genetic correlation and heritability between a single supplementary trait and the target trait by implementing this classical result to many secondary traits, it was better than single-trait GP employing criticized selection indices (Lopez-Cruz and De Los Campos 2021). Additionally, several kernels have been employed to model the impact of secondary traits and SNPs (Schrag et al. 2018). In a similar vein, a number of authors have forecasted yield without marker data using omics, environmental, or management data. Despite their inability to be used for GS, these models can give farmers and policymakers useful tools for making decisions.

The use of causal inference algorithms is a recent advancement that is pertinent to secondary traits as well as G × E modeling. These algorithms can incorporate known functional relations and suggest causal relationships between phenotypes that most closely match the data. These methods have the advantage of allowing for the prediction of the impact of interventions, such as what will occur when a gene is silenced or under various conditions or management (Meinshausen et al. 2016; Peters et al. 2017).

11.6 ML Challenges

The use of ML in plant research presents a various difficulty. Traditionally, statistical methods—which have shown great success over the past years—are used to predict genotype-phenotype. In order to make decisions, researchers and practitioners typically rely on model interpretation and confidence measures. ML techniques' data-driven adaptability may be more advantageous than frequently more exacting statistical methods, but there may be a price for this. First off, ML techniques typically don't carry out explicit inference and don't offer reliable estimates of prediction confidence.

Second, a lot of the more effective ML techniques make it difficult for users to understand the reasoning behind a particular prediction. Third, while ML techniques can be modified to take into account existing information about the issue at hand (like causal relationships), doing so usually takes more work than using certain statistical techniques. However, mainstream ML research also addresses these problems, and plant research is starting to adopt solutions for them (Azodi et al. 2020).

Developing (novel) ML techniques for problems where the information is too complicated, diverse, and fluctuating for existing procedures to manage is the most intriguing scientific challenge. ML is frequently used to incorporate data in one study or analyze individual biochemical data sets. However, it has yet to be extensively utilized to extract and integrate the known scientific data and knowledge, which is frequently less organized. Given the speed at which scientific literature and information are being gathered, such integrative analyses will become more and more significant. Similarly, physiological phenotyping has frequently been successfully applied to relatively simple plants under carefully monitored circumstances.

When faced with physically complicated plants in more real, varied environments, the methods underlying these applications will not work. Developing strategies to address occlusion—which occurs when a different surface obscures an area of the subject of interest—as well as phenotypic and environmental variation still requires a lot of work. By separating complex traits like growth into more readily measurable elements, known as secondary characters, ML may be helpful in modeling genotype–environment interactions and relating genotype to phenotype.

Future plant research decision-making could be aided by ML in a number of ways, from using genetic modification (GM) techniques to determine which portions of the genes must be changed to achieve a desired characteristics in order to measure agricultural yield in vivo in the field or greenhouse and ensure optimal local growth conditions. Even though these are essentially engineering problems, effective ML techniques will provide researchers with useful tools, especially as they improve their ability to interpret their findings. In this sense, ML can assist in tackling the difficulties we encounter in guaranteeing food security for expanding populations in quickly evolving environments.

11.7 Summary

As the volume of inherited and physical information collected using technologies that are highly efficient increases, data-driven solutions that use ML to identify and select superior germplasm can be very beneficial to plant breeding. The implementation of the main supervised type of ML algorithms is summarized in this manuscript. The prediction models are investigated through a variety of CV techniques, which can be made to reveal the advantages and disadvantages of various designs.

References

Abdollahi-Arpanahi, R., Gianola, D., and Peñagaricano, F. (2020). Deep learning versus parametric and ensemble methods for genomic prediction of complex phenotypes. *Genetics Selection Evolution* 52: 1–15.

Austin, P.C. and Brunner, L.J. (2004). Inflation of the type I error rate when a continuous confounding variable is categorized in logistic regression analyses. *Statistics in Medicine* 23 (7): 1159–1178.

Azodi, C.B., Bolger, E., McCarren, A. et al. (2019). Benchmarking parametric and machine learning models for genomic prediction of complex traits. *G3: Genes, Genomes, Genetics* 9 (11): 3691–3702.

Azodi, C.B., Pardo, J., VanBuren, R. et al. (2020). Transcriptome-based prediction of complex traits in maize. *The Plant Cell* 32 (1): 139–151.

Bali, N. and Singla, A. (2022). Emerging trends in machine learning to predict crop yield and study its influential factors: A survey. *Archives of Computational Methods in Engineering* 29 (1): 95–112.

Bernardo, R. (2008). Molecular markers and selection for complex traits in plants: Learning from the last 20 years. *Crop Science* 48: 1649–1664.

Boris, H. (2019). Universal function approximation by deep neural nets with bounded width and ReLU activations. *Mathematics* 7 (10): 992.

Chen, T. and Carlos, G. (2016). XGBoost: A scalable tree boosting system. *Proceedings of the 22nd ACMSIGKDD International Conference on Knowledge Discovery and Data Mining*, 785–794.

Christina, L., Wimmer, V., Albrecht, T. et al. (2013). Sensitivity to prior specification in Bayesian genome-based prediction models. *Statistical applications in genetics and molecular biology* 12 (3): 375–391.

Costa-Neto, G., Fritsche-Neto, R. and Crossa, J. (2021). Nonlinear kernels, dominance, and envirotyping data increase the accuracy of genome-based prediction in multi-environment trials. *Heredity* 126: 92–106.

Crossa, J., Pérez-Rodríguez, P., Cuevas, J. et al. (2017). Genomic selection in plant breeding: methods, models, and perspectives. *Trends in plant science* 22 (11): 961–975.

Daetwyler, H.D., Calus, M.P., Pong-Wong, R. et al. (2013). Genomic prediction in animals and plants: simulation of data, validation, reporting, and benchmarking. *Genetics* 193 (2): 347–365.

de Los Campos, G.U.S.T.A.V.O., Gianola, D., Rosa, G.J. et al. (2010). Semi-parametric genomic-enabled prediction of genetic values using reproducing kernel Hilbert spaces methods. *Genetics research* 92 (4): 295–308.

Deisenroth, M.P., Faisal, A.A., and Ong, C.S. (2020). *Mathematics for Machine Learning*. Cambridge University Press.

Gazestani, V.H. and Lewis, N.E. (2019). From genotype to phenotype: Augmenting deep learning with networks and systems biology. *Current Opinion in Systems Biology* 15: 68–73.

González-Recio, O., Rosa, G.J.M, and Gianola, D. (2014). Machine learning methods and predictive ability metrics for genome-wide prediction of complex traits. *Livestock Science* 166: 217–231.

Goodfellow, I., Bengio, Y., Courville, A. et al. (2016). *Deep learning* (Vol. 1, No. 2). Cambridge: MIT press.

Habier, D., Fernando, R.L, and Garrick, D.J. (2013). Genomic BLUP decoded: A look into the black box of genomic prediction. *Genetics* 194 (3): 597–607.

Hastie, T., Tibshirani, R., and Friedman, J. (2008). Unsupervised learning. In: *The Elements of Statistical Learning: Data mining, Inference, and Prediction* (pp. 485–585). New York, NY: Springer New York.

Hayes, B., Goddard, M., and Goddard, M. (2001). Prediction of total genetic value using genome-wide dense marker maps. *Genetics* 157: 1819–1829.

He, T. and Li, C. (2020). Harness the power of genomic selection and the potential of germplasm in crop breeding for global food security in the era with rapid climate change. *The Crop Journal* 8 (5): 688–700.

Heffner, E.L., Sorrells, M.E., and Jannink, J.L. (2009). Genomic selection for crop improvement. *Crop Science* 49 (1): 1–12.

Heslot, N. Yang, H.P., Sorrells, M.E. et al. (2012). Genomic selection in plant breeding: a comparison of models. *Crop Science* 52 (1): 146–160.

Hope, T.M. (2020). Linear regression. In: *Machine Learning*. Academic Press. 67–81.

Kecman, V. (2005). Support vector machines–an introduction. In: *Support Vector Machines: Theory and Applications*. Berlin, Heidelberg: Springer Berlin Heidelberg 1–47.

Khaki, S. and Lizhi, W. (2019). Crop yield prediction using deep neural networks. *Frontiers in Plant Science* 10: 621.

Kramer, O. (2013). K-nearest neighbors. In: *Dimensionality Reduction with Unsupervised Nearest Neighbors*. Berlin, Heidelberg: Springer Berlin Heidelberg 51: 13–23.

Kuhn, M. and Johnson, K. (2013). *Applied Predictive Modeling*. Springer.

Legarra, A., Robert-Granié, C., Manfredi, E. et al. (2008). Performance of genomic selection in mice. *Genetics* 180 (1): 611–618.

Leo, B. (2001). Random forests. *Machine Learning* 45: 5–32.

Lin, E. and Lane, H.Y. (2017). Machine learning and systems genomics approaches for multi-omics data. *Biomarker Research* 5: 1–6.

Littell, R.C., Milliken, G.A., and Stroup, W.W. et al. (2006). *SAS for Mixed Models*, 2nd ed. SAS Institute.

Lopez-Cruz, M. and De Los Campos, G. (2021). Optimal breeding-value prediction using a sparse selection index. *Genetics* 218 (1): iyab030.

Mahood, E.H., Kruse, L.H., and Moghe, G.D. (2020). Machine learning: A powerful tool for gene function prediction in plants. *Applications in Plant Sciences* 8 (7): e11376.

Maronna, R. (2011). Alan Julian Izenman (2008): Modern multivariate statistical techniques: Regression, classification and manifold learning. *Statistical Papers* 52 (3): 733.

Maulud, D. and Abdulazeez, A.M. (2020). A review on linear regression comprehensive in machine learning. *Journal of Applied Science and Technology trends* 1 (2): 140–147.

Meinshausen, N., Hauser, A., Mooij, J.M. et al. (2016). Methods for causal inference from gene perturbation experiments and validation. *Proceedings of the National Academy of Sciences* 113 (27): 7361–7368.

Meuwissen, T.H.E, Hayes, B.J., and Goddard, M.E. (2001). Prediction of total genetic value using genome-wide dense marker maps. *Genetics* 157 (4): 1819–1829.

Misztal, I. and Legarra, A. (2017). Invited review: Efficient computation strategies in genomic selection. *Animal* 11 (5): 731–736.

Montesinos-López, O. A., Montesinos-López, A., Crossa, J. et al. (2018). Multi-trait, multi-environment deep learning modeling for genomic-enabled prediction of plant traits. *G3: Genes, genomes, genetics* 8 (12): 3829–3840.

Montesinos-López, O.A., Montesinos-López, A., Pérez-Rodríguez, P. et al. (2021). A review of deep learning applications for genomic selection. *BMC Genomics* 22: 1–23.

Navya, Y.S., Pranathi, K., Srija, G. (2025). Spam detection using machine learning: a logistic regression approach. *Advancement of Computer Technology and Its Applications* 8 (3): 1–10.

Nayak, J., Naik, B., and Behera. H.S. (2015). A comprehensive survey on support vector machine in data mining tasks: applications & challenges. *International Journal of Database Theory and Application* 8 (1): 169–186.

Perakakis, N., Yazdani, A., and Karniadakis, G.E. et al. (2018). Omics, big data and machine learning as tools to propel understanding of biological mechanisms and to discover novel diagnostics and therapeutics. *Metabolism-Clinical and Experimental* 87: A1–A9.

Pérez-Enciso, M., and Zingaretti, L.M. (2019). A guide on deep learning for complex trait genomic prediction. *Genes* 10 (7): 553.

Peters, J., Janzing, D., and Schölkopf, B. (2017). *Elements of Causal Inference: Foundations and Learning Algorithms*. The MIT Press.

Prasath, N., Sreemathy, J., Krishnaraj, N. et al. (2023). Analysis of crop yield prediction using random forest regression model. *Information Systems for Intelligent Systems: Proceedings of ISBM 2022*. Singapore: Springer Nature Singapore, 239–249.

Pszczola, M. and Calus, M.P.L. (2016). Updating the reference population to achieve constant genomic prediction reliability across generations. *Animal* 10 (6): 1018–1024.

Qaim, M. (2020). Role of new plant breeding technologies for food security and sustainable agricultural development. *Applied Economic Perspectives and Policy* 42 (2): 129–150.

Rincent, R., Laloë, D., Nicolas, S. et al. (2012). Maximizing the reliability of genomic selection by optimizing the calibration set of reference individuals: comparison of methods in two diverse groups of maize inbreds (Zea mays L.). *Genetics* 192 (2): 715–728.

Salcedo-Sanz, S., Ghamisi, P., Piles, M. et al. (2020). Machine learning information fusion in Earth observation: A comprehensive review of methods, applications and data sources. *Information Fusion* 63: 256–272.

Salman, H.A., Kalakech, A., and Steiti, A. (2024). Random forest algorithm overview. *Babylonian Journal of Machine Learning* 2024: 69–79.

Scheben, A., Yuan, Y., and Edwards, D. (2016). Advances in genomics for adapting crops to climate change. *Current Plant Biology* 6: 2–10.

Schrag, T.A., Westhues, M., Schipprack, W. et al. (2018). Beyond genomic prediction: combining different types of omics data can improve prediction of hybrid performance in maize. *Genetics* 208 (4): 1373–1385.

Selby, J. (2019). The Trump presidency, climate change, and the prospect of a disorderly energy transition. *Review of International Studies* 45 (3): 471–490.

Shahhosseini, M., Hu, G., and Huber, I. et al. (2021). Coupling machine learning and crop modeling improves crop yield prediction in the US corn belt. *Scientific Reports* 11: 1–15.

Signoretto, M. and Suykens, J.A.K. (2015). Kernel methods. In: *Springer Handbook of Computational Intelligence*, 577–605. Springer.

Song, Y.Y. and Lu, Y. (2015). Decision tree methods: applications for classification and prediction. *Shanghai Arch Psychiatry* 27 (2):130–135.

Van Dijk, A.D.J., Kootstra, G., and Kruijer, W. et al. (2021). Machine learning in plant science and plant breeding. *iScience* 24(1): 101890. doi:10.1016/j.isci.2020.101890

Xavier, A., Muir, W.M., and Rainey, K.M. (2020). bWGR: Bayesian Whole-Genome Regression. *Bioinformatics* 36 (6): 1957–1959.

Xu, Y., Liu, X., and Fu, J. et al. (2020). Enhancing genetic gain through genomic selection: From livestock to plants. *Plant Communications* 1: 100005.

Zampieri, G., Vijayakumar, S., and Yaneske, E. et al. (2019). Machine and deep learning meet genome-scale metabolic modeling. *PLoS Computational Biology* 15: c1007084.

Zhan, Y., Li, J., Sun, J., Zhang, W. et al. (2019). The impact of chronic heat stress on the growth, survival, feeding, and differential gene expression in the sea urchin Strongylocentrotus intermedius. *Frontiers in Genetics* 10: 301.

12

Machine Learning-Assisted Plant Systems Biology

Haragopal Dutta[1], Suman Dutta[2], Sudhir Kumar[3]*

[1] *Division of Genetics, ICAR–Indian Agricultural Research Institute, New Delhi, India*
[2] *School of Agriculture and Rural Development, Ramakrishna Mission Vivekananda Educational and Research Institute, Belur Math, Howrah, West Bengal*
[3] *Division of Plant Physiology, ICAR-IARI, New Delhi, India*

*Corresponding author: duttasumansiliguri@gmail.com

12.1 Introduction

Plant systems biology aims to understand the complex interactions and networks governing plant growth, development, and responses to environmental factors. The integration of diverse biological data, including genomics, transcriptomics, proteomics, metabolomics, and phenomics, allows for a more comprehensive understanding of plant functions at the molecular level (Gain et al. 2025). However, managing and analyzing this vast amount of data poses significant challenges. Machine learning (ML), with its ability to detect patterns, make predictions, and identify correlations in large datasets, has emerged as a powerful tool in plant systems biology, revolutionizing how researchers explore plant functions and interactions (Dutta et al. 2024).

Plant systems biology focuses on elucidating the intricate interactions and networks that govern plant growth, development, and responses to a multitude of environmental factors. The field integrates diverse biological data streams, including genomics, transcriptomics, proteomics, metabolomics, and phenomics, to foster a comprehensive understanding of plant functions at the molecular level (Tardieu et al. 2017; Roy et al. 2024). However, managing and analyzing the extensive datasets generated from these sources presents significant challenges. Thus, the integration of advanced computational techniques, specifically ML, has emerged as an innovative approach that transforms research methodologies in plant systems biology. ML excels in detecting patterns, making predictions, and identifying correlations within large datasets (Dutta et al. 2024). Its application to plant sciences allows researchers to leverage complex, high-dimensional data effectively. For instance, ML algorithms such as support vector machines (SVM) and Random Forests have been increasingly utilized for classifying plant traits and elucidating genetic markers associated with desired phenotypic outcomes (Parmley et al. 2019; Rodriguez et al. 2022). These tools enable the analysis of vast quantities of phenotypic data collected via high-throughput phenotyping techniques, leading to significant improvements in plant breeding decisions (Araus et al. 2022; Parmley et al. 2019). Such advancements empower breeders to make informed selections based on predictive models that incorporate a multitude of traits, ultimately enhancing genetic gains in crop improvement programs. Additionally, plant phenomics—the study of phenotypic traits through advanced imaging techniques—benefits immensely from ML's capacity for handling diverse data types. This integration facilitates the identification of key traits that correlate with essential agricultural performance metrics, such as yield stability in varying environmental conditions. Moreover, the computational power of ML allows researchers to perform multivariate analyses without necessitating prior explicit selection of variables, thus capturing complex interactions among traits that may otherwise remain obscured (Table 12.1). Furthermore, the application of ML in plant systems biology spans other domains, including stress responses, metabolic engineering, and ecosystem modeling (Hesami et al. 2022). ML methodologies have been increasingly tailored to process multi-omics data, merging genomics, transcriptomics, and proteomics in predictive models that can reveal crucial biomarkers for specific plant

Machine Learning for Plant Biology, First Edition. Edited by Jen-Tsung Chen.

Table 12.1 Machine learning in plant biology system.

Crop	AI Technique	Algorithm(s)	ICQP Role	Application	Dataset	Plant Biology/Traits	References
Tomato	Deep Learning (DL)	YOLOv4 with DenseNet, PANet, SPP	Identification, Classification	Real-time detection of diseases in tomato plants under various challenging environments	Custom dataset with images of tomato plant diseases	Disease detection (e.g., early blight, late blight)	Roy et al. (2021)
Apple, Maize, Rice	Explainable DL	Vision Transformer (PlantXViT)	Identification, Classification	Detection and classification of plant diseases across multiple crops	Five publicly available datasets for Apple, Maize, and Rice	Disease identification with high accuracy	Thakur et al. (2022)
Potato	DL-based Object Detection	Retina-UNet-Ag	Identification, Classification	Detection of crop stress (e.g., drought stress) in aerial images of potato fields	Aerial images captured using Parrot Sequoia camera with annotated bounding boxes	Stress identification (e.g., premature senescence)	Butte et al. (2021)
Maize	Deep Neural Networks (DNN)	DNN with feature selection	Prediction	Prediction of maize yield based on genotype and environmental factors	Syngenta Crop Challenge dataset (2267 hybrids across 2247 locations)	Yield prediction considering genotype–environment interactions	Khaki and Wang (2019)
Wheat, Barley, Soybean	Convolutional Neural Networks (CNN)	ResNet-50, ResNetXt-101, AlexNet, DenseNet, VGG-16, YOLOv5	Identification, Classification, Quantification	Analysis of plant traits such as root segmentation, leaf counting, and stress detection	Various datasets including RGB images and hyperspectral data	Trait analysis (e.g., spike number, stress responses)	Eladl et al. (2024)
Rice	Machine Learning (ML)	Random Forest (RF), Decision Tree (DT), Naive Bayes, Bagging	Classification	Classification of rice species and detection of weed farms using UAV images	UAV image datasets: WeedNet, Rice Seedling, Rice Varieties	Species classification, weed detection	Farooq et al. (2024)
Cotton	Ensemble DL Models	CNN-based ensemble models	Classification	Classification of cotton crop health and detection of diseases	Image datasets of cotton fields with annotations	Disease detection (e.g., cotton leaf curl virus)	Shahid et al. (2024)

responses (Lewis and Kemp 2021). These advancements in data-driven agriculture are transformative, allowing for rapid iterative testing and development of strategies that enhance resistance to diseases and adverse climatic conditions.

12.2 ML Algorithms in Plant Systems Biology

12.2.1 Supervised Learning

Algorithms like random forests, SVMs, and deep learning models are widely used for trait prediction, classification of plant phenotypes, and regulatory network inference. These models learn from labeled datasets and are particularly useful for genomic selection and trait prediction in breeding programs (Table 12.2). Supervised learning has become an essential technique in plant systems biology, playing a crucial role in the analysis of complex, high-dimensional biological data generated from various omics technologies. The fundamental principle of supervised learning is the utilization of labeled datasets to train algorithms capable of predicting outcomes or classifying data points based on input features. In the context of plant systems biology, this involves correlating phenotypic traits with underlying genomic data, thereby facilitating genomic prediction in plant breeding applications (Hu et al. 2018; Xavier 2021). Supervised learning encompasses several algorithms, including linear regression, SVM, decision trees, and neural networks (Xu and Jackson 2019). These models have been increasingly integrated into plant breeding pipelines to streamline the identification of desirable traits. In particular, deep learning approaches, such as convolutional neural networks (CNNs), have demonstrated robustness in tasks like plant species identification and disease classification, achieving superior performance compared to traditional methods (An et al. 2019; Homan and Preez n.d.; Singh et al. 2018). The advantages of deep learning in handling large volumes of image data have made it a prevalent choice in studies focused on phenotyping under biotic and abiotic stress conditions (Singh et al. 2018). A significant challenge associated with supervised learning is the requirement for extensive labeled datasets, which can be labor-intensive and resource-demanding to produce. The efficiency and accuracy of supervised algorithms are intricately linked to the quality and quantity of annotated training data (Yan and Wang 2022). Alternative methodologies, including semi-supervised and self-supervised learning, have been explored as potential solutions to mitigate the scarcity of labeled data while leveraging vast amounts of unlabeled data available in plant phenotyping contexts (Guo et al. 2021; Kefato and Girdzijauskas 2021). For instance, recent advancements in self-supervised learning techniques promise improvements in model performance by using contrastive learning strategies to better understand unlabeled datasets (Ogidi et al. 2023). Moreover, supervised learning has not only advanced phenotypic trait identification but also revolutionized how researchers approach trait prediction from genomic data. The integration of omics technologies, alongside ML, allows for the real-time assessment of agronomic traits, enhancing the precision of breeding programs. With emerging computational tools and increasing accessibility to genomic data, supervised learning methodologies can synergistically improve the precision of predictions in plant biology, ultimately fostering more resilient agricultural systems.

12.2.2 Unsupervised Learning

Clustering algorithms (e.g., K-means, hierarchical clustering) and dimensionality reduction techniques (e.g., PCA, t-SNE) help in identifying patterns and relationships in large multi-omics datasets without predefined labels. These methods are essential for data exploration and hypothesis generation in systems biology (Table 12.2). Unsupervised learning is a foundational paradigm in ML that is increasingly leveraged in plant systems biology, particularly through clustering algorithms and dimensionality reduction techniques. Clustering methods, such as K-means and hierarchical clustering, are essential for discerning patterns within complex, high-dimensional multi-omics datasets without requiring predefined labels. These methods enable researchers to identify natural groupings among data samples based on their inherent features, which can be crucial for understanding biological processes and generating hypotheses in the context of systems biology (Qiu et al. 2024). For instance, the MoClust method represents an advancement in clustering single-cell multi-omics data, utilizing modality-specific autoencoders to process various omics layers effectively (Yuan et al. 2023; Zhang et al. 2024). Such methods exemplify how clustering can facilitate the integration of diverse biological data forms, allowing for a more comprehensive exploration of biological variability and relationships across different omic levels (Cui et al. 2024; Zhao et al. 2023). Additionally, approaches like GSTRPCA provide refined techniques for clustering by applying tensor singular value decomposition, thus enhancing the integrative analysis of single-cell multi-omics. Dimensionality reduction techniques, such as principal component analysis (PCA) and t-distributed stochastic neighbor embedding (t-SNE), complement

Table 12.2 Machine learning algorithms in plant biology systems.

Crop	Gene Regulatory Network	Supervised Learning	Unsupervised Learning	Reinforcement Learning	Platforms	ML Algorithms	Trait of Interests	References
Rice (*Oryza sativa*)	Co-expression networks derived from transcriptomic data	Random Forests for yield prediction	K-means clustering for genotype classification	Limited application in irrigation optimization	TensorFlow, R	Random Forest, SVM	Yield, drought tolerance	Yoosefzadeh Najafabadi et al. (2023)
Maize (*Zea mays*)	GRNs inferred using SCENIC	Support Vector Machines for disease resistance prediction	Hierarchical clustering for population structure analysis	Emerging use in breeding strategy optimization	Python, scikit-learn	SVM, Decision Trees	Disease resistance, biomass	van Dijk et al. (2021)
Wheat (*Triticum aestivum*)	GRNs constructed from gene expression profiles	Gradient Boosting for grain quality prediction	PCA for dimensionality reduction in genomic data	Experimental use in adaptive breeding	R, caret package	Gradient Boosting, k-NN	Grain quality, stress tolerance	Isewon et al. (2022)
Soybean (*Glycine max*)	GRNs modeled using I Regulon	Neural Networks for protein content prediction	SOMs for metabolic pathway analysis	Preliminary studies in trait selection	Keras, TensorFlow	Neural Networks, SOM	Protein content, oil yield	Chen et al. (2024)
Arabidopsis thaliana	GRNs analyzed with DeepMEL	Deep Learning for gene expression prediction	t-SNE for visualization of gene clusters	Conceptual frameworks in developmental studies	PyTorch, R	Deep Learning, t-SNE	Gene expression, developmental traits	Hrdlickova et al. (2017)
Tomato (*Solanum lycopersicum*)	GRNs derived from cis-regulatory elements	Logistic Regression for fruit ripening stages	DBSCAN for clustering gene expression data	Exploratory use in growth modeling	MATLAB, Orange	Logistic Regression, DBSCAN	Fruit ripening, shelf life	Chao et al. (2025)

clustering by transforming high-dimensional data into a lower-dimensional space, retaining essential structures and relationships that can be cognitively and computationally managed more easily. PCA is utilized extensively for reducing the complexity of datasets while preserving variance (Kemmo Tsafack et al. 2024). t-SNE and its parametric variants have been particularly effective in visualizing single-cell transcriptomic data, allowing for an elucidation of complex relationships in multi-omics datasets (Kobak and Berens 2019; Trosten et al. 2021). This capacity to visualize high-dimensional data greatly aids in the exploratory phase of systems biology research, enabling the generation of new biological insights. Moreover, the integration of deep learning in these unsupervised learning frameworks further enhances their efficacy. For example, deep models like scMDC not only streamline the clustering of multi-omics data but also learn latent representations of the data, which enhances the robustness of the clustering outcomes (Lin et al. 2022; Yang et al. 2021). The interplay between dimensionality reduction and clustering techniques thus plays a critical role in revealing the biological insights embedded in multi-omics datasets, offering pathways toward understanding the multifaceted nature of plant biology and related systems (Leng et al. 2022; Nguyen and Wang 2020).

12.2.3 Reinforcement Learning

Though less commonly applied, reinforcement learning (RL) has potential in optimizing experimental design in plant systems biology, where the goal is to maximize some objective (e.g., plant growth or yield) by sequentially choosing the best experimental conditions. RL has emerged as a promising approach to optimize experimental design in plant systems biology, particularly aimed at maximizing objectives such as growth and yield by dynamically selecting the most effective experimental conditions (Table 12.2). This adaptive methodology leverages the principles of RL, where an agent learns to navigate its environment through trial-and-error interactions to optimize desired outcomes, such as resource allocation in plant growth scenarios. Recent advancements highlight the applicability of RL in agricultural optimization. For instance, Han et al. propose a hybrid deep RL algorithm designed for the joint optimization of power and channel in agricultural wireless sensor networks, which supports the notion that RL can effectively adapt strategies under varying environmental conditions, thereby successfully optimizing resource consumption (Han et al. 2021). This adaptability is central in plant systems biology, where conditions can vary significantly between environments and developmental stages. Furthermore, the implementation of RL techniques can significantly enhance the efficiency of experimental setups. Unlike the claim in the original text, Rodríguez et al. emphasize the capability of RL to learn optimal policies based solely on system interactions in a different context; therefore, the cited work does not directly support the assertion about plant systems biology (Rodriguez et al. 2013). A more appropriate reference is needed to support this claim about optimizing plant experimental processes without requiring pre-existing models. Moreover, the synergy between RL and other ML forms, such as model predictive control (MPC), reinforces its potential utility. As noted by Pan et al., RL can be integrated with MPC to enhance control strategies in dynamic systems, suggesting that similar methodologies could be adapted for optimizing growth conditions in agricultural contexts (Pan et al. 2022). By employing online data to tailor control strategies, researchers can navigate the complexities of plant responses to varying environmental factors effectively. The use of RL in plant systems biology also benefits from advancements in perception models and data classification methodologies. Yan and Wang discuss unsupervised learning techniques that enhance data processing capabilities in plant systems, which can complement RL approaches in extracting meaningful insights from high-throughput biological data (Wang et al. 2022). The integration of these methodologies supports a more holistic approach to optimizing plant growth conditions.

12.3 Key Applications of ML in Plant Systems Biology

12.3.1 Multi-omics Data Integration

Plant systems biology often deals with complex datasets from different "omics" technologies (e.g., genomics, transcriptomics, proteomics, metabolomics) (Figure 12.1). ML algorithms, especially unsupervised learning techniques like clustering and dimensionality reduction, help integrate and analyze these data, revealing hidden relationships between genes, proteins, metabolites, and other biological factors (Table 12.3). For instance, PCA and t-SNE are commonly used to reduce the dimensionality of omics datasets, making it easier to identify critical pathways and interactions in plants. The integration of multi-omics data in plant systems biology has emerged as a vital approach for understanding the complex interactions between various biological molecules, including genes, proteins, and metabolites. With the advent of

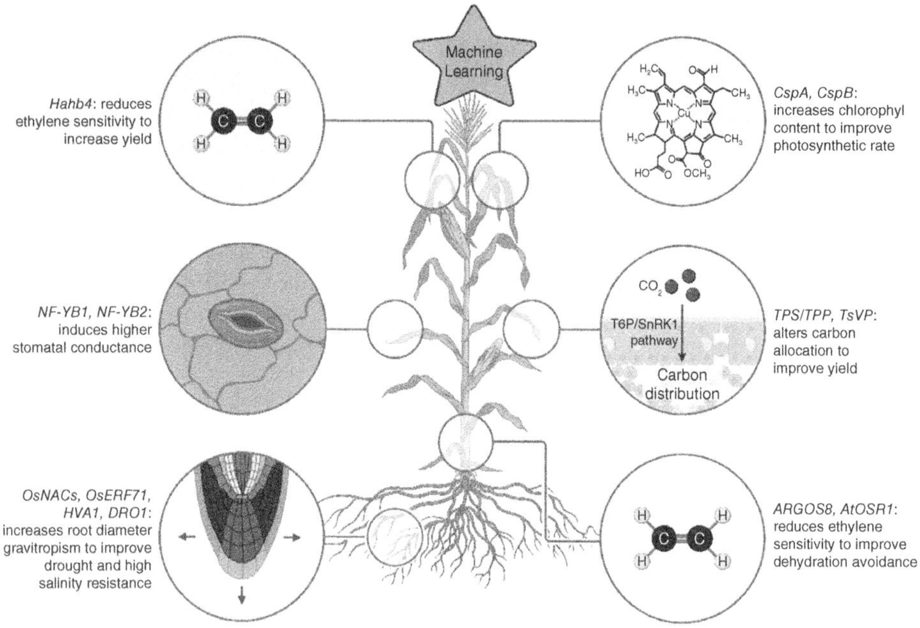

Figure 12.1 Integrated study of plant biology through multi-omics approach.

Table 12.3 Multi-omics data integration through multi-omics approach.

Model Crops	Genomics	Transcriptomics	Proteomics	Metabolomics	ML Algorithms	Crop Trait Studied	References
Maize	ZmDREB2A, ZmNAC111	Drought-responsive transcripts	Stress-induced proteins	Osmoprotectants (e.g., proline)	Random Forest, SVM	Drought tolerance	ScienceDirect
Rice	OsSPL14, OsNRT1.1B	Nitrogen metabolism transcripts	Nitrogen assimilation enzymes	Amino acids, organic acids	Deep Neural Networks	Nitrogen use efficiency	Frontiers
Wheat	TaGW2, TaCKX6	Grain development transcripts	Grain filling proteins	Carbohydrates, phytohormones	Multi-Task Learning	Grain yield and size	Plant Methods
Sorghum	SbMYB60, SbNAC2	Phenolic biosynthesis transcripts	Tannin biosynthetic enzymes	Phenolic compounds	Deep Learning (CNN)	Tannin content prediction	Nature
Tomato	SlGLK2, SlMYB12	Flavonoid pathway transcripts	Flavonoid biosynthetic enzymes	Flavonoids, carotenoids	Support Vector Machines	Fruit quality traits	MDPI
Soybean	GmSWEET10a, GmSWEET10b	Seed development transcripts	Seed storage proteins	Isoflavones, saponins	Random Forest, SVM	Seed composition and yield	ScienceDirect

several omics' technologies such as genomics, transcriptomics, proteomics, and metabolomics, researchers now confront the challenge of integrating high-dimensional datasets that possess heterogeneous features. ML algorithms, particularly unsupervised learning techniques, play a critical role in addressing these challenges by analyzing and revealing underlying biological relationships that are not immediately apparent from individual omics studies. One of the primary methods employed in multi-omics integration is the Multi-Omics Factor Analysis (MOFA), which offers a statistical framework for the unsupervised integration of multi-omics datasets. This approach builds on group factor analysis principles and adapts them to the varied requirements inherent in multi-omics research (Argelaguet et al. 2018). MOFA and its successor,

MOFA+, have been shown to effectively distil complex datasets into latent factors that represent shared biological signals across different biological contexts, thus enabling the identification of key pathways and protein interactions within plant systems. Furthermore, the application of clustering algorithms has been a prominent strategy, highlighted by advancements that facilitate the integration and analysis of multi-omics data (Rappoport et al. 2020). These clustering methods not only enhance the accuracy of the analysis but also minimize the effects of noise typical of high-dimensional datasets, thereby allowing a more nuanced understanding of plant biological systems (Tsafack et al. 2024). Dimensionality reduction techniques such as PCA and t-SNE have also become fundamental tools in the analysis of multi-omics data. PCA captures maximum variance in the data through linear dimensionality reduction, while t-SNE is more adept at preserving the local structure of data in a low-dimensional space, making it particularly useful for visualizing complex datasets (Tian and Tao 2020). This capability is critical for identifying cellular states and their transitions, a task that is increasingly important in plant biology, where complex signaling networks occur (Dutta et al. 2025). As shown in studies of plant responses to environmental stressors, applying these techniques can lead to insights into adaptive mechanisms and the underlying genetic and metabolic pathways affecting a plant's resilience (Gechev et al. 2021). Moreover, the landscape of ML for multi-omics data integration has been expanded by the emergence of novel algorithms that leverage joint probabilistic modeling frameworks. For instance, the totalVI framework capitalizes on the integration of transcriptomic and proteomic data to reveal the relationships at the single-cell level, highlighting the unifying latent structures behind multimodal data types (Gayoso et al. 2021). This integration facilitates a detailed understanding of the intricate biological networks at play, enabling researchers to draw connections between genomic variations and phenotypic expressions in plants.

12.3.2 Gene Regulatory Network

Inference ML is pivotal in reconstructing gene regulatory networks (GRNs), which describe how genes interact and regulate each other. Techniques like random forests, SVM, and neural networks are applied to infer these networks from transcriptomics data. Understanding GRNs enables researchers to identify key regulatory genes and pathways controlling essential processes such as stress responses, development, and metabolic functions in plants. GRNs are complex architectures that delineate how genes interact and regulate each other within biological systems, particularly in plants (Table 12.4). The reconstruction of these networks has become increasingly feasible due to advancements in ML techniques that facilitate the analysis of vast transcriptomic datasets. Approaches such as random forests, SVM, and

Table 12.4 Multi-omics data integration through multi-omics approach.

Application Area	Machine Learning Methods	Purpose in GRN Inference	Example/Case Study	References
GRN Inference from Transcriptomic Data	Support Vector Machines (SVM), Random Forests, Bayesian Networks	Predict regulatory relationships between transcription factors and target genes. Construct GRNs from gene expression profiles	The Beacon GRN tool utilized SVMs to predict GRNs during Arabidopsis embryo development	Tucci et al. (2024)
Time-Series Expression Analysis	Recurrent Neural Networks (RNN), Dynamic Bayesian Networks	Model temporal gene expression changes. Identify dynamic regulatory interactions over time	GRN inference in Mimulus species using time-series RNA-seq data	Mishra et al. (2019)
Integration of Multi-Omics Data	Deep Learning, Ensemble Learning	Combine transcriptomic, proteomic, and metabolomic data. Enhance accuracy of GRN predictions	Systems biology approaches integrating multi-omics data for plant–pathogen interaction studies	Van den Broeck et al. (2020)
Phenotype Prediction from Genotype	Supervised Learning. Graph Neural Networks	Predict phenotypic traits based on GRNs. Assist in crop improvement strategies	Use of GRADIS, a supervised learning approach, to reconstruct GRNs and predict phenotypes	Wang et al. (2018)
Cross-Species GRN Modeling	Transfer Learning. Comparative Genomics	Apply GRN models from model organisms to crops. Identify conserved regulatory mechanisms	Multi-species GRN models to enhance genomic selection and gene editing in crops	Ni et al. (2016)

neural networks are prevalent for inferring GRNs from gene expression data, thereby enabling a deeper understanding of the regulatory dynamics underlying critical processes in plant biology, such as development, metabolic functions, and stress responses. The application of random forest algorithms, exemplified by the GENIE3 method, has shown efficacy in identifying regulatory modules in plant systems. This algorithm utilizes transcriptomic data to associate transcription factors (TFs) with their respective target genes, relying on the presence of specific transcription factor binding motifs (TFBMs) within target gene promoters (Robertson and Wilkins 2023). By constructing GRNs from such correlations, researchers can identify essential regulatory genes, which are vital for understanding plant responses to environmental stimuli and developmental cues (Banf and Rhee 2017). The integration of transcriptomic data with additional layers of biological information, such as chromatin accessibility and protein interactions, further enhances the predictive accuracy of GRN constructs, as demonstrated through methods like SCENIC+ (Aibar et al. 2017; González-Blas et al. 2022). Moreover, ML strategies extend beyond mere inference toward functional elucidation within the GRNs. Studies have demonstrated the utility of weighted gene co-expression network analysis (WGCNA) in reconstructing GRNs and identifying gene modules associated with specific traits or responses in plants. Such frameworks facilitate the identification of biomarkers and provide insights into gene interactions necessary for metabolic processes and stress adaptations. Techniques that combine gene expression profiling with perturbation-based approaches, such as CRISPR integration, represent an innovative frontier in validating these inferred regulatory relationships (Marbach et al. 2016; D. van Dijk et al. 2018). The focus on integrating multiple dimensions of biological data—both transcriptomic and epigenomic—underscores the potential of ML to unravel the complexities of gene regulation in plants.

12.3.3 Trait Prediction and Breeding

Predicting complex traits, such as yield, disease resistance, and drought tolerance, is critical for crop improvement. ML algorithms like deep learning and gradient-boosted decision trees (GBDT) are increasingly used in genomic selection, where they predict phenotypic traits based on genotypic data. This allows for more accurate selection of plants with desired traits, accelerating the breeding process (Table 12.5). For example, CNNs have been applied to high-throughput phenotyping data to predict plant traits with remarkable accuracy. Trait prediction and breeding using ML algorithms in plant systems biology has become essential for enhancing crop improvement, particularly in addressing challenges such as yield optimization, drought tolerance, and disease resistance. The advent of advanced genomic selection methods allows for the efficient prediction of complex phenotypic traits using comprehensive genotypic data. Techniques such as deep learning and GBDT have shown significant potential in accurately predicting desired traits, thus accelerating the breeding processes in various crops (Krannich et al. 2015; Pazhamala et al. 2021). The implementation of ML algorithms, particularly CNNs, represents a notable advancement in high-throughput phenotyping. These algorithms analyze extensive datasets from field trials and laboratory studies to predict plant traits with high accuracy. This approach has been demonstrated in predicting drought tolerance in crops like rice and wheat, where complex traits are informed by both genomic data and phenomics metrics, underscoring the integration of multiple data sources crucial for effective trait selection (Seeli et al. 2021; Zhang et al. 2018). Such sophisticated models are pivotal for understanding how stress conditions, such as drought, influence physiological responses and ultimately crop yield (Li et al. 2024b; Morsy et al. 2022). Moreover, genomic

Table 12.5 Trait prediction and breeding through machine learning.

Crop	Trait of Interest	ML Algorithm	Platform/Tool	References
Maize	Yield, Days to Pollen	Compositional Autoencoder (CAE)	Custom ML Framework	Powadi et al. (2024)
Soybean	Relative Maturity	CNN-LSTM	UAV Imagery	Crossa et al. (2024)
Lettuce	Biomass, Leaf Area	Deformable CNN Regression	RGB and Depth Sensors	Raja et al. (2021)
Rice	Yield, Plant Height	Learnable Group Transform (LGT)	Rice529 Dataset	Dong et al. (2025)
Multiple Crops	Genomic Estimated Breeding Value (GEBV)	Genomic Prediction (GP) Models	Genotypic and Phenotypic Data	van Dijk et al. (2021)
Orphan Crops	Disease Resistance, Yield	Various ML Techniques	Genomic Data Analysis	MacNish et al. (2025)

selection techniques combined with ML strategies enable breeders to manage large datasets involving various traits and genotypes. For example, PCA has been employed to enhance selection efficiency in complex breeding programs, particularly in drought-tolerant maize accessions (Javaid et al. 2023; Maazou et al. 2016). As these models evolve, they facilitate the identification of candidate genes associated with desirable traits. The availability of databases like DroughtDB enhances understanding of the genetic mechanisms related to drought tolerance, significantly contributing to the development of resilient crop cultivars (Alter et al. 2015; Maazou et al. 2016). Operationalizing ML in plant breeding also necessitates a nuanced understanding of the underlying genetic architecture. Research has indicated a correlation between deep rooting systems and drought resilience in crops, demonstrating how root architecture is influenced by genetic traits. Selection for root depth and architecture is increasingly recognized as a key factor in breeding strategies (Al-Naggar et al. 2018; Tomar et al. 2016). Additionally, contemporary studies highlight the importance of epigenetic factors in breeding for drought tolerance, illustrating how multi-generational exposure to drought can lead to advantageous traits being inherited through epigenetic mechanisms (Shih et al. 2016; Zheng et al. 2017).

12.3.4 Metabolic Pathway Analysis

ML techniques are also used to model and predict metabolic pathways in plants. By analyzing metabolomics data, ML algorithms can predict the production of specific metabolites and how they are regulated in different conditions. This is particularly useful in metabolic engineering, where understanding the regulation of metabolic pathways can help optimize the production of valuable compounds such as biofuels, pharmaceuticals, and nutritional components. One notable approach in this field combines network analysis with ML to predict metabolic pathways based on empirical data. For instance, Toubiana et al. (2019) successfully employed this method to analyze metabolomics data from tomato plants, demonstrating the effectiveness of ML techniques in identifying 169 known metabolic pathways while also expanding knowledge on uncharacterized pathways. This combinatory analysis not only highlights the predictive power of ML but also underscores its role in enriching our understanding of plant metabolic networks. Additionally, the use of genome-scale metabolic models (GEMs) further enhances our ability to model plant metabolism (Tafur Rangel et al. 2021). These comprehensive models encompass a wide array of metabolic reactions proposed by genome annotation, thus allowing for predictive analysis that can inform metabolic pathway regulation. The advancement of hybrid modeling techniques that intertwine mechanistic approaches with ML frameworks has shown promise in prediction accuracy and model robustness (Merzbacher et al. 2025; Zhang et al. 2020). Such methodologies pave the way for system-level insights and targeted metabolic engineering interventions. Moreover, recent studies have highlighted the utility of ML techniques in addressing specific questions related to metabolite production and regulatory control. For example, Maruenda and Toubiana (2025) highlighted the activation of rosmarinic acid biosynthesis in potato tubers when subjected to water recovery treatments by employing correlation-based network analysis combined with ML. This indicates the potential to manipulate metabolic pathways in response to environmental triggers, showcasing the practical applications of ML in plant systems biology. The challenges of data complexity and scalability in biological networks have led to the development of specialized algorithms capable of predicting pathway involvement based on chemical structure and metabolic features (Huckvale and Moseley 2024). Such advancements aim to overcome issues like class imbalance in various metabolic data applications, thus improving the reliability of predictions concerning metabolite associations with specific pathways. Noteworthy computational strategies, including the use of graph attention mechanisms for multi-class prediction, further illustrate the potential for sophisticated ML models in advancing pathway prediction accuracy in plant systems (Yang et al. 2020).

12.3.5 Stress Response and Adaptation

Plants encounter various biotic (e.g., pathogens) and abiotic (e.g., drought, salinity) stresses. ML models are used to predict plant responses to these stresses by analyzing transcriptomic, proteomic, and phenotypic data. By identifying the key regulatory genes and pathways involved in stress tolerance, ML assists in designing more resilient crops (Table 12.6). For example, ML models have been used to predict drought tolerance in crops by analyzing gene expression profiles and phenotypic traits under water-limited conditions. Plants are increasingly subjected to various stresses, both biotic (such as pathogens) and abiotic (including drought and salinity) (Kumar et al. 2024). Understanding plant stress responses and their subsequent adaptations is essential for enhancing crop resilience through ML models. These models leverage data from

Table 12.6 Stress response and adaptation through machine learning.

ML Application	Stress Type	Data Inputs	ML Techniques	Key Outcomes	References
Early Detection of Water Stress in Chickpea	Water Deficiency	Time-series images of plant shoots	CNN-LSTM	Achieved over 98% accuracy in classifying stress stages	Lee et al. (2024)
Classification of Stress Symptoms in Cucumber Seedlings	Environmental Stress	RGB images capturing color, texture, and morphology	Various ML models	Accurate classification of stress levels under varied conditions	Mohammadi and Asefpour Vakilian (2023)
Detection of Salt and Drought Stress in Cucumber	Salinity and Drought	Morphological features, physiological/biochemical markers, miRNA concentrations	ML models with metaheuristic optimization	High accuracy in predicting stress type and severity; R^2 up to 0.99 using miRNA data	Shi et al. (2023)
Prediction of Abiotic Stress-Responsive Transcription Factors	Various Abiotic Stresses	Genomic sequences	ML frameworks	Identification of key transcription factors associated with stress responses	Pradhan et al. (2024)
Discrimination of Wheat Yellow Rust and Nitrogen Deficiency	Biotic and Nutrient Stress	Sentinel-2 time-series data	Fast Fourier Convolutional Neural Network (FFDNN)	Accurate and explainable detection of stresses with similar symptoms	Azimi et al. (2021)

diverse biological levels—transcriptomic, proteomic, and phenotypic—to predict plant responses to stresses and to identify key regulatory genes and pathways facilitating stress tolerance. The initial phase of the stress response typically involves an alarm stage where physiological mechanisms are activated upon stress detection, followed by acclimation, maintenance, and potentially exhaustion, leading to recovery or death (Suvarna et al. 2024). Proteomic analyses have illustrated that changes in the protein composition of plants are critical in adapting to stress conditions. For example, studies highlight those proteins associated with signaling and stress Défense, such as the 14-3-3 family, are responsive both to biotic and abiotic stresses (Aghaei and Komatsu 2013; Chen et al. 2018). Additionally, alternative splicing of stress-responsive genes contributes significantly to the plant's ability to adapt, revealing a complex network that is under continuous scrutiny for its potential applications in crop resilience (Li et al. 2018). In terms of ML applications, the integration of phenotypic and transcriptomic data has proven effective in predicting stress tolerance. For instance, research has demonstrated that the combination of metabolic profiling and dynamic phenotyping can successfully classify different ecotypes of *Brachypodium distachyon* into drought tolerance categories (Guo et al. 2016). Such integrated approaches enhance the understanding of how specific genetic traits translate into observable phenotypic changes under stress conditions, paving the way for targeted breeding programs aimed at improving resistance to stresses like drought and salinity (Fisher et al. 2016). Moreover, ML models validate their predictions by analyzing gene expression patterns in response to stress—evident in studies showing that specific genes, when upregulated under salt stress, contribute to enhanced tolerance (Gong et al. 2015; Soren et al. 2020). The development of quantitative trait loci (QTL) for identifying genotypes with better stress performance allows researchers to map stress tolerance traits accurately against controlled conditions, ensuring that future crop varieties exhibit sustainability and resilience in varied environmental scenarios (Li et al. 2024a; Lv et al. 2014). Emerging techniques like quantitative proteomics further elucidate the mechanisms underlying plant responses to stress, offering insights into how plants employ complex signaling networks to cope with adverse conditions (Kakar et al. 2019). For example, the role of heat shock proteins and other stress-responsive proteins has been critical in modulating plant responses and enhancing tolerance—not just to thermal stress but extending to conditions like drought and salinity by stabilizing cellular functions (Ps et al. 2017; Rodziewicz et al. 2014).

12.3.6 Predictive Modeling of Plant–Environment Interactions

Understanding how plants interact with their environment is essential for sustainable agriculture. ML models, particularly neural networks and random forests are used to predict plant growth and productivity under different environmental conditions. These models can incorporate environmental data such as soil conditions, climate variables, and nutrient availability, helping researchers and farmers optimize plant management strategies to enhance crop yields. The interaction between plants and their environment plays a critical role in agricultural productivity and sustainability. As the complexity of these interactions increases, so does the necessity for precise predictive modeling to inform agricultural practices. Recent advancements in ML, particularly through the application of artificial neural networks (ANNs) and random forests, have enabled the creation of sophisticated models that can predict the growth and productivity of plants under varying environmental conditions. These models utilize data encompassing soil characteristics, climate variables, and nutrient levels to help optimize management strategies for crops. Research has demonstrated the efficacy of ANNs in plant growth predictions. For instance, a study focused on the growth of red chili (*Capsicum annuum* L.) reported that an ANN model achieved an R^2 value close to 1, indicating its high accuracy in predicting plant growth under manipulated environmental conditions (Putra et al. 2023). This aligns with findings from another study that employed ANNs to predict lettuce crop growth in a hydroponic system, emphasizing how the model accounted for environmental parameters such as root zone characteristics (Ruslan et al. 2024). The predictive capability of ML models, particularly ANNs, can significantly impact crop yield outcomes and decision-making in practices such as precision agriculture. Random forests, a robust ensemble learning technique, have also been utilized effectively in plant systems biology. These models can manage diverse datasets, perform feature selection, and yield accurate predictions regarding plant responses to environmental changes. For instance, random forests have been employed for disease detection, enhancing understanding of plant health dynamics and improving yield predictions (Sudhakar and Reddy 2023). Furthermore, by harnessing high-throughput omics data, random forests facilitate better insights into genetic and phenotypic variations in plant systems, allowing for improved management and breeding strategies (Yan and Wang 2022). Incorporating high-dimensional environmental data into predictive models is another essential component for optimizing agricultural outputs. By leveraging ML algorithms to assess various factors—such as soil moisture levels, temperature, and nutrient availability—researchers can achieve more comprehensive predictions for plant growth (Benos et al. 2021; Williamson et al. 2023). This approach facilitates the development of tailored management practices that align with real-time environmental conditions, thereby maximizing productivity while maintaining sustainable practices. Moreover, the fusion of genetic information with ML models enhances their predictive accuracy for specific traits in crops. For example, some researchers have suggested integrating genetic markers with quantum ML techniques for marker-assisted plant selection, leading to optimized breeding pathways and disease resistance predictions (Albert et al. 2023). This integrated approach streamlines the breeding process and equips farmers with actionable insights into how genetic factors influence plant responses under various environmental scenarios.

12.3.7 Identification of Key Regulators in Synthetic Biology

In synthetic biology, ML helps in designing genetic circuits and metabolic pathways by identifying key regulators and interactions within the plant genome. By predicting the outcomes of gene modifications, ML tools allow for the rational design of plants with enhanced characteristics, such as increased photosynthetic efficiency or improved nutrient use efficiency. In synthetic biology, the integration of ML presents transformative potential for the design and optimization of genetic circuits and metabolic pathways in plant systems. The identification of key regulators within plant genomes is critical for enhancing various characteristics such as photosynthetic efficiency and nutrient utilization. ML algorithms allow researchers to predict the effects of genetic modifications by analyzing genetic interaction networks and identifying influential genes that modulate desired traits. Recent advancements have led to the development of genetically encoded synthetic circuits using Boolean logic to integrate multiple signals from plant phenotypic data, which can significantly reshape plant morphology and physiology. For example, the creation of synthetic circuits capable of performing logical operations has shown potential in reprogramming root architectures, as discussed by Ferreira and Antunes, although specific examples related to root development were not detailed in their work (Ferreira and Antunes 2024). Additionally, as highlighted by Yokobayashi et al. (2002), the process of directed evolution has proven effective for solving complex biological design problems in synthetic biology, enabling the refinement of individual proteins and entire metabolic pathways through iterative testing of genetic modifications. These findings underscore the importance of identifying key genetic components at both the transcriptional and translational levels, which ML can facilitate by analyzing vast datasets related to gene expression and regulatory interactions. The capabilities of ML in identifying key regulators are evident in the classification of genetic parts critical for the construction of synthetic regulatory circuits. While the reference by Young et al. does not exist, Lu et al. (2020) emphasize that modern ML techniques can efficiently classify high-dimensional data, allowing for the identification of key genetic elements that significantly influence plant metabolic traits. Notably, various ML strategies, including ensemble models, have been shown to improve prediction accuracy; however, only references related to gene expression and regulatory interactions were found relevant for this statement and should be cited appropriately (Lu et al. 2020, 2021). Furthermore, the use of deep learning has emerged as a powerful tool in plant systems biology, as it allows the analysis of large and heterogeneous datasets often collected in high-throughput genotyping and phenotyping efforts. Singh et al. emphasize that deep learning applications have demonstrated efficacy in predicting plant stress responses by assimilating extensive data from multiple phenotypic measurements (Singh et al. 2018). Insights gained from such analyses facilitate a rational design of plants with enhanced attributes.

12.4 Challenges and Future Directions

Despite its transformative potential, several challenges hinder the full integration of ML in plant systems biology:

- **Data Quality and Availability**: High-quality, well-annotated datasets are crucial for training reliable ML models. However, biological datasets are often noisy, incomplete, or imbalanced, which can affect model performance.
- **Interpretability**: ML models, particularly deep learning models, are often seen as "black boxes," making it difficult to interpret their predictions. Developing interpretable models is essential for understanding biological mechanisms and guiding experimental validation.
- **Integration Across Scales**: Integrating data across different biological scales (e.g., from molecular to ecosystem level) remains challenging. Future work must focus on developing ML methods capable of handling multi-scale data to provide a holistic view of plant systems.
- **Cross-disciplinary Collaboration**: Progress in ML-assisted plant systems biology requires collaboration between biologists, computer scientists, and data scientists. Effective communication and interdisciplinary training are necessary to bridge gaps and develop solutions that are both biologically meaningful and computationally feasible.

The application of ML algorithms in plant systems biology holds tremendous promise, yet several challenges persist that inhibit its full integration. These challenges involve data quality and availability, model interpretability, multi-scale integration, and the need for cross-disciplinary collaboration. Addressing these issues is vital for the future directions of ML in this field. One of the primary obstacles is the quality and availability of biological data. High-quality, well-annotated datasets are essential for training robust ML models; however, biological datasets frequently suffer from issues such as noise, incompleteness, and imbalances across classes, which can adversely affect model performance (Dale et al. 2021). For instance, in the context of gene expression profiles, the dependency on high-throughput sequencing data can exacerbate these challenges, as these datasets are often fraught with variability and require considerable preprocessing to ensure reliability (Camacho et al. 2018). Novel methodologies that enhance dataset curation and integration are needed, as demonstrated by graph-linked embedding approaches in integrating multi-omics data (Cao and Gao 2022), which can facilitate the collection of more coherent datasets. Interpretability remains another significant hurdle in the application of ML within plant systems biology. Many advanced ML models, particularly deep learning frameworks, are often criticized for their "black box" nature, making it difficult for researchers to understand the biological implications of their predictions (Weiskittel et al. 2021). This lack of transparency can impede experimental validation, as it becomes challenging to correlate model outputs with underlying biological mechanisms (Seyed Tabib et al. 2020). To counter this, efforts towards developing more interpretable models and identifying biological insights from ML predictions are crucial. For example, research into "white box" models has shown promise, as they offer improved interpretability, allowing for better elucidation of the biological significance of the results. Integrating multi-scale biological data—ranging from molecular levels to whole-ecosystem dynamics—adds another layer of complexity. The challenge lies in creating ML methodologies capable of synthesizing and interpreting vast and varied data types effectively. Notably, the proposed systematic multi-omics integration approach seeks to facilitate this process by structuring integration at various levels, which may enable a more comprehensive understanding of plant systems biology. Developing ML methods that can manage such integration effectively will yield richer insights into plant biology and ecology. Cross-disciplinary collaboration is vital for overcoming these challenges and fostering effective integration of ML into plant systems biology. This field requires the joint efforts of biologists, computer scientists, and data scientists to streamline communication, methodologies, and tools (Auslander et al. 2021; Carbonell et al. 2019). Such collaboration can promote interdisciplinary training and knowledge exchange, necessary for addressing complex biological questions using ML (Jamil et al. 2020). Enhancing collaborative frameworks and encouraging the sharing of expertise will be pivotal as the utilization of ML continues to evolve within the context of plant systems biology.

12.5 Conclusion

ML is playing an increasingly vital role in plant systems biology, offering new avenues for understanding complex biological systems, predicting plant traits, and optimizing breeding programs. By integrating diverse data types and uncovering previously hidden biological patterns, ML is transforming how we study plant systems and unlocking new opportunities for crop improvement and sustainable agriculture. As ML algorithms continue to evolve, their application in plant systems biology will likely expand, offering more refined insights into plant–environment interactions and enabling the design of more resilient, productive, and sustainable crops.

The integration of ML technologies into plant systems biology heralds a transformative era in the study and management of crop systems. By facilitating the analysis of complex biological data, ML enables researchers to uncover intricate patterns and relationships among various biological variables. This capability is pivotal for advancing crop yield prediction and optimizing breeding programs through more refined trait identification and genetic analysis. Recent studies have demonstrated the potential of ML algorithms, such as Long Short-Term Memory (LSTM) networks, in accurately predicting crop types and assessing yield based on diverse datasets, including remote sensing data from Unmanned Aerial Vehicles. The capacity of these algorithms to classify and predict outcomes based on environmental and genetic variables has underscored their importance in developing strategic agricultural practices that enhance productivity and sustainability. The application of hybrid ML models, incorporating techniques like Particle Swarm Optimization, has also been highlighted as an effective way to improve prediction accuracy and model performance within agricultural contexts. The continuous advancement of ML algorithms is expected to yield deeper insights into plant–environment interactions, potentially allowing for the design of resilient crops better equipped to withstand climate change pressures. For instance, incorporating genomic data into ML workflows for crop improvement has enabled the identification of valuable alleles associated with stress tolerance, driving forward the development of high-yield, resilient crops. Furthermore, big data analytics powered by ML methodologies enhance farmers' capabilities by providing actionable insights that aid in decision-making regarding crop management and breeding strategies, thereby directly impacting food security and sustainability. As ML continues to evolve, its application in plant systems biology is poised to expand considerably, fostering further innovations in sustainable agriculture. This growth can lead to more intelligent and automated agricultural systems, promising substantial advancements in crop production efficiency while addressing the challenges posed by biotic and abiotic stresses. Thus, the synergy between ML and plant systems biology represents a fertile ground for future research and development, where interdisciplinary collaboration can pave the way for groundbreaking discoveries in agriculture.

References

Aghaei, K., and Komatsu, S. (2013). Crop and medicinal plants proteomics in response to salt stress. *Frontiers in Plant Science* 4 (JAN). https://doi.org/10.3389/FPLS.2013.00008

Aibar, S., González-Blas, C.B., Moerman, T. et al. (2017). SCENIC: Single-cell regulatory network inference and clustering. *Nature Methods* 14 (11): 1083–1086. https://doi.org/10.1038/NMETH.4463

Albert, E.T.A., Bille, N.H., Martin, B.J. et al. (2023). Integrating genetic markers and adiabatic quantum machine learning to improve disease resistance-based marker assisted plant selection. *Journal of Scientific Agriculture*: 63–76. https://doi.org/10.25081/JSA.2023.V7.8556

Al-Naggar, A.M.M., Shafik, M.M., and Elsheikh, M.O.A. (2018). The effects of genotype, soil water deficit and their interaction on agronomic, physiologic and yield traits of zea mays L. *Annual Research; Review in Biology* 29 (5): 1–17. https://doi.org/10.9734/ARRB/2018/45369

Alter, S., Bader, K.C., Spannagl, M. et al. (2015). DroughtDB: An expert-curated compilation of plant drought stress genes and their homologs in nine species. *Database* 2015. https://doi.org/10.1093/DATABASE/BAV046

An, J., Li, W., Li, M. et al. (2019). Identification and classification of maize drought stress using deep convolutional neural network. *Symmetry* 11 (2). https://doi.org/10.3390/SYM11020256

Araus, J.L., Kefauver, S.C., Vergara-Díaz, O. et al. (2022). Crop phenotyping in a context of global change: What to measure and how to do it. *Journal of Integrative Plant Biology* 64 (2): 592–618. https://doi.org/10.1111/JIPB.13191

Argelaguet, R., Velten, B., Arnol, D. et al. (2018). Multi-Omics Factor Analysis—A framework for unsupervised integration of multi-omics data sets. *Molecular Systems Biology* 14 (6). https://doi.org/10.15252/MSB.20178124

Auslander, N., Gussow, A.B., and Koonin, E.V. (2021). Incorporating machine learning into established bioinformatics frameworks. *International Journal of Molecular Sciences* 22 (6): 1–19. https://doi.org/10.3390/IJMS22062903

Azimi, S., Wadhawan, R., and Gandhi, T.K. (2021). Intelligent monitoring of stress induced by water deficiency in plants using deep learning. *IEEE Transactions on Instrumentation and Measurement* 70. https://doi.org/10.1109/TIM.2021.3111994

Banf, M., and Rhee, S.Y. (2017). Computational inference of gene regulatory networks: Approaches, limitations and opportunities. *Biochimica Et Biophysica Acta (BBA): Gene Regulatory Mechanisms* 1860 (1): 41–52. https://doi.org/10.1016/J.BBAGRM.2016.09.003

Benos, L., Tagarakis, A.C., Dolias, G. et al. (2021). Machine learning in agriculture: A comprehensive updated review. *Sensors* 21 (11). https://doi.org/10.3390/S21113758

Butte, S., Vakanski, A., Duellman, K. et al. (2021). Potato crop stress identification in aerial images using deep learning-based object detection. *Agronomy Journal* 113 (5), 3991–4002. https://doi.org/10.1002/agj2.20841

Camacho, D.M., Collins, K.M., Powers, R.K. et al. (2018). Next-generation machine learning for biological networks. *Cell* 173 (7): 1581–1592. https://doi.org/10.1016/J.CELL.2018.05.015

Cao, Z.J., and Gao, G. (2022). Multi-omics single-cell data integration and regulatory inference with graph-linked embedding. *Nature Biotechnology* 40 (10): 1458–1466. https://doi.org/10.1038/S41587-022-01284-4

Carbonell, P., Radivojevic, T., and García Martín, H. (2019). Opportunities at the intersection of synthetic biology, machine learning, and automation. *ACS Synthetic Biology* 8 (7): 1474–1477. https://doi.org/10.1021/ACSSYNBIO.8B00540

Chao, D., Wang, H., Wan, F. et al. (2025). MtCro: Multi-task deep learning framework improves multi-trait genomic prediction of crops. *Plant Methods* 21 (1): 1–14. https://doi.org/10.1186/S13007-024-01321-0/TABLES/3

Chen, L., Liu, G., and Zhang, T. (2024). Integrating machine learning and genome editing for crop improvement. *ABIOTECH* 5 (2): 262. https://doi.org/10.1007/S42994-023-00133-5

Chen, Z., Zhu, D., Wu, J. et al. (2018). Identification of differentially accumulated proteins involved in regulating independent and combined osmosis and cadmium stress response in *Brachypodium* seedling roots. *Scientific Reports* 8 (1). https://doi.org/10.1038/S41598-018-25959-8

Crossa, J., Montesinos-Lopez, O.A., Costa-Neto, G. et al. (2024). Machine learning algorithms translate big data into predictive breeding accuracy. *Trends in Plant Science* 30 (2): 167–184. https://doi.org/10.1016/J.TPLANTS.2024.09.011/ASSET/2CEB69A9-AE59-4012-A0F8-0A99983BE0FD/MAIN.ASSETS/GR4.JPG

Cui, L., Guo, G., Ng, M.K. et al. (2024). GSTRPCA: irregular tensor singular value decomposition for single-cell multi-omics data clustering. *Briefings in Bioinformatics* 26 (1). https://doi.org/10.1093/BIB/BBAE649

Dale, R., Oswald, S., Jalihal, A.P. et al. (2021). Overcoming the challenges to enhancing experimental plant biology with computational modeling. *Frontiers in Plant Science* 12: 1266. https://doi.org/10.31219/OSF.IO/JFDB9

Dong, Y., Chen, S., and Zhang, J. (2025). Learnable group transform: enhancing genotype-to-phenotype prediction for rice breeding with small, structured datasets. *ArXiv*. https://arxiv.org/abs/2503.11180v1

Dutta, S., Islam, Z., Das, S. et al. (2025). Harmonizing plant resilience: unveiling the symphony of membrane lipid dynamics in response to abiotic stresses: A review. *Discover Plants* 2 (1): 61. https://doi.org/10.1007/s44372-025-00152-0

Dutta, S., Zunjare, R.U., Sil, A. et al. (2024). Prediction of matrilineal specific patatin-like protein governing in-vivo maternal haploid induction in maize using support vector machine and di-peptide composition. *Amino Acids* 56 (1): 20. https://doi.org/10.1007/s00726-023-03368-0

Eladl, S.G., Haikal, A.Y., Saafan, M.M. et al. (2024). A proposed plant classification framework for smart agricultural applications using UAV images and artificial intelligence techniques. *Alexandria Engineering Journal* 109: 466–481. https://doi.org/10.1016/J.AEJ.2024.08.076

Farooq, M.A., Gao, S., Hassan, M.A. et al. (2024). Artificial intelligence in plant breeding. *Trends in Genetics*. https://doi.org/10.1016/J.TIG.2024.07.001

Ferreira, S.S., and Antunes, M.S. (2024). Genetically encoded Boolean logic operators to sense and integrate phenylpropanoid metabolite levels in plants. *New Phytologist* 243 (2): 674–687. https://doi.org/10.1111/NPH.19823

Fisher, L.H.C., Han, J., Corke, F.M.K. et al. (2016). Linking dynamic phenotyping with metabolite analysis to study natural variation in drought responses of *Brachypodium distachyon*. *Frontiers in Plant Science* 7 (November 2016). https://doi.org/10.3389/FPLS.2016.01751

Gain, N., Chhabra, R., Muthusamy, V. et al. (2025). Molecular delineation and haplotype analysis of domain membrane protein (DMP) gene influencing in-vivo haploid induction in maize. *Scientific Reports* 15 (1): 7697. https://doi.org/10.1038/s41598-025-91031-x

Gayoso, A., Steier, Z., Lopez, R. et al. (2021). Joint probabilistic modeling of single-cell multi-omic data with totalVI. *Nature Methods* 18 (3): 272–282. https://doi.org/10.1038/S41592-020-01050-X

Gechev, T., Lyall, R., Petrov, V. et al. (2021). Systems biology of resurrection plants. *Cellular and Molecular Life Sciences* 78 (19–20): 6365–6394. https://doi.org/10.1007/S00018-021-03913-8

Gong, F., Hu, X., and Wang, W. (2015). Proteomic analysis of crop plants under abiotic stress conditions: where to focus our research? *Frontiers in Plant Science* 6 (June). https://doi.org/10.3389/FPLS.2015.00418

González-Blas, C.B., Winter, S. De, Hulselmans, G. et al. (2022). SCENIC+: Single cell multiomic inference of enhancers and gene regulatory networks. *Nature Methods* 20 (9): 1355–1367. https://doi.org/10.1101/2022.08.19.504505

Guo, M., Liu, J.H., Ma, X. et al. (2016). The plant heat stress transcription factors (HSFs): Structure, regulation, and function in response to abiotic stresses. *Frontiers in Plant Science* 7 (FEB 2016). https://doi.org/10.3389/FPLS.2016.00114

Guo, X., Qiu, Y., Nettleton, D. et al. (2021). KAT4IA: K-means assisted training for image analysis of field-grown plant phenotypes. *Plant Phenomics* 2021. https://doi.org/10.34133/2021/9805489

Han, X., Wu, H., Zhu, H. et al. (2021). Joint power and channel optimization of agricultural wireless sensor networks based on hybrid deep reinforcement learning. *Processes* 9 (11). https://doi.org/10.3390/PR9111919

Hesami, M., Alizadeh, M., Jones, A.M.P. et al. (2022). Machine learning: Its challenges and opportunities in plant system biology. *Applied Microbiology and Biotechnology* 106 (9): 3507–3530. https://doi.org/10.1007/s00253-022-11963-6

Homan, D., and Preez, J.A. du. (n.d.). *Automated Feature-Specific Tree Species Identification from Natural Images using Deep Semi-Supervised Learning*. https://doi.org/10.48550/ARXIV.2110.03994

Hrdlickova, R., Toloue, M., and Tian, B. (2017). RNA-Seq methods for transcriptome analysis. *Wiley Interdisciplinary Reviews: RNA* 8 (1): e1364. https://doi.org/10.1002/wrna.1364

Hu, H., Scheben, A., and Edwards, D. (2018). Advances in integrating genomics and bioinformatics in the plant breeding pipeline. *Agriculture* 8 (6). https://doi.org/10.3390/AGRICULTURE8060075

Huckvale, E.D., and Moseley, H.N.B. (2024). *Predicting The Pathway Involvement of Metabolites Based on Combined Metabolite and Pathway Features*. https://doi.org/10.1101/2024.04.01.587582

Isewon, I., Apata, O., Oluwamuyiwa, F. et al. (2022). Machine learning algorithms: Their applications in plant omics and agronomic traits' improvement. *F1000Research* 11: 1256. https://doi.org/10.12688/f1000research.125425.1

Jamil, I.N., Remali, J., Azizan, K.A. et al. (2020). Systematic multi-omics integration (MOI) approach in plant systems biology. *Frontiers in Plant Science* 11. https://doi.org/10.3389/FPLS.2020.00944

Javaid, A., Razzaq, H., Khan, F.A. et al. (2023). Evaluation of maize accessions for drought tolerance through principal component analysis. *SABRAO Journal of Breeding and Genetics* 52 (2): 476–484. https://doi.org/10.54910/SABRAO2023.55.2.19

Kakar, N., Jumaa, S.H., Redoña, E.D. et al. (2019). Evaluating rice for salinity using pot-culture provides a systematic tolerance assessment at the seedling stage. *Rice* 12 (1). https://doi.org/10.1186/S12284-019-0317-7

Kefato, Z.T., and Girdzijauskas, Š. (2021). *Self-supervised graph neural networks without explicit negative sampling. ArXiv*, https://doi.org/10.48550/ARXIV.2103.14958

Kemmo Tsafack, U., Ahn, K.W., Kwitek, A.E. et al. (2024). Meta-analytic gene-clustering algorithm for integrating multi-omics and multi-study data. *Bioengineering* 11 (6). https://doi.org/10.3390/BIOENGINEERING11060587

Khaki, S., and Wang, L. (2019). Crop yield prediction using deep neural networks. *Frontiers in Plant Science* 10. https://doi.org/10.3389/fpls.2019.00621

Kobak, D., and Berens, P. (2019). The art of using t-SNE for single-cell transcriptomics. *Nature Communications* 10 (1). https://doi.org/10.1038/S41467-019-13056-X

Krannich, C.T., Maletzki, L., Kurowsky, C. et al. (2015). Network candidate genes in breeding for drought tolerant crops. *International Journal of Molecular Sciences* 16 (7): 16378–16400. https://doi.org/10.3390/IJMS160716378

Kumar, P., Paul, D., Jhajhriya, S. et al. (2024). Understanding heat-shock proteins' abundance and pivotal function under multiple abiotic stresses. *Journal of Plant Biochemistry and Biotechnology*: 1–22. https://doi.org/10.1007/s13562-024-00932-x

Lee, K.H., Samsuzzaman, Reza, M.N., Islam, S. et al. (2024). Evaluation of machine learning models for stress symptom classification of cucumber seedlings grown in a controlled environment. *Agronomy* 15 (1): 90. https://doi.org/10.3390/AGRONOMY15010090

Leng, D., Zheng, L., Wen, Y. et al. (2022). A benchmark study of deep learning-based multi-omics data fusion methods for cancer. *Genome Biology* 23 (1). https://doi.org/10.1186/S13059-022-02739-2

Lewis, J.E., and Kemp, M.L. (2021). Integration of machine learning and genome-scale metabolic modeling identifies multi-omics biomarkers for radiation resistance. *Nature Communications* 12 (1). https://doi.org/10.1038/S41467-021-22989-1

Li, H., Liu, Y., Wu, J. et al. (2024a). Overexpression of *GhGSTF9* enhances salt stress tolerance in transgenic *Arabidopsis*. *Genes* 15 (6). https://doi.org/10.3390/GENES15060695

Li, Y., Tao, F., Hao, Y. et al. (2024b). Identifying high-yielding and drought-tolerant wheat cultivars based on ideotypic traits and yield responses to stress. *Journal of Agronomy and Crop Science* 210 (4). https://doi.org/10.1111/JAC.12738

Li, Y.F., Zheng, Y., Vemireddy, L.R. et al. (2018). Comparative transcriptome and translatome analysis in contrasting rice genotypes reveals differential mRNA translation in salt-tolerant Pokkali under salt stress. *BMC Genomics* 19 (S10). https://doi.org/10.1186/S12864-018-5279-4

Lin, X., Tian, T., Wei, Z. et al. (2022). Clustering of single-cell multi-omics data with a multimodal deep learning method. *Nature Communications* 13 (1). https://doi.org/10.1038/S41467-022-35031-9

Lu, H., Yuan, G., Strauss, S.H. et al. (2020). Reconfiguring plant metabolism for biodegradable plastic production. *BioDesign Research* 2020. https://doi.org/10.34133/2020/9078303

Lu, Y., Ni, Y., Wang, Q. et al. (2021). Effectiveness of sexual health influencers identified by an ensemble machine learning model in promoting secondary distribution of HIV self-testing among men who have sex with men in China: study protocol for a quasi-experimental trial. *BMC Public Health* 21 (1). https://doi.org/10.1186/S12889-021-11817-2

Lv, D.W., Subburaj, S., Cao, M. et al. (2014). Proteome and phosphoproteome characterization reveals new response and defense mechanisms of *Brachypodium distachyon* leaves under salt stress. *Molecular & Cellular Proteomics* 13 (2): 632–652. https://doi.org/10.1074/MCP.M113.030171

Maazou, A.S., Tu, J., Qiu, J. et al. (2016). Breeding for drought tolerance in maize (*Zea mays* L.). *American Journal of Plant Sciences* 07 (14): 1858–1870. https://doi.org/10.4236/AJPS.2016.714172

MacNish, T.R., Danilevicz, M.F., Bayer, P.E. et al. (2025). Application of machine learning and genomics for orphan crop improvement. *Nature Communications* 16 (1): 1–10. https://doi.org/10.1038/s41467-025-56330-x

Marbach, D., Lamparter, D., Quon, G. et al. (2016). Tissue-specific regulatory circuits reveal variable modular perturbations across complex diseases. *Nature Methods* 13 (4): 366–370. https://doi.org/10.1038/NMETH.3799

Maruenda, H. and Toubiana, D. (2025). Correlation-based network analysis combined with machine learning techniques highlights rosmarinic acid biosynthesis activation in tuber of potato plants corresponding to water recovery treatment. *Potato Research*. https://doi.org/10.21203/RS.3.RS-3388393/V1

Merzbacher, C., Aodha, O. Mac, and Oyarzún, D.A. (2025). Modelling dynamic host-pathway interactions at the genome scale with machine learning. *Metabolic Engineering*. https://doi.org/10.1101/2025.04.09.588720

Mishra, B., Kumar, N., and Mukhtar, M.S. (2019). Systems biology and machine learning in plant–pathogen interactions. *Molecular Plant-Microbe Interactions* 32 (1), 45–55. https://doi.org/10.1094/MPMI-08-18-0221-FI/ASSET/IMAGES/LARGE/MPMI-08-18-0221-FI_F4-1544772603415.JPEG

Mohammadi, P., and Asefpour Vakilian, K. (2023). Machine learning provides specific detection of salt and drought stresses in cucumber based on miRNA characteristics. *Plant Methods* 19 (1): 1–12. https://doi.org/10.1186/S13007-023-01095-X/FIGURES/4

Morsy, S.M., Elbasyoni, I.S., Abdallah, A.M. et al. (2022). Imposing water deficit on modern and wild wheat collections to identify drought-resilient genotypes. *Journal of Agronomy and Crop Science* 208 (4): 427–440. https://doi.org/10.1111/JAC.12493

Nguyen, N.D., and Wang, D. (2020). Multiview learning for understanding functional multiomics. *PLOS Computational Biology* 16 (4). https://doi.org/10.1371/JOURNAL.PCBI.1007677

Ni, Y., Aghamirzaie, D., Elmarakeby, H. et al. (2016). A machine learning approach to predict gene regulatory networks in seed development in *Arabidopsis*. *Frontiers in Plant Science* 7 (December 2016): 1936. https://doi.org/10.3389/FPLS.2016.01936/FULL

Ogidi, F.C., Eramian, M.G., and Stavness, I. (2023). Benchmarking self-supervised contrastive learning methods for image-based plant phenotyping. *Plant Phenomics* 5. https://doi.org/10.34133/PLANTPHENOMICS.0037

Pan, X., Chen, X., Zhang, Q. et al. (2022). Model predictive control: A reinforcement learning-based approach. *Journal of Physics: Conference Series* 2203 (1). https://doi.org/10.1088/1742-6596/2203/1/012058

Parmley, K., Nagasubramanian, K., Sarkar, S. et al. (2019). Development of optimized phenomic predictors for efficient plant breeding decisions using phenomic-assisted selection in soybean. *Plant Phenomics* 2019. https://doi.org/10.34133/2019/5809404

Parmley, K.A., Higgins, R.H., Ganapathysubramanian, B. et al. (2019). Machine learning approach for prescriptive plant breeding. *Scientific Reports* 9 (1). https://doi.org/10.1038/S41598-019-53451-4

Pazhamala, L.T., Kudapa, H., Weckwerth, W. et al. (2021). Systems biology for crop improvement. *The Plant Genome* 14 (2). https://doi.org/10.1002/TPG2.20098

Powadi, A., Jubery, T.Z., Tross, M.C. et al. (2024). Disentangling genotype and environment specific latent features for improved trait prediction using a compositional autoencoder. *ArXiv*, https://arxiv.org/abs/2410.19922v1

Pradhan, U.K., Mahapatra, A., Naha, S. et al. (2024). ASPTF: A computational tool to predict abiotic stress-responsive transcription factors in plants by employing machine learning algorithms. *Biochimica et Biophysica Acta (BBA): General Subjects* 1868 (6): 130597. https://doi.org/10.1016/J.BBAGEN.2024.130597

Ps, S., Mithra, S.V., Prakash, C. et al. (2017). High resolution mapping of QTLs for heat tolerance in rice using a 5K SNP array. *Rice* 10 (1): 28. https://doi.org/10.1186/S12284-017-0167-0

Putra, G., Sutiarso, L., Nugroho, A. et al. (2023). Plant growth prediction model of red chili (*Capsicum annuum L.*) by different manipulation environment. *Advances in Biological Sciences Research* 26. https://doi.org/10.2991/978-94-6463-086-2_4

Qiu, Y., Guo, D., Zhao, P. et al. (2024). scMNMF: A novel method for single-cell multi-omics clustering based on matrix factorization. *Briefings in Bioinformatics* 25 (3). https://doi.org/10.1093/BIB/BBAE228

Raja, P., Olenskyj, A., Kamangir, H. et al. (2021). Simultaneously predicting multiple plant traits from multiple sensors via deformable CNN regression. *Arxiv*, https://arxiv.org/abs/2112.03205v1

Rappoport, N., Safra, R., and Shamir, R. (2020). MONET: Multi-omic module discovery by omic selection. *PLOS Computational Biology* 16 (9). https://doi.org/10.1371/JOURNAL.PCBI.1008182

Robertson, S.M., and Wilkins, O. (2023). Spatially resolved gene regulatory networks in Asian rice (*Oryza sativa* cv. Nipponbare) leaves. *The Plant Journal* 116 (1): 269–281. https://doi.org/10.1111/TPJ.16375

Rodriguez, A., Vrancx, P., Nowé, A. et al. (2013). Model-free learning of wire winding control. *Proceedings of the 2013 9th Asian Control Conference (ASCC), 1–6*. https://doi.org/10.1109/ASCC.2013.6606290

Rodriguez, D.F.C., Urban, M.O., Santaella, M. et al. (2022). Using phenomics to identify and integrate traits of interest for better-performing common beans: A validation study on an interspecific hybrid and its Acutifolii parents. *Frontiers in Plant Science* 13. https://doi.org/10.3389/FPLS.2022.1008666

Rodziewicz, P., Swarcewicz, B., Chmielewska, K. et al. (2014). Influence of abiotic stresses on plant proteome and metabolome changes. *Acta Physiologiae Plantarum* 36 (1): 1–19. https://doi.org/10.1007/S11738-013-1402-Y

Roy, A., Dutta, S., Das, S. et al. 2024. Next-generation sequencing in the development of climate-resilient and stress-responsive crops: A review. *The Open Biotechnology Journal* 18 (1). https://doi.org/10.2174/0118740707301657240517063244

Roy, A.M., Bose, R., and Bhaduri, J. (2021). A fast accurate fine-grain object detection model based on YOLOv4 deep neural network. *Neural Computing and Applications* 34 (5): 3895–3921. https://doi.org/10.1007/s00521-021-06651-x

Ruslan, Suhardiyanto, H., and Supriyanto. (2024). An artificial neural network model for predicting the growth of lettuce crops in hydroponic system with root zone cooling. *IOP Conference Series: Earth and Environmental Science* 1359 (1). https://doi.org/10.1088/1755-1315/1359/1/012038

Seeli, F.D.P., Manonmani, S., Raveendran, M. et al. (2021). Development of drought tolerant rice breeding lines of an elite Indian rice (*Oryza sativa* L.) variety improved white Ponni through molecular breeding. *International Journal of Plant and Soil Science*: 447–457. https://doi.org/10.9734/IJPSS/2021/V33I2430800

Seyed Tabib, N.S., Madgwick, M., Sudhakar, P. et al. (2020). Big data in IBD: Big progress for clinical practice. *Gut* 69 (8): 1520–1532. https://doi.org/10.1136/GUTJNL-2019-320065

Shahid, M.F., Khanzada, T.J.S., Aslam, M.A. et al. (2024). An ensemble deep learning models approach using image analysis for cotton crop classification in AI-enabled smart agriculture. *Plant Methods* 20 (1): 1–22. https://doi.org/10.1186/S13007-024-01228-W/FIGURES/18

Shi, Y., Han, L., González-Moreno, P. et al. (2023). A fast Fourier convolutional deep neural network for accurate and explainable discrimination of wheat yellow rust and nitrogen deficiency from Sentinel-2 time-series data. *Arxiv*, https://arxiv.org/abs/2306.17207v1

Shih, P.M., Vuu, K., Mansoori, N. et al. (2016). A robust gene-stacking method utilizing yeast assembly for plant synthetic biology. *Nature Communications* 7 (1). https://doi.org/10.1038/NCOMMS13215

Singh, A.K., Ganapathysubramanian, B., and Sarkar, S. (2018). Deep learning for plant stress phenotyping: Trends and future perspectives. *Trends in Plant Science* 23 (10): 883–898. https://doi.org/10.1016/J.TPLANTS.2018.07.004

Soren, K.R., Madugula, P., Kumar, N. et al. (2020). Genetic dissection and identification of candidate genes for salinity tolerance using Axiom®CicerSNP array in chickpea. *International Journal of Molecular Sciences* 21 (14): 1–17. https://doi.org/10.3390/IJMS21145058

Sudhakar, C.V., and Reddy, G. (2023). Multi-descriptor enabled leaf disease detection using machine learning methods. *International Journal for Multidisciplinary Research* 5 (3). https://doi.org/10.36948/IJFMR.2023.V05I03.4008

Suvarna, Y.R., Sagar, S., Kuchanur, P.H. et al. (2024). Plant proteome in response to abiotic stresses. In: *Abiotic Stress in Crop Plants: Ecophysiological Responses and Molecular Approaches. Intech Open*. https://doi.org/10.5772/INTECHOPEN.114297

Tafur Rangel, A.E., Ríos, W., Mejía, D. et al. (2021). In silico design for systems-based metabolic engineering for the bioconversion of valuable compounds from industrial by-products. *Frontiers in Genetics* 12: 633073. https://doi.org/10.3389/FGENE.2021.633073

Tardieu, F., Cabrera-Bosquet, L., Pridmore, T. et al. (2017). Plant phenomics, from sensors to knowledge. *Current Biology* 27 (15): R770–R783. https://doi.org/10.1016/J.CUB.2017.05.055

Thakur, P.S., Khanna, P., Sheorey, T. et al. (2022). Explainable vision transformer enabled convolutional neural network for plant disease identification: PlantXViT. *ArXiv*, https://arxiv.org/abs/2207.07919v1

Tian, H., and Tao, P. (2020). ivis Dimensionality Reduction Framework for Biomacromolecular Simulations. *Journal of Chemical Information and Modeling* 60 (10): 4569–4581. https://doi.org/10.1021/ACS.JCIM.0C00485

Tomar, R.S.S., Tiwari, S., Vinod, Naik, B.K. et al. (2016). Molecular and morpho-agronomical characterization of root architecture at seedling and reproductive stages for drought tolerance in wheat. *Plos One* 11 (6). https://doi.org/10.1371/JOURNAL.PONE.0156528

Toubiana, D., Puzis, R., Wen, L. et al. (2019). Combined network analysis and machine learning allows the prediction of metabolic pathways from tomato metabolomics data. *Communications Biology* 2 (1). https://doi.org/10.1038/S42003-019-0440-4

Trosten, D.J., Løkse, S., Jenssen, R. et al. (2021). Reconsidering representation alignment for multi-view clustering. In: *Proceedings of the 2021 IEEE/CVF Conference on Computer Vision and Pattern Recognition (CVPR)*, 1255–1265. IEEE. https://doi.org/10.1109/CVPR46437.2021.00131

Tucci, A., Flores-Vergara, M.A., and Franks, R.G. (2024). Machine learning inference of gene regulatory networks in developing mimulus seeds. *Plants* 13 (23): 3297. https://doi.org/10.3390/PLANTS13233297/S1

Van den Broeck, L., Gordon, M., Inzé, D. et al. (2020). Gene regulatory network inference: Connecting plant biology and mathematical modeling. *Frontiers in Genetics* 11. https://doi.org/10.3389/FGENE.2020.00457

van Dijk, A.D.J., Kootstra, G., Kruijer, W. et al. (2021). Machine learning in plant science and plant breeding. *iScience* 24 (1). https://doi.org/10.1016/J.ISCI.2020.101890

van Dijk, D., Sharma, R., Nainys, J. et al. (2018). Recovering gene interactions from single-cell data using data diffusion. *Cell* 174 (3): 716–729.e27. https://doi.org/10.1016/J.CELL.2018.05.061

Wang, H., Gu, L., Zhang, X. et al. (2018). Global transcriptome and weighted gene co-expression network analyses reveal hybrid-specific modules and candidate genes related to plant height development in maize. *Plant Molecular Biology* 98 (3): 187–203. https://doi.org/10.1007/S11103-018-0763-4

Wang, Y., Wang, K., An, T. et al. (2022). Genetic dissection of branch architecture in oilseed rape (*Brassica napus* L.) germplasm. *Frontiers in Plant Science* 13. https://doi.org/10.3389/FPLS.2022.1053459

Weiskittel, T.M., Correia, C., Yu, G.T. et al. (2021). The trifecta of single-cell, systems-biology, and machine-learning Approaches. *Genes* 12 (7). https://doi.org/10.3390/GENES12071098

Williamson, H.F., Brettschneider, J., Caccamo, M. et al. (2023). Data management challenges for artificial intelligence in plant and agricultural research. *F1000Research* 10. https://doi.org/10.12688/F1000RESEARCH.52204.2

Xavier, A. (2021). Technical nuances of machine learning: implementation and validation of supervised methods for genomic prediction in plant breeding. *Crop Breeding and Applied Biotechnology* 21 (spe). https://doi.org/10.1590/1984-70332021V21SA15

Xu, C., and Jackson, S.A. (2019). Machine learning and complex biological data. *Genome Biology* 20 (1). https://doi.org/10.1186/S13059-019-1689-0

Yan, J., and Wang, X. (2022). Unsupervised and semi-supervised learning: The next frontier in machine learning for plant systems biology. *The Plant Journal* 111 (6): 1527–1538. https://doi.org/10.1111/TPJ.15905

Yang, H., Chen, R., Li, D. et al. (2021). Subtype-GAN: A deep learning approach for integrative cancer subtyping of multi-omics data. *Bioinformatics* 37 (16): 2231–2237. https://doi.org/10.1093/BIOINFORMATICS/BTAB109

Yang, Z., Liu, J., Wang, Z. et al. (2020). Multi-class metabolic pathway prediction by graph attention-based deep learning method. *Proceedings of the 2020 IEEE International Conference on Bioinformatics and Biomedicine (BIBM)*, 126–131. https://doi.org/10.1109/BIBM49941.2020.9313298

Yokobayashi, Y., Weiss, R., and Arnold, F.H. (2002). Directed evolution of a genetic circuit. *Proceedings of the National Academy of Sciences* 99 (26): 16587–16591. https://doi.org/10.1073/PNAS.252535999

Yoosefzadeh Najafabadi, M., Hesami, M., and Eskandari, M. (2023). Machine learning-assisted approaches in modernized plant breeding programs. *Genes* 14 (4): 777. https://doi.org/10.3390/GENES14040777

Yuan, M., Chen, L., and Deng, M. (2023). Clustering single-cell multi-omics data with MoClust. *Bioinformatics* 39 (1). https://doi.org/10.1093/BIOINFORMATICS/BTAC736

Zhang, J., Petersen, S.D., Radivojevic, T. et al. (2020). Combining mechanistic and machine learning models for predictive engineering and optimization of tryptophan metabolism. *Nature Communications* 11 (1). https://doi.org/10.1038/S41467-020-17910-1

Zhang, J., Zhang, S., Cheng, M. et al. (2018). Effect of drought on agronomic traits of rice and wheat: A meta-analysis. *International Journal of Environmental Research and Public Health* 15 (5). https://doi.org/10.3390/IJERPH15050839

Zhang, T., Zhang, H., Ren, J. et al. (2024). scDRMAE: Integrating masked autoencoder with residual attention networks to leverage omics feature dependencies for accurate cell clustering. *Bioinformatics* 40 (10). https://doi.org/10.1093/BIOINFORMATICS/BTAE599

Zhao, J., Zhao, B., Song, X. et al. (2023). Subtype-DCC: Decoupled contrastive clustering method for cancer subtype identification based on multi-omics data. *Briefings in Bioinformatics* 24 (2). https://doi.org/10.1093/BIB/BBAD025

Zheng, X., Chen, L., Xia, H. et al. (2017). Transgenerational epimutations induced by multi-generation drought imposition mediate rice plant's adaptation to drought condition. *Scientific Reports* 7 (1). https://doi.org/10.1038/SREP39843

13

Machine Learning-Driven Precision Plant Breeding

*Krishna Kumar Rai**

Department of Biotechnology, Amity School of Biological Sciences, Amity University Punjab, Mohali, Punjab, India

*Corresponding author: raikrishna16@gmail.com; kkrai@pb.amity.edu

13.1 Introduction

The conjunction of expanding global populations, rapid urbanization, and escalating climate instability has triggered a crisis in agriculture, threatening humanity's ability to sustainably nourish its people (Das et al. 2024). By 2050, the global population is projected to exceed 9.7 billion, necessitating a 60–70% increase in food production (Bren d'Amour et al. 2017). However, urbanization and land degradation are diminishing arable land at alarming rates up to 12 million hectares annually while climate change exacerbates droughts, floods, and temperature extremes, destabilizing agricultural systems (Bren d'Amour et al. 2017; Reynolds et al. 2020). These challenges demand urgent innovation in crop breeding to enhance yield resilience and resource efficiency. Plant performance is governed by complex genotype–environment (G×E) interactions, where genetic potential is modulated by external conditions to produce observable traits, or phenotypes (Xiong et al. 2020). For instance, a drought-tolerant maize genotype may thrive in arid regions but underperform in well-watered environments, highlighting the context-dependent nature of genetic traits. Climate change intensifies these interactions, altering pest dynamics, soil health, and growing seasons (Tesfaye et al. 2018). Rising temperatures alone could reduce global wheat yields by 6% per 1 °C increase, jeopardizing staple crops (Xiong et al. 2020). Understanding these dynamics is critical for developing climate-resilient crops.

Traditional breeding, the backbone of agricultural progress for millennia, relies on crossing plants with desirable traits and selecting superior offspring over generations. This approach underpinned the Green Revolution, which doubled global cereal production between 1960 and 2000 through semi-dwarf wheat and rice varieties (Chawla et al. 2023). Despite its successes, conventional breeding has its own limitations. Identifying beneficial quantitative trait loci (QTLs) genomic regions associated with traits like yield or disease resistance is labor-intensive, requiring years of field trials (Behera et al. 2020). Moreover, "linkage drag," where undesirable genes near target QTLs are inadvertently inherited, can compromise crop quality (Behera et al. 2020). Breeding cycles spanning 7–9 years further slow progress, lagging behind the accelerating pace of environmental change (Lamichhane and Thapa 2022). The advent of molecular markers in the late 20th century marked a paradigm shift. These DNA sequences, such as single-nucleotide polymorphisms (SNPs) or microsatellites, serve as genetic landmarks/signposts, enabling breeders to screen seedlings for target traits within weeks rather than years (Amiteye 2021). Marker-assisted selection (MAS) allows precise introgression of traits like blast resistance in rice or beta-carotene enrichment in maize without lengthy phenotypic evaluations. Genome-wide association studies (GWAS) further accelerated trait discovery by linking genomic diversity to agronomic performance across diverse populations, revealing loci for drought tolerance in sorghum and nutrient efficiency in soybean (Bohra et al. 2020).

The integration of genomics with distinct "omics" disciplines perse., transcriptomics, metabolomics, and phenomics has unlocked functional insights into plant biology (Huang 2024). High-throughput phenotyping platforms, using drones and sensors, quantify traits like canopy temperature or photosynthetic efficiency across thousands of plants daily, bridging the gap between genotype and phenotype (Arya et al. 2022). Genomic-enabled breeding, powered by next-generation

Machine Learning for Plant Biology, First Edition. Edited by Jen-Tsung Chen.

sequencing, enables the assembly of haplotype maps to prioritize optimal crosses. For example, the 3000 Rice Genomes Project identified novel alleles for grain quality and yield, facilitating targeted breeding (Yang et al. 2025). Such integration has slashed breeding cycles by 30–50%, a critical advance in the race against climate change. Speed breeding, which uses controlled environments with extended photo periods and LED lighting, enables up to six generations of wheat annually as compared to two generations in the field (Varshney et al. 2021). Coupled with CRISPR-Cas9 gene editing, breeders can now engineer disease resistance or nutrient profiles with the utmost precision thus reducing linkage drag (Pandey et al. 2022). Furthermore, artificial intelligence (AI) coupled with machine learning (ML) models trained on omics datasets predict trait inheritance, optimize crossing schemes, and even design "synthetic" genomes tailored to specific environments (Rai 2022). For instance, IBM's Watson Decision Platform for Agriculture integrates weather data and satellite imagery to recommend planting strategies, enhancing climate adaptation (Zhai et al. 2020).

This chapter explores the significance of traditional agricultural methods and their integration with molecular tools, omics technologies, and AI-driven platforms, which are revolutionizing agricultural growth and productivity. Additionally, it provides a comprehensive understanding of advanced breeding tools that can empower smallholder farmers in climate-vulnerable regions. Finally, this chapter critically discusses how the fusion of biology and technology can be used to ensure agricultural sustainability.

13.2 Conventional Breeding Approaches

Conventional plant breeding has shaped the genetic architecture of modern crops through millennia of natural selection (Borlaug 1983). Early farmers domesticated wild species by preferentially propagating plants with desirable traits, larger grains, reduced seed shattering, or drought resilience laying the foundation for systematic crop improvement (Caruso et al. 2020). For instance, teosinte, a wild grass with tiny, inedible kernels, was transformed into modern maize through iterative selection over 9000 years, indicating the potential for genetic advancement (Caruso et al. 2020). In the modern era, conventional breeding has been used to hybridize wild relatives and cultivated varieties to introduce traits such as disease resistance, nutritional quality, and abiotic stress tolerance (Salonia et al. 2020). This approach underpinned landmark achievements like the Green Revolution, where semi-dwarf wheat and rice varieties, bred through targeted crosses, averted famine for millions (Acquaah 2015). Despite the rise of molecular techniques, conventional methods remain indispensable, particularly in resource-limited regions, due to their cost-effectiveness and reliance on phenotypic observation (Table 13.1) (Smith 2021). Self-pollinating species, such as wheat, rice, and barley, present unique opportunities and constraints for breeders. These crops naturally inbreed, resulting in homozygous lines with stable trait expression across generations (Begna 2021). Three primary strategies dominate their improvement such as (i) Mass selection, the oldest method, involves selecting superior plants from a heterogeneous population and bulking their seeds to form the next generation which is effective for improving seed size or uniformity in landraces (Labroo et al. 2021). (ii) Pedigree breeding, by contrast, tracks the lineage of individual plants over multiple generations, enabling precise selection for complex traits such as yield stability or disease resistance. For example, the development of the high-yielding rice variety IR8 involved meticulous pedigree selection to combine dwarfing genes with pest resistance (Acquaah 2015). (iii) Single seed descent (SSD), a more recent innovation, accelerates homozygosity by advancing generations via single-seed propagation under controlled conditions, bypassing the need for laborious field evaluations at early stages (Labroo et al. 2021). These methods are often integrated with molecular tools; for instance, marker-assisted backcrossing has refined pedigree breeding by reducing linkage drag when introgressing traits from wild relatives into elite cultivars (Lema 2018).

Cross-pollinated crops, such as maize, alfalfa, and rye, demand distinct approaches due to their heterozygosity and genetic diversity (Saxena et al. 2025). Recurrent selection, i.e., repeated cycles of crossing selected individuals and evaluating progeny, is a cornerstone for improving polygenic traits like yield or forage quality (Lamichhane and Thapa 2022). In alfalfa, maternal selection, a form of recurrent selection, enhanced dry matter yield by 12–15% over four cycles by prioritizing families with superior combining ability (Annicchiarico et al. 2021). Mutation breeding, another key strategy, induces genetic variation through chemical mutagens (e.g., ethyl methanesulfonate) or radiation, followed by large-scale screening for beneficial mutants. This approach identified blast-resistant wheat mutants in Ethiopia, where irradiation of seeds yielded lines with durable resistance to *Magnaporthe oryzae* without compromising grain quality (Goddard et al. 2020). Similarly, colchicine-induced polyploidy in watermelon created seedless triploid varieties, revolutionizing consumer markets. However, mutation breeding's randomness necessitates screening thousands of individuals, a resource-intensive process mitigated by modern phenomics platforms.

Table 13.1 Recent studies involving conventional and molecular breeding techniques in crop improvement.

Techniques	Crops	Mapping Population	Type of Mapping	DNA Markers Used	Improved Traits	References
Conventional breeding						
Phenotypic selection	*Solanum melongena*	F1-F2	-	-	Fruit quality and yield	Solieman et al. (2012)
	Fagopyrum esculentum Moenc	F4-F5	Linkage mapping	Genome wide marker	Growth and yield	Yabe et al. (2018)
	Picea glauca	F1-F2	Linkage mapping	SNP	Agronomic traits	Lenz et al. (2020)
Inbred line selection	*Sesamum indicum*	F1, F2, F3	-	-	Oil quality and yield	Baydar et al. (2008)
	Noccaea caerulescens	F1, F2, F3	-	-	Phytoremediation	Sterckeman et al. (2019)
	Zea mays L.	F8	QTL mapping	SSR	Yield and agronomic traits	Mukri et al. (2022)
Pedigree method	*Dioscorea* spp.	F2-F3	-	-	Tuber quality and yield	Asfaw et al. (2020)
	Vicia faba	F2-F6	-	-	Yield and agronomic traits	Arya et al. (2022)
	Vigna unguiculata	Mutant populations M3-M4	-	-	Seed quality and yield	Metwally et al. (2021)
Bulk segregation	*Triticum* spp.	F2, F3	Linkage mapping	KASP	Rust resistance	Klymiuk et al. (2022)
	Pennisetum glaucum	F2, F3	QTL mapping	SNP	Yield and agronomic traits	de la Fuente Cantó and Vigouroux (2021)
	Chenopodium quinoa	F2:6	Linkage disequilibrium	SNPs and SilicoDArT	Disease resistance	Fondevilla et al. (2024)
Backcross method	*Zea mays* L.	BC1F1	-	-	Disease resistance	Sunitha et al. (2020)
	Oryza sativa L.	BC1F1-BC3F1	-	-	High temperature resistance	Malumpong et al. (2020)
	Brassica napus	BC4F2	QTL mapping	SNP	Seed oil quality and yield	Wang et al. (2023)
Molecular breeding						
Marker-assisted selection	*Oryza sativa* L.	BC2F2	QTL mapping	SNP	Grain quantity and yield	Gouda et al. (2020)
	Triticum aestivum	F2	Linkage mapping	SSR	Disease resistance	Wang et al. (2021)
	Brassica napus L.	F5	-	CAPS	Oil content and yield	Spasibionek et al. (2020)
Genotyping by sequencing	Glycine Max	F8:11	QTL mapping	SNP	Phytophthora-resistance	Jiang et al. (2020)
	Zea mays L.	Double haploid	Association mapping	rAmpSeq	Enhanced Zn in kernel	Guo et al. (2020)

(Continued)

Table 13.1 (Continued)

Techniques	Crops	Mapping Population	Type of Mapping	DNA Markers Used	Improved Traits	References
	Zea mays L.	RILs	QTL mapping	SNP	Cob color	Guo et al. (2021)
Genome-wide association analysis	*Oryza sativa*	RILs	QTL mapping	SNP	Salt resistance	Lv et al. (2022)
	Glycine max	RILs	QTL mapping	SNP	Growth and yield	Wang et al. (2022)
	Triticum aestivum	RILs	Association mapping	SNP	Leaf rust resistance	Vikas et al. (2022)
Tilling	*Vigna radiata*	M1	-	-	Plant growth	Varadaraju et al. (2021)
	Brassica rapa	M1	-	-	Stress tolerance	Navarro-Leon et al. (2020)
	Glycine Max	M2	-	-	Seed oil content and yield	Lakhssassi et al. (2021)

A central challenge in conventional breeding lies in progeny evaluation, which traditionally required multi-year, multi-location field trials to assess traits like yield, flowering time, or stress tolerance (Begna 2021). For example, developing a new wheat cultivar involves evaluating up to 100,000 genotypes annually across diverse agroclimatic zones (Rieseberg et al. 1993). The integration of molecular markers since the 1990s has streamlined this process, for example, restriction fragment length polymorphisms (RFLPs) and simple sequence repeats (SSRs) enabled breeders to screen seedlings for target loci, eliminating the need of phenotypic evaluation (Anand et al. 2023). In sunflower, MAS accelerated the introgression of *Pl* genes conferring resistance to downy mildew, reducing the breeding cycle by three years (Rieseberg et al. 1993). Advances in genotyping-by-sequencing (GBS) now allow genome-wide profiling at minimal cost, facilitating the selection of multiple traits simultaneously. For instance, GBS identified QTLs associated with iron bioavailability in pearl millet, enabling the rapid development of biofortified varieties for iron-deficient populations (Rayaprolu et al. 2022). Despite its successes, conventional breeding faces systemic constraints such as limited genetic diversity in elite gene pools because of intensive selection practices, particularly in self-pollinating crops (Pootakham 2023). For example, 60% of modern wheat varieties trace their ancestry to just four parental cultivars, heightening vulnerability to emerging pathogens (Lamichhane and Thapa 2022). Lengthy breeding timelines, often exceeding a decade from the cross to commercialization, further affect their growth under climate-driven crises like shifting pest zones or abrupt rainfall patterns (Raffo and Jensen 2023). Additionally, traits governed by epistatic interactions or genotype–environment (G×E) effects such as drought tolerance in maize require large-scale, replicated trials, and are time-consuming and laborious (Viana 2023). Mutation breeding, while valuable, is constrained by its reliance on random mutagenesis, with less than 1% of induced mutations yielding agronomically useful variants (Goddard et al. 2020).

13.3 Molecular Breeding Innovations

The advent of molecular breeding has revolutionized agricultural biotechnology, offering unprecedented precision in crop improvement (Aziz and Masmoudi 2025). Central to this transformation is next-generation sequencing (NGS) platforms, that has enabled the rapid identification of molecular markers such as single nucleotide polymorphisms (SNPs) and insertions-deletions (InDels) (Amiteye 2021). These markers serve as chromosomal landmarks, allowing breeders to pinpoint loci associated with agronomically valuable traits (McGowan et al. 2021). For instance, SNPs have been instrumental in dissecting complex traits like drought tolerance in maize and disease resistance in tomatoes, facilitating marker-assisted selection (MAS) with 90% accuracy in predicting phenotypic outcomes (Jaganathan et al. 2020). Similarly, InDels, which involve small DNA sequence insertions or deletions, have been leveraged to improve grain size in rice, directly linking genetic variation to phenotypic expression (Jain et al. 2019). Another approach is genotyping-by-sequencing (GBS) and GWAS have emerged as potential tools of high-resolution QTL mapping (Rani et al. 2023). GBS, a cost-effective method for genome-wide SNP discovery, has been pivotal in crops with complex genomes. In cotton, GBS identified 15 QTLs associated with fiber strength and length, enabling the development of elite cultivars with a 20% improvement in textile quality (Sabev et al. 2020). GWAS, which correlates genomic diversity with trait variation across diverse populations, has unlocked genetic insights in soybean. A 2022 study linked *GmNARK* gene polymorphisms to enhanced nitrogen fixation, paving the way for varieties that reduce synthetic fertilizer dependency by 30% (Shingote et al. 2022). These tools not only accelerate trait discovery but also refine genomic selection models, narrowing the gap between genetic potential and field performance (Table 13.1).

While transgenic technologies face ethical and regulatory issues due to public skepticism toward foreign DNA integration, CRISPR-Cas9 has emerged as a game-changer in precision breeding (Ma et al. 2016). This RNA-guided genome-editing tool enables targeted modifications such as knockouts, knock-ins, or base edits without introducing foreign DNA (Belhaj et al. 2015). In rice, CRISPR-mediated editing of the *OsERA1* gene enhanced drought tolerance by modulating abscisic acid sensitivity, resulting in a 25% yield increase under drought conditions (Zhang et al. 2019). Similarly, wheat engineered with edited *TaDREB2* alleles exhibited improved heat tolerance, sustaining grain filling at temperatures up to 38 °C (Demirer et al. 2021). Such advancements underscore CRISPR's potential to bypass the lengthy timelines of conventional breeding, with edited lines reaching field trials in just 2–3 years (Belhaj et al. 2015). Nonetheless, molecular breeding faces significant barriers such as high costs of NGS infrastructure and bioinformatics expertise limiting accessibility in developing regions (Liu et al. 2017). For example, a single GWAS analysis can be very expensive, excluding recurrent phenotyping expenses (Tibbs Cortes et al. 2021). Furthermore, ethical debates persist, particularly around off-target effects in CRISPR applications and intellectual property rights over edited genomes (Liu et al. 2023).

The future of molecular breeding lies in integrating multi-omics data like transcriptomics, proteomics, and metabolomics to unravel the molecular networks underpinning agriculturally important traits (Muduli et al. 2021). ML algorithms are being trained on these datasets to predict optimal gene-editing targets or hybrid combinations, exemplified by IBM's AgriTech platform, which reduced breeding cycles by 40% in pilot trials (Yan and Wang 2023). International collaborations, such as the African Orphan Crops Consortium, aim to democratize access to molecular tools for underutilized crops like fonio and baobab (Yan and Wang 2023). Concurrently, ethical guidelines emphasizing equity and sustainability, as outlined by the FAO, are critical to ensuring these innovations benefit smallholder farmers and ecosystems alike.

13.4 Speed Breeding and AI Integration

The escalating pressures of population growth, climate volatility, and land degradation demand transformative solutions to accelerate crop improvement (Rai 2022). Speed breeding, a groundbreaking methodology that manipulates environmental conditions to reduce plant life cycles, has emerged as a cornerstone of modern agricultural innovation (Table 13.2) (Wanga et al. 2021). Under controlled environments, AI coupled with ML can be used to reduce breeding timelines from years to months, offering a lifeline for global food security (Farooq et al. 2024). At its core, speed breeding exploits the physiological plasticity of plants by optimizing photoperiods, temperatures, and light spectra to accelerate growth and flowering (Sun et al. 2024). Traditional breeding cycles, constrained by seasonal photoperiods and climatic limitations, often yield only 1–2 generations annually for crops like wheat (*Triticum aestivum*) or barley (*Hordeum vulgare*) (Sun et al. 2024). Speed breeding natural circadian rhythm of plants by extending daily light exposure to 22 hours using energy-efficient LED systems, which emit specific wavelengths (e.g., red and blue spectra) to maximize photosynthetic efficiency and suppress shade-avoidance responses (Ghosh et al. 2018). Concurrently, temperatures are maintained at 22–25 °C during the day and 17 °C at night to balance metabolic activity and reduce heat stress (Watson et al. 2018). These conditions mimic perpetual spring, tricking plants into rapid maturation. For instance, protocols developed by the University of Queensland achieve six generations of wheat annually a threefold increase over conventional field breeding (Jahne et al. 2020). The true power of speed breeding lies in its integration with multi-omics technologies and AI-driven analytics (Watson et al. 2018). High-throughput phenotyping platforms capture real-time data on plant physiology, while genomic sequencing identifies stress-responsive genes and metabolic pathways (Jahne et al. 2020). AI algorithms, trained on these multi-dimensional datasets, predict gene-trait associations and prioritize candidate lines for further breeding. For example, Wang et al. (2024) combined transcriptomic profiles of drought-stressed wheat with ML models to pinpoint TaNAC29, a transcription factor

Table 13.2 Recent studies utilizing speed breeding and artificial intelligence to accelerate crop improvement.

Crops	Speed Breeding	AI and ML Approach	Generation Advanced/year	Trait Improved	References
Glycine max, Oryza sativa, Amaranthus spp	light-emitting diodes (LEDs)	_	4–6	Agronomic traits	Jahne et al. (2020)
Glycine max	-	Convolutional neural networks, RetinaNet, Mask-RCNN, and YOLO	-	Relative maturity	Moeinizade et al. (2022)
Triticum aestivum	-	CropQuant-Air	-	Yield- and performance-related traits	Chen et al. (2023)
Brassica rapa, Brassica oleracea, Medicago truncatula, T. Aestivum, T. Durum, C. arietinum	LEDs or SVLs	-	4–6	Yield- and performance-related traits	Ghosh et al. (2018)
Solanum lycopersicum	RO-LUXF36W/GRO	-	3–4	Agronomic traits	Gimeno-Páez et al. (2024)
Multiple plants	-	Black-box optimization algorithm, StoSOO	-	Breeding cycle	Hamazaki and Iwata (2024)
Capsicum annuum	Fluorescent lamp	-	2–3	Flowering gene	Choi et al. (2023)
Triticum durum	Fluorescent lamp	-	4–6	Agronomic traits	Alhamad et al. (2018)
Pisum sativum	Fluorescent lamp		4–6	Agronomic traits	Cazzola et al. (2020)
Solanum lycopersicum	-	Flower pollination algorithm (FPA) and support vector machine (SVM), and a convolutional neural network (CNN)	-	Disease pattern	Yag and Altan (2022)
Musa paradisiaca	-	UAV, machine learning models	-	Banana bunchy top disease	Selvaraj et al. (2020)
Cucurbita pepo	-	UAV, radial basis function	-	Powdery mildew disease	Abdulridha et al. (2020)

regulating root architecture and water-use efficiency. Such insights enable breeders to bypass trial-and-error selection, focusing resources on genetically optimized progeny.

Convolutional neural networks (CNNs), a class of deep learning algorithms, further enhance precision by analyzing hyperspectral or RGB images to quantify traits like leaf area, chlorophyll content, or disease severity (Table 13.2) (Shi et al. 2023). In soybean, CNNs trained on 50,000 leaf images achieved 98% accuracy in detecting *Phakopsora pachyrhizi* (soybean rust) infections before visual symptoms emerged (Karlekar and Seal 2020). The scalability of speed breeding depends on automation, with robotics addressing labor-intensive tasks such as planting, phenotyping, and harvest (Wanga et al. 2021). The BoniRob V2, a modular agricultural robot, have been extensively used for various agricultural work shifting manual seed sowing, planting and harvesting to automated robotics (Ruckelshausen et al. 2009). Furthermore, it is also equipped with LiDAR, multispectral cameras, and robotic arms, enabling easy and autonomous navigation from growth chambers to fields, capturing millimeter-scale measurements of plant height, biomass, and root architecture (Lottes et al. 2020). In trials at the Leibniz Institute, BoniRob monitored 10,000 maize plants daily, detecting nitrogen deficiency 10 days earlier than manual scouts (Xu and Li 2022). Such systems not only reduce human error but also generate standardized datasets to automate hydroponic and aeroponic systems in controlled environments (Krishnan and Swarna 2020). Vertical farms employing these technologies report 70% reductions in water and fertilizer use compared to traditional greenhouses, while accelerating seedling growth by 30% (Ghosh et al. 2018). When paired with CRISPR-Cas9 gene editing, these platforms enable "closed-loop" breeding cycles, where AI identifies target genes and speed breeding rapidly advances edited lines to maturity (Ellery 2021).

Despite immense potential, speed breeding faces significant barriers such as capital costs of light-emitting diode (LED)-equipped growth chambers and AI infrastructure limit accessibility for public-sector institutions in developing nations (Sharma et al. 2022). In addition, energy consumption remains a concern, with 24-hour lighting demanding 2–3 MW annually per 1000 m^2 facility (Watson et al. 2018). Furthermore, data management poses another hurdle for example, a single speed breeding trial can generate 10 TB of genomic and phenotypic data, necessitating cloud computing resources and bioinformatics expertise often unavailable to smallholder cooperatives (Pandey et al. 2022). Ethical debates also persist as rapid deployment of gene-edited crops, accelerated by speed breeding, raises questions about biodiversity loss and corporate monopolization of genetic resources (Sharma et al. 2022). The future of speed breeding relies on equalizing access and enhancing sustainability of open-source platforms like Speed Breeder OS aim to reduce costs by providing blueprints for low-cost LED chambers and AI models tailored to regional crops (Potts et al. 2023). International consortia, such as the CGIAR Excellence in Breeding Platform, are piloting shared-speed breeding hubs in Kenya and India, focusing on drought-tolerant sorghum and heat-resistant chickpeas.

13.5 Challenges and Future Directions

While AI and speed breeding reduce breeding cycles, challenges include data complexity, high costs, and algorithm limitations thus accentuating the need of self-supervised ML models for unlabeled datasets and gap validation (Bhat et al. 2023). Future efforts must prioritize affordable AI tools for small-scale farmers and robust data integration to ensure food security (Kundu 2024). Recognizing these limitations, breeders increasingly hybridize conventional methods with cutting-edge technologies such as genomics-assisted breeding to predict the performance of untested crosses thus optimizing parental selection (Gupta and Verma 2024). In rice, the *Green Super Rice* initiative combined pedigree breeding with genomic selection to develop varieties yielding 20% more under drought conditions (Labroo et al. 2021). Similarly, *CRISPR-Cas9* gene editing is being deployed to refine traits introgressed via conventional crosses, as seen in tomato, where editing of *SP5G* and *SP* genes enhanced yield without altering decades of pedigree-based improvements in fruit quality (Jeon et al. 2023). These synergies underscore that conventional breeding is not obsolete but evolving, exploiting molecular tools to enhance precision while retaining its foundational emphasis on phenotypic relevance and adaptability (Jeon et al. 2023).

13.6 Conclusion

Molecular breeding innovations represent a paradigm shift in agriculture, merging genetic precision with scalability to address food security and climate challenges. While cost and ethical barriers persist, ongoing advancements in technology and global cooperation promise to equal access and refine regulatory landscapes. As the field evolves, the synergy between cutting-edge science and inclusive policy will determine its efficacy towards sustainable agriculture. Speed breeding, in conjunction with AI and robotics, represents a paradigm shift in agriculture. By reducing breeding cycles and enhancing precision, it addresses the limitation of conventional breeding for developing climate-resilient crops. However, realizing its full potential requires overcoming economic, technical, and ethical challenges through innovation and inclusive governance. As these technologies improve, their integration into holistic, sustainable farming systems will determine their role in ensuring food security for future generations. Overall, AI and speed breeding represent transformative tools for developing climate-resilient crops. By merging OMICS data with predictive algorithms, these technologies address global agricultural challenges. However, advancing computational efficiency and cost-effectiveness will determine their scalability, ensuring equitable benefits across diverse farming systems.

Suggested Further Readings

Kaya, C. (2025). Optimizing crop production with plant phenomics through high-throughput phenotyping and AI in controlled environments. *Food and Energy Security* 14(1): e70050.

Maß, V., Alirezazadeh, P., Seidl-Schulz, J., et al. (2025). Annotated image dataset with different stages of European pear rust for UAV-based automated symptom detection in orchards. *Data in Brief* 58: 111271.

Vieira, R.A., Nogueira, A.P.O., and Fritsche-Neto, R. (2025). Optimizing the selection of quantitative traits in plant breeding using simulation. *Frontiers in Plant Science* 16: 1495662.

Yuan, X., Hu, J., Marcelis, L.F., et al. (2025). Advanced technologies in plant factories: Exploring current and future economic and environmental benefits in urban horticulture. *Horticulture Research* 12: uhaf024.

References

Acquaah, G., (2015). Conventional plant breeding principles and techniques. *Advances in Plant Breeding Strategies: Breeding, Biotechnology and Molecular Tools*: 115–158.

Amiteye, S., (2021). Basic concepts and methodologies of DNA marker systems in plant molecular breeding. *Heliyon* 7 (10).

Anand, A., Subramanian, M., and Kar, D., (2023). Breeding techniques to dispense higher genetic gains. *Frontiers in Plant Science* 13: 1076094.

Annicchiarico, P., Nazzicari, N., Notario, T. et al. (2021). Pea breeding for intercropping with cereals: variation for competitive ability and associated traits, and assessment of phenotypic and genomic selection strategies. *Frontiers in Plant Science* 12: 731949.

Arya, R.K., Hooda, J.S., Kaushik, S.K. et al. (2022). Development of new promising varieties of faba bean through traditional pedigree method for commercial cultivation in plain zone of India. *Indian Journal of Traditional Knowledge (IJTK)* 21 (4): 895–904.

Arya, S., Sandhu, K.S., Singh, J. et al. (2022). Deep learning: As the new frontier in high-throughput plant phenotyping. *Euphytica* 218 (4): 47.

Asfaw, A., Aderonmu, D.S., Darkwa, K. et al. (2021). Genetic parameters, prediction, and selection in a white Guinea yam early-generation breeding population using pedigree information. *Crop Science* 61 (2): 1038–1051.

Aziz, M.A. and Masmoudi, K., (2025). Molecular breakthroughs in modern plant breeding techniques. *Horticultural Plant Journal* 11 (1): 15–41.

Baydar, H.A.S.A.N., Marquard, R., and Turgut, I., (1999). Pure line selection for improved yield, oil content and different fatty acid composition of sesame, Sesamum indicum. *Plant Breeding* 118 (5): 462–464.

Begna, T., (2021). Conventional breeding methods widely used to improve self-pollinated crops. *International Journal of Research* 7 (1): 1–16.

Behera, T.K., Jat, G.S., and Pathak, M. (2020). Classical genetics and traditional breeding. In: *The Bitter Gourd Genome*: 91–104. Cham: Springer International Publishing.

Belhaj, K., Chaparro-Garcia, A., Kamoun, S. et al. (2015). Editing plant genomes with CRISPR/Cas9. *Current Opinion in Biotechnology* 32: 76–84.

Bhat, J.A., Feng, X., Mir, Z.A. et al. (2023). Recent advances in artificial intelligence, mechanistic models, and speed breeding offer exciting opportunities for precise and accelerated genomics-assisted breeding. *Physiologia Plantarum* 175 (4): e13969.

Bohra, A., Chand Jha, U., Godwin, I.D. et al. (2020). Genomic interventions for sustainable agriculture. *Plant Biotechnology Journal* 18 (12): 2388–2405.

Borlaug, N.E., (1983). Contributions of conventional plant breeding to food production. *Science* 219 (4585): 689–693.

Bren d'Amour, C., Reitsma, F., Baiocchi, G. et al. (2017). Future urban land expansion and implications for global croplands. *Proceedings of the National Academy of Sciences* 114 (34): 8939–8944.

Caruso, M., Smith, M.W., Froelicher, Y. et al. (2020). Traditional breeding. In: *The Genus Citrus*, 129–148. Woodhead Publishing.

Chawla, R., Poonia, A., Samantara, K. et al. (2023). Green revolution to genome revolution: Driving better resilient crops against environmental instability. *Frontiers in Genetics* 14: 1204585.

Das, S., Choudhury, M.R., Chatterjee, B. et al. (2024). Unraveling the urban climate crisis: Exploring the nexus of urbanization, climate change, and their impacts on the environment and human well-being: A global perspective. *AIMS Public Health* 11 (3): 963.

de la Fuente Cantó, C., and Vigouroux, Y., (2022). Evaluation of nine statistics to identify QTLs in bulk segregant analysis using next generation sequencing approaches. *BMC Genomics* 23 (1): 490.

Demirer, G.S., Silva, T.N., Jackson, C.T. et al. (2021). Nanotechnology to advance CRISPR–Cas genetic engineering of plants. *Nature Nanotechnology* 16 (3): 243–250.

Efficiency of Mass Selection and Selfing with Selection Breeding Methods on Improving Some Important Characters of Three Eggplant Cultivars

Ellery, A., (2021). Opening closed loop ecological life support systems with in-situ resources on the moon.

Farooq, M.A., Gao, S., Hassan, M.A. et al. (2024). Artificial intelligence in plant breeding. *Trends in Genetics* 40 (10): 891–908.

Fondevilla, S., Calderón-González, Á., Rojas-Panadero, B. et al. (2024). Genome-wide association study, combined with bulk segregant analysis, identify plant receptors and defense related genes as candidate genes for downy mildew resistance in quinoa. *BMC Plant Biology* 24 (1): 594.

Ghosh, S., Watson, A., Gonzalez-Navarro, O.E. et al. (2018). Speed breeding in growth chambers and glasshouses for crop breeding and model plant research. *Nature Protocols* 13 (12): 2944–2963.

Goddard, R., Steed, A., Chinoy, C. et al. (2020). Dissecting the genetic basis of wheat blast resistance in the Brazilian wheat cultivar BR 18-Terena. *BMC Plant Biology* 20: 1–15.

Gouda, G., Gupta, M.K., Donde, R. et al. (2020). Marker-assisted selection for grain number and yield-related traits of rice (*Oryza sativa* L.). *Physiology and Molecular Biology of Plants* 26: 885–898.

Guo, R., Dhliwayo, T., Mageto, E.K. et al. (2020). Genomic prediction of kernel zinc concentration in multiple maize populations using genotyping-by-sequencing and repeat amplification sequencing markers. *Frontiers in Plant Science* 11: 534.

Guo, Z., Yang, Q., Huang, F. et al. (2021). Development of high-resolution multiple-SNP arrays for genetic analyses and molecular breeding through genotyping by target sequencing and liquid chip. *Plant Communications* 2 (6): 100230.

Gupta, D., and Verma, T., (2024). Advanced technologies in plant breeding: From genome editing to machine learning. *Molecular Plant Breeding: Principles and Practices*: 193.

Huang, W., (2024). The current situation and future of using GWAS strategies to accelerate the improvement of crop stress resistance traits. *Molecular Plant Breeding* 15: 52–62.

Jaganathan, D., Bohra, A., Thudi, M. et al. (2020). Fine mapping and gene cloning in the post-NGS era: Advances and prospects. *Theoretical and Applied Genetics* 133 (5): 1791–1810.

Jahne, F., Hahn, V., Würschum, T. et al. (2020). Speed breeding short-day crops by LED-controlled light schemes. *Theoretical and Applied Genetics* 133 (8): 2335–2342.

Jain, A., Roorkiwal, M., Kale, S. et al. (2019). InDel markers: An extended marker resource for molecular breeding in chickpea. *PLoS One* 14 (3): e0213999.

Jeon, D., Kang, Y., Lee, S. et al. (2023). Digitalizing breeding in plants: A new trend of next-generation breeding based on genomic prediction. *Frontiers in Plant Science* 14: 1092584.

Jiang, B., Cheng, Y., and Cai, Z. (2020). Fine mapping of a Phytophthora-resistance locus RpsGZ in soybean using genotyping-by-sequencing. *BMC Genomics* 21: 1–11.

Karlekar, A., and Seal, A. (2020). SoyNet: Soybean leaf diseases classification. *Computers and Electronics in Agriculture* 172: 105342.

Klymiuk, V., Chawla, H.S., Wiebe, K. et al. (2022). Discovery of stripe rust resistance with incomplete dominance in wild emmer wheat using bulked segregant analysis sequencing. *Communications Biology* 5 (1): 826.

Krishnan, A., and Swarna, S. (2020). Robotics, IoT, and AI in the automation of agricultural industry: A review. *2020 IEEE Bangalore Humanitarian Technology Conference (B-HTC)*, 1–6 (8 October 2020, Banglore). IEEE.

Kundu, V. (2024). Artificial intelligence (AI): Cultivating innovation in crop breeding. *Food and Scientific Reports* 5 (3): 21–27.

Labroo, M.R., Studer, A.J., and Rutkoski, J.E. (2021). Heterosis and hybrid crop breeding: A multidisciplinary review. *Frontiers in Genetics* 12: 643761.

Lakhssassi, N., Lopes-Caitar, V.S., Knizia, D. et al. (2021). TILLING-by-sequencing+ reveals the role of novel fatty acid desaturases (GmFAD2-2s) in increasing soybean seed oleic acid content. *Cells* 10 (5): 1245.

Lamichhane, S., and Thapa, S. (2022). Advances from conventional to modern plant breeding methodologies. *Plant Breeding in Biotechnology* 10 (1): 1–4.

Lema, M., (2018). Marker assisted selection in comparison to conventional plant breeding. *Agricultural Research and Technology* 14: 555914.

Lenz, P.R., Nadeau, S., Azaiez, A. et al. (2020). Genomic prediction for hastening and improving efficiency of forward selection in conifer polycross mating designs: An example from white spruce. *Heredity* 124 (4): 562–578.

Liu, X., Tian, D., Li, C. et al. (2023). GWAS Atlas: An updated knowledgebase integrating more curated associations in plants and animals. *Nucleic Acids Research* 51 (D1): D969–D976.

Liu, X., Wu, S., Xu, J. et al. (2017). Application of CRISPR/Cas9 in plant biology. *Acta Pharmaceutica Sinica B* 7 (3): 292–302.

Lv, Y., Ma, J., Wei, H. et al. (2022). Combining GWAS, genome-wide domestication and a transcriptomic analysis reveals the loci and natural alleles of salt tolerance in rice (*Oryza sativa* L.). *Frontiers in Plant Science* 13: 912637.

Ma, X., Zhu, Q., Chen, Y. et al. (2016). CRISPR/Cas9 platforms for genome editing in plants: Developments and applications. *Molecular plant* 9 (7): 961–974.

Malumpong, C., Buadchee, R., Thammasamisorn, B. et al. (2020). Backcross breeding for improvement of heat tolerance at reproductive phase in Thai rice (*Oryza sativa* L.) varieties. *The Journal of Agricultural Science* 158 (6): 496–510.

Metwally, E., Sharshar, M., Masoud, A. et al. (2021). Development of high yielding cowpea [*Vigna unguiculata* (L.) walp.] lines with improved quality seeds through mutation and pedigree selection methods. *Horticulturae* 7 (9): 271.

Mukri, G., Patil, M.S., Motagi, B.N. et al. (2022). Genetic variability, combining ability and molecular diversity-based parental line selection for heterosis breeding in field corn (*Zea mays* L.). *Molecular Biology Reports* 49 (6): 4517–4524.

Navarro-León, E., López-Moreno, F.J., Atero-Calvo, S. et al. (2020). CAX1a tilling mutations modify the hormonal balance controlling growth and ion homeostasis in Brassica rapa plants subjected to salinity. *Agronomy* 10 (11): 1699.

Pandey, S., Singh, A., Parida, S.K. et al. (2022). Combining speed breeding with traditional and genomics-assisted breeding for crop improvement. *Plant Breeding* 141 (3): 301–313.

Pootakham, W., (2023). Genotyping by sequencing (GBS) for genome-wide SNP identification in plants. In: *Plant Genotyping: Methods and Protocols*, 1–8. Springer US: New York, NY.

Potts, J., Jangra, S., Michael, V.N. et al. (2023). Speed breeding for crop improvement and food security. *Crops* 3 (4): 276–291.

Raffo, M.A., and Jensen, J., (2023). Gene× gene and genotype× environment interactions in wheat. *Crop Science* 63 (4): 1779–1793.

Rai, K.K., (2022). Integrating speed breeding with artificial intelligence for developing climate-smart crops. *Molecular Biology Reports* 49 (12): 11385–11402.

Rani, K., Kumar, M., Razzaq, A. et al. (2023). Recent advances in molecular marker technology for QTL mapping in plants. *QTL Mapping in Crop Improvement*: 1–15.

Rayaprolu, L., Deshpande, S.P., and Gupta, R. (2022). Genotyping-by-sequencing (GBS) method for accelerating marker-assisted selection (MAS) program. In: *Genomics of Cereal Crops*, 245–257. Springer US: New York, NY.

Rieseberg, L.H., Choi, H., Chan, R. et al. (1993). Genomic map of a diploid hybrid species. *Heredity* 70 (3): 285–293.

Ruckelshausen, A., Biber, P., Dorna, M. et al. (2009). BoniRob: An autonomous field robot platform for individual plant phenotyping. In: *Precision Agriculture 09*, 841–847. Wageningen Academic.

Sabev, P., Valkova, N., and Todorovska, E.G. (2020). Molecular markers and their application in cotton breeding: progress and future perspectives. *Bulgarian Journal of Agricultural Science* 26 (4): 816–828.

Salonia, F., Ciacciulli, A., Poles, L. et al. (2020). New plant breeding techniques in citrus for the improvement of important agronomic traits. A review. *Frontiers in Plant Science* 11: 1234.

Saxena, K.B., Dalvi, V.A., Pandey, M. et al. (2025). Sybrid population: A new breeding method for often cross-pollinated legumes (Leguminocea). *Plant Breeding*.

Sharma, S., Kumar, A., Dhakte, P. et al. (2022). Speed breeding opportunities and challenges for crop improvement. *Journal of Plant Growth Regulation*: 1–14.

Shi, T., Liu, Y., Zheng, X. et al. (2023). Recent advances in plant disease severity assessment using convolutional neural networks. *Scientific Reports* 13 (1): 2336.

Shingote, P.R., Gotarkar, D.N., Kale, R.R. et al. (2022). Recent advances and applicability of GBS, GWAS, and GS in soybean. *Genotyping by Sequencing for Crop Improvement*: 218–249.

Smith, C.M., (2021). Conventional breeding of insect-resistant crop plants: Still the best way to feed the world population. *Current Opinion in Insect Science* 45: 7–13.

Solieman, T.H., Abdel-Razzak, H.S., Doss, M.M. et al. (2012). Efficiency of mass selection and selfing with selection breeding methods on improving some important characters of three eggplant cultivars. *American Eurasian Journal of Agricultural and Environmental Sciences* 12 (3): 342–351.

Spasibionek, S., Mikołajczyk, K., Ćwiek-Kupczyńska, H. et al. (2020). Marker assisted selection of new high oleic and low linolenic winter oilseed rape (*Brassica napus* L.) inbred lines revealing good agricultural value. *PloS One* 15 (6): e0233959.

Sterckeman, T., Cazes, Y., and Sirguey, C. (2019). Breeding the hyperaccumulator Noccaea caerulescens for trace metal phytoextraction: First results of a pure-line selection. *International Journal of Phytoremediation* 21 (5): 448–455.

Sun, L., Lai, M., Ghouri, F. et al. (2024). Modern plant breeding techniques in crop improvement and genetic diversity: from molecular markers and gene editing to artificial intelligence—A critical review. *Plants* 13 (19): 2676.

Sunitha, N.C., Gangappa, E., Gowda, R.V. et al. (2020). Effectiveness of one cycle of phenotype-based backcross breeding for resistance to late wilt disease in maize (*Zea mays* L.). 54 (1): 47–50.

Tesfaye, K., Kruseman, G., Cairns, J.E. et al. (2018). Potential benefits of drought and heat tolerance for adapting maize to climate change in tropical environments. *Climate Risk Management* 19: 106–119.

Tibbs Cortes, L., Zhang, Z., and Yu, J., (2021). Status and prospects of genome-wide association studies in plants. *The Plant Genome* 14 (1): e20077.

Varadaraju, A., Ramadoss, B.R., Gurunathan, S. et al. (2021). TILLING by sequencing (TbyS) reveals mutations in flowering control genes that are associated with altered plant architecture in Mungbean (*Vigna radiata* (L.) R. Wilczek). *Genetic Resources and Crop Evolution* 68 (3): 849–864.

Varshney, R.K., Bohra, A., Roorkiwal, M. et al. (2021). Fast-forward breeding for a food-secure world. *Trends in Genetics* 37 (12): 1124–1136.

Viana, J.M.S., (2023). The impact of epistasis in the heterosis and combining ability analyses. *Frontiers in Plant Science* 14: 1168419.

Vikas, V.K., Pradhan, A.K., Budhlakoti, N. et al. (2022). Multi-locus genome-wide association studies (ML-GWAS) reveal novel genomic regions associated with seedling and adult plant stage leaf rust resistance in bread wheat (*Triticum aestivum* L.). *Heredity* 128 (6): 434–449.

Wang, D., Zhang, X., Cao, Y. et al. (2024). TabHLH27 orchestrates root growth and drought tolerance to enhance water use efficiency in wheat. *Journal of Integrative Plant Biology* 66 (7): 1295–1312.

Wang, J., Hu, B., Jing, Y. et al. (2022). Detecting QTL and candidate genes for plant height in soybean via linkage analysis and GWAS. *Frontiers in Plant Science* 12: 803820.

Wang, M., King, G.J., et al. (2023). Genome-wide recombination variation in biparental segregating and reciprocal backcross populations provides information for introgression breeding in Brassica napus. *The Crop Journal* 11 (1): 208–219.

Wang, W., He, H., Gao, H. et al. (2021). Characterization of the powdery mildew resistance gene in wheat breeding line KN0816 and its evaluation in marker-assisted selection. *Plant Disease* 105 (12): 4042–4050.

Wanga, M.A., Shimelis, H., Mashilo, J. et al. (2021). Opportunities and challenges of speed breeding: A review. *Plant Breeding* 140 (2): 185–194.

Watson, A., Ghosh, S., Williams, M.J. et al. (2018). Speed breeding is a powerful tool to accelerate crop research and breeding. *Nature Plants* 4 (1): 23–29.

Xiong, W., Reynolds, M.P., Crossa, J. et al. (2020). Climate change has increased genotype-environment interactions in wheat breeding. https://doi.org/10.21203/rs.3.rs-69475/v1

Xu, R., and Li, C. (2022). A review of high-throughput field phenotyping systems.

Yabe, S., Hara, T., Ueno, M. et al. (2018). Potential of genomic selection in mass selection breeding of an allogamous crop: An empirical study to increase yield of common buckwheat. *Frontiers in Plant Science* 9: 276.

Yan, J., and Wang, X. (2023). Machine learning bridges omics sciences and plant breeding. *Trends in Plant Science* 28 (2): 199–210.

Yang, L., He, W., Zhu, Y. et al. (2025). GWAS meta-analysis using a graph-based pan-genome enhanced gene mining efficiency for agronomic traits in rice. *Nature Communications* 16 (1): 3171.

Zhai, Z., Martínez, J.F., Beltran, V. et al. (2020). Decision support systems for agriculture 4.0: Survey and challenges. *Computers and Electronics in Agriculture* 170: 105256.

Zhang, Z., Hua, L., Gupta, A. et al. (2019). Development of an Agrobacterium-delivered CRISPR/Cas9 system for wheat genome editing. *Plant Biotechnology Journal* 17 (8): 1623–1635.

14

Machine Learning-Driven Smart Agriculture

*Rajdeep Mohanta, Soumik Dey Roy, Sahely Kanthal, Sanjay Mochary, Subhadwip Ghorai, Soham Hazra**

Department of Agriculture, Brainware University, Barasat, Kolkata, West Bengal, India

*Corresponding author: sohamhazra94@gmail.com

14.1 Introduction to Smart Agriculture

The United Nations Food and Agricultural Organization (FAO) projects that the world's population may surpass 9 billion people by 2050, resulting in a 50% increase in agricultural demand compared to 2013 levels (FAO 2017). Furthermore, climate change is making matters worse by increasing the frequency of extreme weather events, which have a devastating effect on global agriculture and, more frequently, the food security of developing countries (Fedoroff et al. 2010; FAO 2018). Furthermore, the 2019 COVID-19 pandemic increased agricultural system uncertainty and threatened global food security. In order to lower down or completely eradicate the detrimental effects of climate change in agriculture sector, degradation of land, biodiversity, water, and energy resources, food production systems must be regulated by sustainable management of agricultural areas (Gomiero et al. 2011).

14.1.1 Importance of Smart Agriculture in Modern Era

In India, the main source of income is agriculture and its related industries. The main source of work for almost 80% of the population is agriculture or its byproducts. In order to make agriculture sustainable and profitable, government policymakers must encourage smart farming given the population's rapid expansion. In order to improve the agricultural sector, it is imperative that novel technology be integrated more quickly, particularly in developing nations like India (González-Orozco et al. 2020). Crop cutting and manual observations were the traditional methods used in India to estimate crops and track their condition. In order to increase crop productivity, farmers use fertilizer, irrigation, pesticides, and herbicides on their properties. However, this procedure requires a lot of time and resources, which raises the cost of agricultural inputs. In order to overcome this urgent problem, real-time crop growth monitoring in a variety of settings and with different environmental circumstances is desperately needed. In addition to improving management practices and saving time, real-time crop monitoring helps to manage the extreme weather events and lessen their effects on the world food chain (Wheeler et al. 2013).

14.1.2 The Role of Technology in Modern Agriculture

Since remote sensing (RS) provides a nondestructive way to gather data at local to global sizes, it is a crucial tool for addressing the aforementioned needs. It has shown itself to be an effective instrument for quickly obtaining large amounts of crop data and producing useful information for crop managers (Kasampalis et al. 2018). Characterizing the spatiotemporal variability within a certain area is made possible by RS. Recent years have seen tremendous improvements in our

Machine Learning for Plant Biology, First Edition. Edited by Jen-Tsung Chen.

ability to collect RS data, bringing in a new era marked by the growth of massive databases known as "big data." Big data, which is frequently summarized as the five Vs, includes not only the vast amount of data but also the difficulties brought on by its diversity, dependability, and intrinsic worth. In the age of smart agriculture, traditional processing methods cannot keep up with the ever-increasing needs. In light of this, machine learning (ML) has become a potent method for effectively obtaining information from a variety of sources.

The application of ML algorithms to global agriculture management is the subject of several thorough reviews (Chlingaryan et al. 2018; Liakos et al. 2018; Jha et al. 2019; Diaz-Gonzalez et al. 2022). Chlingaryan et al. (2018) examined the application of ML methods to forecast agricultural production and nitrogen status. A few global RS tool contributions were incorporated by Liakos et al. (2018). Jha et al. (2019) also examined the automation methods used in agricultural systems. Both ML and RS approaches were employed in Diaz-Gonzalez et al.'s (2022) work, but solely to review the research on global soil quality evaluation. Every review listed above concentrated on studies conducted on a worldwide scale (Diaz-Gonzalez et al. 2022). Thus, the goal is to talk about how RS and ML might be used to solve issues in Indian agriculture. The aim to highlight the application of RS satellites and ML techniques in the agriculture field in this work. Based on the publications that is investigated; it is also provided an overview of the difficulties and potential applications of integrating ML technology with RS data for Indian agriculture.

14.1.3 Introduction to ML and AI in Agriculture

Data analysis, classification, and pattern recognition are the primary applications of ML, a multidisciplinary technique that integrates computer science, engineering, and mathematics. Various statistical tools are used to assess the performance of ML models and algorithms. ML learns from experience and makes predictions, just like humans do. As ML-based models continue to learn from previous events, their performance gradually gets better. To understand input-output relations from past events, computers and machines must be trained using historical data on events or experiences (Liakos et al. 2018). Following training, the ML-based models are applied to additional data (testing data) for categorization or prediction. Numerous new ML algorithms have been developed to address a range of issues as a result of the development of computing and information technology.

14.2 The Role of ML in Agriculture

Modern agriculture faces numerous challenges, driven by the escalating demand for food due to rapid global population growth, climate change (Thayer et al. 2020), depletion of natural resources (Nassani et al. 2019), shifts in dietary preferences (Conrad et al. 2018), and concerns related to food safety and human health (Benos et al. 2020). In response to these pressing issues, which exert substantial pressure on the agricultural sector, there is an urgent need to enhance the efficiency of agricultural practices while concurrently mitigating their environmental impact. These two fundamental requirements have been key drivers in the transition of agriculture toward precision farming. This advancement in agricultural practices holds significant potential to ensure sustainability, maximize productivity, and promote environmental safety (Lampridi et al. 2019). Broadly, smart farming is founded on four core principles to address the growing demands: (i) efficient management of natural resources, (ii) preservation of ecosystem integrity, (iii) development of appropriate agricultural services, and (iv) integration of modern technological innovations (Benos et al. 2021).

In numerous countries, farmers continue to depend on traditional agricultural practices, which are primarily guided by the advice and experience of elders. However, this approach leaves farmers vulnerable to unpredictable climatic conditions, which have become increasingly erratic due to global warming and irregular rainfall patterns (Sharma et al. 2020). Traditional methods, often reliant on human intuition and assumptions, lack systematic strategies to effectively mitigate these challenges. Plant nutrition, health, disease diagnosis, and growth assessment often require precise treatments and targeted interventions to ensure optimal plant health (Mirani et al. 2021). Large-scale improvements in crop yield can be achieved through the application of ML techniques. ML, a subset of artificial intelligence (AI), has demonstrated significant potential in advancing various components of Agriculture 4.0 (Araújo et al. 2023). The integration of ML in agricultural sciences has enhanced data-driven predictive modeling, improving the accuracy of future forecasts and decision-making processes.

The integration of ML into modern agriculture represents a growing trend, facilitating the adoption of smart farming through advanced technological approaches. Recent methodologies in agriculture utilize ML for various applications, including soil moisture assessment based on crop type, efficient water management, disease detection, pesticide selection through pattern recognition, and overall crop management. The ML framework (Figure 14.1) in agriculture comprises five key components: (i) data collection from farms, (ii) data storage, (iii) data pre-processing, (iv) model training, and (v) performance evaluation (Veeragandham and Santhi 2020).

Within the framework of Agriculture 4.0, several widely adopted ML algorithms have gained prominence. These include established techniques such as Random Forest (RF), Support Vector Machine (SVM), Artificial Neural Network (ANN), and various Deep Learning (DL) models (Araújo et al. 2021). These algorithms are instrumental in transforming the agricultural sector, driving innovation and enhancing operational efficiency.

ML algorithms are generally categorized into three types: supervised learning (SL), unsupervised learning (UL), and semi-supervised learning (SSL), with the primary distinction being the presence or absence of labeled training data (Yan and Wang 2022) (Figure 14.2). In SL, the model is trained using a labeled dataset, allowing it to make predictions on new, unlabeled data. Conversely, UL utilizes an unlabeled dataset for both model training and identifying patterns or relationships within the data. Supervised learning (SL) is broadly divided into two main approaches: classification and regression, while unsupervised learning (UL) encompasses clustering and dimensionality reduction (DR) techniques (Petegrosso et al. 2020). In semi-supervised learning (SSL), only a subset of the training data is labeled, and the model leverages both labeled and unlabeled data to make predictions or decisions on test samples. A notable example of SSL is positive-unlabeled (PU) learning, which utilizes partially labeled datasets for model training (Li et al. 2022).

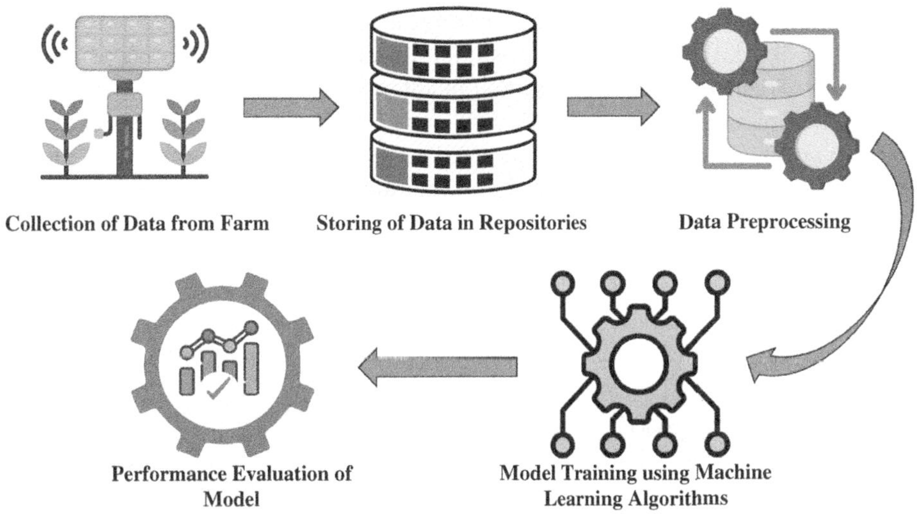

Figure 14.1 General framework of machine learning in agriculture.

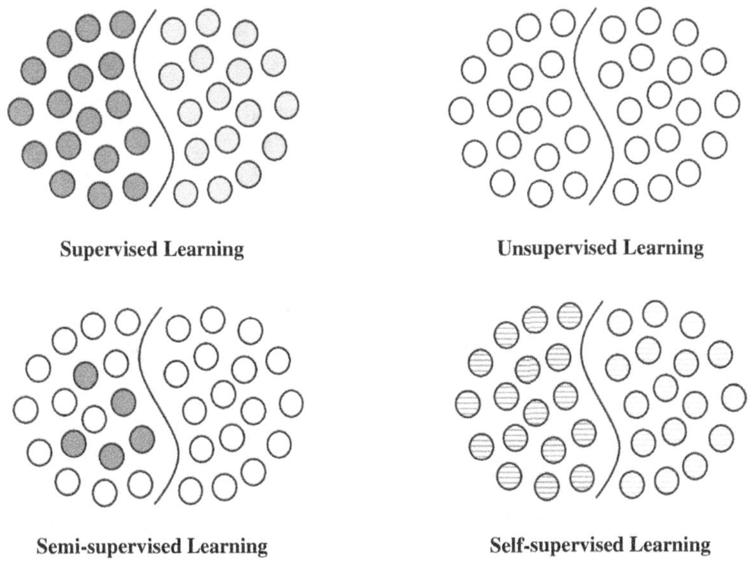

Figure 14.2 Illustration of four primary machine learning approaches (Yan and Wang 2022/John Wiley & Sons): White circles denote unlabeled data points, while shaded circles signify labeled data points belonging to different classes. Striped circles represent initially unlabeled data points that acquire labels during model training. The black boundary lines indicate decision divisions between classes. The semi-supervised learning (SSL) diagram highlights positive-unlabeled (PU) learning, where only a small subset of data points (blue) are labeled as positive samples, while the majority remain unlabeled.

In supervised learning (SL), a guiding mechanism oversees the training process, where the model learns from labeled training examples and subsequently applies this knowledge to derive accurate conclusions for new data (Mishra et al. 2016). Common SL algorithms include Artificial Neural Networks, Bayesian Networks, Decision Trees, Support Vector Machines, ID3, k-Nearest Neighbors, and Hidden Markov Models. In contrast, unsupervised learning (UL) involves providing a large dataset to the model, allowing it to independently identify patterns and relationships within the data. This approach enables the discovery of hidden structures without predefined labels (Ouf 2018). Examples of UL algorithms include k-Nearest Neighbors, Self-Organizing Maps, Partition-Based Clustering, Hierarchical Clustering, and K-means Clustering.

14.3 Key Applications of ML in Smart Agriculture

14.3.1 Precision Farming

Precision agriculture revolutionizes modern farming by leveraging data-driven technologies to boost crop yields and minimize waste. It integrates advanced technologies like AI, IoT, and ML to optimize resource use and enhance productivity. ML models, such as support vector regression and decision trees, analyze complex datasets to predict yields and classify diseases, while algorithms recommend suitable crops based on environmental data to improve yield and reduce environmental impact (Bachu et al. 2024; Manikandababu et al. 2024) (Figure 14.3).

Techniques like convolutional neural networks detect early signs of pests and diseases, enabling timely interventions (Lakshmi et al. 2024). Deep learning is used for soil classification, facilitating real-time monitoring and optimal fertilizer application (Triantafyllou et al. 2019). Soil monitoring systems provide critical data for crop management, while data warehouse frameworks enhance decision-making and crop quality (Abioye et al. 2020; Wakhare et al. 2020). Intelligent irrigation systems use ML to predict water needs, ensuring precise delivery and reducing waste (Karim and Karim 2017; Divyajyothi et al. 2024; Fotabong et al. 2024; Korlepara et al. 2024).

14.3.2 Pest and Disease Detection

Image-based pest identification using ML is a crucial aspect of precision agriculture, enhancing accuracy and efficiency in pest management (Figure 14.3). Convolutional Neural Networks (CNNs) are widely employed for their ability to extract

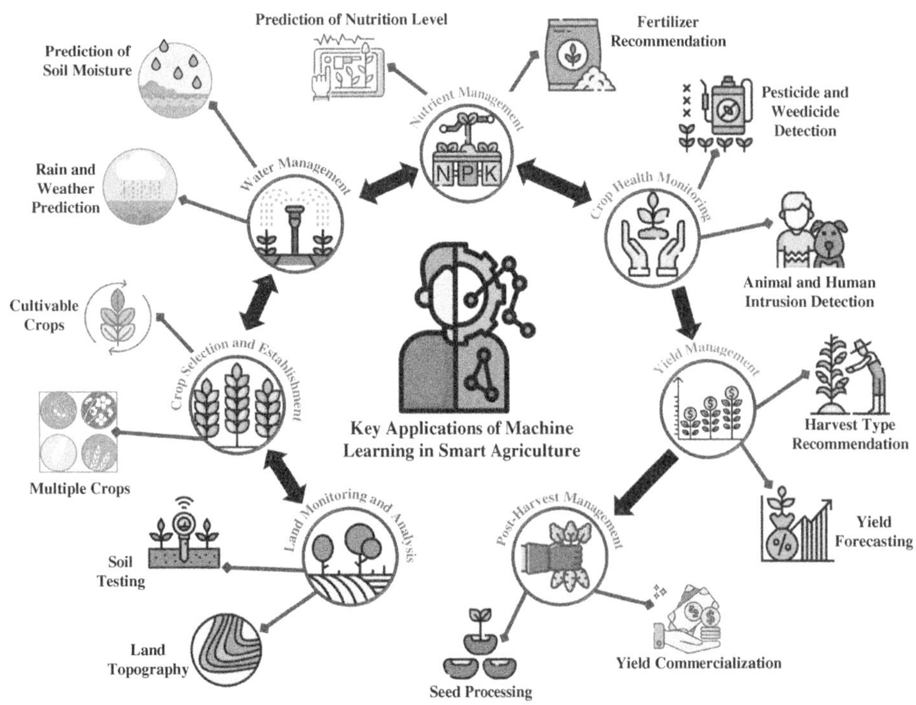

Figure 14.3 Key applications of machine learning in smart agriculture. Adapted from Rani et al. 2023.

and classify features from images. For example, a CNN-based model achieved 97.5% accuracy in identifying pests in guava orchards by utilizing features like Haralick features and color histograms (Ahmed et al. 2024). The MobileENet architecture further improved accuracy to 98.83% by optimizing computational efficiency (Aldosari 2024).

Transfer learning models, such as the Transfer-Driven Self-Adaptive Learning Model (TSLM), leverage pre-trained networks to enhance pest classification by adapting to specific agricultural contexts, thereby improving precision and efficiency (Guruprakash et al. 2024). Hybrid models combining ML and deep learning achieve over 95% accuracy in pest detection, particularly effective in early identification and preventing extensive crop damage (Zarboubi et al. 2024).

14.3.2.1 Disease Prediction and Control Measures

ML has revolutionized the prediction and control of plant diseases, enhancing agricultural productivity and food security. Convolutional Neural Networks (CNNs) play a key role in real-time disease detection through image processing, enabling early diagnosis and management by extracting features from images of diseased plants (Ahmed and Yadav 2023; Sivaprakash et al. 2023) (Figure 14.3). Hybrid models combining CNNs and Long Short-Term Memory (LSTM) networks further optimize detection accuracy, as seen in pomegranate fruit disease classification (Sivaprakash et al. 2023). Additionally, ML models use environmental data like temperature and humidity to predict disease outbreaks, such as the Cotton Leaf Curl Virus (CLCV) model (Metagar and Walikar 2024). Soil condition analysis also leverages ML to predict disease likelihood, offering a sustainable alternative to chemical treatments (Awais et al. 2023).

14.3.3 Yield Prediction and Crop Management

14.3.3.1 Predictive Models for Yield Based on Various Parameters (Soil, Weather, and Crop Type)

Predictive models for crop yield estimation are becoming increasingly sophisticated by incorporating diverse parameters such as soil characteristics, weather conditions, and crop types (Figure 14.3). By integrating temperature, rainfall, and humidity data, models can significantly enhance accuracy, as seen in a study achieving a correlation value of 0.761 (Ahmed et al. 2024).

Ensemble-based approaches like the Random Forest Extreme Gradient (RFXG) model have shown improved accuracy in predicting cotton yields by combining multiple ML techniques (Van et al. 2020). Furthermore, the Crop Yield Prediction Algorithm (CYPA) leverages IoT and big data to analyze extensive datasets, enabling precise yield predictions and informed decision-making (Talaat 2023).

14.3.3.2 Crop Rotation and Field Management Insights

ML models like Random Forest, Decision Tree, and K-Neighbors Classifier are being used to predict suitable crops for rotation, enhancing yield and sustainability by accurately analyzing historical data (Hageltoum et al. 2024). Building on this, Recurrent Neural Networks, particularly LSTMs in Seq2Seq architectures, are employed to predict crop rotation scenarios. This approach aids decision-making in sustainable agriculture by balancing profitability and environmental impact, further supporting informed choices for farmers (Gupta et al. 2024).

14.3.4 Climate Prediction and Weather Forecasting

14.3.4.1 Predictive Analytics for Extreme Weather Events

The integration of GIS spatial analysis, numerical weather models, historical data, and RS creates a powerful toolkit for predicting extreme weather events, enabling detailed risk mapping and temperature forecasting critical for resilience planning (Akana et al. 2024). Within this framework, ML plays a pivotal role: Long Short-Term Memory (LSTM) models, for instance, excel in handling complex, nonlinear rainfall patterns. With an R^2 of 0.235 and MSE of 92.18, these models enhance flood prediction accuracy, empowering decision-makers in agriculture and disaster management to act proactively (Sarwade et al. 2024).

14.3.4.2 Mitigating the Impact of Climate Change on Agriculture

Innovative systems like Renewable Energy-based Smart Agriculture Systems (RESAS) and crop simulation models are transforming agriculture's resilience to climate change. RESAS harnesses solar and wind power to drive real-time data analysis, leveraging AI for enhanced forecasting and resource optimization, thereby reducing carbon footprints compared to traditional systems (Savitha et al. 2024). Crop simulation models help evaluate farm-level adaptations, promoting sustainable practices (Annie et al. 2023). Meanwhile, agricultural biotechnology develops climate-adaptive crops through gene editing, enhancing resilience against extreme weather conditions and mitigating climate change impacts (Bonciu et al. 2020).

14.4 Data Collection and Preprocessing in Agriculture

14.4.1 Sources of Data in Agriculture

ML-driven smart agriculture relies on diverse data sources to optimize farming practices and enhance productivity. Key data sources include:

14.4.1.1 Role of Technology in Agricultural Data Collection

14.4.1.1.1 Soil Sensors

In SAS, many kinds of soil moisture sensors are used to investigate variables including conductivity and pH. By indirectly inferring soil organic matter and soil texture, soil conductivity maps aid in crop yield prediction. The availability of water and the existence of possible weeds are indicated by these two criteria. Therefore, the measurement of soil electrical conductivity is also used to estimate the amounts of herbicides that have been applied to the soil. Comparably, very acidic soils (pH 4.0 to 5.0) may contain large concentrations of soluble iron, manganese, and aluminum, which can be detrimental to certain plant growth, making soil pH essential for healthy crops. The pH range of approximately 6 to 7 is where plant nutrients are most effective (Qazi et al. 2022).

14.4.1.1.2 Satellite and Drone Imagery

In this article, the word "remote sensing" refers to the use of satellite-based sensors that can measure the reflected and emitted radiation from a specific location in order to identify and track its physical features. It is typically used to contrast with local observation. Although RS methods are primarily employed in earth scientific fields (geology, ecology, meteorology, etc.), they can also yield valuable data for the agricultural industry, such as Vegetation Indices (VIs). In order to increase the contribution of vegetation qualities, a VI is a spectral transformation of two or more electromagnetic spectrum spectral bands that are measured as reflectance from the Earth's surface.

Galford et al. (2008) provide an example of a satellite image, illustrating a pseudo-color infrared MODIS image. The image's various colors are used to convey different types of vegetation: the white areas on the map indicate bare agricultural fields, the bright turquoise blues show row-crop agriculture, the dark greens suggest Cerrado native vegetation, and the bright red areas represent Cerrado savanna native vegetation.

Drone technology is becoming more and more important in big data applications, particularly in agriculture. First, drones can quickly and effectively gather a variety of data related to agricultural output because they are outfitted with a variety of sensors and cameras. Infrared cameras, for example, can be used to track the growth state of crops, and multispectral sensors can yield valuable data like vegetation indices. By collecting data at a high frequency and with high resolution, agricultural decision-makers can gain a more thorough understanding of farming conditions and quickly spot problems like fertilizer deficits or plant diseases (Ray et al. 2024; Ghorai and Mochary 2025). Second, drone technology allows for accurate surveillance of vast agricultural areas. By use of automated flight and RS technology, drones are able to quickly cover large agricultural fields and provide precise geographic data. The sustainability and profitability of agricultural production can be improved by using this effective tool to help agricultural practitioners create better planting plans and irrigation schemes (Zhang 2024).

14.4.1.1.3 Internet of Things (IoT) Devices

The IoT is being used by farmers and the agriculture sector to boost analytics and production capacities. According to (Tzounis et al. 2017), the management and analysis of Internet of Things data, or "Big Data," can improve a number of activities, forecast outcomes, and automate procedures—all in real time. The Internet of Things (IoT) makes use of a variety of sensors that are made to perform particular tasks that will improve farming in the future. By 2020, there will be 75 million agricultural IoT device installs, a 20% yearly growth, predicts Business Insider Intelligence (Tang et al. 2021) Even if the 4G IoT network offers dependable connectivity and fast speeds, it is not feasible to integrate all of the smart farming equipment in remote areas at a cost that is affordable for both installation and upkeep. This endeavor has gotten easier for effectively exchanging resources and information since the introduction of 5G mobile networks and ultra-high-speed networks (Li et al. 2018).

IoT technology is growing in popularity in a number of industries, such as food processing, agricultural, security surveillance, and environmental monitoring (Da Xu et al. 2014).

The Industrial Internet of Things (IIoT), sometimes referred to as Industry 4.0, is what happens when IoT is used for industrial reasons. It combines advanced analytics, automation driven by AI, and intelligent manufacturing equipment to assist increase productivity for all workers and factories. Because it can store agricultural data and facilitate resource sharing at a reduced cost, cloud computing technology is another example of industry 4.0 (Rajeswari et al. 2017).

IoT and cloud computing integration in agriculture enables access to shared resources (Farooq et al. 2019).

14.4.1.1.4 Smartphone Applications

Cell phones are equipped with three different kinds of sensors: motion, environmental, and location. The first kind of motion sensor measures acceleration and rotational forces. Motion sensors include gyroscopes, accelerometers, rotating vector sensors, and gravity sensors. The second group includes environmental sensors, which take measurements of the environment. A thermometer can measure the ambient air temperature, a barometer can measure the pressure, and a photometer can measure the illumination. Position sensors, the third kind, measure the actual location of the gadget. These sensors include orientation sensors, GPS, and magnetometers (Liu et al. 2015). Communication channels (such as WIFI, Bluetooth, 3G, and NFC) and display channels are not included in the list of typical smartphone sensors for the purposes of this evaluation.

14.4.1.1.5 Digital Platforms

Digital platforms are currently reshaping global commercial and socioeconomic structures (de Reuver et al. 2018). This has sparked conversations on digital entrepreneurship, frugal innovations, data rights and privacy, the growth of innovation platform ecosystems in resource-constrained environments, and the digital economy and decent work (Mann 2018). Digital platforms, due to their inherent characteristics that distinguish them from their non-digital counterparts, provide affordance (both enabling and constraining) that can change, reorganize, or even demolish the systems they are introduced into (Kazan et al. 2014; Koskinen et al. 2018). Furthermore, these platforms' governance greatly influences how accessible, open, and inclusive they are, all of which have wider development-related ramifications.

14.4.2 Types of Data

Traditional farming practices have changed as a result of the incorporation of ML with agricultural data. Farmers can enhance yield efficiency, forecast environmental effects, and optimize resource utilization by examining a variety of variables.

14.4.2.1 Key Data Types in Smart Agriculture

14.4.2.1.1 Soil Data

When choosing the best times to plant and apply fertilizer, soil characteristics like pH, nutrient content, and moisture content are essential. There are several different types of soil sensors that monitor vital soil characteristics like temperature, PH, moisture content, and nutrient levels (nitrogen, potassium, and phosphorus). In order to prevent resource waste and guarantee the best possible nutrient availability for crops, these sensors continuously assess soil conditions to direct precise irrigation and fertilization techniques (Vuran et al. 2018).

Keswani et al. (2019) employed a Wireless Sensor Network (WSN) framework enabled by the Internet of Things (IoT) that comprised CO_2 sensors, light intensity sensors, ambient sensors, and a range of soil sensors to gather real-time agricultural data. These IoT-enabled WSN nodes collected and stored data, and then applied a neural network-based prediction model to optimize irrigation by regulating water valves based on estimates of soil water requirements. The Variable Learning Rate Gradient Descent (VLRGD) fared better in soil moisture prediction than Gradient Descent (GD), according to the study's comparative examination of optimization methodologies. The study also created a comprehensive map of soil moisture requirements utilizing interpolation and a Structural Similarity Index (SSIM)-based method to satisfy consistent irrigation needs under a variety of weather circumstances. This led to fuzzy logic weather modeling-based irrigation valve control directives that worked well.

14.4.2.2 Weather Data

Real-time weather information, such as temperature, humidity, wind speed, and precipitation, is provided by meteorological and environmental sensors in the field. Environmental factors like solar radiation and atmospheric pressure are also measured by these sensors.

They are crucial for predicting and controlling weather-related hazards, maximizing irrigation, and guaranteeing the overall health of crops. Crop management has advanced significantly with the introduction of wireless sensor networks in greenhouses. By giving plants an artificial habitat, greenhouses minimize risk and maximize crop production by successfully shielding them from weather, pests, and disease.

A wireless sensor network was put in place by (Savitha et al. 2022) to track variables like soil moisture, temperature, and humidity, guaranteeing automated climate control. Every sensor node sends its data to the cloud so that it may be tracked and examined.

A major technical advancement that simplifies agricultural procedures is the employment of sophisticated sensor systems in contemporary agriculture, especially in greenhouses.

According to (Touhami et al. 2018), this system uses a wireless sensor network, which comprises the DHT11 temperature and humidity sensor, and an intelligent LoWPAN gateway to connect the Zigbee network to the Internet. This makes it possible to transfer and analyze data continuously.

The DHT11 sensor is a useful tool in greenhouse settings because of its exceptional speed, precision, and high resolution in sensing temperature and humidity. These sensors gather data, which is subsequently relayed to an intelligent gateway via the ZigBee network and eventually to a web service via mobile data connection.

By analyzing this data, intelligent software programs allow actuators to respond automatically and adjust the greenhouse's environment. At a time when technology is transforming agriculture, this creative technique promotes higher-quality production while increasing greenhouse farming's efficiency.

14.4.2.3 Crop Health Data

Crop health is monitored with the help of high-resolution pictures from satellites and drones as well as information from ground-based sensors. Crop health sensors make it easier to identify diseases, nutritional shortages, and stressors early on, allowing for quicker action and higher yields. These sensors use methods like the Normalized Difference Vegetation Index (NDVI) to analyze variables including canopy temperature, vegetation indices, and leaf chlorophyll content in order to determine agricultural viability (Tripodi et al. 2018; Sun et al. 2022).

The use of RS methods and unmanned aerial vehicles (UAVs) in South African PA to enhance crop monitoring for smallholder farmers was the main emphasis of (Brewer et al. 2022). The study measures the amount of chlorophyll in maize at various stages of growth using multispectral UAV data and a random forest ML method. tent using red-edge and near-infrared wavelength bands, along with derived vegetation indices, giving vital information on the condition of maize during phenological stages. Particularly in the early vegetative, late vegetative, and early reproductive development stages, the random forest model operated at its best. The findings demonstrate the regional variability of chlorophyll in the maize field, offering useful data for agricultural management and decision-making. This study demonstrates how crucial it is to incorporate ML algorithms and drone imagery in PA to support small-scale farmers in preserving crop productivity and health.

Alam et al. (2022) introduced a real-time computer vision system for variable rate pesticide spraying that addresses issues like overdosing and underdosing that arise with conventional spraying techniques. A Random Forest classifier is used by the system to detect and categorize weeds and crops. Before the classification model is put into use for real-world testing, it is trained using a specially created data set.

Under the direction of a vision-based feedback system, the agrochemical spraying process guarantees precise application of the required agrochemical quantities through the use of a fluid flow control system based on pulse-width modulation (PWM). The results of several field tests confirm the real-time precision of the proposed vision-based pesticide spraying system.

14.4.2.4 Livestock Monitoring Data

It promotes real-time animal health and growth monitoring to guarantee maximum productivity. A variety of animal monitoring sensors, in contrast to crop monitoring sensors, are affixed on animals in order to track their health and performance. Additionally, farmers can use wearable technology and detectors to collect data, such as smart sensors that are affixed to an animal's neck, and transmit that data to a cloud server.

Livestock monitoring accounts for 6.66% of all studies (6 articles out of 90), which is significantly less than crop monitoring, irrigation, and disease detection. We can conclude from reading these articles that farmers can use RFID to track the location of individual animals and deter animal theft (Garg and Alam 2020). To guarantee improved animal health, producers may make decisions more easily and conduct ongoing monitoring thanks to advanced technologies (Mekala and Viswanathan 2017).

Farmers can use the Internet of Things (IoT) to track livestock using a variety of sensors that can regulate temperature, heart rate, and digestion, among other animal ailments (wa Maina 2017; Walter et al. 2017). Using ML methods, wa Maina (2017) investigated sensors to track the health of cows and dairy animals. Farms commonly use animal behavior recognition to evaluate the health of their animals, which increases overall profits.

Shakeel et al. (2022) propose the K-nearest neighbor (KNN) method as a way to improve recognition accuracy. The categorization process makes use of the random forest (RM) technique. The KNN-RM fusion model improves overall accuracy while reducing system computing costs when compared to other recognition models.

14.4.3 Data Preprocessing Steps and Challenges in Agricultural Data

Technology is helping agriculture more and more, especially ML, which improves precision farming, insect detection, soil health analysis, and crop production forecasting. However, preprocessing is an essential step before deploying ML models because agricultural data is frequently unstructured, noisy, and gathered from a variety of sources. Liakos et al. (2018) claim that ML applications in agriculture are greatly enhancing productivity and judgment. Agricultural data handling issues and necessary pretreatment methods are covered in this chapter.

14.4.3.1 Data Preprocessing Steps

14.4.3.1.1 Data Collection

Agricultural data is collected from a variety of sources, including sensors, satellites, drones, Internet of Things devices, and handwritten recordkeeping. The data collected often includes temperature, humidity, soil moisture, crop pictures, and weather patterns. The first step in preprocessing is to ensure that high-quality data is obtained. As Kamilaris and Prenafeta-Boldú (2018) point out, to improve agricultural insights, deep learning techniques are increasingly being used on these datasets.

i. **Data Cleaning**

   Data processing and cleaning are essential processes in ensuring the quality and usability of data when using big data in smart agriculture. To guarantee accuracy and completeness, data cleaning first entails filtering, eliminating outliers, and adding missing values to the raw data that has been gathered.

   Environmental influences and sensor equipment constraints can introduce noise or outliers into agricultural data; data cleaning successfully increases the data's reliability (Zhang 2024). Second, data processing is the process of arranging, calculating, and modifying cleaned data in order to provide higher-level information. In order to provide scientific support for agricultural decision-making, data processing in smart agriculture involves employing statistical techniques, ML algorithms, etc. to extract useful information from vast amounts of data, such as crop growth patterns and pest predictions.

   The dependability of ensuing data analysis and applications is closely correlated with how well data is cleaned and processed. Practitioners can improve the scientific and operational components of decision making by obtaining more accurate and dependable data through standardized data cleaning and processing techniques. Furthermore, a high-quality data foundation for agricultural big data visualization and model construction is provided by efficient data cleaning and processing (Zhang 2024). Thus, data processing and cleaning are essential to smart agriculture because they provide a strong basis for the comprehensive use of big data.

ii. **Data Transformation**

   Transformation processes include:

   - **Normalization:** Scaling data to a specific range (e.g., 0 to 1)
   - **Standardization:** Converting data to have zero mean and unit variance
   - **Encoding Categorical Variables:** Using techniques like one-hot encoding or label encoding for qualitative attributes such as crop types or soil classifications Schmidhuber (2015) highlights that deep learning models perform better when properly transformed and pre-processed datasets are used.

iii. **Feature Selection and Extraction**
The most crucial phases in the ML implementation procedure are feature selection and extraction as features have to be accessible at the time of predicting. Only the training dataset is used to incorporate feature selection and resampling techniques. The selection procedure is carried out at every internal resampling iteration since, in the context of food security, some properties (such as vegetation, climate, etc.) are significant, and food security data can already contain useful information when incorporated into ML models. Lastly, the forecasting outcomes are influenced by the data attributes.

iv. **Data Integration**
A lot of agricultural data comes from many sources; therefore, it needs to be integrated to make sure it is consistent and useful. Duplicate records, inconsistent data, and format mismatches are obstacles in this procedure that can compromise data accuracy. In order to remove redundancy, entity resolution—a crucial integration technique—finds and combines matched entries from several datasets. Schema alignment is another important strategy that creates a consistent format by combining data structures. Saiz-Rubio and Rovira-Más (2020) assert that combining various agricultural datasets is essential to improving smart farming decision-making and facilitating more accurate and data-driven farming methods.

v. **Data Augmentation**
For image-based agricultural ML applications, data augmentation techniques (e.g., flipping, rotation, contrast adjustments) can improve model robustness. Schmidhuber (2015) suggests that augmentation is particularly useful in deep learning applications for agriculture, where labeled data is scarce.

vi. **Data Splitting**
Data is separated into discrete subsets for model validation and training in order to facilitate efficient learning and assessment. The model learns patterns from the data by being trained on the training set. The validation set offers an intermediate assessment during training, which aids in adjusting model parameters and avoiding overfitting. The test set is used for the last evaluation to gauge how well the model generalizes to new inputs. Cross-validation methods, such k-fold cross-validation, make sure that performance indicators are independent of a single data split, which improves the resilience of model evaluation (Liakos et al. 2018).

14.4.3.2 Challenges in Agricultural Data Preprocessing

i. **Data Heterogeneity**
Standardization is difficult since agricultural data comes from a variety of sources with varying formats, resolutions, and time periods. Interoperability solutions are necessary to manage heterogeneous data, according to Yang et al. (2021).

ii. **Missing and Noisy Data**
Data gaps occur due to sensor failures, manual input errors, or adverse weather conditions. Effective imputation methods and noise reduction techniques are essential.

iii. **Labeling Issues**
Getting labeled data for supervised ML models is expensive and time-consuming. An important obstacle still exists in expert-driven annotation of agricultural datasets (Schmidhuber 2015).

iv. **Data Imbalance**
Limited data representation of specific crop diseases or meteorological circumstances may result in class imbalance. Oversampling, undersampling, and synthetic data generation techniques (like SMOTE) can all be beneficial (Liakos et al. 2018).

v. **Scalability**
Preprocessing large agricultural datasets demands a substantial amount of computing power. Scalability problems can be managed with the use of cloud computing technologies or effective data handling frameworks like Apache Spark (Liu et al. 2022).

vi. **Privacy and Security Concerns**
Protecting sensitive farm data is necessary to avoid abuse. Access management, data encryption, and adherence to agricultural data policies are essential (Kamilaris and Prenafeta-Boldú 2018).

14.5 ML Techniques in Agriculture

A subfield of AI and data science, ML is primarily concerned with the theory, performance, and features of learning systems and algorithms. AI, statistics, engineering, mathematics, information theory, cognitive science, and numerous other scientific and technical fields are all significantly impacted by this clever instrument (Cherkassky et al. 2007). Generally speaking, ML aims to maximize task performance by drawing on prior knowledge. Based on prior experiences, ML can efficiently create associations between data inputs and analyze vast amounts of data (Benos et al. 2021). In order to make judgments and predictions, it entails building mathematical models on sample data sets known as "training datasets" (Bal and Kayaalp 2021). The model or algorithm was trained first, and then it was classified or forecasted using ML.

14.5.1 Categorization of ML Techniques

Supervised Learning (SL) and Unsupervised Learning (UL) are the two main types of ML algorithms (Figure 14.4).

14.5.1.1 SL (Supervised Learning)

For out-of-sample data, SL trains a model to predict the target variable using a known set of labeled data (Barber 2012). Regression and classification methods are frequent uses of supervised learning. Table 14.1 shows a list of popular algorithms that fit within the various categories.

14.5.1.2 UL (Unsupervised Learning)

Conversely, UL uses data's inherent structures or hidden patterns to infer conclusions from unlabeled data. It is helpful in exploratory applications where the information contained in the data is unclear or there is no defined aim. Additionally, it works perfectly as a dimensionality reduction strategy for data with several features. Under this category of learning, clustering is the most widely used learning model. It can be used for exploratory data analysis, including object identification and gene sequencing (Bishop 2006). The choice of algorithm is determined by the amount of data, kind, and anticipated level of insight. Nevertheless, there is no set formula for choosing an algorithm; it is typically a trial-and-error process. In a variety of disciplines, IoT smart data analysis makes substantial use of both supervised and unsupervised learning approaches (Mahdavinejad et al. 2018).

WSN (Wireless Sensor Networks) and IoT-enabled smart farming is one of the areas where the approaches of ML are being developed to measure and understand the huge amount of data in this field. ML applications in precision agriculture (PA) can be categorizes into: soil management (Chlingaryan et al. 2018; Padarian et al. 2019), water management (Dou and Yang

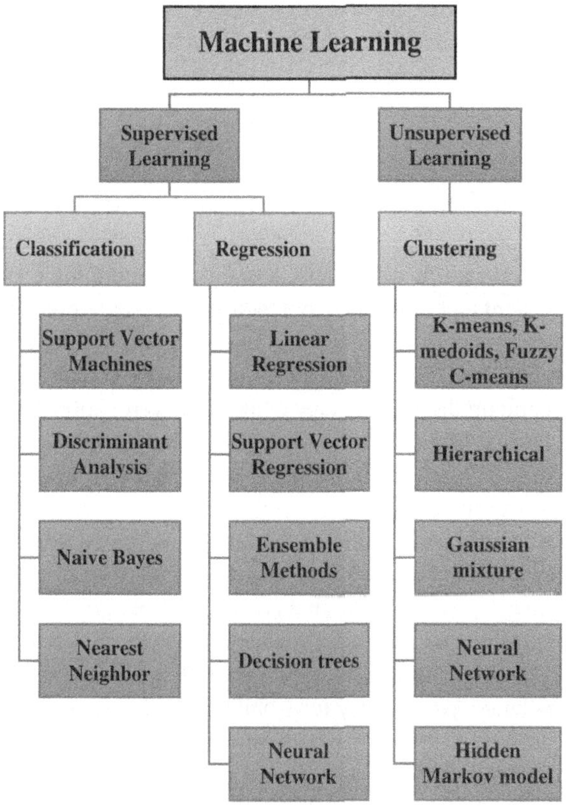

Figure 14.4 Algorithms of machine learning.

Table 14.1 Different ML techniques.

Techniques	Unsupervised Learning	Supervised Learning
Uses	• Clustering • Dimensionality reduction	• Classification • Regression • Estimation
Algorithms	• K-means • X-means • Anomaly detection • Principal component analysis • Independent component analysis • Gaussian mixture models	• Bayesian networks • Support vector machine • Random forest • Decision tree • Neural networks • Hidden Markov model • Naive Bayes

2018; Feng et al. 2017), livestock management (Gokul and Tadepalli 2017), or crop management (Chlingaryan et al. 2018; Mann et al. 2019; Pantazi et al. 2019; Irmak et al. 2006). The use of ML in crop management includes phenotype categorization (Ashapure et al. 2019), disease identification (Ramcharan et al. 2017), weed detection (Tiwari et al. 2019), and yield prediction (Chlingaryan et al. 2018; Mann et al. 2019; Pantazi et al. 2016; Irmak et al. 2006).

14.6 Challenges and Limitations of ML in Agriculture

The integration of ML in agriculture presents significant potential; however, it is accompanied by several challenges. Agricultural data often exhibit high variability, noise, and scarcity, posing difficulties in generating high-quality datasets for effective model training (Garg and Alam 2020). Furthermore, insufficient connectivity and underdeveloped technological infrastructure in rural areas create barriers to efficient data collection and model implementation (Balducci et al. 2018). Moreover, the adaptation of ML models to diverse agricultural environments necessitates a flexible approach that accounts for variations in climate, soil properties, and farming practices to ensure reliability and effectiveness (Araújo et al. 2023).

Effective data management is crucial, as delays in data retrieval can hinder real-time decision-making in agricultural applications. Many rural regions face challenges due to inadequate internet infrastructure, which limits the feasibility of IoT-driven data collection and cloud-based ML/DL services (Aker et al. 2016). Maintaining data accuracy and consistency is essential, as erroneous or incomplete datasets can introduce biases, compromise model reliability, and result in suboptimal farm management decisions. Maintaining contextual relevance is essential, as utilizing data from different environmental conditions may result in inaccurate predictions (Araújo et al. 2023). Additionally, ensuring data security and privacy is critical, requiring robust cybersecurity measures to protect sensitive information related to farm operations and crop yields (Garg and Alam 2020). Striking a balance between real-time and historical data usage is fundamental for optimizing predictive accuracy while effectively capturing long-term agricultural trends (Araújo et al. 2021).

Human expertise remains essential in the application of ML in agriculture, as AI should complement rather than replace traditional farming knowledge. Encouraging collaboration between technology and agricultural stakeholders is crucial for effective integration. The interpretability of ML models plays a key role in fostering trust, particularly among farmers with limited technical proficiency. Many farmers face challenges in adopting ML, deep learning (DL), and the Internet of Things (IoT) due to a lack of necessary skills and resources, underscoring the importance of targeted education and training initiatives to support adoption (Fabregas et al. 2019). Enhancing transparency in AI-driven decision-making will strengthen credibility and facilitate broader acceptance of technology-based agricultural solutions (Araújo et al. 2021).

The substantial costs associated with ML, deep learning (DL), and Internet of Things (IoT) solutions present a significant challenge, particularly for smallholder farmers with constrained financial resources. Ensuring affordability and demonstrating a clear return on investment (ROI) are critical factors for widespread adoption (Garg and Alam 2020; Benti et al. 2024). Developing cost-effective approaches, such as lightweight ML models optimized for low-power devices and the use of edge computing, can help mitigate resource limitations. Addressing the digital divide through targeted training programs and the development of scalable, low-cost AI-driven tools will be essential for fostering sustainable agricultural innovation.

14.7 Case Studies and Real-world Applications

Modern agriculture is undergoing a digital revolution, with ML and deep learning (DL) at its core. Below in Table 14.2, there are key case studies and applications demonstrating how these technologies are transforming farming practices.

14.8 Future Prospects and Emerging Trends

14.8.1 The Potential Impact of ML on Global Food Security Through Smart Agriculture

As the world's population is expected to reach 9.7 billion by 2050, maintaining global food security remains a top priority (FAO 2017). Ensuring food supply, affordability, and accessibility requires innovative thinking. ML-powered smart

Table 14.2 Real-world application of machine learning in agriculture.

Application Area	Case Study/ Technology	Key Features	Impact/Outcome	Key Takeaways and Future Directions	Research References
Crop Disease Detection	Innovatics' Deep Learning Solution	CNNs, MobileNetV2, Transfer Learning	High accuracy in disease diagnosis, reduced pesticide use	Hybrid models combining CNNs and LSTMs enhance disease classification accuracy.	Ahmed and Yadav (2023); Sivaprakash et al. (2023)
Weed Management	Blue River Technology's See and Spray	Computer Vision, Deep Learning	90% reduction in herbicide use, cost savings	AI-powered robots like Carbon Robotics' laser weeder reduce chemical usage.	Sawhney (2023)
Soil Health and Nutrient Optimization	Trace Genomics	ML for Soil Analysis, k-Nearest Neighbors	Tailored fertilization plans, improved yields	Machine learning models optimize fertilization based on soil nutrient analysis.	Rani et al. (2023)
Predictive Analytics for Yield	Cropin's AI-Driven Precision Farming	Satellite Imagery, Weather Data, ML Models	30% reduction in water use, improved yields	Models like RFXG and CYPA improve yield predictions using weather and soil data.	Jha (2025)
Climate Resilience and Adaptive Farming	Smart Agriculture Systems (RESAS)	Renewable Energy, Real-Time Data Analysis	Reduced carbon footprint, improved drought resilience	RESAS integrates renewable energy and ML for sustainable farming practices.	Savitha et al. (2024); Bonciu et al. (2020)

agriculture has emerged as a revolutionary approach to addressing these problems. By using data analytics, predictive modeling, and automation, ML boosts agricultural productivity, optimizes resource utilization, and lowers losses. This chapter highlights the primary applications, benefits, challenges, and possibilities of ML as it relates to global food security.

14.8.1.1 The Role of ML in Smart Agriculture

i. **Crop Yield Prediction**

Crop yield forecasting is essential to ensuring food availability and achieving food security. Performing this work with great precision at the regional and national levels is also preferable for quick decision-making (Shahhosseini 2021; Ruan et al. 2022). For example, decision makers can use precise crop yield projections to support the import/support decision. Because it is impacted by many different kinds of data (such as climate and oil data), this process is quite complicated. According to Tian et al. (2022), ML and DL techniques are commonly employed to predict agricultural productivity through the integration of multi-spectral and multi-temporal input data with ground observed data.

ii. **Precision Farming**

Which entails applying what is necessary when and where it is required, has further enhanced farm management efficiency with the integration of data-based digital systems that expand producers' understanding of their fields; this is referred to as Agriculture 4.0 or Digital Farming. The term Agriculture 5.0 refers to the general notion that is used when these data-driven farms integrate robotics with AI algorithms into their systems. According to several research, agricultural robots that incorporate AI are capable of performing some jobs more quickly than humans (Walch 2019).

Robotics is a rapidly expanding industry with a wide range of possible uses in agriculture, despite other research that challenges this claim (Saiz-Rubio and Rovira-Más 2020).

iii. **Pest and Disease Detection**

Reducing crop losses requires early disease and pest identification. ML-based image recognition systems use drones, smartphone applications, and satellite data to assess the health of plants. Precision agriculture improves farming's ability to control pests and diseases by utilizing cutting-edge technologies and data management strategies (Getahun et al. 2024). Through the integration of technologies like GPS, RS, and GISs, farmers can precisely monitor and control crop health (Sood et al. 2015). By offering comprehensive geographical and temporal data, these technologies enable the early diagnosis of disease outbreaks and pest infestations (Zhang et al. 2019). RS, for example, may detect minute variations in crop color and vigor that may be invisible to the human eye but are suggestive of stress or illness (Franzen et al. 2021). With the use of this data, farmers may carry out focused interventions, which lessens the need for general pesticide applications and encourages more environmentally friendly agricultural methods. The capacity to administer treatments more effectively and efficiently is one important advantage of precision agriculture in the management of pests and diseases (Roberts et al. 2021). VRT makes it possible to apply fertilizer and insecticides precisely, meeting the unique requirements of various fields (Pawase et al. 2023). This focused strategy reduces soil contamination and chemical runoff, optimizing resource usage and minimizing environmental effect. Additionally, predictive modeling, which forecasts pest and disease threats using historical data and present circumstances, may be integrated into precision agriculture (Ukhurebor et al. 2022). This enables farmers to prevent any problems before they worsen by taking preventative actions, such changing planting dates or choosing resistant crop types.

iv. **Soil Health Monitoring:**
Food security is threatened by soil degradation since it reduces the amount of arable land. To suggest the best land management techniques, ML algorithms examine the nutrient concentration, moisture content, and soil composition. Predictive analytics aids in preserving fertility and halting soil erosion (Lehmann et al. 2022).

v. **Climate Adaptation and Risk Assessment**
With its capacity to enhance crop yields and optimize resource usage, precision agriculture is becoming more widely acknowledged as an essential component of agricultural climate adaption methods (Lindblom et al. 2017). Changes in precipitation, temperature, and extreme weather occurrences are just a few of the ways that climate change is affecting weather patterns (Ummenhofer et al. 2017). PATs that provide real-time monitoring and data collecting capabilities, such as sensors, drones, and satellite imaging, let farmers better comprehend and react to these shifting conditions (Sishodia et al. 2020). Farmers may save water and reduce the danger of drought by modifying irrigation schedules to correspond with shifting precipitation patterns, for example, thanks to sensors that measure soil moisture levels.

vi. **Supply Chain Optimization**
By forecasting demand, streamlining logistics, and lowering post-harvest losses, ML simplifies agricultural supply chains (Liakos et al. 2018). Food waste is decreased, fair pricing is ensured, and market forecasting is aided by predictive analytics.

14.8.1.2 Benefits of ML-driven Smart Agriculture

- **Enhanced Productivity:** Higher agricultural yields result from more efficient farming that is data-driven (Schmidhuber 2015).
- **Resource Optimization:** By minimizing input waste, ML lowers expenses and its negative effects on the environment (Liu et al. 2020).
- **Food Waste Reduction:** Overproduction and post-harvest losses can be avoided with predictive analytics (Liakos et al. 2018).
- **Improved Farmer Decision-Making:** AI-powered insights provide farmers with practical, real-time advice (Kamilaris and Prenafeta-Boldú 2018).
- **Sustainability:** Eco-friendly agricultural methods are promoted by less reliance on chemical inputs (FAO 2018).

14.8.1.3 Challenges in Implementing ML in Agriculture

- **Data Availability and Quality:** Agricultural data frequently needs a great deal of preparation since it is unstructured and fragmented.
- **High Implementation Costs:** Training and infrastructure investments are necessary for the use of ML technology.
- **Limited Digital Literacy:** Many farmers lack the technical know-how to use ML-based solutions, particularly in underdeveloped nations (Schmidhuber 2015).
- **Ethical and Privacy Concerns:** To increase stakeholder confidence, data security and ownership concerns must be resolved (FAO 2018).
- **Scalability:** Smallholder farmers may find it difficult to integrate ML, even when it performs well on large farms with ample data (Liakos et al. 2018).

14.8.1.4 Future Prospects and Recommendations

With developments in edge computing, blockchain integration, and AI, ML in agriculture appears to have a bright future. To guarantee broader adoption, governments and the business sector should work together to offer financing, infrastructural assistance, and training. User-friendly ML technologies and open-access agricultural datasets will further democratize technology.

14.8.2 Future Trends: Future Trends in ML-driven Smart Agriculture: AI Integration, Robotics, and Precision Breeding

Agriculture has been transformed by ML, which increases sustainability, productivity, and resource optimization. Precision breeding, robotics, and AI integration are anticipated to be the main focuses of future developments in ML-driven smart agriculture as technology develops. Increased automation, better decision-making, and increased global food security are all benefits of these innovations. New developments in ML-driven agriculture are examined in this chapter, with an emphasis on the difficulties and possible effects.

14.8.2.1 AI Integration in Agriculture

i. **Autonomous Decision-making**
Real-time decision-making is made possible by AI-powered systems that analyze enormous volumes of agricultural data. Deep learning models are used by AI-driven systems to forecast weather patterns, plan irrigation schedules, and recommend optimum farming techniques.

ii. **Predictive Analytics for Crop Management**
Predictive analytics models powered by AI assist farmers in anticipating changes in yield, insect infestations, and crop illnesses. AI improves proactive farm management and risk assessment by utilizing real-time satellite images and IoT sensor data.

iii. **AI-Powered Smart Irrigation**
By examining weather patterns, crop needs, and soil moisture content, AI systems optimize water use. Irrigation systems that operate automatically save water and improve resource efficiency.

14.8.2.2 Robotics in Smart Agriculture

i. **Autonomous Farming Machinery**
Precision farming is made possible by robotics driven by ML algorithms, which enable autonomous harvesters, drones, and tractors. These devices facilitate large-scale farming operations, boost productivity, and lower labor expenses (Schmidhuber 2015).

ii. **Robotic Weed and Pest Control**
Herbicides are used less frequently when AI-powered robots detect and eradicate weeds. Similarly, vision-based ML models are used by robotic pest management systems to identify and target pests with little harm to the environment (Liakos et al. 2018).

iii. **Automated Harvesting Systems**
To assess fruit maturity and optimize harvesting schedules, robotic harvesting systems employ ML-driven image recognition. In addition to addressing labor shortages, this technology increases productivity and decreases food waste (Kamilaris and Prenafeta-Boldú 2018).

14.8.2.3 Precision Breeding with ML

i. **Genetic Data Analysis**
Large genetic datasets are analyzed using ML models to find desired crop attributes. Crop varieties that are disease-, climate-, and yield-resistant can be created through precision breeding.

ii. **AI-assisted Crop Improvement**
Plant breeding procedures can be accelerated by scientists using AI and ML systems (Figure 14.5). AI makes it possible to find genetic markers for desired features, which makes hybridization and selection processes more effective.

iii. **Livestock Genetic Optimization**
By evaluating genetic data to enhance characteristics like growth rates, disease resistance, and milk output, ML is also revolutionizing cattle breeding. AI-powered breeding initiatives improve the production and health of animals (FAO 2018).

14.8.2.4 Challenges and Ethical Considerations

i. **High Implementation Costs:** The adoption of AI-driven robotics and precision breeding requires significant investment in infrastructure and expertise.

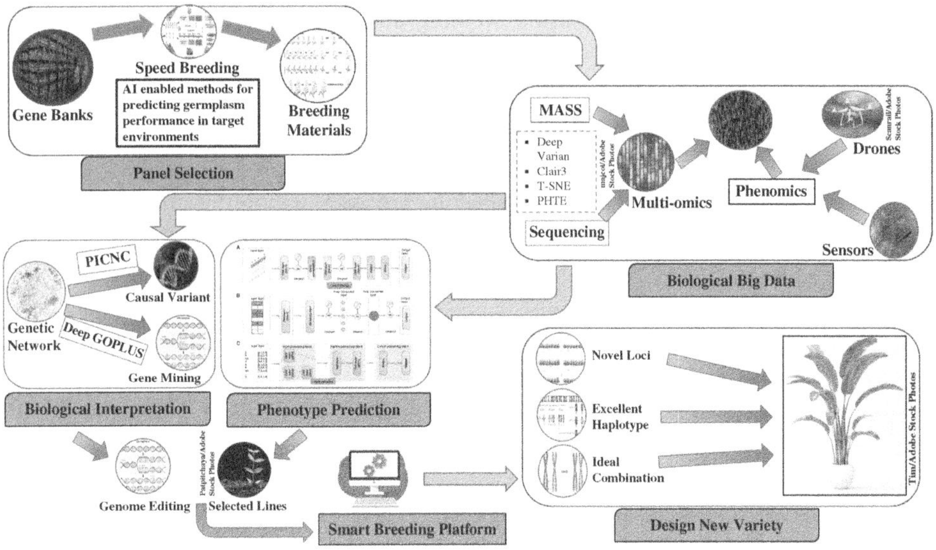

Figure 14.5 A strategic framework for AI-driven plant breeding.

ii. **Data Privacy and Security:** Ensuring the protection of genetic and agricultural data is crucial to prevent misuse.
iii. **Technological Accessibility:** Small-scale farmers may face barriers to accessing advanced AI and robotic technologies.
iv. **Environmental and Ethical Concerns:** The ethical implications of genetic modification and AI-driven automation need careful regulation and monitoring.

14.8.2.5 Future Outlook and Recommendations

The world's food supply could be completely transformed by combining agriculture with robotics, AI, and precision breeding. To maximize benefits, governments, educational institutions, and the private sector must collaborate by:

- Providing funding for AI-driven agricultural research and development.
- Promoting policies that ensure fair access to ML technologies for smallholder farmers.
- Investing in digital education and training programs for farmers.
- Encouraging ethical AI practices to ensure sustainable and responsible agricultural advancements.

14.8.3 Sustainability Implications and Environmental Impact of ML-driven Smart Agriculture

Agriculture needs to be sustainable to address food security and slow down environmental degradation. By reducing waste, enhancing ecological balance, and optimizing resources, ML-powered smart agriculture holds promise for enhancing sustainability. However, issues with energy use, e-waste, and technological inequality are also brought up by its adoption. This chapter examines the environmental effects and sustainability implications of ML-driven agriculture, stressing both the advantages and difficulties.

14.8.3.1 Environmental Benefits of ML-driven Smart Agriculture

i. **Resource Efficiency and Conservation**

Important agricultural resources like water, fertilizer, and pesticides are used as efficiently as possible by ML models. IoT sensors, RS, and predictive analytics are used in precision agricultural approaches to make sure:

- **Efficient Water Use:** Water waste is decreased via AI-powered irrigation systems that modify water distribution based on real-time soil moisture data (Liakos et al. 2018).
- **Fertilizer and Pesticide Optimization:** By anticipating nutrient shortages and insect infestations, ML algorithms allow for tailored administration and minimize chemical runoff (Kamilaris and Prenafeta-Boldú 2018).

ii. **Reduction in Greenhouse Gas Emissions**

A substantial portion of the world's greenhouse gas (GHG) emissions come from agriculture. ML-based methods reduce emissions in the following ways:

- **Improving Livestock Management:** AI lowers methane emissions from cattle by optimizing feeding procedures and monitoring animal health.
- **Enhancing Soil Carbon Sequestration:** Regenerative farming methods are made easier by ML-driven soil health assessments, which increases the soil's potential to store carbon (Schmidhuber 2015).
- **Optimizing Supply Chains:** Analytics for prediction minimize food waste and streamline operations to cut down on emissions from food transportation (Liakos et al. 2018).

iii. **Biodiversity and Soil Health Preservation**
ML-driven smart agriculture promotes sustainable land use by:
- **Preventing Soil Degradation:** Early indicators of soil erosion and nutrient depletion are detected by AI-powered monitoring systems, enabling prompt action (Kamilaris and Prenafeta-Boldú 2018).
- **Protecting Biodiversity:** By using fewer pesticides, precision farming protects the soil's microbial diversity and beneficial insects (Liu et al. 2022).

14.8.3.2 Environmental Challenges of ML-driven Smart Agriculture

i. **Energy Consumption and Carbon Footprint**
ML model implementation needs a significant amount of processing power, leading to:
- **High Energy Demand:** Electricity used to train deep learning models for agricultural applications is substantial and frequently comes from fossil fuels (Schmidhuber 2015).
- **Data Center Emissions:** Emissions of CO_2 are caused by the processing and storage of huge agricultural datasets (Liu et al. 2020).

ii. **E-Waste and Hardware Dependency**
The increasing reliance on digital agriculture generates electronic waste due to:
- **Short Hardware Lifespan:** The short operating lifespan of IoT sensors, drones, and automated machinery results in frequent replacements and disposal problems (FAO 2018).
- **Disposal of Obsolete Technology:** The quick development of AI-powered farming equipment leads to obsolete equipment that adds to the buildup of e-waste (Kamilaris and Prenafeta-Boldú 2018).

iii. **Digital Divide and Sustainability Disparities**
Although ML-driven agriculture improves sustainability, not everyone can profit from it because:
- **High Implementation Costs:** Small-scale farmers find it difficult to purchase AI-based agricultural solutions, especially in developing nations (Liakos et al. 2018).
- **Infrastructure Challenges:** The broad use of ML technology is hampered by poor internet access and a lack of technical know-how (FAO 2018).

14.8.3.3 Strategies for Sustainable Implementation

These tactics should be used to minimize adverse effects and optimize the sustainability advantages of ML-driven smart agriculture:

i. **Promoting Renewable Energy for AI Operations:** The carbon footprint of ML applications can be decreased by promoting the use of wind energy and solar-powered sensors in data centers (Liu et al. 2020).
ii. **Developing Sustainable Hardware:** The problem of e-waste can be solved by creating IoT gadgets that are both biodegradable and energy-efficient (Schmidhuber 2015).
iii. **Enhancing Accessibility and Affordability:** For smallholder farmers, governments and nonprofits should fund digital literacy initiatives and reasonably priced AI solutions (FAO 2018).
iv. **Encouraging Circular Economy Practices:** E-waste may be reduced by recycling electronic parts and encouraging the appropriate disposal of outdated equipment (Kamilaris and Prenafeta-Boldú 2018).

14.9 Conclusion

The Indian agriculture sector has yet to fully embrace advanced technologies, despite challenges such as land degradation, water scarcity, and climate change. RS and ML offer significant advantages but face limitations, including the unavailability of optical RS data on cloudy days and difficulties in real-time crop monitoring. Additionally, UAV usage remains limited, and the high cost of ML-based cognitive solutions hinders widespread adoption. Affordable, open-source platforms could bridge this gap, making smart farming accessible to more farmers. Decision support systems (DSSs) equipped with advanced algorithms can efficiently process vast agricultural datasets, yet high infrastructure costs often prevent adoption. To sustain India's large agricultural population, integrating sensors, automated data recording, ML, satellite and UAV datasets, and DSSs is crucial for enhancing crop productivity and quality. A holistic framework combining RS and ML can drive sustainable agriculture by providing data-driven insights and optimizing resource utilization. This systematic review serves as a valuable guide for policymakers, researchers, developers, and farmers, offering a comprehensive analysis of RS and ML advancements in Indian agriculture. By leveraging these technologies, the future of Indian agriculture can move toward greater efficiency, resilience, and sustainability.

References

Abioye, E.A., Abidin, M.S.Z., Mahmud, M.S.A. et al. (2020). A review on monitoring and advanced control strategies for precision irrigation. *Computers and Electronics in Agriculture* 173: 105441. https://doi.org/10.1016/j.compag.2020.105441

Ahmed, F.U., Das, A., and Zubair, M. (2024). A machine learning approach for crop yield and disease prediction integrating soil nutrition and weather factors. *Proceedings of the International Conference on Advances in Computing, Communication, Electrical, and Smart Systems (iCACCESS)*, Dhaka, Bangladesh (08–09 March 2024). Dhaka, Bangladesh: IEEE

Ahmed, I. and Yadav, P.K. (2023). Plant disease detection using machine learning approaches. *Expert Systems* 40 (5): e13136. https://doi.org/10.1111/exsy.13136

Ahmed, S., Marwat, S.N.K., Brahim, G.B. et al. (2024). IoT based intelligent pest management system for precision agriculture. *Scientific Reports* 14 (1): 31917. https://doi.org/10.1038/s41598-024-83012-3

Akana, C.M.V.S., Mishra, N., Kumar, K.S. et al. (2024). An in-depth analysis of advanced data analytics for accurate assessment and mitigation of extreme weather events. *Proceedings of the 1st International Conference on Innovative Sustainable Technologies for Energy, Mechatronics, and Smart Systems (ISTEMS)*, Dehradun, India (26–27 April 2024). Dehradun, India: IEEE.

Aker, J.C., Ghosh, I., and Burrell, J. (2016). The promise (and pitfalls) of ICT for agriculture initiatives. *Agricultural Economics* 47 (S1): 35–48. https://doi.org/10.1111/agec.12301

Alam, M., Alam, M.S., Roman, M. et al. (2020). Real-time machine-learning based crop/weed detection and classification for variable-rate spraying in precision agriculture. *Proceedings of the 7th International Conference on Electrical and Electronics Engineering (ICEEE)*, Antalya, Turkey (14–16 April 2020). Antalya, Turkey: IEEE.

Aldosari, H. (2024). An expert model using deep learning for image-based pest identification with the TSLM approach for enhancing precision farming. *Journal of Wireless Mobile Networks* 15 (3): 160–183. https://doi.org/10.58346/jowua.2024.i3.012

Annie, M., Pal, R.K., Gawai, A.S. et al. (2023). Assessing the impact of climate change on agricultural production using crop simulation model. *International Journal of Environment and Climate Change* 13: 538–550.

Araújo, S.O., Peres, R.S., Barata, J. et al. (2021). Characterising the agriculture 4.0 landscape—Emerging trends, challenges and opportunities. *Agronomy* 11 (4): 667. https://doi.org/10.3390/agronomy11040667

Araújo, S.O., Peres, R.S., Ramalho, J.C. et al. (2023). Machine learning applications in agriculture: current trends, challenges, and future perspectives. *Agronomy* 13 (12): 2976. https://doi.org/10.3390/agronomy13122976

Ashapure, A., Oh, S., Marconi, T.G. et al. (2019). Unmanned aerial system-based tomato yield estimation using machine learning. In: *Autonomous Air and Ground Sensing Systems for Agricultural Optimization and Phenotyping IV*, 11008: 110080O. Maryland, USA. https://doi.org/10.1117/12.2519129

Awais, M., Naqvi, S.M.Z.A., Zhang, H. et al. (2023). AI and machine learning for soil analysis: an assessment of sustainable agricultural practices. *Bioresources and Bioprocessing* 10 (1): 90. https://doi.org/10.1186/s40643-023-00710-y

Bachu, L., Kandibanda, A., Grandhi, N. et al. (2024). Machine learning for enhanced crop management and optimization of yield in precision agriculture. *Proceedings of the 8th International Conference on I-SMAC (IoT in Social, Mobile, Analytics and Cloud) (I-SMAC)*, Kirtipur, Nepal (03–05 October 2024). Kirtipur, Nepal: IEEE.

Bal, F. and Kayaalp, F. (2021). Review of machine learning and deep learning models in agriculture. *International Advanced Researches and Engineering Journal* 5: 309–323. https://doi.org/10.35860/iarej.848458

Balducci, F., Impedovo, D., and Pirlo, G. (2018). Machine learning applications on agricultural datasets for smart farm enhancement. *Machines* 6 (3): 38. https://doi.org/10.3390/machines6030038

Benos, L., Bechar, A., and Bochtis, D. (2020). Safety and ergonomics in human-robot interactive agricultural operations. *Biosystems Engineering* 200: 55–72. https://doi.org/10.1016/j.biosystemseng.2020.09.009

Benos, L., Tagarakis, A.C., Dolias, G. et al. (2021). Machine learning in agriculture: A comprehensive updated review. *Sensors* 21 (11): 3758. https://doi.org/10.3390/s21113758

Benti, N.E., Chaka, M.D., Semie, A.G. et al. (2024). Transforming agriculture with machine learning, deep learning, and IoT: Perspectives from Ethiopia—Challenges and opportunities. *Discover Agriculture* 2 (1): 63. https://doi.org/10.1007/s44279-024-00066-7

Bishop, C.M. (2006). *Pattern Recognition and Machine Learning*. Springer.

Bonciu, E. (2020). The climate change mitigation through agricultural biotechnologies. *Cadastre Series* 49 (1): 36–43.

Brewer, K., Clulow, A., Sibanda, M. et al. (2022). Predicting the chlorophyll content of maize over phenotyping as a proxy for crop health in smallholder farming systems. *Remote Sensing* 14 (3): 518. https://doi.org/10.3390/rs14030518

Cherkassky, V. and Mulier, F.M. (2007) *Learning from Data: Concepts, Theory, and Methods*. John Wiley and Sons.

Chlingaryan, A., Sukkarieh, S., and Whelan, B. (2018). Machine learning approaches for crop yield prediction and nitrogen status estimation in precision agriculture: A review. *Computers and Electronics in Agriculture* 151: 61–69. https://doi.org/10.1016/j.compag.2018.05.012

Conrad, Z., Niles, M.T., Neher, D.A. et al. (2018). Relationship between food waste, diet quality, and environmental sustainability. *PLOS One* 13 (4): e0195405. https://doi.org/10.1371/journal.pone.0195405

Da Xu, L., He, W., and Li, S. (2014). Internet of things in industries: A survey. *IEEE Transactions on Industrial Informatics* 10 (4): 2233–2243.

De Reuver, M., Sørensen, C., and Basole, R.C. (2018). The digital platform: A research agenda. *Journal of Information Technology* 33 (2): 124–135. https://doi.org/10.1057/s41265-016-0033-3

Diaz-Gonzalez, F.A., Vuelvas, J., Correa, C.A. et al. (2022). Machine learning and remote sensing techniques applied to estimate soil indicators: Review. *Ecological Indicators* 135: 108517. https://doi.org/10.1016/j.ecolind.2021.108517

Divyajyothi, M.G., Jopate, R., and Albalushi, R.A.A. (2024). AI precision for irrigation, crop management, and pest control for sustainable agriculture in Oman. *IOP Conference Series: Earth and Environmental Science* 1401 (1): 012005. https://doi.org/10.1088/1755-1315/1401/1/012005

Dou, X. and Yang, Y. (2018). Evapotranspiration estimation using four different machine learning approaches in different terrestrial ecosystems. *Computers and Electronics in Agriculture* 148: 95. https://doi.org/10.1016/j.compag.2018.03.010

Fabregas, R., Kremer, M., and Schilbach, F. (2019). Realizing the potential of digital development: The case of agricultural advice. *Science* 366 (6471): eaay3038.

FAO. (2017). *The Future of Food and Agriculture, Trends and Challenges*. FAO.

FAO. (2018). *Impact of Disasters and Crises on Agriculture and Food Security*. FAO.

Farooq, M.S., Riaz, S., Abid, A. et al. (2019). A survey on the role of IoT in agriculture for the implementation of smart farming. *IEEE Access* 7: 156237–156271.

Fedoroff, N.V., Battisti, D.S., Beachy, R.N. et al. (2010). Radically rethinking agriculture for the 21st century. *Science* 327: 833–834.

Feng, Y., Peng, Y., Cui, N. et al. (2017). Modeling reference evapotranspiration using extreme learning machine and generalized regression neural network only with temperature data. *Computers and Electronics in Agriculture* 136: 71. https://doi.org/10.1016/j.compag.2017.01.027

Fotabong, T.F., Muhammad, A., Keisham, M. et al. (2024). Soil analysis using deep learning for precision agriculture. *Proceedings of the First International Conference on Pioneering Developments in Computer Science and Digital Technologies (IC2SDT)*, Delhi, India (02–04 August 2024). Delhi, India: IEEE.

Franzen, D.W., Miao, Y., Kitchen, N.R. et al. (2021). Sensing for health, vigour and disease detection in row and grain crops. In: *Sensing Approaches for Precision Agriculture* (ed. Kerry, R., and Escolà, A.), 159–193. Springer: Cham. https://doi.org/10.1007/978-3-030-78431-7_6

Galford, G.L., Mustard, J.F., Melillo, J. et al. (2008). Wavelet analysis of MODIS time series to detect expansion and intensification of row-crop agriculture in Brazil. *Remote Sensing of Environment* 112 (2): 576–587. https://doi.org/10.1016/j.rse.2007.05.017

Garg, D. and Alam, M. (2020). Deep learning and IoT for agricultural applications. In: *Internet of Things (IoT)* (ed. Alam, M., Shakil, K.A., and Khan, S.), 273–284. Springer: Cham. https://doi.org/10.1007/978-3-030-37468-6_14

Getahun, S., Kefale, H., and Gelaye, Y. (2024). Application of precision agriculture technologies for sustainable crop production and environmental sustainability: A systematic review. *The Scientific World Journal* 2024 (1): 2126734. https://doi.org/10.1155/2024/2126734

Ghorai, S. and Mochary, S. (2025). Use of agricultural drone: Solving the farmers problems. *The Agriculture Magazine* 4 (5): 417–421.

Gokul, V. and Tadepalli, S. (2017). Implementation of smart infrastructure and non-invasive wearable for real time tracking and early identification of diseases in cattle farming using IoT. *Proceedings of the International Conference on I-SMAC (IoT in Social, Mobile, Analytics and Cloud) (I-SMAC)*, Palladam, India (10–11 February 2017). Palladam, India: IEEE.

Gomiero, T., Pimentel, D., and Paoletti, M.G. (2011). Environmental impact of different agricultural management practices: Conventional vs. organic agriculture. *Critical Reviews in Plant Sciences* 30: 95–124. https://doi.org/10.1080/07352689.2011.554355

González-Orozco, C.E., and Porcel, M. et al. (2020). Extreme climate variability weakens a major tropical agricultural hub. *Ecological Indicators* 111: 106015. https://doi.org/10.1016/j.ecolind.2019.106015

Gupta, C. (2024). Crop management using predictive analytics. In: *Exploring Ethical Dimensions of Environmental Sustainability and Use of AI* (ed. A. Ade-Ibijola, H. Kannan, Z.Z. Paprika, and R.V. Rodriguez), 106–125. IGI Global: New York.

Guruprakash, K.S., Siva Karthick, P., Ramachandran, A. et al. (2024). Crop pest identification using deep network based extracted features and MobileENet in smart agriculture. *Land Degradation and Development* 35 (11): 3642–3652. https://doi.org/10.1002/ldr.5157

Hageltoum, I.A., Abdallah, E.K., Alfeel, M.I. et al. (2024). Predictive modelling of crop rotation using data mining approaches. *Journal of Artificial Intelligence and Computational Technology* 1 (1): 32–46. https://doi.org/10.70274/jaict.2024.1.1.35

Irmak, A., Jones, J.W., Batchelor, W.D. et al. (2006). Artificial neural network model as a data analysis tool in precision farming. *Transactions of the ASABE* 49 (6): 2027.

Jha, K., Doshi, A., Patel, P. et al. (2019). comprehensive review on automation in agriculture using artificial intelligence. *Artificial Intelligence in Agriculture* 2: 1–12. https://doi.org/10.1016/j.aiia.2019.05.004

Jha, M. (2025). Driving innovation in agriculture: 3 key AgTech trends for 2025. https://www.cropin.com/blogs/agri-trends-2025.html (accessed 03 March 2025).

Kamilaris, A. and Prenafeta-Boldú, F.X. (2018). Deep learning in agriculture: A survey. *Computers and Electronics in Agriculture* 147: 70–90. https://doi.org/10.1016/j.compag.2018.02.016

Karim, F. and Karim, F. (2017). Monitoring system using web of things in precision agriculture. *Procedia Computer Science* 110: 402–409. https://doi.org/10.1016/j.procs.2017.06.083

Kasampalis, D.A., Alexandridis, T.K., Deva, C. et al. (2018). Contribution of remote sensing on crop models: A review. *Journal of Imaging* 4: 52. https://doi.org/10.3390/jimaging4040052

Kazan, E., Tan, C.W., and Lim, E.T. (2014). Towards a framework of digital platform disruption: A comparative study of centralized and decentralized digital payment providers. *Proceedings of the 25th Australasian Conference on Information Systems: ACIS 2014*, Auckland, New Zealand (8–10 December 2024). Auckland, New Zealand: Australasian Association for Information Systems (AAIS).

Keswani, B., Mohapatra, A.G., Mohanty, A. et al. (2019). Adapting weather conditions based IoT enabled smart irrigation technique in precision agriculture mechanisms. *Neural Computing and Applications* 31: 277–292. https://doi.org/10.1007/s00521-018-3737-1

Korlepara, N.P., Raju, V.N., Satyanarayana, P.V.V. et al. (2024). Real-time precision irrigation system for optimal crop yield and water conservation. *IOP Conference Series: Earth and Environmental Science* 1375 (1): 012019. https://doi.org/10.1088/1755-1315/1375/1/012019

Koskinen, K., Bonina, C., and Eaton, B. (2019). Digital platforms in the global south: Foundations and research agenda. *Proceedings of the Information and Communication Technologies for Development. Strengthening Southern-Driven Cooperation as a Catalyst for ICT4D: 15th IFIP WG 9.4 International Conference on Social Implications of Computers in Developing Countries (ICT4D 2019)*, Dares Salaam, Tanzania (1–3 May 2019). Cham, Switzerland: Springer.

Lakshmi, G., Brindha, S., Sathish, V. et al. (2024). Advancing precision agriculture: AI, IoT, and machine learning solutions for optimal crop yield. *Proceedings of the International Conference on Communication, Computing and Internet of Things (IC3IoT)*, Chennai, India (17–18 April 2024). Chennai, India: IEEE.

Lampridi, M.G., Sørensen, C.G., and Bochtis, D. (2019). Agricultural sustainability: A review of concepts and methods. *Sustainability* 11 (18): 5120. https://doi.org/10.3390/su11185120

Lehmann, J., Bossio, D.A., Kögel-Knabner, I. et al. (2020). The concept and future prospects of soil health. *Nature Reviews Earth and Environment* 1 (10): 544–553. https://doi.org/10.1038/s43017-020-0080-8

Li, F., Dong, S., Leier, A. et al. (2022). Positive-unlabeled learning in bioinformatics and computational biology: A brief review. *Briefings in Bioinformatics* 23 (1): bbab461. https://doi.org/10.1093/bib/bbab461

Li, S., Da Xu, L., and Zhao, S. (2018). 5G Internet of Things: A survey. *Journal of Industrial Information Integration* 10: 1–9. https://doi.org/10.1016/j.jii.2018.01.005

Liakos, K.G., Busato, P., Moshou, D. et al. (2018). Machine learning in agriculture: A review. *Sensors* 18 (8): 2674. https://doi.org/10.3390/s18082674

Lindblom, J., Lundström, C., Ljung, M. et al. (2017). Promoting sustainable intensification in precision agriculture: Review of decision support systems development and strategies. *Precision Agriculture* 18: 309–331. https://doi.org/10.1007/s11119-016-9491-4

Liu, X., Liu, J., and Wang, W. (2015). Exploring sensor usage behaviors of android applications based on data flow analysis. *Proceedings of the IEEE 34th International Performance Computing and Communications Conference (IPCCC)*, Nanjing, China (14–16 December 2015). Nanjing, China: IEEE.

Liu, Z., Hu, H., Lin, Y. et al. (2022). Swin transformer v2: Scaling up capacity and resolution. *Proceedings of the IEEE/CVF Conference on Computer Vision and Pattern Recognition (CVPR)*, New Orleans, USA (19–24 June 2022). California, USA: IEEE CPS.

Mahdavinejad, M.S., Rezvan, M., Barekatain, M. et al. (2018). Machine learning for internet of things data analysis: A survey. *Digital Communications and Networks* 4 (3): 161. https://doi.org/10.1016/j.dcan.2017.10.002

Manikandababu, C.S., Preethi, V., Kanna, M.Y. et al. (2024). Enhancing crop yield prediction with IoT and machine learning in precision agriculture. *Proceedings of the International Conference on Advances in Computing, Communication and Applied Informatics (ACCAI)*, Chennai, India (09–10 May 2024). Chennai, India: IEEE.

Mann, L. (2018). Left to other peoples' devices? A political economy perspective on the big data revolution in development. *Development and Change* 49 (1): 3–36. https://doi.org/10.1111/dech.12347

Mann, M.L., Warner, J.M., and Malik, A.S. (2019). Predicting high-magnitude, low-frequency crop losses using machine learning: An application to cereal crops in Ethiopia. *Climatic Change* 154: 211–227. https://doi.org/10.1007/s10584-019-02432-7

Mekala, M.S. and Viswanathan, P. (2017). A survey: Smart agriculture IoT with cloud computing. *Proceedings of the International Conference on Microelectronic Devices, Circuits and Systems (ICMDCS)*, Vellore, India (10–12 August 2017). Vellore, India: IEEE.

Metagar, S.M. and Walikar, G.A. (2024). Machine learning models for plant disease prediction and detection: A review. *Agricultural Science Digest* 44 (4): 591–602.

Mirani, A., Memon, M.S., Chohan, R. et al. (2021). Machine learning in agriculture: A review. *International Journal of Scientific and Technology Research* 10 (5): 229–234.

Mishra, S., Mishra, D., and Santra, G.H. (2016). Applications of machine learning techniques in agricultural crop production: A review paper. *Indian Journal of Science and Technology* 9 (38): 1–14.

Nassani, A.A., Awan, U., Zaman, K. et al. (2019). Management of natural resources and material pricing: Global evidence. *Resources Policy* 64: 101500. https://doi.org/10.1016/j.resourpol.2019.101500

Ouf, N.S. (2018). A review on the relevant applications of machine learning in agriculture. *Methods* 6 (8): 1–17.

Padarian, J., Minasny, B., and McBratney, A.B. (2019). Using deep learning to predict soil properties from regional spectral data. *Geoderma Regional* 16: e00198. https://doi.org/10.1016/j.geodrs.2018.e00198

Pantazi, X.E., Moshou, D., Alexandridis, T. et al. (2016). Wheat yield prediction using machine learning and advanced sensing techniques. *Computers and Electronics in Agriculture* 121: 57. https://doi.org/10.1016/j.compag.2015.11.018

Pawase, P.P., Nalawade, S.M., Bhanage, G.B. et al. (2023). Variable rate fertilizer application technology for nutrient management: A review. *International Journal of Agricultural and Biological Engineering* 16 (4): 11–19.

Petegrosso, R., Li, Z., and Kuang, R. (2020). Machine learning and statistical methods for clustering single-cell RNA-sequencing data. *Briefings in Bioinformatics* 21 (4): 1209–1223. https://doi.org/10.1093/bib/bbz063

Qazi, S., Khawaja, B.A., and Farooq, Q.U. (2022). IoT-equipped and AI-enabled next generation smart agriculture: A critical review, current challenges and future trends. *IEEE Access* 10: 21219–21235.

Rajeswari, S.K.R.K.A., Suthendran, K., and Rajakumar, K. (2017). A smart agricultural model by integrating IoT, mobile and cloud-based big data analytics. *Proceedings of the International Conference on Intelligent Computing and Control (I2C2)*, Coimbatore, India (23–24 June 2017). Coimbatore, India: IEEE.

Ramcharan, A., Baranowski, K., McCloskey, P. et al. (2017) Deep learning for image-based cassava disease detection. *Frontiers in Plant Science* 8: 1852. https://doi.org/10.3389/fpls.2017.01852

Rani, S., Mishra, A.K., Kataria, A. et al. (2023). Machine learning-based optimal crop selection system in smart agriculture. *Scientific Reports* 13 (1). https://doi.org/10.1038/s41598-023-42356-y

Ray, J., Ghosh, P.K., Banik, S. et al. (2024). Exploring agriculture in the age of drones: A comprehensive review. *African Journal of Biological Sciences* 6 (3): 216–229.

Roberts, D.P., Short, N.M., Sill, J. et al. (2021). Precision agriculture and geospatial techniques for sustainable disease control. *Indian Phytopathology* 74 (2): 287–305. https://doi.org/10.1007/s42360-021-00334-2

Ruan, G., Li, X., Yuan, F. et al. (2022). Improving wheat yield prediction integrating proximal sensing and weather data with machine learning. *Computers and Electronics in Agriculture* 195: 106852. https://doi.org/10.1016/j.compag.2022.106852

Saiz-Rubio, V. and Rovira-Más, F. (2020). From smart farming towards agriculture 5.0: A review on crop data management. *Agronomy* 10 (2): 207. https://doi.org/10.3390/agronomy10020207

Sarwade, M.P., Shinde, S.A., and Patil, V.S. (2024). Predictive modeling of extreme weather forecasting events: An LSTM approach. *Current World Environment* 19 (1): 185. http://dx.doi.org/10.12944/CWE

Savitha, A.C., Aravind, H.S., Jayaram, M.N. et al. (2022). Monitoring plant growth in a greenhouse using IoT with the energy-efficient wireless sensor network. In: *Recent Innovations in Artificial Intelligence and Smart Applications.* (ed. Al-Emran, M., and Shaalan, K.), 85–104. Springer: Cham. https://doi.org/10.1007/978-3-031-14748-7_5

Savitha, A.C., DSouza, S., Shanthakumar, S. et al. (2024). Renewable energy-based smart agriculture systems for climate change prediction and impact mitigation. *Proceedings of the Second International Conference on Networks, Multimedia and Information Technology (NMITCON)*, Bengaluru, India (09–10 August 2024). Bengaluru, India: IEEE.

Sawhney, D. (2023). Machine learning in agriculture: Use cases and applications. https://www.itransition.com/machine-learning/agriculture (accessed 28 February 2025).

Schmidhuber, J. (2015). Deep learning in neural networks: An overview. *Neural networks* 61: 85–117. https://doi.org/10.1016/j.neunet.2014.09.003

Shahhosseini, M., Hu, G., Huber, I. et al. (2021). Coupling machine learning and crop modeling improves crop yield prediction in the US Corn Belt. *Scientific Reports* 11 (1): 1606. https://doi.org/10.1038/s41598-020-80820-1

Shakeel, P.M., bin Mohd Aboobaider, B., and Salahuddin, L.B. (2022). A deep learning-based cow behavior recognition scheme for improving cattle behavior modeling in smart farming. *Internet of Things* 19: 100539. https://doi.org/10.1016/j.iot.2022.100539

Sharma, A., Jain, A., and Gupta, P. (2020). Machine learning applications for precision agriculture: A comprehensive review. *IEEE Access* 9: 4843–4873.

Sishodia, R.P., Ray, R.L., and Singh, S.K. (2020). Applications of remote sensing in precision agriculture: A review. *Remote Sensing* 12 (19): 3136. https://doi.org/10.3390/rs12193136

Sivaprakash, P., Sankar, M., Chithambaramani, R. et al. (2023). Crop disease prediction using machine learning. *Proceedings of the International Conference on System, Computation, Automation and Networking (ICSCAN)*, Puducherry, India (17–18 November 2023). Puducherry, India: IEEE.

Sood, K., Singh, S., Rana, R.S. et al. (2015). Application of GIS in precision agriculture. *Paper presented as lead lecture in national seminar on precision farming technologies for high Himalayas*, Jammu and Kashmir, India (October 2015).

Sun, H., Li, M., and Zhang, Q. (2022). Crop sensing in precision agriculture. In: *Soil and Crop Sensing for Precision Crop Production* (ed. Li, M., Yang, C., and Zhang, Q.), 251–293. Springer Nature: Switzerland. https://doi.org/10.1007/978-3-030-70432-2_8

Talaat, F.M. (2023). Crop yield prediction algorithm (CYPA) in precision agriculture based on IoT techniques and climate changes. *Neural Computing and Applications* 35 (23): 17281–17292. https://doi.org/10.1007/s00521-023-08619-5

Tang, Y., Dananjayan, S., Hou, C. et al. (2021). A survey on the 5G network and its impact on agriculture: Challenges and opportunities. *Computers and Electronics in Agriculture* 180: 105895. https://doi.org/10.1016/j.compag.2020.105895

Thayer, A.W., Vargas, A., Castellanos, A.A. et al. (2020). Integrating agriculture and ecosystems to find suitable adaptations to climate change. *Climate* 8 (1): 10. https://doi.org/10.3390/cli8010010

Tian, H., Wang, P., Tansey, K. et al. (2021). An LSTM neural network for improving wheat yield estimates by integrating remote sensing data and meteorological data in the Guanzhong Plain, PR China. *Agricultural and Forest Meteorology* 310: 108629. https://doi.org/10.1016/j.agrformet.2021.108629

Tiwari, O., Goyal, V., Kumar, P. et al. (2019). An experimental set up for utilizing convolutional neural network in automated weed detection. *Proceedings of the 2019 4th International Conference on Internet of Things: Smart Innovation and Usages (IoT-SIU)*, Ghaziabad, India (18–19 April 2019). Ghaziabad, India: IEEE.

Touhami, A., Benahmed, K., and Bounaama, F. (2018). Monitoring of greenhouse based on internet of things and wireless sensor network. In: *Proceedings of the 8th International Conference on Sciences of Electronics, Technologies of Information and Telecommunications (SETIT'18)* (ed. Bouhlel, M.S. and Rovetta, S.), 281–289, Springer: Cham.

Triantafyllou, A., Sarigiannidis, P., and Bibi, S. (2019). Precision agriculture: A remote sensing monitoring system architecture. *Information* 10 (11): 348. https://doi.org/10.3390/info10110348

Tripodi, P., Massa, D., Venezia, A. et al. (2018). Sensing technologies for precision phenotyping in vegetable crops: Current status and future challenges. *Agronomy* 8 (4): 57. https://doi.org/10.3390/agronomy8040057

Tzounis, A., Katsoulas, N., Bartzanas, T. et al. (2017). Internet of Things in agriculture, recent advances and future challenges. *Biosystems Engineering* 164: 31–48. https://doi.org/10.1016/j.biosystemseng.2017.09.007

Ukhurebor, K.E., Adetunji, C.O., Olugbemi, O.T. et al. (2022). Precision agriculture: Weather forecasting for future farming. In: *AI, Edge and IoT-based Smart Agriculture* (ed. Abraham, A., Dash, S., Rodrigues, J.J.P.C. et al.), 101–121. Academic Press. https://doi.org/10.1016/B978-0-12-823694-9.00008-6

Ummenhofer, C.C. and Meehl, G.A. (2017). Extreme weather and climate events with ecological relevance: A review. *Philosophical Transactions of the Royal Society B: Biological Sciences* 372 (1723): 20160135. https://doi.org/10.1098/rstb.2016.0135

Van Klompenburg, T., Kassahun, A., and Catal, C. (2020). Crop yield prediction using machine learning: A systematic literature review. *Computers and Electronics in Agriculture* 177: 105709. https://doi.org/10.1016/j.compag.2020.105709

Veeragandham, S. and Santhi, H. (2020). A review on the role of machine learning in agriculture. *Scalable Computing: Practice and Experience* 21 (4): 583–589. https://doi.org/10.12694/scpe.v21i4.1699

Vuran, M.C., Salam, A., Wong, R. et al. (2018). Internet of underground things in precision agriculture: Architecture and technology aspects. *Ad Hoc Networks* 81: 160–173. https://doi.org/10.1016/j.adhoc.2018.07.017

wa Maina, C. (2017). IoT at the grassroots—Exploring the use of sensors for livestock monitoring. *Proceedings of the IST-Africa Week Conference (IST-Africa)*, Windhoek, Namibia (30 May 2017–02 June 2017). Windhoek, Namibia: IEEE.

Wakhare, P.B., Neduncheliyan, S., and Sonawane, G.S. (2020). Automatic irrigation system based on Internet of Things for crop yield prediction. *Proceedings of the International Conference on Emerging Smart Computing and Informatics (ESCI)*, Pune, India (12–14 March 2020). Pune, India: IEEE.

Walch, K. (2019). How AI is transforming agriculture. https://www.forbes.com/sites/cognitiveworld/2019/07/05/how-ai-is-transforming-agriculture (accessed 28 February 2025).

Walter, A., Finger, R., Huber, R. et al. (2017). Smart farming is key to developing sustainable agriculture. *Proceedings of the National Academy of Sciences* 114 (24): 6148–6150. https://doi.org/10.1073/pnas.1707462114

Wheeler, T. and Braun, V.J. (2013). Climate change impacts on global food security. *Science* 341: 508–513.

Yan, J. and Wang, X. (2022). Unsupervised and semi-supervised learning: The next frontier in machine learning for plant systems biology. *The Plant Journal* 111 (6): 1527–1538. https://doi.org/10.1111/tpj.15905

Yang, C., Wang, Q., Xu, M. et al. (2021). Characterizing impacts of heterogeneity in federated learning upon large-scale smartphone data. *Proceedings of the Web Conference 2021*, Ljubljana, Slovenia (19–23 April 2021). New York, USA: Association for Computing Machinery.

Zarboubi, M., Bellout, A., Chabaa, S. et al. (2024). IoT-based pest detection in agriculture using Raspberry Pi and YOLOv10m for precision farming. *Transactions of International Conference on Mobility, Artificial Intelligence and Health (MAIH2024)*, Marrakesh, Morocco (20–22 November 2024). Marrakesh, Morocco: ITM Web of Conferences. https://doi.org/10.1051/itmconf/20246904017

Zhang, J., Huang, Y., Pu, R. et al. (2019). Monitoring plant diseases and pests through remote sensing technology: A review. *Computers and Electronics in Agriculture* 165: 104943. https://doi.org/10.1016/j.compag.2019.104943

Zhang, Y. (2024). Application of big data in smart agriculture. *Advances in Resources Research* 4 (2): 221–230. https://doi.org/10.50908/arr.4.2_221

15

Plant Leaf Disease Detection and Classification Using Convolutional Neural Networks

A. V. Senthil Kumar[1], N. Abinesh[1], Shanmugasundaram Hariharan[2], Kyla L. Tennin[3]*

[1] *Hindusthan College of Arts & Science, Coimbatore, India*
[2] *Department of Artificial Intelligence and Data Science, Vardhaman College of Engineering, India*
[3] *United Nations Peace Ambassador Foundation, Southaven, USA*

*Corresponding author: avsenthilkumar@yahoo.com

15.1 Introduction

The global agricultural business is persistently dealing with challenges to satisfy the increasing needs. The global agricultural sector faces continual pressure to satisfy the increasing demands for food, fuel, and raw materials. As climate change, environmental degradation, and limited arable land increasingly impact crop production, the need for effective and sustainable farming practices has never been more critical. Among the myriad issues confronting farmers, plant diseases pose a substantial risk to crop output, quality, and overall agricultural productivity. These infections, caused by various pathogens including fungi, bacteria, viruses, and nematodes, can result in significant losses if not swiftly recognized and addressed. Conventional techniques for identifying plant diseases typically depend on visual assessments conducted by experienced specialists, which can be labor-intensive, subjective, and susceptible to inaccuracies, especially when the disease manifestations are not readily visible or are subtle.

The emergence of technology has created new opportunities for improving plant disease management. One notable breakthrough is the utilization of machine learning (ML) techniques, which can analyze extensive datasets and discern patterns that are frequently undetectable to human observation. Machine learning, a branch of artificial intelligence (AI), facilitates the automation of the identification and classification of plant diseases, making the process more rapid, precise, and scalable. Machine learning algorithms can be trained to identify nuanced variations in plant health and detect early disease indicators by employing extensive datasets of plant photos, environmental variables, and disease symptoms, even prior to visual manifestation.

The integration of ML into plant disease detection and classification systems is transforming the way farmers and agricultural professionals approach crop management. Instead of relying solely on the expertise of agronomists, farmers can now use mobile apps, drones, and remote sensing technologies equipped with ML algorithms to diagnose plant diseases in real time, often in the field. These tools not only enable more precise identification of diseases but also offer predictive capabilities, allowing farmers to implement preventive measures before a disease spreads and causes widespread damage. As a result, ML-based systems hold the promise of increasing agricultural productivity, reducing crop loss, and minimizing the use of harmful pesticides, leading to more sustainable farming practices.

ML methodologies for the identification of plant diseases can be broadly categorized as supervised and unsupervised learning techniques. Monitored learning algorithms such as convolutional neural networks (CNNs), decision trees, and support vector machines (SVMs) necessitate labeled datasets to train the model for the identification of specific illness patterns. These algorithms are particularly effective in scenarios when disease kinds and symptoms are clearly delineated. Conversely, unsupervised learning techniques, such as clustering and anomaly detection, are advantageous in scenarios where disease traits are poorly understood or when identifying novel plant diseases. Deep learning, a subset of ML, has

Machine Learning for Plant Biology, First Edition. Edited by Jen-Tsung Chen.

demonstrated significant potential in the identification of plant diseases using images, owing to its capacity to learn intricate characteristics and hierarchies from extensive image datasets.

Considering the considerable capability of ML in identifying plant diseases, numerous problems persist. A primary impediment is the necessity for high-quality labeled data, which is crucial for training precise models. Furthermore, environmental factors, including illumination conditions, plant species, and developmental stages, might affect the manifestation of foliar disease symptoms, complicating the establishment of generalized models applicable across diverse geographical areas and plant taxa. Moreover, guaranteeing the resilience and comprehensibility of ML models is essential for their effective implementation by farmers and agricultural stakeholders.

In this chapter, the various ML algorithms that have been applied to plant disease detection and classification, emphasizing their strengths, limitations, and potential for real-world implementation, were explored. Recent advancements in neural network techniques and how they have revolutionized the field of plant pathology were discussed. Additionally, we will analyze case studies and research findings that illustrate the efficacy of ML models in detecting illnesses in diverse crops. By comprehending the possibilities and constraints of these technologies, we seek to offer insights into their integration into contemporary agricultural practices to improve disease management, promote food security, and support more sustainable farming systems.

Ultimately, the use of ML in plant disease detection offers an exciting opportunity to revolutionize the way diseases are diagnosed and managed in agriculture. Ongoing enhancements in processing capacity, data accessibility, and algorithmic refinement position ML as a crucial instrument in the worldwide initiative to address plant diseases and provide a stable and resilient food supply for the future.

15.2 Plant Disease Challenges and Issues

The incorporation of ML algorithms into plant disease and pest detection has shown great promise in revolutionizing the way agricultural practitioners approach crop management. By enabling accurate, timely, and scalable diagnosis of plant diseases and pests, ML offers a powerful tool to minimize crop losses and optimize yield. However, while these technologies bring substantial benefits, several significant challenges and issues must be addressed before they can be fully realized in real-world agricultural practices. These challenges span data quality and availability, model robustness, environmental variability, and the need for continuous updates and adaptation to new diseases and pest species. In this section, we will explore some of the key challenges that affect the use of ML for plant disease and pest detection.

15.2.1 Data Quality and Availability

One of the foundational requirements for any ML model is the presence of reliable, high-quality data in the context of plant disease and pest detection, this typically involves large datasets of images, sensor readings, and environmental data that can be used to train algorithms. Unfortunately, acquiring high-quality labeled data for plant diseases and pests is often difficult and expensive. Labeling images of plant diseases requires expertise in plant pathology, as diseases can have similar or overlapping symptoms, making it difficult to distinguish between different conditions. Furthermore, accurate data collection may be hindered by limited access to field sites, time constraints, and seasonal variations in crop conditions.

Additionally, the problem of imbalanced datasets arises when certain diseases or pests are underrepresented in the data. For example, rare or newly emerging diseases may not have sufficient data available to train an effective ML model, resulting in suboptimal performance when trying to detect these conditions. Similarly, some plant species may not be well-represented in existing datasets, leading to a model that is biased towards more common crops and diseases.

15.2.2 Environmental Variability and Conditions

Plants are influenced by various environmental elements, including fluctuations in temperature, humidity, soil quality, and light exposure, which can markedly impact the manifestation of disease symptoms. These environmental factors introduce a major challenge when using ML for plant disease detection because they can cause significant variability in the way diseases manifest. For example, the same disease may look different under varying environmental conditions, complicating the task for ML algorithms trying to classify and detect plant diseases accurately.

Lighting conditions, in particular, can significantly influence the quality of image data utilized for training ML models. A disease that appears visibly distinct under bright sunlight may be harder to detect under cloudy or low-light conditions. Similarly, different growing stages of plants, such as seedlings versus mature plants, can display different symptoms, requiring ML models to be highly adaptable to these changes. Diversity in plant species, developmental phases, and environmental factors hinders the model's capacity to generalize across varied environments.

15.2.3 Model Generalization Across Geographic Regions

ML algorithms for the detection of plant diseases and pests frequently depend on data gathered from particular geographic areas. However, plant diseases and pests vary not only based on the species but also according to regional factors such as climate, soil conditions, and farming practices. A model based on data from one region may not function as efficiently when utilized in another region with varying variables. This geographic limitation presents a challenge when deploying ML models globally or across diverse agricultural zones.

For instance, a model developed in temperate climates may struggle to recognize diseases in tropical regions, where different environmental factors influence disease development and symptom expression. This issue highlights the importance of ensuring that ML models are trained with diverse and region-specific datasets to improve their ability to generalize across varying conditions.

15.2.4 Complexity of Disease Symptom Expression

Many plant diseases exhibit complex and sometimes subtle symptoms that can be difficult to distinguish from other environmental stressors or natural plant variations. Symptoms such as yellowing of leaves, wilting, or spotting can be indicative of multiple possible conditions, including diseases, pests, or nutrient deficiencies. This ambiguity poses a significant challenge to ML algorithms, as they must learn to distinguish between these overlapping symptoms.

Additionally, some plant diseases manifest very early or in hidden locations (such as on the undersides of leaves), making them harder to detect through image-based machine learning models, which plays a key role in early detection, which is essential for reducing the impact of diseases, but many models rely on visible symptoms that may only appear once the disease has progressed to a later stage. Developing models that can detect diseases at early stages remains a significant challenge.

15.2.5 Limited Generalization to New or Unknown Diseases

ML models for plant disease detection are typically trained on known diseases that have been well-documented. However, the emergence of new and unknown plant diseases poses a challenge for these models. When a new disease appears, the algorithm may not recognize the novel symptoms, as it has not been trained on examples of that disease. This lack of adaptability can hinder the real-time detection of new threats, which is especially concerning given the fast pace at which plant diseases can spread and cause damage.

Furthermore, while supervised learning techniques can be effective for well-defined diseases, they may struggle with the detection of previously unobserved pests or diseases that lack a clear, structured pattern. In such cases, unsupervised learning methods, including clustering and anomaly detection, could offer an advantage by identifying potential outbreaks based on unusual patterns, but these methods are still in their developmental stages and face their own set of challenges.

15.2.6 Model Interpretability and Trust

A major obstacle to the implementation of ML in agriculture is the opaque character of numerous ML models, particularly those based on deep learning. Although these models can attain great accuracy, their decision-making processes are frequently unidentified, complicating users' comprehension of how a specific conclusion was reached. This lack of interpretability can be a deterrent for farmers and agricultural professionals, who may be reluctant to trust a model if they cannot understand its reasoning process.

In agricultural settings, farmers and practitioners must have clear explanations for why a specific diagnosis was made, particularly when it involves costly decisions regarding pesticide application or crop management. Developing models with better transparency and explainability will be essential to fostering trust and encouraging the widespread adoption of ML tools in plant disease and pest management.

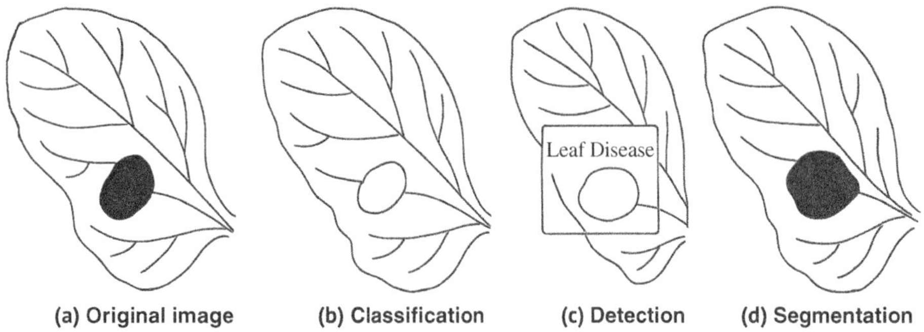

Figure 15.1 Steps for automatic plant disease detection, which is useful in precision agriculture to monitor and manage plant health efficiently. (a) Original image: An image of a leaf with a visible disease spot. It serves as the input for analysis. (b) Classification: In this step, the model classifies the disease affecting the leaf. The label "Leaf Disease" is displayed at the bottom, indicating that the model has identified the disease type. (c) Detection: The diseased region is detected and enclosed within a bounding box with the label "Leaf Disease." This highlights the area of interest where the disease is present. (d) Segmentation: The affected area is segmented and marked in black. This step isolates the diseased portion, helping in precise analysis and potential treatment recommendations.

15.2.7 Scalability and Real-time Implementation

For ML models to be useful in large-scale agricultural settings, they must be able to operate in real time and be scalable across different environments and devices. Many current models are developed and tested in controlled environments, but applying them in the field requires overcoming challenges related to hardware constraints, connectivity issues, and the integration of various data sources such as drones, sensors, and mobile devices. Figure 15.1a–d presents different stages of plant disease detection and classification using ML techniques. Table 15.1 highlights key challenges in plant disease prediction using ML. It explains issues like limited labeled data, symptom complexity, plant variability, image quality, environmental factors, and dataset imbalance. These challenges impact model accuracy, making disease detection and classification more difficult. Addressing them requires better datasets and robust algorithms.

15.3 Plant Disease Detection and Classification

Plant diseases, caused by pathogens such as fungi, bacteria, viruses, and nematodes, pose significant threats to global food production and security. Prompt and precise identification of these diseases is essential for reducing crop losses, protecting food security, and fostering sustainable agricultural practices. Traditionally, plant disease identification relied heavily on manual methods, including visual inspections by agricultural experts who often rely on these methods, but they tend to be time-consuming, subjective, and not scalable, particularly in large-scale operations agricultural settings. In recent years, the application of ML algorithms has emerged as a promising solution for automating and enhancing the plant disease detection and classification process. ML approaches, especially when integrated with image recognition and deep learning, provide more efficient, accurate, and scalable solutions for tackling the issues of plant disease management.

This literature review aims to examine the current landscape of research and development in applying ML to plant disease detection and classification. We will examine different ML techniques, review relevant studies, and assess the challenges and opportunities within this domain.

15.3.1 ML Techniques for Plant Disease Detection

ML algorithms are generally categorized into supervised and unsupervised learning types techniques, each with its own advantages and limitations in plant disease detection.

15.3.2 Supervised Learning Approaches

Supervised learning techniques require labeled data to train models, making them highly effective for plant disease detection, where diseases can often be identified by their characteristic visual symptoms. In the realm of plant disease identification, prevalent supervised learning methods encompass CNNs, SVMs, Decision Trees, and Random Forests (RFs).

Table 15.1 Key challenges in plant disease prediction using machine learning.

Challenge/Issue	Description	Impact on Detection and Classification
Limited Labeled Data	Plant disease datasets often lack a sufficient number of labeled images for training machine learning models.	Machine learning models require large, diverse datasets for accurate classification, leading to poor model performance with small datasets.
Complexity of Disease Symptoms	Symptoms of plant diseases can differ markedly based on the species, disease progression, and environmental conditions.	This complexity can make it difficult for algorithms to generalize across different diseases and plant species.
Variations in Plant Characteristics	Plants of the same species can exhibit different leaf shapes, colors, and textures due to environmental factors or genetic variation.	Variability can complicate the model's capacity to accurately identify diseases among diverse kinds.
Low Image Quality	Images captured in noncontrolled environments can have varying lighting, angles, and resolutions, affecting classification accuracy.	Poor-quality images reduce the reliability and accuracy of disease-detection models.
Environmental Factors	The appearance of plant diseases can be influenced by weather, soil conditions, or pests, adding noise to image data.	Environmental variability can confuse models and reduce accuracy when they encounter data from different conditions.
Imbalanced Dataset	Many disease datasets are imbalanced, where certain diseases are underrepresented, making it harder to classify them accurately.	An imbalanced dataset leads to biased models that perform well on the more common diseases but poorly on rarer ones.
Overfitting and Underfitting	Machine learning models may overfit or underfit data, especially with small datasets or complex models.	Overfitting results in poor generalization of new data, while underfitting leads to inaccurate predictions.
Real-Time Classification	Real-time disease detection requires fast and efficient processing of images, which	High computational requirements can delay real-time detection, making it
Generalization Across Species	Disease symptoms may vary greatly across different plant species, making it challenging for the model to generalize effectively.	Models trained on a limited range of species may fail to detect diseases accurately in new, unseen plant species.
Lack of Robustness in Adverse Conditions	Disease detection models may struggle under adverse conditions like poor lighting, low-resolution images, or background clutter.	These factors can reduce the effectiveness of the model in real-world agricultural scenarios, leading to inconsistent performance.

- **Convolutional Neural Networks (CNNs):** Deep learning methodologies, especially in image classification for plant disease diagnosis, have demonstrated significant potential. CNNs excel in autonomously extracting hierarchical features from images, rendering them optimal for detecting nuanced patterns in plant disease symptoms. Research has highlighted the success of CNNs in accurately classifying over 50 plant diseases using leaf images. CNNs can achieve high classification accuracy, even for complex diseases with similar symptoms.
- **SVMs:** SVMs are also commonly used in plant disease detection. SVMs are particularly effective when the data set is smaller or less complex. For example, the use of SVMs for identifying tomato plant diseases using features extracted from leaf images. The method showed strong results in terms of accuracy but required careful selection of kernel functions and parameters.
- **Decision Trees and RF:** Decision trees and RF have been widely applied for plant disease classification due to their ability to handle nonlinear relationships and provide interpretable models. RF have been utilized to classify multiple plant diseases based on spectral features from remote sensing data, demonstrating the flexibility of these models across various types of data.

15.3.3 Unsupervised Learning Approaches

Unsupervised learning approaches, including clustering and anomaly detection, are useful when labeled data is scarce or when attempting to discover unknown disease patterns.

- **K-means Clustering and Hierarchical Clustering:** K-means clustering has been applied to group plant leaf images based on disease symptoms, allowing for the identification of clusters that correspond to particular diseases.

Clustering techniques, while less accurate than supervised models, can offer valuable insights when disease data is limited or when studying new diseases that are not yet well-documented.

- **Anomaly Detection:** Anomaly detection methods focus on identifying abnormal plant conditions that may not match any known disease model. These methods have been applied in cases where new diseases or pest infestations emerge. Anomaly detection has been used to identify unusual patterns in plant images, providing an approach for the early detection of emerging diseases or pest attacks.

15.3.4 Deep Learning Approaches

Deep learning, a branch of ML, has lately acquired significance in the identification of plant diseases, partially due to its capacity to process extensive datasets and discern complex patterns from raw data without requiring manual feature extraction. CNNs, as mentioned earlier, are a significant part of deep learning techniques applied to plant disease detection, but there are also other deep learning models that are being explored:

- **Recurrent Neural Networks (RNNs):** RNNs have been explored for plant disease detection in time-series data, such as monitoring plant health over time through sensors and camera systems. RNNs have been used to track plant health and detect diseases based on temporal patterns of disease progression.
- **Transfer Learning:** Transfer learning is a method that entails modifying pre-trained models for new tasks, proving advantageous in situations when labeled data for certain diseases is limited. Ferentinos (2018) applied transfer learning to plant disease detection by leveraging pre-trained CNNs for classifying plant disease images, achieving high accuracy even with relatively small datasets.

15.4 Data Sources and Feature Extraction

The efficacy and performance of ML models in plant disease detection predominantly rely on data quality. Images from various sources, such as field cameras, drones, and satellites, are commonly used to capture disease symptoms.

15.4.1 Image-based Data

Most plant disease detection systems use *image-based data* to identify diseases. Leaf images are the most common, but other plant parts, such as stems, fruits, and flowers, are also used. High-resolution leaf images have been used to classify various plant diseases, demonstrating that image-based classification can outperform traditional diagnostic methods.

15.4.2 Multispectral and Hyperspectral Data

Beyond visible light, multispectral and hyperspectral imaging technologies provide additional insights into plant health by capturing spectral information from different wavelengths. Multispectral imaging has been explored to detect early signs of fungal diseases in crops. These imaging techniques can help improve disease detection, especially in cases where visible symptoms are minimal or not easily observable.

15.4.3 Feature Extraction Techniques

The efficacy of feature extraction is vital to the overall performance of ML models. Methods including Histogram of Oriented Gradients (HOG), Scale-Invariant Feature Transform (SIFT), and Local Binary Patterns (LBP) have been employed to extract significant features from plant pictures. HOG has been utilized for extracting features from leaf images to classify plant diseases, achieving promising results.

15.5 Challenges in Plant Disease Detection Using ML

Despite the considerable progress in ML applications for plant disease diagnosis, numerous hurdles persist that hinder the extensive implementation of these technologies.

15.5.1 The Reliability and Accessibility of Data

High-quality labeled data is essential for training accurate ML models. Nonetheless, obtaining extensive datasets can prove to be difficult. Acquiring accurately classified illness occurrences can be challenging. In many cases, plant diseases are not well-documented or may present symptoms that overlap with other conditions, which can make it challenging to acquire sufficient training data. This issue is compounded by the need for expert knowledge to label disease images correctly, which can be time-consuming and costly.

15.5.2 Model Generalization and Overfitting

ML models, especially deep learning models, sometimes want substantial data volumes to generalize proficiently. Overfitting is a prevalent issue wherein a model excels on the training dataset yet fails to generalize to new, unseen data. This is a substantial challenge in plant disease identification, as the manifestation of disease symptoms can fluctuate markedly based on environmental variables, such as light, humidity, and plant species.

15.5.3 Environmental Variability

Environmental elements, such as soil composition, climatic circumstances, and the growth stage of plants, can profoundly affect the manifestation of disease symptoms. As a result, ML models may struggle to detect diseases under varying conditions, which poses a challenge for developing robust models that can be used across different regions and crops. Ensuring that models generalize well across a wide range of environments remains a critical challenge in this field.

15.5.4 Interpretability and Trust

The absence of interpretability in certain ML models, particularly deep learning models, may diminish their acceptability in practical applications. Farmers and agricultural experts often require an understanding of how a model arrived at a particular diagnosis. Without this interpretability, there may be hesitance in trusting ML-based systems for plant disease detection, especially in high-stakes agricultural environments. Figure 15.2 presents images of different plant leaves, showcasing both healthy and diseased samples. Each labeled image represents a specific condition of a plant leaf, which is crucial for plant disease detection using ML.

15.6 Algorithm Description

15.6.1 Algorithm Overview

The general approach for plant disease detection and classification consists of several key steps:

1. Data Collection: Gathering relevant datasets of plant images, sensor data, or other features.
2. Data Preprocessing: Preparing the data required to train the ML model.
3. Feature Extraction: Identifying and extracting useful features from raw data, particularly images.
4. Model Selection and Training: Selecting a suitable ML model and training it on the dataset.
5. Model Evaluation: Assessing the model's performance through various evaluation metrics.
6. Deployment: Implementing the trained model for real-time detection or classification in a real-world agricultural setting.

15.6.2 Data Collection and Preprocessing

The first step in any ML-based plant disease detection system begins with the gathering of data, which can be divided into two main categories:

- Image Data: This often comprises high-resolution photos of plant leaves, fruits, or stems obtained using cameras, drones, or satellites. The dataset should cover a wide range of diseases and plant species to ensure robust training.
- Sensor Data: Data from environmental sensors, such as temperature, humidity, soil conditions, and other factors, can also be useful in conjunction with images to enhance disease prediction.

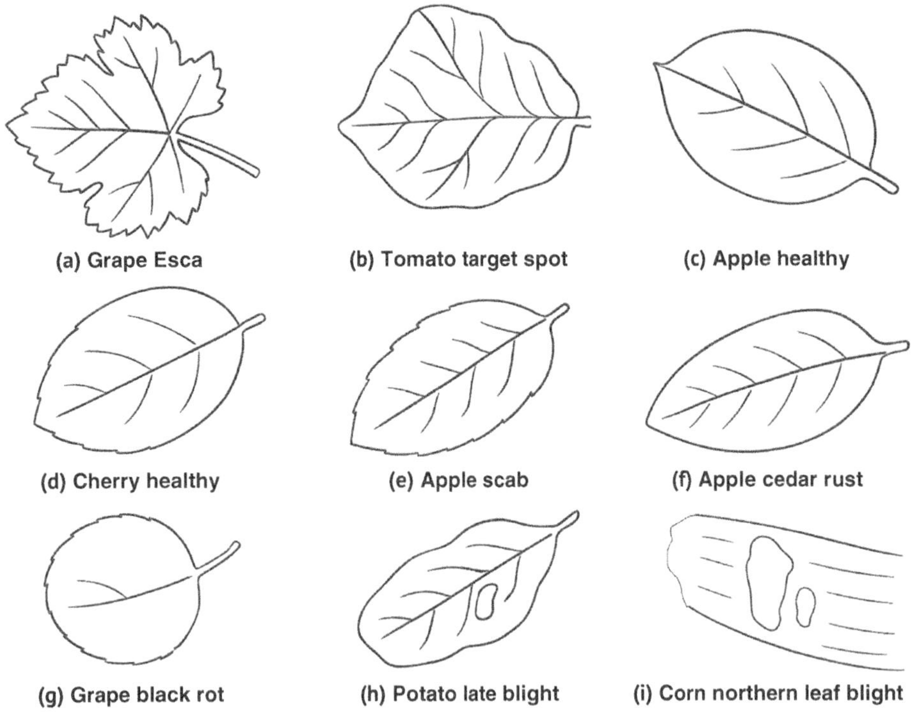

Figure 15.2 Different types of plant diseases. These images are essential for training ML models to differentiate between healthy and diseased plant leaves, aiding in early disease detection and prevention. (a) Grape Esca: A diseased grape leaf showing symptoms of Esca, a fungal disease that causes discoloration and decay. (b) Tomato target spot: A tomato leaf affected by target spot disease, which results in circular necrotic lesions. (c) Apple healthy: A healthy apple leaf with no visible signs of disease. (d) Cherry healthy: A healthy cherry leaf that appears normal, used as a reference for disease detection models. (e) Apple scab: an apple leaf with black, scabby spots, which is characteristic of the apple scab disease caused by a fungus. (f) Apple cedar rust: A diseased apple leaf with rust-colored spots, indicating infection by cedar-apple rust fungus. (g) Grape black rot: A grape leaf showing symptoms of black rot, a fungal infection that leads to dark, sunken lesions. (h) Potato late blight: a potato leaf affected by late blight, a serious fungal disease causing brown and water-soaked lesions. (i) Corn northern leaf blight: A corn leaf with large, elongated, grayish-brown lesions, which is characteristic of Northern leaf blight.

15.6.3 Data Preprocessing Involves Several Steps

1. **Image Resizing and Normalization:** Image data is frequently scaled to a standardized dimension to facilitate effective processing by the model. Furthermore, pixel values are generally normalized (scaled to a range of 0 to 1) to enhance model performance and convergence during training.
2. **Data Augmentation:** In instances of small or imbalanced datasets, data augmentation techniques including rotation, flipping, zooming, and shifting are employed to artificially enlarge the dataset and enhance generalization.
3. **Noise Reduction:** Image preprocessing techniques, including filtering, denoising, and contrast enhancement, can be utilized to eliminate extraneous noise from the image data and improve the visibility of disease symptoms.

15.6.4 Feature Extraction

Feature extraction denotes the procedure of identifying and isolating pertinent attributes from raw data (such as photographs) that ML algorithms might employ to differentiate between various plant diseases. For image-based datasets, feature extraction typically involves detecting specific patterns related to disease symptoms such as leaf spots, lesions, color changes, and textures.

- **Traditional Feature Extraction:** Earlier methods rely on extracting hand-crafted features such as:
  - **Color Histograms:** Capturing the distribution of pixel colors in the image, which can be indicative of disease symptoms (e.g., yellowing or browning of leaves).

  - **Texture Features:** Using techniques like LBP or Gray Level Co-occurrence Matrix (GLCM) to capture the texture of the plant surface, which is often altered by the presence of diseases.
  - **Shape Features:** Analyzing the shape and size of spots or lesions on the leaves.
- **Deep Learning Feature Extraction:** Modern approaches, particularly deep learning methods, automate feature extraction through CNNs that automatically learn spatial feature hierarchies from raw images. In CNNs, the feature extraction process involves convolutional layers that progressively extract low-level (edges, textures) and high-level (disease patterns) features from input images.

15.6.5 Model Selection and Training

The selection of a ML model is contingent upon the characteristics of the data and the specific task involved. In plant disease detection, common model types include the following.

15.6.5.1 Convolutional Neural Networks

CNNs have emerged as the preeminent method for image-based plant disease identification, owing to their robust capacity to autonomously learn hierarchical features from images. CNNs consist of multiple layers:

- **Convolutional Layer:** Utilizes a series of filters on the input image to identify features including edges, corners, and textures.
- **Activation Function:** Rectified Linear Units (ReLU) are commonly employed as activation functions to incorporate nonlinearity into the model, enabling the learning of more intricate patterns.
- **Pooling Layer:** Minimizes the spatial dimensions of the feature maps, hence decreasing computational requirements and mitigating overfitting.
- **Fully Connected Layer:** After feature extraction, the final layers of the CNN combine the extracted features and produce predictions for classifying plant diseases.

15.6.5.2 Support Vector Machine

SVMs are a widely utilized option for the classification of plant diseases. SVMs operate by identifying the best hyperplane that maximally distinguishes between various data classes. When using SVM for image classification, features are typically extracted beforehand (using methods like HOG or CNN), and the SVM model is then trained on these features. SVMs are very effective for binary classification tasks, but they may also be adapted for multi-class issues through methods like the one-vs-all strategy.

15.6.5.3 Decision Trees and RF

Decision Trees are a category of supervised ML algorithms that construct a tree-like framework to facilitate decision-making based on attributes derived from the data. A RF is an ensemble learning technique that generates numerous decision trees and consolidates their predictions to enhance accuracy and mitigate overfitting. These models can be effective when dealing with structured data, such as sensor readings or tabular data, but can also be applied to image-based data once features are extracted.

15.6.5.4 Transfer Learning

Transfer learning is a methodology that utilizes pre-trained models, usually developed on extensive datasets such as ImageNet, and modifies them for new applications. Rather than developing a deep learning model from the ground up, a pre-trained CNN is optimized on a plant disease dataset, conserving time and processing resources. Transfer learning is especially beneficial when labeled data for plant diseases is limited.

15.6.5.5 Model Evaluation and Tuning

Following finalizing the model training, it is essential to assess its performance using a distinct test dataset that was not utilized during the training phase. Common evaluation metrics include:

- **Accuracy**: The ratio of accurately classified instances to the total number of instances.
- **Precision, Recall, and F1-Score**: These metrics are especially beneficial in imbalanced datasets when one class, such as a particular disease, may have a lower frequency of incidents compared to others.

- **Confusion Matrix**: A confusion matrix offers a comprehensive analysis of categorization efficacy, delineating the true positive, false positive, true negative, and false negative rates for each disease category.
- **Cross-Validation**: Cross-validation methods, like k-fold cross-validation, are employed to verify that the model generalizes effectively and does not overfit the training data.

Hyperparameter tuning, which entails modifying parameters like as learning rate, batch size, number of layers, and filter sizes in CNNs, is a crucial step in enhancing the model's performance.

15.7 Deployment and Real-time Inference

Once the model is trained and evaluated, it can be deployed for real-time disease detection in the field. Deployment strategies include:

- **Mobile Applications:** Mobile apps equipped with the trained model can be used by farmers to capture plant images and receive disease classifications in real time.
- **Drones and Remote Sensing:** Drones or satellites equipped with cameras can capture large-scale images of crops, and the model can process these images to identify diseases in the field.
- **Automated Systems:** In large-scale farms, automated systems can capture images of plants and immediately classify the health of the crop, allowing for precision agriculture practices like targeted pesticide application.

Table 15.2 outlines the key steps in implementing a plant disease detection project using ML. It details each stage, from model training and evaluation to deployment on cloud and edge devices, API serving, and real-time inference. Various tools and technologies such as TensorFlow, Docker, Flask, and cloud platforms are used to enhance model performance, deployment, and monitoring.

15.8 Challenges and Future Directions

While ML provides promising solutions in plant disease detection, numerous challenges still persist:

- Data Scarcity and Labeling: Labeled datasets for plant diseases are often small and imbalanced, which can affect model accuracy and generalization. Data augmentation and synthetic data generation methods are being explored to address this issue.

Table 15.2 Key steps in implementing a plant disease detection.

Stage	Action	Tools/Technologies
Model Training	Data collection (plant images), data preprocessing (image augmentation), model architecture design	TensorFlow. Pyloric, OpenCV
Model Evaluation	Test the model using evaluation metrics	Scikit-learn, TensorFlow, Py Torch
Containerization	Package the trained model for deployment	Docker, TensorFlow Serving, Torch Serve
Cloud Deployment	Deploy model to cloud infrastructure	AWS, GCP, Azure, Docker
Edge Device	Deploy model to edge devices like Raspberry Pi or Nvidia Jetson	TensorFlow Lite, ONNX, Raspberry Pi, Jetson, Docker
API Serving	Serve the model as an API to handle image inputs and return predictions	Flask, Fast API. REST API, Tensorflow Serving
Real-time Inference	Capture real-time images and send them to the API for prediction	Mobile apps, Web apps, API integration with edge devices
Model Monitoring	Monitor the model's performance and accuracy and retrain if needed	Prometheus, Grafana, CloudWatch

- Environmental Variability: Elements such as lighting, phases of plant growth, and geographical positioning can profoundly influence the efficacy of disease detection models. The integration of multi-modal data, such as combining pictures with environmental information, is a burgeoning field of research.
- Model Interpretability: Although deep learning models like CNNs can attain elevated accuracy, their opaque nature renders them challenging to analyze. Research into explainable AI (XAI) aims to provide transparency in how models make predictions.

15.9 Proposed Methodology

15.9.1 Overview of the Methodology

The proposed approach comprises six essential stages: data collection, data preprocessing, feature extraction, developing models and training, evaluation, and deployment. Each stage plays a critical role in ensuring the accuracy, reliability, and real-world applicability of the plant disease detection system.

Step 1: Data Collection
The initial and essential phase in the proposed methodology is data collection. The quality and variety of the dataset profoundly influence the efficacy of ML models. The collected data should represent various plant species, diseases, and environmental conditions.

15.9.1.1 Image Data Collection

The primary source of data for plant disease detection is image data. High-resolution images of plant leaves, stems, fruits, or flowers are typically used to identify visible disease symptoms. These images can be captured through:

- **Field cameras:** Stationary or mobile cameras set up in crop fields.
- **Drones:** Drones equipped with high-resolution cameras for capturing images of large agricultural areas.
- **Smartphones:** Portable image-capture devices used by farmers to take pictures of plant diseases.

15.9.1.2 Sensor Data Collection

Besides images, sensor data can enhance illness detection. Diverse environmental conditions, including temperature, humidity, soil moisture, and light intensity, can affect plant health and the advancement of diseases. Sensors placed in the field or greenhouse can collect this data, which may be incorporated into the ML model for a more robust prediction.

15.9.1.3 Disease Annotation and Labeling

A critical part of the data collection process is the annotation and labeling of disease data. Expert plant pathologists must label the images with the correct disease class. This step ensures that the model has the correct ground truth for supervised learning. Disease classes should cover a range of conditions, from common fungal, bacterial, and viral diseases to rare or emerging diseases.

Step 2: Data Preprocessing
Data preprocessing entails the cleansing and preparation of raw data for utilization in the ML model. This step is essential for improving model accuracy and reducing errors.

15.9.2 Image Preprocessing

- **Resizing:** All images must be scaled to a uniform resolution to guarantee efficient processing by the ML model. It is customary to scale images to dimensions of 224×224 or 256×256 pixels.
- **Normalization:** Pixel values are frequently normalized to a range of [0, 1] to standardize inputs and enhance the convergence rate of deep learning models.
- **Data Augmentation:** To mitigate the problem of insufficient data, techniques for data augmentation, including rotation, flipping, scaling, and cropping, may be utilized. Augmentation facilitates the creation of a broader array of training samples from the original dataset, hence mitigating overfitting and enhancing model generalization.

15.9.2.1 Sensor Data Preprocessing

- **Feature Engineering:** Raw sensor data may need to be processed into more useful features such as moving averages, trends, or time-based aggregates.
- **Normalization:** Like image data, sensor data should be standardized so that all factors have similar ranges. This keeps one feature from controlling the learning process of the model.

15.9.2.2 Data Splitting

There are training, validation, and test sets in the collection. Most of the time, 70% is used for teaching, 15% for validation, and 15% for testing. The training set teaches the model what to do, the validation set checks the model's accuracy and makes any necessary changes to the hyperparameters, and the test set checks how well the model performed in its final phase.

Step 3: Feature Extraction

To make sure that the ML model can correctly discover disease trends in the raw data, feature extraction is a very important step. Traditional feature extraction and deep learning-based feature extraction are the two methods that will be looked at in the present research.

15.9.3 Traditional Feature Extraction

When computational resources are constrained or deep learning is not appropriate, conventional image processing methods can be applied to extract features from plant images. These features may include:

- **Texture Features:** Features such as the LBP or GLCM can capture surface textures, which are often altered by diseases.
- **Color Features:** Color histograms or color moments can be extracted to capture color changes associated with disease symptoms (e.g., chlorosis or necrosis).
- **Shape Features:** Shape-based features, such as the contour or area of lesions, can be used to characterize specific symptoms of a disease.

15.9.4 Deep Learning-based Feature Extraction

CNNs are commonly used in image-based disease detection tasks because they learn the most relevant features from raw images during training. CNNs can find complex patterns in the data that might not be evident using conventional feature extraction techniques by stacking several layers of convolutions and pooling.

For transfer learning, CNN architectures such as Res Net, VGG, and Inception are frequently utilized. This type of learning involves the process of fine-tuning models that have been pre-trained on large image datasets (such as ImageNet) using plant disease datasets.

Step 4: Model Development and Training

The next stage involves choosing a ML algorithm and training the model after the data has been preprocessed and pertinent features have been extracted. The complexity of the task and the type of data determine which algorithm is best.

15.9.4.1 Model Selection

Several ML algorithms can be used for plant disease detection, including:

- One of the most effective deep learning models for image classifications is the CNN. Numerous studies on plant disease detection have shown that these models are very accurate and automatically learn the pertinent features for disease detection.
- **SVMs:** SVMs can be used for classification tasks where features have been extracted from images or sensor data. SVMs are effective in situations where data is limited or when the disease classes are well-separated.
- **RF:** Multiple decision trees are combined in the ensemble learning technique known as RF. In contrast to single decision trees, it is less likely to overfit and works well with both structured (sensor) and unstructured (image) data.
- **Transfer Learning:** The plant disease dataset is used to refine pre-trained models like ResNet50, VGG16, or InceptionV3. By utilizing the insights gathered from extensive image datasets, transfer learning helps address the issue of sparsely labeled data.

15.9.4.2 Model Training

- **Training:** The training dataset is used to train the selected model. The model gains knowledge of the connections between the disease classes and the extracted features (or raw images) during training.
- **Hyperparameter Tuning:** Learning rate, batch size, number of epochs, and dropout rate are among the hyperparameters chosen to raise model performance and minimize overfitting.
- **Validation:** The model's performance is consistently assessed on the validation dataset to evaluate its generalization capacity.

Step 5: Model Evaluation

After training, the model undergoes evaluation on the test dataset to determine its accuracy and overall efficacy. Common evaluation metrics include:

- **Accuracy:** The ratio of accurately categorized cases to the total number of occurrences.
- **Precision, Recall, and F1-Score:** These metrics are especially beneficial when dealing with imbalanced datasets. Precision is the comparison of the number of true positives to the total number of predicted positives. The F1-score is the harmonic mean of accuracy and memory, recall quantifies the percentage of true positives among true positives, and expected positives.
- **Confusion Matrix:** A confusion matrix delineates true positives, false positives, true negatives, and false negatives for each disease category.
- **Cross-validation:** K-fold cross-validation is employed to enhance the model's robustness and reduce overfitting.

Step 6: Deployment

After the accomplishment of training and evaluation, the model is prepared for use in practical agricultural environments. The deployment process may include:

- Integration with Mobile Applications: The trained model can be deployed as part of a mobile app, allowing farmers to upload images of their plants and receive instant disease classification results.
- Integration with Drones and Remote Sensing Platforms: Drones or satellite platforms can capture large-scale images of crops, and the trained model can process these images in real-time to detect plant diseases.
- Real-time Disease Monitoring: The approach may be implemented in the field using IoT devices or cameras to continually monitor plant health, providing warnings for the early identification of illnesses and insect infestations.

15.10 Conclusion

Modern agriculture has seen a revolution with the use of ML algorithms for plant disease detection and classification. Utilizing data, especially image and sensor data, ML algorithms may accurately and efficiently diagnose and categorize plant diseases at an unachievable scale by conventional approaches. Timely and precise disease identification is essential for reducing the effects of plant diseases on crop productivity and quality, hence promoting sustainable agricultural practices and food security.

ML models, particularly deep learning techniques like CNNs, have shown remarkable performance in classifying plant diseases across a variety of species and environmental conditions through the proposed methodology, which integrates various stages such as data collection, preprocessing, feature extraction, model development, and deployment. Disease classification is made easier even with fewer datasets by using transfer learning to use pre-trained models, which also improves model accuracy and lessens the difficulties caused by a lack of labeled data.

However, despite the promising outcomes, challenges remain, including data scarcity, environmental variability, and the need for model interpretability. These challenges can be addressed through continuous advancements in data collection, model optimization, and the integration of diverse data sources like environmental sensors. Moreover, as ML models evolve, they will need to become more adaptable to real-time conditions in the field, where image quality and environmental factors may vary.

The future of plant disease detection using ML holds great potential. Enhanced model accuracy, real-time performance, and integration with future technologies such as drones and IoT devices may enable farmers to make trained actions, minimize crop losses, and implement more sustainable agricultural practices. Through the ongoing enhancement of these systems, ML may significantly contribute to addressing global agricultural concerns, ensuring food security, and promoting the sustained health of our agricultural.

References

Basu, S. and Pradeep, K. (2015). Machine learning techniques for plant disease detection and classification. *International Journal of Advanced Research in Computer Science and Software Engineering* 5 (2): 27–34.

Chandrashekar, P.P. and Chandrashekar, H.P. (2017). Machine learning approaches for plant disease detection. *International Journal of Computer Applications* 159 (5): 21–27.

Choudhary, S. and Gupta, A. (2020). Hybridization of machine learning algorithms for the detection of plant diseases. *Journal of Computing and Technology* 27 (2): 115–123.

Chouhan, M. and Patil, A. (2018). Machine learning models for plant disease detection and diagnosis. *Journal of Agricultural Informatics* 10 (3): 23–31.

Ferentinos, K.P. (2018). Deep learning models for plant disease detection and classification. *Computers and Electronics in Agriculture* 145: 311–318.

Gupta, A. and Gupta, P. (2020). Plant disease classification and detection using convolutional neural networks. *Journal of Computational Biology and Bioinformatics* 12 (1): 1–9.

Gupta, D. and Pratap, R. (2020). Hybrid models for early detection of plant diseases using machine learning techniques. *Journal of Advanced Computational Intelligence* 12 (5): 123–131.

Jayalakshmi, M. and Karthik, S. (2019). A novel approach for plant disease detection using deep learning techniques. *Journal of Computing and Information Technology* 27 (3): 145–154.

Jha, P. and Soni, P. (2021). Improved plant disease detection using hybrid feature extraction and classification models. *Journal of Plant Pathology* 103 (4): 845–853.

Kaur, P. and Mittal, P. (2020). Plant disease detection using convolutional neural networks. *Computers and Electronics in Agriculture* 176: 105653.

Kaur, R. and Gupta, P. (2021). Plant disease identification using artificial intelligence and machine learning techniques. *Journal of Agricultural Engineering and Technology* 39 (4): 23–32.

Kavitha, K. and Sriram, M. (2019). Detection of tomato plant diseases using deep convolutional neural networks. *Advances in Computer Vision and Machine Learning* 5 (4): 101–109.

Khan, M.A. and Ahmad, S. (2019). Application of image processing and machine learning for detection of plant diseases. *Computers and Electronics in Agriculture* 160: 35–42.

Kumar, A. and Choudhary, A. (2019). A survey on machine learning approaches in plant disease detection. *Proceedings of the 2019 IEEE 5th International Conference on Signal and Image Processing*: 200–205.

Liu, C. and Wang, M. (2021). Deep convolutional neural networks for plant disease detection and classification. *Frontiers in Plant Science* 12: 655–664.

Mohan, S. and Kumar, B. (2020). Evaluation of machine learning approaches for plant disease detection. *International Journal of Computer Science in Agriculture* 8 (2): 215–224.

Nagarajan, N. and Ramesh, V. (2021). Efficient plant disease detection via hybrid machine learning algorithms. *Computers in Agriculture and Food Processing* 45 (6): 143–152.

Nayak, S. and Mondal, M. (2018). Application of support vector machines for plant disease classification. *Journal of Machine Learning and Applications* 19 (2): 149–158.

Nielsen, A. and Taylor, J. (2021). Hybrid deep learning for accurate plant disease identification. *International Journal of Agricultural Technology* 17 (2): 987–996.

Patil, S. and Rani, A. (2020). Detection of plant diseases using neural networks. *Journal of Computational Intelligence in Agriculture* 5 (4): 185–196.

Rajendran, M. and Baskar, S. (2021). Plant disease detection and classification using deep neural networks. *Advances in Intelligent Systems and Computing* 1367: 147–156.

Ramya, R. and Suresh, R. (2020). A comparative study of machine learning algorithms for plant disease classification. *Journal of Computer Science and Technology* 35 (4): 629–638.

Ranjan, P. and Mishra, S. (2020). Using deep learning for predicting plant diseases: A review. *International Journal of Computer Science and Applications* 17 (3): 89–96.

Saha, P. and Sharma, A. (2020). Data augmentation for plant disease classification using convolutional neural networks. *Journal of Digital Agriculture* 23 (2): 119–127.

Singh, M. and Singh, R. (2019). Predictive models for plant disease diagnosis: A survey of machine learning approaches. *Journal of Artificial Intelligence Research* 60 (1): 245–255.

Tavares, M.G. and Rodrigues, M.R. (2021). A survey on machine learning approaches for plant disease detection from leaf images. *Computers and Electronics in Agriculture* 183: 106089.

Teng, S.W. and Chen, J.M. (2019). A new machine learning approach for plant disease classification using image recognition. *Journal of Computer Science* 47 (8): 785–792.

Thakur, M. and Sharma, A. (2020). Automated plant disease classification using hybrid machine learning techniques. *Procedia Computer Science* 168: 264–272.

Thangavel, K. and Sivaprasad, V. (2021). Hybrid machine learning model for plant disease detection using RGB images. *Computers, Materials & Continua* 68 (1): 1–11.

Vasudevan, R. and Raj, A. (2020). Detection and classification of plant diseases using machine learning algorithms. *Journal of Artificial Intelligence in Agriculture* 5 (1): 55–62.

Zhang, S. and Zhong, Z. (2020). A deep learning approach for plant disease classification using convolutional neural networks. *IEEE Access* 8: 213445–213456.

Zhao, Z. and Li, Y. (2021). Deep learning for plant disease recognition: A survey and case study. *Computers* 10 (4): 60–71.

16

The Future Farming: Machine Learning and Crop Health

Sadhana Veeramani[1], Jeya Rani Maria Michael[2], Kalaichelvi Kalaignan[3], Ehab A A Salama[4], Annasamy Kaliyan[5], Thiruveni Thangaraj[6], Anantha Raju Pokkaru[7], Raveena Ravi[8], Karthiba Loganathan[9], Murali Sankar Perumal[10], Manasa Samuthiravelu[11]*

[1] *Department of Agricultural Entomology, Aravindhar Agricultural Institute of Technology, Tiruvannamalai, India*
[2] *Department of Agronomy, Pushkaram College of Agriculture Sciences, Pudukkottai, India*
[3] *Department of Agronomy, Krishi Vigyan Kendra, Virddhachalam, India*
[4] *Department of Agricultural Botany (Genetics), Saba Basha (Faculty of Agriculture), Alexandria University, Alexandria, Egypt*
[5] *Department of Horticulture, Pushkaram College of Agriculture Sciences, Pudukkottai, India*
[6] *Department of Agricultural Entomology, Nammazhvar College of Agriculture and Technology, Kamuthy, India*
[7] *Department of Plant Breeding and Genetics, Cotton Research Station, Veppanthattai, India*
[8] *Department of Environmental Sciences, Tamil Nadu Agricultural University, Coimbatore, India*
[9] *Department of Plant Pathology, Tamil Nadu Agricultural University, Coimbatore, India*
[10] *Department of Plant Pathology, Pushkaram College of Agriculture Sciences, Pudukkottai, India*
[11] *Department of Seed Science and Technology, TNAU, Coimbatore, India*

*Corresponding author: muralisankar58@gmail.com

16.1 Introduction

In world economics, agriculture is regarded as a crucial foundation of the global economy and fulfills the fundamental needs of food. In several countries, it is viewed as the primary source of employment and livelihood (Meshram et al. 2021). Mostly, all over the world, conventional or traditional farming practices are followed by expert farmers (Paustian and Thevusen 2016). However, these approaches lack precision, resulting in significant labor, time consumption, and expenditure (Nowak 2021). Furthermore, it is projected that by the year 2050, the global population will grow from the present 7.7 billion to 9.2 billion. Additionally, the urban population is expected to increase by 66%, while the amount of arable land is likely to decrease up to 50 million ha. (Abebe 2024). Significantly, the anthropogenic factors of rapid growth of population, transformation of rural to urbanization, growth and development of industrialization, decreasing the efficiency of farmland use, scarcity of water sources, and poor environmental management and conservation were noticed in recent decades, which have led to a heightened demand for agricultural products and continuously provides poor performance in outcome (Abbasi et al. 2022). Additionally, in farming, the lack of knowledge about information of soil types, cropping systems, weather factors, improper purpose of using pesticides, irrigation issues, poor harvesting, crop failures, and lack of poor knowledge of market trends consequently resulted in a substantial expansion of farming activities (Kumar et al. 2024). In sustainable agriculture, a food chain is depicted by an organized road map of stages, namely, pre-field (seed sources, field geography, selection of crops, climatic factors, and demand of products); in-field (cropping to harvesting); and post-field (harvesting to dining) (Jaya Singh et al. 2024). All these stages achieved through several practices, such as ploughing, sowing, irrigating, fertigating, protecting, harvesting, handling, preserving, marketing, and distributing (from field to dine), contributed to better outcomes of basics in agricultural products (Gamage et al. 2023). In farming, initial to these practices are employed through the enormous potential of labor and economic investment along with time requirements (Giller et al. 2021). Under field conditions, emission of GHG and climate changes provide favorable to 50% (incidence) increase in pest (70$ billion/year) and disease (220$ billion/year) loss (Lahlali et al. 2024; Domingues et al. 2022) and its impact to 20% of food production (Duchenne-Moutien and Neetoo 2021) and estimated rising of food demand by 59–98% with insecurity of food safety, availability, and supplies (Tchonkouang et al. 2024). In these situations, the applications of new cutting-edge

Machine Learning for Plant Biology, First Edition. Edited by Jen-Tsung Chen.

technologies are playing a pivotal role in sustainable agriculture and food security (Kumar et al. 2022). Under these circumstances, using technologies were embedded with smart farming approaches of minimal needs, optimized utilization of resources management and eco-conservation, frequent developments, and periodical updating in scientific approaches and modern technologies (Choudhary et al. 2025). In these situations, the high-throughput modern technologies such as artificial intelligence (AI), deep learning (DL), unmanned aerial vehicles (UAVs), information and communication technology (ICT), internet of things (IoT), sensors, cameras, and drones were emerged and introduced in agriculture and allied sectors (Komatineni et al. 2024) for their enhancement and beneficial to sustainable livelihood for future generations without zero poverty and hunger index (Islam et al. 2021). In this context, machine learning (ML) and its approaches, timeline challenges or issues, and future prospective are broadly discussed further.

16.2 Background

In day-to-day life, food is playing a vital role in all living things (Manoogian et al. 2019). With the evolution of civilization, agriculture plays a crucial role in humans' livelihood and it provides socioeconomic growth and sustainable development in all sectors such as science, education, national GDP, and political empowerment globally (Wijerathna-Yapa and Pathirana 2022). Generally, the development of modern farming practices and their production attained several tragic challenges such as the expansion of population, climate changes, deforestation, urbanization, changes in food choices, depletion of natural resources, and natural hazards during the incorporation of novel technologies in SDGs of agriculture 4.0 (Saleem et al. 2024). These situations highly pose novel applications in agriculture for the development and enhancement of production without loss (Viana et al. 2022). During a period, the entering agriculture through digital-based scientific approaches such as augmented reality (AR), autonomous robotic systems (ARS), AI, additive manufacturing (AM), big data analytics (BDA), cloud computing (CC), cyber-physical systems (CPS), drones, deep learning (DL), digital twin (TW), internet of things (IoT), ML, unmanned vehicles (UMVs), and wireless sensor networks (WSN) enable the agriculture to digitalization (Singh et al. 2022). By integrating these technologies, sparkling and agricultural practices are being transformed into sustainable, smart, and digital systems—commonly referred to as "smart or digital farming" (Fawzy and El-Ramady 2022).

16.3 Significance

Technology breakthroughs and the promising demand for sustainable practices are rapid growth in the agricultural sectors (Gamage et al. 2024). ML is one of these extremely revolutionary innovations, specifically when it comes to crop health. It is categorized by enhancing production, encouraging sustainable practices, optimizing resource use, and improving crop health (Padhiary et al. 2024) through periodical monitoring accompanied by surrounding factors (Mana et al. 2024). It has passed multidimensional ways (Durai and Shamili 2022) of *approach 1*: *promoting decision-making* by utilization of algorithms that study the extensive data from different sources like weather patterns, soil conditions (biophysicochemical properties), soil maps, history of previous crops (growth and yield) those are contributed to farmers' decision for avoiding unnecessary issues and loss rapidly (Nde et al. 2024); *approach 2*: *early detection of pest and disease incidences* through large data set of imagery collections from drones, satellites, and analysis through artificial neural algorithms, which reduces the interventions of new and existing pathogens or pests occurrences (Aziz et al. 2025); *approach 3*: *resource optimization* by ML algorithms can enhance water and soil nutrition efficiency by assessing crop requirements, time, and agro-climatic factors. This leads to considerable economic reductions and balances the ecological niche in agriculture (Awais et al. 2023). For example, ML can enable precise irrigation, guaranteeing that plants get the correct amount and requirement of water under the appropriate circumstances (Xing and Wang 2024), thereby conserving water resources and enhancing crop health (Abioye et al. 2022); *approach 4*: *in sustainable farming yield prediction and optimization* by ML algorithms make remarkably accurate predictions about future yields by studying and analyzing of past data of yield and the environment (Abu Jabed and Azrifah Azmi Murad 2024). Improved execution, efficient allocation of resources, and enhanced techniques in harvesting and post harvesting progresses are made all possible, which helps to yield better outcome with least per cent of loss (Nath et al. 2024); *approach 5*: *at crop management handled through data-driven technologies* such as predictive analytics of climate change, cropping patterns, crop selection, environmental adapting resilient farming practices, and information about market demands are promising, which helps the farmers for challenges in crop production and avoiding unfortunate losses (Javaid et al. 2022); *approach 6: even practices ensure eco-safe and economic viability* by application of ML-based instrumentation and neural network organization setups give a cloud knowledge in marketing and profiting from products with day-by-day demands in agriculture (Pokhariyal et al. 2023) and *approach of knowledge sharing and capacity building* through implementation of ML in farming boosts the exchange of scientific knowledge within farmers, researchers, and policy makers. Additionally, the whole farming community benefits from these cooperative environments, fostering inclusive growth in agriculture and livelihood (Aijaz et al. 2025).

16.4 Understanding ML

ML is one of the sub-fields of AI that aims to learn from the collection, analysis, and prediction of data without a planned or designed programmed one, and it improves their accuracy and performance from experience of learning with the help of enabled data and core models of algorithms (Sarker 2021). It is bearing a multi-discipline way of AI, cognitive science, engineering, mathematics, information technology, statistics, and others. Furthermore, it creates potential bonding between the volume of data inputs from the past and can be analyzed with the current (Sahu et al. 2023). When trying to make assessments and predictions, it requires developing mathematical models on sample data sets known as "*training datasets.*" Using an ML method, the model or algorithm was first trained before being categorized or predicted (Taye 2023). The testing dataset, which is also used to refine the model and achieve precise classification and prediction, is used to generally test the correctness of the ML model (Pudjihartono et al. 2022). ML is a core objective algorithm model categorized into three various types such as unsupervised learning (UL), supervised learning (SL), and reinforced learning (RL) (Barbierato and Gatti 2024). Furthermore, it discussed their basic working principles of algorithms and functions described in Table 16.1.

Table 16.1 Detailed algorithms of machine learning and their uses in precision farming and food security.

Types of Machine Learning	Algorithms and Their Uses		References
	Algorithms	Uses	
Unsupervised Learning (UL)	Anomaly detection (AD), component analysis (CA), independent component analysis (ICA), Gaussian mixture models, K-means, principal component analysis (PCA), and X-means.	Soil health, crop health, yield prediction, and marketing analysis.	(Marian and Tremmel 2021; Garcia-Vera et al. 2024; Kuswidiyanto et al. 2022; Perez-Patricio et al. 2024; Ahmad Effendi et al. 2021; Elbasi et al. 2023; Mary John et al. 2023).
Supervised Learning (SL)	Linear regression, logistic regression, decision trees, random forest, support vector machines (SVM), K-nearest neighbors (KNN), Naive bayes (NB), gradient boosting machines (GBM), extreme gradient boosting (XGBoost), LightGBM, artificial neural networks (ANNs), convolutional neural networks (CNNs), recurrent neural networks (RNNs), elastic net regression and lasso regression.	Yield prediction, soil nutrient analysis, economic forecasting, presence/absence of diseases or pests, farm health, classification of soil types, predictions of crops based on agro-climatic conditions, resource allocation, rapid classification of different crop diseases, soil quality analysis, weather patterns, growth pattern of crop, environmental factors, pest and disease detection, quality assessment, crop yield modeling and selection of crop.	(Murugan et al. 2020; Donatelli et al. 2017; Aworka et al. 2022; Klompenburg et al. 2020; Trentin et al. Javidan et al. 2024; Suljug et al. 2024; Noorunnahar et al. 2023; Zaman et al. 2023; Castillo-Girones et al. 2025).
Reinforced Learning (RL)	Q-Learning, deep Q-networks, policy gradient methods, proximal policy optimization (PPO), trust region policy optimization (TRPO), asynchronous actor-citric agents (A3C), soft actor-critic (SAC), hierarchical reinforcement learning (HRL), multi-agent reinforcement learning (MARL), inverse reinforcement learning (IRL), and curriculum learning in RL.	Irrigation schedule, crop rotation, pest and disease management, autonomous vehicle driving patterns, improving precision farming and crop monitoring, organizing multi-drones, controlling automated vehicles, drones, UMVs, optimal knowledge gaining, and decision-making.	(Alkaff et al. 2025; Li et al. 2024; Feng et al. 2023; Sun et al. 2025; Zhou et al. 2021; Zhao et al. 2022; Martinez-Gil et al. 2025).

16.5 Unsupervised Learning

In ML, UL is one of the most important subset branches and is employed to identify patterns and relationships with untagged data (Wang and Biljecki 2022). Which enables algorithms to innovate elementary structures in data or information without any pre-existing information. Applied in agriculture, it significantly improves productivity and resource optimization and tackles production-related issues in sustainable ways (Benti et al. 2024). It comprised algorithms of anomaly detection (AD), component analysis (CA), independent component analysis (ICA), Gaussian mixture models, K-means, principal component analysis (PCA), and X-means. Analysis of the dimensionality reduction through clustering was performed on both unlabeled and precise farming data (Liakos et al. 2018).

Furthermore, they worked in multifunctional ways of analyzing soil health (pH, nutritional status, microbiome and pollutants load, topography, geographical distribution, and water capabilities were helped to tailor irrigation, fertigation, and optimization resource use efficiency during cultivation) (Marian and Tremmel 2021). In crop health monitoring algorithms of UL exploring the multispectral or hyperspectral images (380/440 nm to 8/12 μm) were used, which have been obtained from satellites and drones (Garcia-Vera et al. 2024). Initially, algorithmic patterns processing images of crop diseases (Kuswidiyanto et al. 2022), disorders, and abiotic stresses (Perez-Patricio et al. 2024) will be progressed through dimensionality reduction clustering techniques of K-means and effectively contributing to reducing farmers' misleading errors in crop health monitoring (Ahmad Effendi et al. 2021). Yield predictions were carried out through past historical data and analysis with similarity and dissimilarity yield patterns and influencing the relationship between productivity enhancement and loss accompanied by crops, varieties, soil health, farming practices, climatic factors, and demands (Elbasi et al. 2023). In marketing, UL clustering algorithms teach farmers through their assumption of trading strategies, namely, demand-related productions, purchasing behavior of consumers, and their preferences, and it meets farmers to enhance skill in production and marketing segment with an investment of smart knowledge based ones (Mary John et al. 2023).

16.6 Supervised Learning

It is a core methodology in ML that involves training algorithms using tagged (labeled) datasets, on each input closely associated with a peculiar output of results. Approaches like data-driven decisions enhance productivity optimization of resource utilization and remodeling or refine crop management based on farming practices, which is mostly helpful to farmers in avoiding unfortunate crop loss and yield also (Pugliese et al. 2021). It includes an assortment of algorithms that can be utilized for classification and relapse assignments. Each algorithm has its qualities, shortcomings, and ranges of application. Underneath could be a full outline of a few of the foremost common supervised learning algorithms, such as *Linear Regression* (yield prediction, soil nutrient analysis, and economic forecasting) (Murugan et al. 2020); *Logistic Regression* (presence/absence of diseases or pests and farm health) (Donatelli et al. 2017); *Decision Trees* (classification of soil types, predictions of crops based on agroclimatic conditions and resource allocation) (Aworka et al. 2022); *Random Forest* (rapid classification of different crop diseases and yield prediction) (Klompenburg et al. 2020); *Support Vector Machines (SVM)* (detecting plant diseases and soil quality analysis) (Islam et al. 2024); *K-Nearest Neighbors (KNN)* (differentiating plant species, yield estimation); *Naive Bayes (NB)* (extracting details of images from captured and infected plants) (Javidan et al. 2024); *Gradient Boosting Machines (GBM)* (yield prediction and weather patterns) (Suljug et al. 2024); *Extreme Gradient Boosting (XGBoost)* (yield forecasting) (Noorunnahar et al. 2023); *LightGBM* (efficiently used to detect pests, yield prediction, and real-time analytics of weather) (Zaman et al. 2023); *Artificial Neural Networks (ANNs), Convolutional Neural Networks (CNNs), and Recurrent Neural Networks (RNNs)* (growth pattern of crop, environmental factors, pest and disease detection, and quality assessment) (Castillo-Girones et al. 2025), *Elastic Net Regression,* and *Lasso Regression* (crop yield modeling and selection of crop) (Trentin et al. 2024). Mostly, the selection of algorithms depends on the specific issue, data availability, and the complexity of the relationship between input and output datasets. In the agriculture domain, the integration of algorithms of remote sensing, IoT, and vast data analytics based on sensors will promote efficiency while employing ML (Wang et al. 2024).

16.7 Reinforced Learning

RL is one of the most important branches or subdivisions of ML dedicated to developing algorithms that enable to make a continuous series of decisions within an environment and to optimize the maximum of returns. It differs from supervised learning by its principle of acquiring knowledge from the previous outcomes of its trial and error, not from labeled data

(Ning and Xie 2024). It can be approached in multi-potential ways, for optimizing decision-making and resource utilization. RL had several different algorithms; they were employed to solve complex problems in farming from field to dine, such as *Q-Learning* (irrigation scheduling, optimize usage of water resources, and crop rotation) (Alkaff et al. 2025), *Deep Q-Networks* (pest and disease management autonomous vehicle driving patterns usage in agricultural operations) (Li et al. 2024), *Policy Gradient Methods* (environmental pollution study and resource allocation) (Feng et al. 2023), *Proximal Policy Optimization (PPO), Trust Region Policy Optimization (TRPO)* (improving precision farming through optimized distribution of resources, automated crop monitoring) (Sun et al. 2025), *Asynchronous Actor-Critic Agents (A3C), Soft Actor-Critic (SAC)* (coordinating functions of numerous drones and monitoring) (Zhou et al. 2021), *Hierarchical Reinforcement Learning (HRL), Multi-Agent Reinforcement Learning (MARL)* (complex farming operations, market distribution, coordinating multiple sensors, drones, and UMVs) (Zhao et al. 2022), *Inverse Reinforcement Learning (IRL),* and *Curriculum Learning in RL* (optimal knowledge gathering from traditional farming, decision-making, real-time tasks of planting, watering, fertigation, and harvesting) (Martinez-Gil et al. 2020).

16.8 Generic Functions of ML in Crop Health

Generally, ML has a wide range of applications in farming and related sectors for improving their efficiency in crop survivability, productivity, decision-making, and sustainability (Dey et al. 2024). Significantly, it improves potential nature in crop monitoring, resource optimization, and economic forecasting, increasing profitability from unique operations like detection of pests and diseases, disease forecasting, improving management by precautionary measures, crop health mapping, and climatic predictive analytics (Mahmoudi et al. 2024) for optimization of yield by time to harvesting and postharvest. Additionally, it works from land to dining at a consumer level (Talaviya et al. 2020). The multi-dimensional functions, such as disease detection, involve identifying foliar diseases based on large image datasets. These images, collected from satellites, sensors, and other imaging tools, are then categorized using CNNs (Sarkar et al. 2023); *pest identification* (by infestations and images collected from monitoring of sensory cameras, drones, and visual recognition of neural networks) (Popescu et al. 2023); *diagnosis of nutrient deficiency* (analysis by phenotypic changes in leaves, texture, and chroma explored by models of ML) (Kadam and Mohite-Patil 2022); *predictive analysis of disease outbreaks* (based on historical and comparison with current data and accompanied risk factors of weather, soil conditions, crops, and forecasting) (Fenu et al. 2021); *crop health monitoring and yield prediction* (consistently monitoring crop health by capturing of hyper-spectral images, past yield data, and predicted yield were analyzed through algorithms of automated sensors and correlation and regression) (Li et al. 2022); *phenotyping, trait analysis, and autonomous decision support systems* (Analyzing a crop's phenotype helps determine if it's healthy. This analysis includes assessing its physiological growth and developmental traits. Using labeled data with AI algorithms, farmers can use this information for future decision-making, such as selecting the best crop varieties) (Gupta et al. 2024); predictive analytics of climate changes and crop health mapping (based on remote sensing and weather patterns, crop health has monitored and their growth and risky factors has been forecasted) (Omia et al. 2023); *crop health recommendations and market trend analysis* (crop monitored and managed by biotic stresses by using crop protectants and its commercial demand and profitability were explored through utilization of ML algorithms) (Chizoba Mba et al. 2025).

16.9 Advantages

In crop health management (CHM), ML grants tremendous advantages in agriculture; during the implementation, ML improves farmers' practices through their potential outputs of early detection of pests and diseases, automated crop monitoring, efficient usage of resource optimization, climate adaptation, rapid and comprehensive analysis of big data, predictive analytics, and customization, which provides a better outcome in right time with zero loss (Benos et al. 2021).

16.10 Challenges

During employment of ML, numerous practical issues by poly-way of manual and technical defaults such as *quality of data and availability* (pure and HQ datasets are required for training models in ML when it was unavailable or poor status it provides failure) (Alzubaidi et al. 2023) have been faced; *complexity in the biological system of crops* (maybe drawback in the accuracy of results and poor yield) (Khaki and Wang 2019); *selection, interpretation, and customization algorithms and models* playing a vital role in breaking the complexity and confusion leads, so when it was a right choice that optimized the potential outcome (Akter et al. 2022); skilled, expertise and application cost of investment are essential to achieving a successful outcome; nature of the environment and well-known organized approaches of traditional farming practices and their integration is necessarily one (Cravero et al. 2022).

16.11 Future Perspectives

The future of ML in crop health is on the edge of significant advancements that will greatly improve precision farming and sustainability (Maryam et al. 2025). It encompasses the growing science of high-resolution satellite imagery, IoT, sensors, and extensive datasets of real-time algorithms and crop health (Sharma et al. 2023). AI incorporation of predictive analytics is more helpful to farmers with a sustainable prediction of weather patterns, soil nutritional updates, natural resource optimization, early detection of diseases, and upsurge of pests infestations, allowing farmers their decision-making. It collaborates with farming and trading and encourages knowledge of demand strategies and profitability point-outs from the field to consumers (Assimakopoulos et al. 2024). Recently, it contributed to climate change mitigations, resilient crop varieties, and modernization of farm practices. In summary, ML is poised to transform crop health management, leading to a more sustainable and food-secure future (Oleiro Araujo et al. 2023).

16.12 Conclusion

The incorporation of ML into crop health management signifies a significant transformation in the future of agriculture, providing innovative solutions to the complex challenges encountered by contemporary farming. As issues such as climate change, increasing population, and limited resources become more pressing, the importance of precision agriculture and data-driven decision-making has reached unprecedented levels. ML technologies enable farmers to monitor crop health in real time, anticipate potential risks, and execute targeted interventions, ultimately resulting in improved productivity and sustainability. Moving forward necessitates not only technological advancements but also collaborative initiatives among agricultural scientists, data specialists, and farmers. By promoting interdisciplinary collaborations, we can create more effective and accessible tools tailored to specific agricultural needs, thereby enhancing resilience within farming systems. As the agricultural sector continues to evolve, it is vital to tackle challenges such as data integration, model interpretability, and technological accessibility to fully harness the capabilities of ML. The potential implementation of ML in crop health management has the potential to revolutionize agriculture, making it a more sustainable and efficient industry. This transformation is essential for ensuring food security for future generations while preserving ecological balance. Embracing this technological advancement is critical for developing a resilient agricultural framework capable of adapting to ongoing environmental changes and fulfilling the needs of a growing global population.

References

Abbasi, R., Martinez, P., and Ahmad, R. (2022). The digitization of agricultural industry: A systematic literature review on agriculture 4.0. *Smart Agricultural Technology* 2: 1–24.

Abebe, M.G. (2024). Impacts of urbanization on food security in Ethiopia: A review with empirical evidence. *Journal of Agriculture and Food Research* 15: 1–8.

Abioye, E.A., Hensel, O., Esau, T.J. et al. (2022). Precision irrigation management using machine learning and digital farming solutions. *AgriEngineering* 4: 70–100.

Abu Jabed, M., and Azrifah Azmi Murad, M. (2024). Crop yield prediction in agriculture: A comprehensive review of machine learning and deep learning approaches, with insights for future research and sustainability. *Heliyon* 10 (24): e40836.

Ahmad Effendi, A.F.A., Md Isa, M.N., Ahmad, M.I. et al. (2021). Image processing for paddy disease detection using K-means clustering and GLCM algorithm. *International Journal of Nanoelectr onics and Materials* 14: 253–263.

Aijaz, N., Lan, H., Raza, T. et al. (2025). Artificial intelligence in agriculture: Advancing crop productivity and sustainability. *Journal of Agriculture and Food Research* 20: 1–9.

Akter, S., Dwivedi, Y.K., Sajib, S. et al. (2022). Algorithmic bias in machine learning-based marketing models. *Journal of Business Research* 144: 201–216.

Alkaff, M., Basuhail, A., and Sari, Y. (2025). Optimizing water use in maize irrigation with reinforcement learning. *Mathematics* 13: 595.

Alzubaidi, L., Bai, J., Al-Sabaawi, A. et al. (2023). A survey on deep learning tools dealing with data scarcity: Definitions, challenges, solutions, tips, and applications. *Journal of Big Data* 10: 1–82.

Assimakopoulos, F., Vassilakis, C., Margaris, D. et al. (2024). Artificial intelligence tools for the agriculture value chain: Status and prospects. *Electronics* 13: 4362.

Awais, M., Abbas Naqvi, S.M.Z., Zhang, H. et al. (2023). AI and machine learning for soil analysis: An assessment of sustainable agricultural practices. *Bioresources and Bioprocessing* 10: 90.

Aworka, R., Saadio Cedric, L., Hamilton Adoni, W.Y. et al. (2022). Agricultural decision system based on advanced machine learning models for yield prediction: Case of East African countries. *Smart Agricultural Technology* 2: 1–9.

Aziz, D., Rafiq, S., Saini, P. et al. (2025). Remote sensing and artificial intelligence: Revolutionizing pest management in agriculture. *Frontiers in Sustainable Food Systems* 9: 1551460.

Barbierato, E. and Gatti, A. (2024). The challenges of machine learning: A critical review. *Electronics* 13 (2): 416.

Benos, L., Tagarakis, A.C., Dolias, G. et al. (2021). Machine learning in agriculture: A comprehensive updated review. *Sensors* 21: 3758.

Benti, N.E., Diro Chaka, M., Gezahegn Semie, A. et al. (2024). Transforming agriculture with machine learning, deep learning, and IoT: Perspectives from Ethiopia—Challenges and opportunities. *Discover Agriculture* 2: 63.

Castillo-Girones, S., Munera, S., Martínez-Sober, M. et al. (2025). Artificial neural networks in agriculture, the core of artificial intelligence: What, when, and why. *Computers and Electronics in Agriculture* 230: 1–19.

Chizoba Mba, P., Njoku, J.N., and Uyeh, D.D. (2025). Enhancing resilience in specialty crop production in a changing climate through smart systems adoption. *Smart Agricultural Technology* 25: 1–45.

Choudhary, V., Guha, P., Pau, G. et al. (2025). An overview of smart agriculture using internet of things (IoT) and web services. *Environmental and Sustainability Indicators* 26: 1–18.

Cravero, A., Pardo, S., Sepulveda, S. et al. (2022). Challenges to use machine learning in agricultural big data: A systematic literature review. *Agronomy* 12: 748.

Dey, B., Ferdous, J., and Ahmaed, R. (2024). Machine learning based recommendation of agricultural and horticultural crop farming in India under the regime of NPK, soil pH and three climatic variables. *Heliyon* 10: e25112.

Domingues, T., Brandao, T., and Ferreira, J.C. (2022). Machine learning for detection and prediction of crop diseases and pests: A comprehensive survey. *Agriculture* 12: 1350.

Donatelli, M., Magarey, R.D., Bregaglio, S. et al. (2017). Modelling the impacts of pests and diseases on agricultural systems. *Agricultural Systems* 155: 213–224.

Duchenne-Moutien, R.A., and Neetoo, H. (2021). Climate change and emerging food safety issues: A review. *Journal of Food Protection* 84 (11): 1884–1897.

Durai, S.K.S. and Shamili, M.D. (2022). Smart farming using machine learning and deep learning techniques. *Decision Analytics Journal* 3: 1–30.

Elbasi, E., Zaki, C., Topcu, A.E. et al. (2023). Crop prediction model using machine learning algorithms. *Applied Sciences* 13: 9288.

Fawzy, Z.F. and El-Ramady, H. (2022). Applications and challenges of smart farming for developing sustainable agriculture. *Environment, Biodiversity and Soil Security* 6: 81–90.

Feng, X., Lie, Y., Chen, X. et al. (2023). Study of the allocation strategy of water pollutant emission permits under a bidirectional ecological compensation mechanism. *Ecological Indicators* 154: 1–13.

Fenu, G. and Maridina Malloci, F. (2021). Forecasting plant and crop disease: An explorative study on current algorithms. *Big Data and Cognitive Computing* 5: 2.

Gamage, A., Gangahagedara, R., Gamage, J. et al. (2023). Role of organic farming for achieving sustainability in agriculture. *Farming System* 1 (1): 1–14.

Gamage, A., Gangahagedara, R., Subasinghe, S. et al. (2024). Advancing sustainability: The impact of emerging technologies in agriculture. *Current Plant Biology* 40: 1–16.

Garcia-Vera, Y.E., Poloche-Arango, A., Mendivelso-Fajardo, C.A. et al. (2024). Hyperspectral image analysis and machine learning techniques for crop disease detection and identification: A review. *Sustainability* 16: 6064.

Giller, K.E., Delaune, T., Vasco Silva, J. et al. (2021). The future farming: Who will produce our food?. *Food Security* 13: 1073–1099.

Gupta, D.K., Pagani, A., Zamboni, P. et al. (2024). AI-powered revolution in plant sciences: Advancements, applications, and challenges for sustainable agriculture and food security. *Exploration of Foods and Foodomics* 2: 443–459.

Islam, N., Rashid, Md., Pasandideh, F. et al. (2021). A review of applications and communication technologies for internet of things (IoT) and unmanned aerial vehicle (UAV) based sustainable smart farming. *Sustainability* 13: 1821.

Islam, S., Samsuzzaman, S., Reza, M. et al. (2024). Image processing and support vector machine (SVM) for classifying environmental stress symptoms of pepper seedlings grown in a plant factory. *Agronomy* 14: 2043.

Javaid, M., Haleem, A., Singh, R.P. et al. (2022). Enhancing smart farming through the applications of agriculture 4.0 technologies. *International Journal of Intelligent Networks* 3: 150–164.

Javidan, S.M., Banakar, A., Rahnama, K. et al. (2024). Feature engineering to identify plant diseases using image processing and artificial intelligence: A comprehensive review. *Smart Agricultural Technology* 8: 1–14

Jaya Singh, D.G., Elangovan, S., Paramasivam, A. et al. (2024). Impact of seed rate and time of sowing on rice fallow crops and their sustainable production system in machine-harvested rice fields. *Frontiers in Sustainable Food Systems* 8: 1–8.

Kadam, V.J. and Mohite-Patil, T.B. (2022). Detection of nutrient deficiencies in crops using support vector machine (SVM). *International Research Journal of Engineering and Technology* 09 (08): 1–4.

Khaki, S. and Wang, L. (2019). Crop yield prediction using deep neural networks. *Frontiers in Plant Science* 10: 621.

Klompenburg, T., Kassahun, A., and Catal, C. (2020). Crop yield prediction using machine learning: A systematic literature review. *Computers and Electronics in Agriculture* 177: 1–18.

Komatineni, B.K., Makam, S., and Meena, S. (2024). A comprehensive review of the functionality and applications of unmanned aerial vehicles (UAVs) in the realm of agriculture. *Journal of Electrical Systems and Information Technology* 11: 1–28.

Kumar, L., Chhogyel, N., Gopalakrishnan, T. et al. (2022). Climate change and future of agri-food production. In: *Future Foods Global Trends, Opportunities and Sustainability Challenges* (ed. Bhat, R.), 49–79. Academic Press: New York, USA.

Kumar, V., Sharma, K.V., Kedam, N. et al. (2024). A comprehensive review on smart and sustainable agriculture using IoT technologies. *Smart Agricultural Technology* 8: 1–22.

Kuswidiyanto, L.W., Noh, H.H., and Han, X. (2022). Plant disease diagnosis using deep learning based on aerial hyperspectral images: A review. *Remote Sensing* 14: 6031.

Lahlali, R., Taoussi, M., Laasli, S.E. et al. (2024). Effects of climate change on plant pathogens and host-pathogen interactions. *Crop and Environment* 3 (3): 159–170.

Li, K.Y., Lima, R.S., Burnside, N.G. et al. (2022). Toward automated machine learning-based hyperspectral image analysis in crop yield and biomass estimation. *Remote Sensing* 14: 1114.

Li, W., Han, X., Lin, Z. et al. (2024). Enhanced pest and disease detection in agriculture using deep learning-enabled drones. *Acadlore Transactions on AI and Machine Learning* 3 (1): 1–10.

Liakos, K.G., Busato, P., Moshou, D. et al. (2018). Machine learning in agriculture: A review. *Sensors* 18: 2674.

Mahmoudi, S., Davar, A., Sohrabipour, P. et al. (2024). Leveraging imitation learning in agricultural robotics: A comprehensive survey and comparative analysis. *Frontiers in Robotics and AI* 11: 1–21.

Mana, A.A., Allouhi, A., Hamrani, A. et al. (2024). Sustainable AI-based production agriculture: Exploring AI applications and implications in agricultural practices. *Smart Agricultural Technology* 7: 1–15.

Manoogian, E.N.C., Chaix, A., and Panda, S. (2019). When to eat: The importance of eating patterns in health and disease. *Journal of Biological Rhythms* 34 (6): 579–581.

Marian, M. and Tremmel, S. (2021). Current trends and applications of machine learning in tribology: A review. *Lubricants* 9 (9): 86.

Martinez-Gil, F., Lozano, M., García-Fernández, I. et al. (2020). Using inverse reinforcement learning with real trajectories to get more trustworthy pedestrian simulations. *Mathematics* 8: 1479.

Mary John, J., Shobayo, O., and Ogunleye, B. (2023). An exploration of clustering algorithms for customer segmentation in the UK retail market. *Analytics* 2: 809–823.

Maryam, G., Khan, N., Sehar, S. et al. (2025). Optimizing crop health with artificial intelligence and machine learning technologies. In: *Ecologically Mediated Development* (ed. Jatav, H.S., Raiput, V.D., and Minkina, T.), 487–509. Springer: New York, USA.

Meshram, V., Patil, K., Meshram, V. et al. (2021). Machine learning in agriculture domain: A state-of-art survey. *Artificial Intelligence in the Life Sciences* 1: 1–11.

Murugan, R., Thomas, F., Geetha Sree, G. et al. (2020). Linear regression approach to predict crop yield. *Asian Journal of Computer Science and Technology* 9 (1): 40–44.

Nath, B., Chen, G., O'Sullivan, C.M. et al. (2024). Research and technologies to reduce grain postharvest losses: A review. *Foods* 13 (12): 1875.

Nde, R.K., Kedieng Fendji, J.L.E., Omer Yenke, B. et al. (2024). Crop selection: A survey on factors and techniques. *Smart Agricultural Technology* 9: 1–18.

Ning, Z. and Xie, L. (2024). A survey on multi-agent reinforcement learning and its application. *Journal of Automation and Intelligence* 3 (2): 73–91.

Noorunnahar, M., Chowdhury, A.H., and Mila, F.A. (2023). A tree based eXtreme gradient boosting (XGBoost) machine learning model to forecast the annual rice production in Bangladesh. *PLoS One* 18 (3): e0283452.

Nowak, B. (2021). Precision agriculture: Where do we stand? A review of the adoption of precision agriculture technologies on field crops farms in developed countries. *Agricultural Research* 10: 515–522.

Oleiro Araujo, S., Peres, R.S., Ramalho, J.C. et al. (2023). Machine learning applications in agriculture: Current trends, challenges, and future perspectives. *Agronomy* 13: 2976.

Omia, E., Bae, H., Park, E. et al. (2023). Remote sensing in field crop monitoring: A comprehensive review of sensor systems, data analyses and recent advances. *Remote Sensing* 15: 354.

Padhiary, M., Saha, D., Kumar, R. et al. (2024). Enhancing precision agriculture: A comprehensive review of machine learning and AI vision applications in all-terrain vehicle for farm automation. *Smart Agricultural Technology* 8: 1–14.

Paustian, M. and Thevusen, L. (2016). Adoption of precision agriculture technologies by German crop farmers. *Precision Agriculture* 18: 701–716.

Perez-Patricio, M., de Jesus Osuna-Coutino, J.A., Rios-Toledo, G. et al. (2024). Plant stress detection using a three-dimensional analysis from a single RGB image. *Sensors* 24: 7860.

Pokhariyal, S., Patel, N.R., and Govind, A. (2023). Machine learning-driven remote sensing applications for agriculture in India: A systematic review. *Agronomy* 13: 1–30.

Popescu, D., Dinca, A., Ichim, L. et al. (2023). New trends in detection of harmful insects and pests in modern agriculture using artificial neural networks: A review. *Frontiers in Plant Science* 14: 1–29.

Pudjihartono, N., Fadason, T., Kempa-Liehr, A.W. et al. (2022). A review of feature selection methods for machine learning-based disease risk prediction. *Frontiers in Bioinformatics* 2: 927312.

Pugliese, R., Regondi, S., and Marini, R. (2021). Machine learning-based approach: Global trends, research directions, and regulatory standpoints. *Data Science and Management* 4: 19–29.

Sahu, S.K., Mokhade, A., and Bokde, N.D. (2023). An overview of machine learning, deep learning, and reinforcement learning-based techniques in quantitative finance: Recent progress and challenges. *Applied Sciences* 13 (3): 1956.

Saleem, A., Anwar, S., Nawaz, T. et al. (2024). Securing a sustainable future: The climate change threat to agriculture, food security, and sustainable development goals. *Journal of Umm Al-Qura University for Applied Sciences*: 1–17.

Sarkar, C., Gupta, U., Hazarika, B.B. et al. (2023). Leaf disease detection using machine learning and deep learning: Review and challenges. *Applied Soft Computing* 145 (11): 110534.

Sarker, I.H. (2021). Machine learning: Algorithms, real-world applications and research directions. *SN Computer Science* 2: 160.

Sharma, A., Sharma, A., Tselykh, A. et al. (2023). Artificial intelligence and internet of things oriented sustainable precision farming: Towards modern agriculture. *Open Life Sciences* 18 (1): 20220713.

Singh, G., Kalra, N., Yadav, N. et al. (2022). Smart agriculture: A review. *Siberian Journal of Life Sciences and Agriculture* 14 (6): 423–454.

Suljug, J., Spisi, J., Grgi, K. et al. (2024). A comparative study of machine learning models for predicting meteorological data in agricultural applications. *Electronics* 13: 3284.

Sun, K., Yang, J., Li, J. et al. (2025). Proximal policy optimization-based hierarchical decision-making mechanism for resource allocation optimization in UAV networks. *Electronics* 14: 747.

Talaviya, T., Shah, D., Patel, N. et al. (2020). Implementation of artificial intelligence in agriculture for optimisation of irrigation and application of pesticides and herbicides. *Artificial Intelligence in Agriculture* 4: 58–73.

Taye, M.M. (2023). Understanding of machine learning with deep learning: Architectures, workflow, applications and future directions. *Computers* 12 (5): 91.

Tchonkouang, R.D., Onyeaka, H., and Nkoutchou, H. (2024). Assessing the vulnerability of food supply chains to climate change-induced disruptions. *Science of the Total Environment* 920: 1–30.

Trentin, C., Ampatzidis, Y., Lacerda, C. et al. (2024). Tree crop yield estimation and prediction using remote sensing and machine learning: A systematic review. *Smart Agricultural Technology* 9: 1–12.

Viana, C.M., Freire, D., Abrantes, P. et al. (2022). Agricultural land systems importance for supporting food security and sustainable development goals: A systematic review. *Science of the Total Environment* 806: 1–13.

Wang, J. and Biljecki, F. (2022). Unsupervised machine learning in urban studies: A systematic review of applications. *Cities* 129: 1–20.

Wang, J., Wang, Y., Li, G. et al. (2024). Integration of remote sensing and machine learning for precision agriculture: A comprehensive perspective on applications. *Agronomy* 14: 1975.

Wijerathna-Yapa, A., and Pathirana, R. (2022). Sustainable agro-food systems for addressing climate change and food security. *Agriculture* 12 (10): 1554.

Xing, Y. and Wang, X. (2024). Precise application of water and fertilizer to crops: Challenges and opportunities. *Frontiers in Plant Science* 15: 1–17.

Zaman, A., Nyarko-Boateng, O., Adekoya, A.F. et al. (2023). A predictive analytics model for crop suitability and productivity with tree-based ensemble learning. *Decision Analytics Journal* 8: 1–11.

Zhao, P., Yuan, Y., and Guo, T. (2022). Extensible hierarchical multi-agent reinforcement-learning algorithm in traffic signal control. *Applied Sciences* 12: 12783.

Zhou, K., Wang, W., Hu, T. et al. (2021). Application of improved asynchronous advantage actor critic reinforcement learning model on anomaly detection. *Entropy* 23: 274.

17

Social Impact of Machine Learning on Agricultural Communities

Atef M. El-Sagheer[1], Eman A. Ahmed[2], Hamdy A. Sayed[3]*

[1] *Agricultural Zoology and Nematology Department, Faculty of Agriculture, Al-Azhar University, Assiut, Egypt*
[2] *Department of Community Organization, Faculty of Social Work, Assiut University, Assiut, Egypt*
[3] *Social Planning Department, Faculty of Social Work, Assiut University, Assiut, Egypt*

*Corresponding author: atefelsagheer@azhar.edu.eg

17.1 Introduction

In recent years, machine learning (ML) has become a transformative tool in plant biology, reshaping how we study plant systems, improve agricultural practices, and manage ecosystems. ML's ability to analyze vast amounts of data and uncover patterns has made it indispensable for predicting plant traits and disease outbreaks and optimizing crop yields. These capabilities are crucial for tackling the modern challenges of plant biology and agriculture, particularly in the face of climate change and resource constraints (Liakos et al. 2018). By integrating ML into fields such as phenotyping, genetics, and breeding, researchers can make faster, more precise discoveries while also supporting sustainable agricultural systems (Kamilaris and Prenafeta-Boldú 2018). ML is also driving advancements in precision agriculture, where real-time data from sensors and drones are used to monitor plant health, improve resource efficiency, and tailor interventions to the specific needs of crops (Van Klompenburg et al. 2020). However, the adoption of ML poses challenges, particularly regarding the impact on traditional farming practices, the management of data ownership, and ensuring equitable access to these technologies (Wolfert et al. 2017). Smallholder farmers and rural communities, in particular, may struggle to integrate these advanced systems, potentially exacerbating existing inequalities in agriculture (Basso and Antle 2020; Birner 2018b).

Therefore, this chapter explores the role of ML in plant biology, focusing on its applications in improving plant productivity, sustainability, and resilience. It also discusses the need for inclusive strategies to ensure that the benefits of ML are widely shared across agricultural communities and that traditional knowledge is integrated with technological innovations. As ML continues to reshape plant biology, striking a balance between technological innovation and sustainable, inclusive practices will be critical for the future of both the science and the agricultural systems it supports.

17.2 Impact of ML on Traditional Farming Practices

ML has transformed many sectors, and agriculture is no exception. Traditionally, farming has depended on manual labor, experience, and techniques that have been passed down through generations (Liakos et al. 2018). With advancements in data collection, computational power, and algorithm development, ML now offers unprecedented possibilities for precision agriculture, enhancing efficiency in farming, pest management, and resource allocation. However, while ML presents numerous opportunities, its introduction also disrupts deeply ingrained cultural, environmental, and local knowledge systems that have defined traditional farming for centuries (Wolfert et al. 2017).

Machine Learning for Plant Biology, First Edition. Edited by Jen-Tsung Chen.

17.2.1 Disruption of Traditional Knowledge Systems

One of the most notable impacts of ML on traditional farming practices is the disruption of long-established knowledge systems. For centuries, farmers have relied on their intimate understanding of local ecosystems, weather patterns, and soil conditions to guide their agricultural decisions (Altieri 2004). This indigenous knowledge has been critical in ensuring productivity and sustainability, especially in regions with highly variable environmental conditions. With the advent of ML, however, data-driven algorithms are increasingly replacing farmer intuition for tasks like optimizing planting schedules, irrigation, and pest control (Gebbers and Adamchuk 2010).

While these advancements have led to increases in yield and reduced resource consumption, they also risk devaluing traditional knowledge. Machine-generated predictions, although efficient, may lead to the gradual erosion of cultural heritage and practices that have been carefully adapted to specific environmental conditions over generations (Cochrane 2019). As a result, there is a growing concern that the introduction of ML could result in a loss of local knowledge systems, which have been crucial for environmental stewardship and long-term resilience (Brondizio and Moran 2008).

17.2.2 Precision Agriculture and Efficiency Gains

ML is a key driver behind the development of precision agriculture, which leverages data from sensors, drones, and satellites to monitor crop health, predict weather, and optimize resource usage (Lamb et al. 2008). These technologies allow for real-time data analysis and decision-making, providing more accurate insights than traditional methods, which often rely on generalized practices (Van Klompenburg et al. 2020). For example, ML models can detect signs of disease in crops earlier than manual visual inspections, allowing for timely interventions and reduced crop loss (Kamilaris and Prenafeta-Boldú 2018).

Additionally, ML enables the precise tailoring of inputs like water and fertilizers to the specific needs of each crop. This reduces waste and minimizes the environmental impact, a significant improvement over conventional farming methods (Wang et al. 2018). The efficiency gains from ML-driven precision agriculture are often difficult, if not impossible, to achieve using traditional farming techniques, which are more labor-intensive and less specific in their resource allocation (Mulla 2013).

17.2.3 Economic Impacts on Smallholder Farmers

The economic effects of ML on smallholder farmers, particularly in developing countries, have been profound. Large-scale farming operations with access to capital are often able to adopt ML technologies, benefiting from increased productivity and efficiency (Rahman 2020). However, smallholder farmers, who constitute a significant portion of the global farming population, may lack access to the necessary technology and expertise, exacerbating existing economic inequalities (Birner 2018a).

Furthermore, as ML-driven technologies replace labor-intensive farming practices, rural employment is likely to decline, directly affecting the livelihoods of smallholder farmers who depend on manual labor (Lowder et al. 2016). Without targeted interventions, such as affordable technology solutions and access to training, ML could widen the economic gap between large-scale and small-scale farmers, threatening the survival of traditional farming communities (Westermann et al. 2018).

17.2.4 Environmental and Sustainability Considerations

Traditional farming systems are often designed to work in harmony with local ecosystems, relying on methods like crop rotation, polyculture, and organic farming to maintain soil health and biodiversity (Altieri 1999). However, ML's application in industrial farming has raised concerns about environmental degradation. Monoculture farming, driven by ML-based efficiency models, can lead to soil depletion and reduced biodiversity, which are common issues in conventional large-scale agriculture (Tsiafouli et al. 2015).

That said, ML has the potential to contribute to sustainable agricultural practices. When used appropriately, ML algorithms can monitor soil conditions, water usage, and pest populations, reducing the need for chemical fertilizers and pesticides (Chlingaryan et al. 2018). Moreover, the integration of traditional ecological knowledge into ML systems may foster a more balanced approach to farming, allowing for sustainable resource use and environmental preservation

(Crossa et al. 2017). For ML to be effective in advancing sustainability, it must be implemented in ways that respect local farming traditions and prioritize long-term environmental health over short-term productivity gains (Pretty 2008).

17.2.5 Adapting Traditional Farmers to the Digital Era

Despite the challenges posed by ML technologies, pathways exist to integrate traditional farmers into the digital agricultural revolution. Key among these is the need for education and digital literacy programs, which are essential for empowering farmers to adopt and benefit from new technologies (Vankov et al. 2020). Training programs that combine ML tools with traditional practices can enhance decision-making without erasing cultural heritage (Singh et al. 2020).

Decision-support systems that provide actionable insights based on ML predictions, when combined with farmer input, can boost both productivity and cultural resilience (Basso and Antle 2020). Furthermore, open-source ML platforms and low-tech mobile applications could help democratize access to advanced technologies for smallholder farmers, enabling them to compete in the modern agricultural landscape (van Es and Woodard 2017).

In conclusion, while ML presents significant advantages in terms of efficiency, precision, and sustainability, its impact on traditional farming practices is complex. The challenges it poses to local knowledge systems, economic equity, and environmental sustainability necessitate a balanced approach that integrates the strengths of traditional practices with technological innovations. By fostering collaboration between traditional farmers and technological innovators, the agricultural sector can leverage ML's potential without sacrificing the values and practices that have sustained farming communities for generations.

17.2.6 Redefining Rural Employment in the Age of AI

Artificial intelligence (AI) and ML are rapidly transforming agriculture, fundamentally altering employment structures in rural areas. Traditionally, rural employment has been centered on manual labor and small-scale farming. However, as AI technologies such as autonomous machinery, smart sensors, and predictive algorithms become more prevalent in agriculture, the labor dynamics within these communities are shifting. While AI presents new opportunities, it also raises concerns about job displacement, skill gaps, and the future economic sustainability of rural areas (Jha et al. 2019).

17.2.7 Automation and Job Displacement

One of the most significant effects of AI in agriculture is the automation of labor-intensive tasks. Autonomous tractors, robotic harvesters, and drone-based crop monitoring systems are becoming more widespread, particularly on larger farms. These technologies increase efficiency, reduce human error, and can operate continuously, thereby reducing the need for human labor (Reichstein 2020). Tasks traditionally performed by large groups of workers, such as planting and harvesting, can now be managed by machines, leading to a reduction in demand for manual labor (Alonso and Calderon 2021).

For rural workers who depend on seasonal agricultural jobs, the increased automation poses a serious threat to their livelihoods. While farmers benefit from lower operational costs and improved productivity, many workers—especially those with low skills—face the risk of job loss (World Bank 2021). The displacement of labor by AI-driven machines may exacerbate unemployment in rural communities, where alternative job opportunities are often scarce (Lowenberg-DeBoer et al. 2020).

17.2.8 Emergence of New Job Roles

Despite the threat of job displacement, AI is also creating new employment opportunities in agriculture. The widespread use of AI and ML requires a workforce skilled in operating, maintaining, and optimizing these systems. Roles in agricultural technology (AgTech) are emerging, including AI developers, data analysts, and technicians specializing in autonomous farming equipment (Schimmelpfennig 2018).

Furthermore, data management has become increasingly important. AI systems generate large datasets from various sources such as sensors, drones, and satellites, requiring professionals who can interpret the data and convert it into actionable insights (van Es and Woodard 2017). However, these new job roles often require a high level of technical proficiency, creating a skills gap in rural areas where access to advanced education is limited (Klerkx and Rose 2020).

17.2.9 Education and Skill Development

The skill gap between traditional agricultural labor and the demands of AI-related roles presents a significant challenge. While traditional agricultural work often requires little formal education, AI-related jobs demand technical proficiency and digital literacy (Rose et al. 2021). Many rural communities lack access to the education and training programs necessary to prepare workers for the emerging AI-driven agricultural sector (Daum 2018).

To address these gaps, governments and educational institutions must develop vocational training programs tailored to the needs of the modern agricultural workforce. Training initiatives focused on AI system operation, data analysis, and robotics can help rural workers transition into new roles created by the rise of AI in agriculture (Schneider 2020). Moreover, integrating digital literacy and science, technology, engineering, and mathematics (STEM) education into rural curricula is essential to ensure future generations are equipped to thrive in a tech-oriented agricultural industry (He and Zhang 2020).

17.2.10 Opportunities for Entrepreneurship

In addition to reshaping traditional employment, AI opens up new avenues for entrepreneurship in rural areas. Farmers and rural entrepreneurs can harness AI technologies to optimize operations, reduce costs, and enhance yields. AI-powered tools enable small-scale farmers to practice precision agriculture, allowing them to compete with larger farming operations (Gebbers and Adamchuk 2010).

Rural entrepreneurs can also develop local AI solutions tailored to the specific needs of their regions. For example, ML algorithms customized for local crops or climates can be developed and sold to local farmers (Brynjolfsson and McAfee 2014). By fostering such entrepreneurial ventures, rural economies can benefit from job creation and new business opportunities in AI-related services (Zeng et al. 2020).

17.2.11 The Role of Policy in Supporting Rural Employment

Governments play a critical role in ensuring that the adoption of AI in agriculture benefits rural communities. Policies that promote equitable access to AI technologies and offer incentives for adopting these tools can help mitigate the risks of job displacement (United Nations 2021). Additionally, governments must strengthen social safety nets and invest in retraining programs for displaced workers to help them transition into new roles in the evolving agricultural landscape (OECD 2019).

Policymakers should also promote inclusive access to education and training, ensuring that rural workers have the opportunity to acquire the skills needed for AI-related jobs (Klerkx and Begemann 2020). In ng so, governments can help bridge the skills gap and support the development of a tech-savvy rural workforce (Schimmelpfennig 2018).

Thus, AI and ML are reshaping rural employment by shifting the focus from manual labor to technology-driven roles. While automation poses a risk of job displacement, it also presents opportunities for new, high-skilled employment and entrepreneurship. Bridging the skills gap through education and training is essential to ensure that rural workers can benefit from these new opportunities. Governments and policymakers must work together to foster an inclusive transition to AI in agriculture, ensuring that rural communities thrive in the age of AI.

17.3 ML and Farmer Autonomy: Decision-making in a Data-driven World

ML has become a critical tool in modern agriculture, offering farmers unprecedented insights into crop management, resource allocation, and disease prediction. By leveraging large datasets from sensors, satellites, and weather forecasts, ML enables more precise decision-making, ultimately boosting productivity and sustainability. However, the increasing reliance on data-driven models raises important questions about farmer autonomy. The shift from intuition-based practices to AI-driven decisions may reduce farmers' control over their operations, creating new dependencies on technology providers and reducing the role of traditional expertise (Lowenberg-DeBoer et al. 2020).

17.3.1 The Role of ML in Decision-making

ML algorithms have transformed agricultural decision-making by offering predictive insights that allow farmers to optimize planting, irrigation, fertilization, and pest control (van Es and Woodard 2017). These algorithms analyze historical data and real-time inputs to predict outcomes with greater accuracy than human judgment alone. For example, ML models can predict disease outbreaks before symptoms are visible, enabling farmers to act swiftly and prevent crop loss (Jha et al. 2019). Similarly, ML-driven precision irrigation systems optimize water use based on crop needs, soil conditions, and weather patterns, resulting in more efficient resource use (Gebbers and Adamchuk 2010).

While these advances undoubtedly improve farm efficiency, they also shift the locus of decision-making away from the farmer to the AI system. Farmers increasingly rely on ML-generated recommendations rather than personal experience or traditional knowledge to make critical decisions, reducing their direct involvement in everyday operations (Rose et al. 2021). This shift toward data-driven decision-making has sparked debate about the potential erosion of farmer autonomy.

17.3.2 Impact on Farmer Autonomy

As ML systems become more central to agricultural management, concerns about farmer autonomy have grown. Autonomy, in the context of agriculture, refers to the ability of farmers to make independent decisions about their operations based on their knowledge, experience, and preferences. ML increasing role in decision-making can challenge this autonomy in several ways (Klerkx and Rose 2020).

First, farmers may become overly dependent on ML tools, particularly when they are designed as black-box systems that provide recommendations without explaining the underlying logic (Doshi-Velez and Kim 2017). The lack of transparency in ML models can undermine farmers' confidence in their decision-making abilities, as they may feel compelled to follow the algorithm's recommendations even when they conflict with their intuition or local knowledge (Floridi et al. 2018).

Second, ML adoption often involves the outsourcing of decision-making to technology companies that own and operate the platforms. This dynamic can create a dependency on external entities for critical farm decisions, potentially limiting farmers' control over their data and operations (Hoffman 2019). Technology providers typically collect, process, and analyze farm data, which can lead to a concentration of decision-making power in the hands of companies rather than the farmers themselves (Wendling et al. 2020). This dependency risks undermining the self-reliance that has traditionally characterized farming communities.

17.3.3 Data Ownership and Control

Data ownership is a critical issue in the discussion of farmer autonomy in a data-driven agricultural landscape. As farmers, increasingly rely on ML platforms to make decisions, large volumes of data—ranging from soil conditions to crop yields—are collected and analyzed by these systems. However, the question of who owns this data and how it is used remains contentious (van der Burg et al. 2019).

In many cases, the ownership of data generated on farms is unclear, with technology providers often retaining control over it. This creates a power imbalance where farmers are dependent on companies for access to the insights generated from their own data (Bronson and Knezevic 2016). Additionally, farmers may lack the legal or technical expertise to negotiate favorable data-sharing agreements, further limiting their autonomy. Policies that ensure farmers retain ownership and control over their data are critical to preserving autonomy in the face of growing reliance on ML technologies (Carolan 2020).

Moreover, the use of aggregated farm data by large corporations can give them a competitive advantage in predicting market trends, potentially disadvantaging individual farmers. Companies with access to vast amounts of agricultural data may use this information to make market decisions, such as adjusting prices or predicting demand, that small-scale farmers cannot compete with. This dynamic risks further marginalizing farmers in the agricultural supply chain (Carbonell 2016).

17.3.4 Maintaining Autonomy Through Explainable AI

One solution to the issue of eroding autonomy is the development of explainable AI (XAI) systems in agriculture. Unlike traditional black-box ML models, XAI systems are designed to provide clear, understandable explanations for their recommendations (Samek et al. 2017). This transparency allows farmers to understand why a particular decision is recommended and how the AI system arrived at its conclusion. By providing farmers with more insight into the decision-making process, XAI can help restore a sense of control and empower them to make more informed choices based on both AI recommendations and their own expertise (Carvalho et al. 2019).

In addition to promoting transparency, XAI can also facilitate a more collaborative approach to decision-making, where farmers use ML as a tool to augment, rather than replace, their knowledge. Farmers who can critically evaluate the recommendations of an XAI system are more likely to maintain autonomy over their operations, balancing AI-driven insights with their own experience and understanding of local conditions (Rudin 2019).

17.3.5 Collaborative Models of AI and Farmer Expertise

Maintaining farmer autonomy does not require rejecting ML technologies but instead calls for a more collaborative model in which farmers and AI systems work together. By integrating farmer knowledge with AI insights, a hybrid decision-making model can emerge that respects both traditional expertise and the benefits of data-driven farming (van der Burg et al. 2019).

For example, decision-support systems (DSS) that incorporate both ML and farmer inputs can allow for a more balanced approach, where the farmer's knowledge is taken into account alongside the algorithm's recommendations. These systems empower farmers by giving them the final say in decision-making while providing them with valuable data-driven insights (Lindblom et al. 2017). Such models offer a way forward in which ML technologies enhance farmer autonomy rather than diminish it.

As a result, ML is transforming agricultural decision-making, offering farmers powerful tools for optimizing operations and increasing productivity. However, the rise of data-driven farming also poses risks to farmer autonomy, particularly through the over-reliance on opaque ML systems and the concentration of power in the hands of technology providers. To maintain autonomy in a data-driven world, it is essential to develop transparent, XAI systems that empower farmers to make informed decisions. Additionally, policies that ensure data ownership and foster collaboration between farmers and AI systems are critical for balancing the benefits of ML with the preservation of farmer independence.

17.4 Cultural Shifts and Acceptance of Technology in Agriculture

Agriculture has long been a sector rooted in tradition, with farming practices often passed down through generations. However, the rise of technology-particularly digital tools like AI, ML, and precision agriculture-has introduced profound changes. These technological advances promise increased efficiency and sustainability but also challenge longstanding cultural norms in farming communities (Rose et al. 2021). As digital technologies become more prevalent, the process of integrating them into agricultural practices is influenced not only by the economic benefits but also by cultural acceptance, resistance, and adaptation (Klerkx and Rose 2020).

17.4.1 Technology Adoption in Agricultural Communities

Technology adoption in agriculture is a complex process influenced by factors beyond economic feasibility. Cultural values, social norms, and historical farming practices play significant roles in determining how and when farmers integrate new technologies (Daum 2018). Research has shown that farmers' personal experiences, community influences, and trust in technology providers heavily impact their willingness to adopt new tools (Caffaro et al. 2021). For example, while precision agriculture and ML offer promising advancements in crop management, many farmers remain hesitant due to concerns about losing control over their traditional practices and the potential complexity of these technologies (Eastwood et al. 2017).

Moreover, the generational divide within farming communities also affects technology acceptance. Younger farmers, who are often more digitally literate, tend to adopt new technologies more readily than older generations, who may be more resistant to changing their methods (Barnes et al. 2019). This generational difference highlights the importance of addressing cultural aspects alongside economic factors when introducing new agricultural technologies (Aubert et al. 2012).

17.4.2 Trust and Technology Providers

The acceptance of technology in agriculture is also linked to the level of trust farmers place in technology providers. Farmers are more likely to adopt technologies when they believe the tools will benefit their operations and that the companies behind them prioritize the farmers' needs and data security. Concerns about data ownership, privacy, and the role of large agribusinesses in controlling technology platforms can lead to resistance, particularly among smaller, family-run farms (Bronson and Knezevic 2016).

Many farmers feel that adopting AI and ML technologies involves relinquishing control over their data and farm management decisions to corporations, a shift that conflicts with traditional values of self-reliance and autonomy (Rotz et al. 2019). Building trust between farmers and technology providers is essential for fostering cultural acceptance. Transparent communication about how data is used, efforts to provide equitable access to technology, and partnerships that include farmers in decision-making processes can help bridge the gap between digital tools and agricultural traditions (Carolan 2020).

17.4.3 Cultural Resistance to Change

Cultural resistance to adopting new technologies is often rooted in the fear of losing traditional knowledge and practices. Agriculture has historically been built on generational wisdom—knowledge of soil health, weather patterns, and pest control passed down through family and community networks (Wheeler et al. 2018). The introduction of ML and AI, which rely on data and algorithms, can challenge these traditional knowledge systems, leading to resistance among farmers who fear that their expertise will be devalued (Klerkx et al. 2019).

In many cases, farmers are reluctant to replace time-honored practices with data-driven models because they feel that local knowledge, tailored to specific environmental conditions, may not be adequately captured by AI algorithms (Vanclay 2004). Additionally, the complexity of some technologies creates a barrier for farmers who may not have the technical skills to fully engage with ML systems, further fueling resistance to change (Eastwood et al. 2017).

Cultural resistance is particularly strong in communities where agriculture is closely tied to identity and heritage. For example, in many developing countries, farming is not only an economic activity but also a cultural tradition intertwined with social values and local customs. In these regions, adopting new technologies may be viewed as a threat to cultural preservation (Salami et al. 2020). Addressing this resistance requires culturally sensitive approaches that respect traditional practices while promoting the benefits of technology.

17.4.4 Facilitating Cultural Shifts

To encourage cultural acceptance of technology in agriculture, several strategies can be employed. One key approach is participatory technology development, where farmers are involved in the design, testing, and implementation of new technologies (Daum and Birner 2020). By engaging farmers early in the development process, technology providers can create tools that are more aligned with farmers' needs and preferences, increasing the likelihood of adoption. Additionally, this collaborative approach helps farmers retain a sense of control and ownership over the technology, which is crucial for cultural acceptance (Darnhofer 2020).

Training and education also play vital roles in facilitating cultural shifts. Providing farmers with accessible, hands-on training in digital tools helps bridge the gap between traditional practices and new technologies (Rose et al. 2021). Demonstrating how AI and ML can complement rather than replace traditional knowledge can help alleviate fears that technology will erode local expertise. Training programs that focus on integrating digital literacy with practical applications in farming contexts can build farmers' confidence in using new tools (Barnes et al. 2019).

Social networks and community influences are also critical in the adoption of technology. Research shows that farmers are more likely to adopt new technologies when they see other community members successfully using them (Wheeler et al. 2018). Extension services and farmer cooperatives that facilitate peer-to-peer learning and showcase technology successes within the community can drive broader cultural acceptance (Lindblom et al. 2017).

Government policies and incentives also have a significant impact on cultural shifts in technology adoption. Policies that provide financial support, subsidies, and technical assistance to farmers adopting new technologies can reduce barriers to entry and make technology adoption more appealing (Salami et al. 2020). Additionally, policies that prioritize equitable access to technology for smallholder farmers and marginalized communities can help ensure that the benefits of technological advancements are widely distributed, further encouraging cultural acceptance.

Overall, the integration of technology in agriculture, particularly AI and ML, presents both opportunities and challenges. While these technologies offer enhanced productivity, efficiency, and sustainability, their successful adoption depends on addressing the cultural barriers and resistance within farming communities. Trust in technology providers, fear of losing traditional practices, and the generational divide all play significant roles in shaping how farmers respond to technological change. By promoting participatory technology development, offering education and training, leveraging community networks, and implementing supportive policies, the agricultural sector can facilitate cultural shifts that embrace technology while respecting traditional farming practices. Ultimately, the future of technology in agriculture will depend on finding the right balance between innovation and cultural preservation.

17.4.5 Socioeconomic Disparities in Access to ML Technologies

ML, a subset of AI, is revolutionizing fields such as healthcare, finance, and education (Jordan and Mitchell 2015). Despite its transformative potential, access to ML technologies remains unevenly distributed, often influenced by socioeconomic factors. Disparities in access can exacerbate existing inequalities, limiting opportunities for disadvantaged groups (O'Neil 2016). This article examines the socioeconomic factors affecting access to ML technologies and discusses potential solutions for mitigating these disparities.

17.5 Economic Factors

Economic resources play a crucial role in determining access to ML technologies. Wealthier individuals and organizations are more likely to afford advanced ML tools and services, whereas those with limited financial means face barriers (Brynjolfsson and McElheran 2016). Companies in low-income regions or sectors may struggle to implement ML solutions due to high costs associated with software, hardware, and skilled personnel (Miller et al. 2018). Furthermore, the high cost of data acquisition and processing can also limit access for economically disadvantaged groups (Zeng et al. 2020).

17.5.1 Educational Disparities

Education is another critical factor influencing access to ML technologies. Individuals with advanced educational backgrounds are better positioned to understand and leverage ML tools (Chui et al. 2018). Educational disparities contribute to unequal access, as those with lower levels of education or training may lack the skills required to utilize ML technologies effectively (Smith et al. 2020). Furthermore, educational institutions in economically disadvantaged areas often have fewer resources and less access to cutting-edge technologies, perpetuating the gap (Bessen 2019).

17.5.2 Infrastructural Challenges

Infrastructure plays a significant role in determining access to ML technologies. Regions with inadequate technological infrastructure—such as limited internet connectivity and outdated hardware-face substantial challenges (Goh et al. 2017). Rural and underdeveloped areas are particularly affected, as they often lack the necessary infrastructure to support ML applications (Hargittai and Hinnant 2008). These infrastructural deficits can hinder the deployment and utilization of ML technologies, further entrenching socioeconomic disparities.

17.5.3 Impact on Different Socioeconomic Groups

The socioeconomic disparities in access to ML technologies have far-reaching implications. For instance, in healthcare, regions with limited access to ML tools may experience slower diagnostic processes and less effective treatments (Rajkomar et al. 2019). In education, students in underserved areas may miss out on the benefits of personalized learning algorithms, widening the achievement gap (Pasek et al. 2018). Additionally, in the labor market, those without access to ML technologies may face disadvantages in job opportunities and career advancement (Bessen 2019).

17.5.4 Strategies for Mitigating Disparities

Addressing socioeconomic disparities in access to ML technologies requires multifaceted strategies. Investments in educational programs that focus on ML and data science can help build the necessary skills in underserved communities (Chui et al. 2018). Governments and organizations can also implement policies to reduce the cost of ML technologies and improve infrastructure in disadvantaged areas (Brynjolfsson and McElheran 2016). Additionally, promoting collaboration between private sector companies and educational institutions can facilitate knowledge transfer and resource sharing (Miller et al. 2018).

Overall, socioeconomic disparities in access to ML technologies present significant challenges that need to be addressed to ensure equitable distribution of these transformative tools. Economic resources, educational background, and infrastructural quality all play critical roles in determining access. By implementing targeted strategies and fostering collaboration, it is possible to reduce these disparities and promote a more inclusive technological future.

17.5.5 Gender and Social Equity in ML-driven Agriculture

ML has become a powerful tool in agriculture, offering innovations that range from precision farming to automated pest control (Liakos et al. 2018). While these advancements promise significant benefits, there is a growing concern about how they affect gender and social equity within agricultural communities (Srinivasan et al. 2019). This article explores the intersection of ML technologies with gender and social equity, focusing on the implications for farmers and agricultural workers from diverse backgrounds.

17.6 Gender Disparities in ML-driven Agriculture

Gender disparities in agriculture are well-documented, with women often facing barriers to accessing resources and technology (FAO 2011). The introduction of ML technologies can either bridge or exacerbate these gaps, depending on how these tools are deployed and adopted. Studies show that women in agriculture are less likely to have access to advanced technologies due to lower levels of education, limited financial resources, and restricted land ownership (Doss 2018; Peterman et al. 2020).

For example, research by Beintema and di Marco (2019) indicates that women are underrepresented in agricultural research and extension services that utilize ML technologies. This underrepresentation can limit women's ability to benefit from innovations such as precision farming, which could otherwise enhance productivity and income. Moreover, gendered patterns in technology adoption suggest that without targeted interventions, ML-driven advancements may not reach or benefit women equally (Alston 2015).

17.6.1 Social Equity and Access to ML Technologies

Social equity in the context of ML-driven agriculture involves ensuring that marginalized and low-income communities also gain access to technological benefits. ML technologies can enhance agricultural productivity and sustainability, but their benefits are often concentrated in more affluent and technologically advanced regions (Bennett et al. 2020). This concentration can exacerbate existing inequalities between high-income and low-income regions. Kumar et al. (2020) highlight that rural and economically disadvantaged areas often lack the infrastructure necessary to support ML technologies, such as high-speed internet and advanced computing facilities. As a result, farmers in these regions may be excluded from the benefits of ML-driven innovations, leading to wider socioeconomic disparities.

17.6.2 Impact of ML Technologies on Gender and Social Equity

The impact of ML technologies on gender and social equity is multifaceted. On one hand, ML can improve decision-making and reduce labor-intensive tasks, potentially benefiting all farmers (Wolfert et al. 2017). On the other hand, the distribution of these benefits is not uniform. For instance, the automation of tasks may disproportionately affect women who are engaged in manual labor, potentially leading to job displacement without adequate reskilling opportunities (Srinivasan et al. 2019).

Additionally, ML-driven data analysis can inform better agricultural practices and resource management, but if data collection and analysis are not inclusive, they may overlook the needs and contributions of marginalized groups (Agarwal et al. 2021). For example, ML models trained on data from predominantly male farmers may not accurately reflect the practices and challenges faced by female farmers (Arora et al. 2022).

17.6.3 Strategies for Promoting Equity in ML-driven Agriculture

To address gender and social equity in ML-driven agriculture, several strategies can be employed. First, increasing the representation of women and marginalized groups in agricultural research and development can ensure that ML technologies are designed to meet their needs (FAO 2020). Programs aimed at providing training and resources to underserved communities can also enhance access to ML technologies and foster greater inclusivity.

Second, policymakers and organizations should work to improve the infrastructure in rural and low-income areas to support the adoption of ML technologies (Kumar et al. 2020). This includes investing in digital connectivity, providing financial support, and ensuring that agricultural policies promote equitable access to technological advancements.

Finally, it is crucial to design ML systems with inclusivity in mind. This involves collecting diverse data, incorporating gender-sensitive variables, and conducting impact assessments to ensure that the benefits of ML technologies are equitably distributed (Bennett et al. 2020).

In light of the above, the gender and social equity are critical considerations in the deployment of ML technologies in agriculture. While ML offers significant potential for improving agricultural practices, its benefits are not automatically equitably distributed. Addressing gender and social disparities requires deliberate efforts to ensure inclusive access and equitable outcomes. By implementing targeted strategies and fostering inclusivity, it is possible to leverage ML technologies to benefit all agricultural stakeholders and promote a more equitable agricultural sector.

17.7 Digital Literacy and Skill Development in Farming Communities

Digital literacy has become increasingly crucial in the agricultural sector as digital technologies, including precision farming tools, remote sensing, and farm management software, play a growing role in optimizing agricultural practices (Rose et al. 2018). However, many farming communities, especially in rural areas, face challenges in adopting these technologies due to varying levels of digital literacy and skills (Zhao et al. 2021). This article examines the importance of digital literacy in agriculture, identifies key barriers to skill development, and proposes solutions to enhance digital capabilities among farmers.

17.7.1 State of Digital Literacy in Farming Communities

Digital literacy encompasses the ability to effectively use digital tools and technologies, including understanding and managing digital information (Hargittai 2020). In farming communities, digital literacy can significantly influence the adoption and utilization of agricultural technologies, affecting productivity and efficiency (Kumar et al. 2019). Studies indicate that while some farmers are adept at using digital tools, a significant portion still lacks the necessary skills to fully leverage these technologies (Barton et al. 2019).

For instance, research by Zhao et al. (2021) found that while farmers in developed regions often have higher levels of digital literacy, those in developing regions face considerable challenges. These include limited access to training resources, inadequate infrastructure, and low levels of prior experience with digital technologies. This disparity underscores the need for targeted educational initiatives to bridge the digital divide in agriculture.

17.7.2 Barriers to Digital Skill Development

Several factors contribute to the barriers faced by farming communities in developing digital skills. Infrastructure limitations, such as unreliable internet access and outdated technology, hinder farmers' ability to engage with digital tools (Smith et al. 2020). Additionally, a lack of formal education and training opportunities exacerbates the skills gap, particularly among older farmers and those in remote areas.

Economic constraints also play a significant role, as many farmers cannot afford the necessary technology or training programs (Kumar et al. 2019). Furthermore, there is often a lack of tailored training programs that address the specific needs of farmers, which can lead to lower engagement and effectiveness of digital skill development initiatives (Barton et al. 2019).

17.7.3 Impact of Digital Literacy on Agricultural Productivity

Digital literacy has been shown to positively impact agricultural productivity by enabling farmers to use technology more effectively (Rose et al. 2018). Technologies such as Geographic Information Systems (GIS), Internet of Things (IoT) sensors, and data analytics tools require a certain level of digital proficiency to be utilized to their full potential. Farmers with higher digital literacy are better positioned to make data-driven decisions, optimize resource use, and improve crop yields.

For example, a study by Bhatia et al. (2020) demonstrated that farmers who engaged with precision agriculture technologies saw significant improvements in yield and efficiency. These improvements were attributed to their ability to interpret and act on data provided by these technologies, which was made possible by their digital literacy.

17.7.4 Strategies for Enhancing Digital Literacy in Farming Communities

To improve digital literacy and skill development among farmers, several strategies can be employed. First, providing accessible and affordable training programs is essential. These programs should be tailored to the specific needs of farmers and delivered through various formats, such as online courses, workshops, and on-site training sessions (Horrigan 2016).

Second, enhancing digital infrastructure in rural areas is crucial for enabling effective use of digital technologies. Investments in reliable internet access, modern computing facilities, and technical support can help overcome infrastructure barriers and support skill development (Smith et al. 2020).

Third, fostering partnerships between agricultural organizations, technology providers, and educational institutions can facilitate knowledge transfer and resource sharing. Collaborative efforts can lead to the development of targeted training programs and the creation of resources that address the unique challenges faced by farmers (Barton et al. 2019).

Finally, promoting digital literacy from an early age through educational initiatives can prepare future generations of farmers for a digitally-driven agricultural landscape. Integrating digital skills into agricultural education curricula can help build a foundation for lifelong learning and adaptation (Hargittai 2020).

To conclude, the digital literacy is a critical component of modern agriculture, influencing the adoption and effective use of digital technologies in farming communities. Addressing the barriers to digital skill development and implementing targeted strategies can enhance productivity, efficiency, and sustainability in agriculture. By improving digital literacy and infrastructure, and fostering collaborative efforts, it is possible to bridge the digital divide and support the growth of a digitally proficient farming sector.

17.8 Balancing Technological Innovation with Social Equity

Technological innovation drives progress across various sectors, promising improvements in efficiency, productivity, and quality of life. However, the rapid pace of technological advancement often raises concerns about social equity, as the benefits of innovation are not always evenly distributed (Brynjolfsson and McElheran 2016). This article explores the challenges of balancing technological innovation with social equity, examining how disparities in access and opportunities can arise and proposing strategies to ensure that the benefits of technology are equitably shared.

17.8.1 Technological Innovation and Its Impact on Equity

Technological innovations, such as AI, ML and advanced automation, have transformative potential but can also exacerbate existing inequalities (O'Neil 2016). For example, while AI-driven tools can enhance productivity and decision-making, they may also lead to job displacement and widen income gaps if certain groups are unable to adapt or access these technologies (Bessen 2019).

The benefits of technological innovations are often concentrated in regions and sectors that are already technologically advanced. High-income countries and large corporations typically have better access to cutting-edge technologies, while low-income regions and smaller enterprises may lag behind (Chui et al. 2018). This disparity can perpetuate and even deepen existing social and economic inequalities (Miller et al. 2018).

17.8.2 Barriers to Equitable Access

Several factors contribute to the uneven distribution of technological benefits. Economic barriers are significant, as the costs associated with acquiring and implementing new technologies can be prohibitive for individuals and organizations with limited financial resources (Zeng et al. 2020). Additionally, infrastructure limitations, such as inadequate internet connectivity and outdated hardware, can hinder access to technological innovations in underserved areas (Goh et al. 2017).

Educational and skill gaps also play a crucial role in determining who can effectively utilize new technologies. Individuals with higher levels of education and specialized skills are more likely to benefit from technological advancements, while those with lower levels of education may struggle to keep up (Smith et al. 2020). This disparity in skill development can lead to unequal opportunities and outcomes.

17.8.3 Case Studies of Technological Disparities

Several case studies illustrate the impact of technological disparities on social equity. In the healthcare sector, the adoption of AI and advanced diagnostics has improved patient outcomes but has also highlighted gaps in access. Research by Rajkomar et al. (2019) shows that while AI can enhance diagnostic accuracy, its benefits are often limited to well-resourced healthcare settings, leaving underserved populations with less access to these advancements.

In education, the digital divide has been a major concern, particularly during the COVID-19 pandemic. Studies have shown that students in low-income households face significant challenges accessing remote learning resources, which can affect their educational outcomes and future opportunities (Pasek et al. 2018). This divide underscores the need for targeted interventions to ensure that technological innovations in education reach all students equitably.

17.8.4 Strategies for Promoting Social Equity

To balance technological innovation with social equity, several strategies can be employed. First, it is essential to invest in infrastructure development to ensure that underserved areas have access to the necessary technological resources (Goh et al. 2017). This includes improving internet connectivity, upgrading hardware, and providing technical support.

Second, educational initiatives that focus on digital literacy and skills development can help bridge the gap between those who have and have not benefited from technological advancements (Hargittai 2020). Training programs should be designed to address the needs of diverse populations, including low-income and marginalized communities.

Third, policymakers and organizations should implement inclusive technology policies that promote equitable access and address barriers to entry. This may involve subsidies or financial assistance for acquiring technology, as well as support for small businesses and startups in underserved regions (Brynjolfsson and McElheran 2016).

Finally, fostering collaboration between technology developers, educational institutions, and community organizations can facilitate the development of solutions that address specific needs and challenges (Miller et al. 2018). By working together, these stakeholders can create programs and resources that ensure a more equitable distribution of technological benefits. It follows that, the balancing technological innovation with social equity is a complex but crucial endeavor. While technological advancements have the potential to drive significant progress, they can also exacerbate existing inequalities if not managed carefully. By addressing barriers to access, investing in infrastructure and education, and promoting inclusive policies, it is possible to ensure that the benefits of technology are distributed more equitably and contribute to a more just and inclusive society.

17.9 The Role of ML in Shaping Rural Economies

ML has become a transformative force across various sectors, including agriculture and rural economies. By leveraging data-driven insights and predictive analytics, ML technologies offer significant potential to enhance productivity, efficiency, and economic growth in rural areas. This article explores the impact of ML on rural economies, examining how these technologies are reshaping agricultural practices, influencing economic development, and addressing the unique challenges faced by rural communities.

17.9.1 ML in Agriculture

Agriculture, a cornerstone of many rural economies, stands to benefit greatly from ML applications. ML technologies can optimize crop management, improve yield prediction, and enhance pest and disease management. For example, precision agriculture uses ML algorithms to analyze data from sensors and satellite imagery to make more informed decisions about planting, irrigation, and fertilization (Liakos et al. 2018). This technology can lead to increased crop yields and more efficient use of resources, directly impacting rural economies by boosting agricultural productivity (Wolfert et al. 2017).

ML-based pest detection systems can help in early identification of pest outbreaks, minimizing crop damage and reducing the need for extensive pesticide use (Kumar et al. 2021). These advancements contribute to greater economic stability and sustainability in rural areas.

17.9.2 Economic Development and Diversification

ML can also facilitate economic development and diversification in rural regions. By enabling more efficient agricultural practices, ML frees up resources that can be redirected towards other economic activities. For instance, improved agricultural efficiency can lead to surplus production, which can stimulate local economies through increased trade and value-added processing (Basso et al. 2016).

Additionally, ML technologies are fostering innovation in sectors beyond agriculture. Rural communities are increasingly adopting ML in areas such as renewable energy, healthcare, and small-scale manufacturing. ML applications in renewable energy, such as predictive maintenance for wind turbines and optimization of solar energy production, can support rural economies by creating new job opportunities and enhancing energy security (Aydin et al. 2021). In healthcare, ML can improve diagnostic accuracy and personalize treatment, addressing health disparities in rural areas and contributing to economic stability by reducing healthcare costs (Rajkomar et al. 2019).

17.9.3 Challenges and Barriers

Despite the potential benefits, the adoption of ML in rural economies faces several challenges. One significant barrier is the lack of infrastructure and digital connectivity in many rural areas. Limited access to high-speed internet and advanced computing resources can hinder the implementation of ML technologies (Smith et al. 2020).

Moreover, there is a need for specialized skills and knowledge to effectively use ML tools. The shortage of data scientists and technical experts in rural areas can limit the ability of local businesses and farmers to harness the full potential of ML (Miller et al. 2018). Additionally, the initial costs associated with implementing ML technologies can be prohibitive for small-scale farmers and businesses (Zeng et al. 2020).

17.9.4 Strategies for Effective Implementation

To overcome these challenges and maximize the benefits of ML for rural economies, several strategies can be employed. First, improving digital infrastructure is crucial. Investments in high-speed internet and advanced computing facilities can enable more widespread adoption of ML technologies and support economic growth (Goh et al. 2017).

Second, providing education and training programs is essential for building the necessary skills and knowledge. Local governments and organizations can partner with educational institutions to offer training in data science and ML applications tailored to the needs of rural communities (Hargittai 2020). These programs can help bridge the skills gap and empower individuals to leverage ML for economic development.

Third, fostering collaboration between technology developers, rural businesses, and agricultural organizations can facilitate the development of tailored ML solutions. Collaborative efforts can lead to the creation of customized tools that address specific challenges faced by rural communities (Brynjolfsson and McElheran 2016).

Finally, financial support mechanisms, such as grants and subsidies, can help offset the initial costs of implementing ML technologies. Targeted financial assistance can make it more feasible for small-scale farmers and rural businesses to invest in ML tools and infrastructure (Chui et al. 2018).

In light of the above, the ML has the potential to significantly shape rural economies by enhancing agricultural productivity, fostering economic diversification, and addressing key challenges. While barriers to adoption exist, strategic investments in infrastructure, education, and collaboration can help overcome these challenges and maximize the benefits of ML technologies. By addressing these issues, rural communities can harness the power of ML to drive economic growth and development.

17.10 Ethical Dilemmas of AI-driven Agriculture in Developing Communities

AI has the potential to revolutionize agriculture by optimizing crop management, enhancing yields, and improving resource efficiency. However, the deployment of AI technologies in developing communities presents a range of ethical dilemmas. These challenges arise from issues related to equity, privacy, data ownership, and the potential exacerbation of existing inequalities. This article explores the ethical considerations surrounding AI-driven agriculture in developing communities, examining both the benefits and the potential pitfalls of these technologies.

17.10.1 Equity and Access Issues

One of the foremost ethical concerns in deploying AI-driven agriculture in developing communities is the issue of equity and access. AI technologies can exacerbate existing disparities if not implemented with careful consideration of local contexts and needs (Aker and Mbiti 2010). For instance, while AI can offer significant productivity gains, these benefits may be unevenly distributed, favoring large-scale or commercially oriented farms over smallholder and subsistence farmers (Chisanga et al. 2020).

Research by Zeng et al. (2020) highlights that access to AI technologies often requires substantial financial investment and infrastructure, which may be beyond the reach of many farmers in developing regions. This disparity can result in a technological divide where only a few benefit from advancements, while the majority are left behind, potentially widening existing economic inequalities (Gordon and Grant 2021).

17.10.2 Privacy and Data Ownership

Another critical ethical issue is the management of data privacy and ownership. AI-driven agriculture relies heavily on data collection, including information on crop performance, soil conditions, and weather patterns (Liakos et al. 2018). In developing communities, there is a risk that data collected from smallholder farms may be exploited without proper consent or compensation, raising concerns about data ownership and control (Miller et al. 2018).

Studies such as those by Smith and Kopp (2020) emphasize the need for clear agreements on data rights and privacy. Farmers must be informed about how their data will be used and have a say in the decisions regarding data sharing. Without appropriate measures, there is a risk that valuable agricultural data could be monopolized by external entities, leading to potential exploitation and loss of local benefits.

17.10.3 Algorithmic Bias and Fairness

AI systems can inadvertently reinforce existing biases or create new forms of discrimination if not carefully designed and implemented. Algorithmic bias arises when AI models produce unfair outcomes due to biased training data or flawed algorithms (O'Neil 2016). In the context of agriculture, biased AI systems could disproportionately benefit certain groups over others, such as favoring crops or farming practices that are more common in wealthier regions (Binns et al. 2018).

Research by Barocas and Selbst (2016) suggests that addressing algorithmic bias requires diverse and representative data, as well as transparency in the development and deployment of AI systems. Ensuring that AI technologies are fair and equitable necessitates ongoing scrutiny and adaptation to address any potential biases that may arise.

17.10.4 Environmental and Social Implications

The environmental and social implications of AI-driven agriculture also present ethical challenges. While AI technologies can enhance efficiency and reduce resource use, there is a risk that they may lead to unintended environmental consequences if not properly managed (Wolfert et al. 2017). For example, the reliance on AI for precision agriculture might encourage the overuse of inputs like fertilizers and pesticides, potentially leading to soil degradation or water contamination (Kumar et al. 2021).

Additionally, the social implications of AI adoption must be considered. The introduction of AI-driven technologies can disrupt traditional farming practices and community structures, potentially leading to job displacement and social unrest (Rajkomar et al. 2019). It is important to balance technological advancements with strategies to support affected communities and mitigate negative impacts.

17.10.5 Strategies for Ethical AI Implementation

To address these ethical dilemmas, several strategies can be employed. First, promoting inclusive access to AI technologies is essential. This includes providing financial and technical support to smallholder farmers and ensuring that AI tools are designed with the needs of diverse farming communities in mind (Chisanga et al. 2020).

Second, establishing robust data governance frameworks is crucial for protecting privacy and ensuring fair data use. Clear agreements on data ownership, consent, and compensation should be put in place to safeguard the rights of farmers and ensure that they benefit from the use of their data (Smith and Kopp 2020). Third, addressing algorithmic bias requires continuous monitoring and evaluation of AI systems. Developing inclusive datasets and involving diverse stakeholders in the design and testing phases can help mitigate biases and ensure that AI technologies are fair and equitable (Barocas and Selbst 2016).

Finally, incorporating environmental and social impact assessments into the development and deployment of AI technologies can help identify potential risks and unintended consequences. Engaging with local communities to understand their needs and concerns is essential for ensuring that AI-driven agriculture contributes positively to both the environment and society (Wolfert et al. 2017).Ultimately, AI-driven agriculture has the potential to bring significant benefits to developing communities, but it also presents a range of ethical dilemmas. Addressing issues related to equity, data privacy, algorithmic bias, and environmental impact is crucial for ensuring that these technologies are implemented in a way that is fair, inclusive, and sustainable. By adopting strategies that promote equitable access, protect data rights, and assess impacts, it is possible to harness the benefits of AI while mitigating potential ethical risks.

17.11 Community Resilience and Adaptation to Technological Change

As technological advancements continue to reshape various sectors, communities must adapt to these changes to ensure resilience and sustainability. Technological change can offer significant benefits but also poses challenges that may disrupt existing social structures, economies, and ways of life. This article explores how communities can build resilience and adapt effectively to technological change, examining strategies, challenges, and case studies to provide a comprehensive understanding of this complex process.

17.11.1 Understanding Community Resilience

Community resilience refers to the capacity of a community to anticipate, prepare for, respond to, and recover from adverse events or changes, including technological disruptions (Norris et al. 2008). Resilience involves not only the ability to withstand shocks but also the capability to adapt and transform in the face of new challenges (Adger 2000). In the context of technological change, resilience entails leveraging technological innovations to enhance community well-being while mitigating potential negative impacts.

17.11.2 Adaptation Strategies for Technological Change

17.11.2.1 Education and Skill Development

One of the fundamental strategies for adapting to technological change is investing in education and skill development. As technologies evolve, the demand for new skills and knowledge increases. Providing training programs and educational opportunities can help individuals and communities acquire the necessary skills to thrive in a changing technological landscape (Hargittai 2020). For example, digital literacy programs can empower community members to use new technologies effectively, reducing the risk of exclusion and enhancing overall resilience (Smith and Kopp 2020).

17.11.2.2 Community Engagement and Participation

Active community engagement and participation are crucial for effective adaptation to technological change. Involving community members in decision-making processes related to technology implementation ensures that solutions are tailored to local needs and preferences (Arnstein 1969). Participatory approaches can foster a sense of ownership and collaboration, which enhances the likelihood of successful adaptation (Pretty 1995). For instance, community-driven technology projects can address specific local challenges and create more sustainable solutions (Berkes and Folke 1998).

17.11.2.3 Building Social Capital

Social capital, defined as the networks, norms, and trust that facilitate collective action, plays a vital role in community resilience (Putnam 2000). Strong social networks can enhance information sharing, support collaborative problem-solving, and facilitate access to resources and opportunities. Building social capital through community organizations, networks, and partnerships can help communities adapt to technological changes more effectively (Adger 2003). For example, local technology hubs and innovation centers can serve as platforms for knowledge exchange and collaborative projects.

17.11.2.4 Infrastructure Development

Developing and upgrading infrastructure is essential for supporting technological adaptation. Adequate infrastructure, including access to high-speed internet, reliable power sources, and modern facilities, can enable communities to leverage new technologies and improve their resilience (Goh et al. 2017). Investments in infrastructure should be prioritized to ensure that all community members have equal access to technological benefits and resources (Chui et al. 2018).

17.11.3 Challenges in Adapting to Technological Change

17.11.3.1 Digital Divide

The digital divide represents a significant challenge to community resilience. Disparities in access to technology, internet connectivity, and digital skills can exacerbate existing inequalities and hinder the ability of some community members to benefit from technological advancements (Van Dijk 2020). Addressing the digital divide requires targeted efforts to improve access and reduce barriers for underserved populations (Warschauer 2003).

17.11.3.2 Resistance to Change

Resistance to technological change is another challenge that communities may face. Resistance can stem from various factors, including fear of the unknown, perceived risks, and cultural values (Rogers 2003). Overcoming resistance involves addressing concerns through education, communication, and demonstrating the benefits of new technologies (Kotter 1996). Engaging community leaders and influencers can also help in fostering acceptance and facilitating smoother transitions.

17.11.3.3 Economic Constraints

Economic constraints can limit the ability of communities to invest in and adopt new technologies. Limited financial resources may restrict access to advanced technologies and hinder the implementation of adaptation strategies (Zeng et al. 2020). Financial support mechanisms, such as grants, subsidies, and public-private partnerships, can help alleviate economic barriers and support community adaptation efforts (Brynjolfsson and McElheran 2016).

17.11.4 Case Studies of Successful Adaptation

17.11.4.1 The Digital Green Initiative

The Digital Green initiative is a notable example of a successful adaptation to technological change in rural communities. By utilizing video technology to disseminate agricultural knowledge, Digital Green has improved farming practices and increased crop yields in several developing countries (Kumar et al. 2011). This initiative demonstrates how technology can be effectively integrated into community-based approaches to enhance resilience and support local development.

17.11.4.2 Smart Cities and Urban Resilience

Smart cities leverage technology to improve urban infrastructure, services, and overall quality of life. Initiatives such as smart grids, intelligent transportation systems, and digital health services have been implemented in cities around the world to enhance resilience and adaptability (Hollands 2008). These technologies contribute to more efficient resource management, improved public services, and increased community engagement, showcasing the potential of technology to foster resilience in urban environments.

Generally, the building community resilience in the face of technological change requires a multifaceted approach that includes education, engagement, social capital, and infrastructure development. While challenges such as the digital divide, resistance to change, and economic constraints may arise, addressing these issues through targeted strategies can facilitate effective adaptation and enhance overall resilience. By leveraging technology in a thoughtful and inclusive manner, communities can navigate technological disruptions and harness their potential for positive change.

17.12 Long-term Social Impacts: Sustainability and Food Security

Sustainability and food security are two intertwined global challenges that have long-term social impacts. As the world faces population growth, climate change, and resource depletion, ensuring food security for all while maintaining environmental sustainability has become a central goal for governments, communities, and international organizations.

This article explores the long-term social implications of sustainability practices and their direct relationship to food security, emphasizing how both concepts influence one another.

17.12.1 Defining Sustainability and Food Security

Sustainability refers to meeting present needs without compromising the ability of future generations to meet their own. It encompasses environmental, social, and economic dimensions (WCED 1987). Sustainability is particularly crucial in the context of agriculture and food systems because unsustainable practices can degrade ecosystems, reduce biodiversity, and compromise future food production (Pretty 2008).

Food security is defined by the Food and Agriculture Organization (FAO) as the state where all people, at all times, have physical, social, and economic access to sufficient, safe, and nutritious food that meets their dietary needs and food preferences (FAO 1996). Achieving food security is not only about food availability but also about access, utilization, and stability over time (Godfray et al. 2010). In this light, food security depends heavily on the sustainability of agricultural practices and the equitable distribution of resources.

17.12.2 The Role of Sustainability in Food Security

17.12.2.1 Sustainable Agricultural Practices

Sustainable agriculture involves practices that improve efficiency and productivity while preserving the environment. Agroecology, organic farming, and conservation agriculture are examples of practices that minimize the use of chemical inputs and enhance ecosystem services such as pollination and soil fertility (Altieri 1995). These methods contribute to long-term food security by maintaining soil health and reducing the dependence on nonrenewable resources (Tittonell and Giller 2013). Moreover, sustainable agriculture enhances the resilience of food systems to climate variability. For example, agroforestry, a method that integrates trees into agricultural landscapes, improves water retention and mitigates the impacts of droughts (Jose 2009). This not only ensures stable food production but also supports biodiversity conservation, which is vital for ecosystem balance and crop pollination.

17.12.2.2 Climate Change Mitigation

Climate change presents one of the greatest threats to global food security. Increased temperatures, erratic rainfall patterns, and more frequent extreme weather events disrupt agricultural production and lead to reduced yields (IPCC 2014). Sustainable practices like carbon sequestration in soil, renewable energy use, and sustainable water management can reduce agriculture's carbon footprint and enhance its capacity to adapt to changing climatic conditions (Lal 2004). Such adaptation is critical for maintaining food production in vulnerable regions, particularly in developing countries where agriculture is the backbone of the economy.

17.12.3 Social Effects of Sustainability and Food Security

17.12.3.1 Equitable Access to Resources

Social equity is a fundamental aspect of both sustainability and food security. Inequitable access to land, water, and agricultural inputs has long been a barrier to achieving food security for marginalized communities (Zimmerer et al. 2018). Ensuring equitable access to resources is essential to sustain livelihoods and reduce poverty. Sustainable land management practices that incorporate social equity—such as community-based resource management—empower local populations and improve food distribution (Leach et al. 1999).

Furthermore, gender equity in agriculture is vital for achieving sustainability and food security. Women, who comprise nearly half of the agricultural workforce in many developing countries, often face significant barriers in accessing land, credit, and agricultural technologies (FAO 2011). Empowering women farmers through education, training, and policy reforms can increase agricultural productivity and improve food security at the household and community levels (Quisumbing et al. 2014).

17.12.3.2 Health and Nutrition

Sustainability directly impacts the health and nutrition of populations. Unsustainable agricultural practices, such as the overuse of chemical fertilizers and pesticides, can contaminate water supplies, degrade soils, and reduce the nutritional quality of food (Horrigan et al. 2002). In contrast, sustainable farming practices tend to produce food that is more nutritious by promoting crop diversity and reducing the reliance on synthetic inputs (IAASTD 2009). Diverse diets that include a variety of fruits, vegetables, and legumes are essential for combating malnutrition and improving overall public health.

Long-term food security also hinges on addressing the "double burden" of malnutrition, which refers to the coexistence of undernutrition and obesity in many parts of the world (Popkin et al. 2012). Sustainable food systems that prioritize the production of nutritious, diverse foods can help mitigate this issue by ensuring that populations have access to both affordable and healthy food.

17.12.3.3 Economic Stability and Livelihoods

Sustainability practices have a profound impact on economic stability and the livelihoods of millions, particularly in rural areas. Agriculture remains a primary source of employment in many developing countries, and sustainable practices can enhance the economic resilience of farming communities by increasing yields and reducing input costs. For example, the use of organic fertilizers and integrated pest management reduces the need for expensive chemical inputs, thus increasing profit margins for smallholder farmers (Rigby and Cáceres 2001).Additionally, sustainable agriculture opens up opportunities for diversification of income sources. By adopting agroecological practices, farmers can engage in agroforestry, organic certification, and the production of high-value crops, which can improve income stability and reduce vulnerability to market fluctuations (Altieri and Toledo 2011).

17.12.4 Long-term Challenges to Sustainability and Food Security

17.12.4.1 Resource Depletion

Despite the clear benefits of sustainability, the overexploitation of natural resources remains a significant barrier to achieving long-term food security. Deforestation, soil degradation, and water scarcity continue to threaten agricultural productivity in many regions (Tilman et al. 2002). Policies that promote sustainable land-use practices, reforestation, and the protection of water sources are essential to addressing these issues and ensuring future food security (Foley et al. 2005).

17.12.4.2 Global Trade and Food Systems

Global trade policies also pose challenges to sustainable food security. The liberalization of agricultural markets can lead to the dominance of large-scale industrial farms that prioritize export crops over food production for local consumption (Patel 2007). This can undermine food security in developing countries by making them dependent on volatile international markets and reducing the availability of staple foods (McMichael 2009). A shift towards localized, sustainable food systems is necessary to enhance food sovereignty and reduce vulnerability to global market shocks.

To conclude, The long-term social impacts of sustainability and food security are vast and complex. Sustainability practices offer a pathway to improving food security by promoting equitable resource access, enhancing health and nutrition, and strengthening the resilience of food systems to climate change. However, significant challenges remain, including resource depletion, global trade imbalances, and the need for inclusive policies that address social and gender equity. To achieve sustainable food security, coordinated efforts at the local, national, and global levels are essential, alongside continued investment in sustainable agriculture and social resilience.

17.13 Collaborative Models: Integrating Local Knowledge with AI Systems

AI has emerged as a transformative tool in various sectors, including agriculture, healthcare, and environmental management. However, integrating AI with local knowledge through collaborative models is critical to ensuring that these technologies are not only effective but also culturally relevant and sustainable. This article discusses the importance of combining local knowledge with AI systems, particularly in developing regions where traditional practices and community-based knowledge play a significant role in resource management and decision-making.

17.13.1 The Significance of Local Knowledge

Local knowledge refers to the skills, practices, and understandings developed by communities over time, often passed down through generations (Berkes 2012). This knowledge is embedded in the local context, accounting for unique environmental, social, and economic conditions. For example, indigenous farming communities often possess deep insights into weather patterns, soil conditions, and crop management, which can be critical for developing climate-resilient agricultural practices (Altieri 2004).

Local knowledge is particularly valuable because it often provides a holistic understanding of complex ecosystems and social systems, integrating environmental observations with social and cultural values (Folke et al. 2005). AI systems, by contrast, excel in processing large datasets and identifying patterns but can struggle to account for local variability and contextual factors. Thus, combining these two knowledge systems can create robust, adaptable models that are grounded in both data-driven insights and contextual understanding.

17.13.2 Collaborative Models in AI and Local Knowledge Integration

17.13.2.1 Participatory AI Development

One of the most effective ways to integrate local knowledge with AI systems is through participatory development approaches. Participatory AI development involves local stakeholders in the design, implementation, and evaluation of AI tools. This approach ensures that AI technologies are aligned with local needs and preferences, fostering trust and cooperation between communities and developers (Rahman 2020).

For instance, in rural agricultural communities, farmers can collaborate with AI researchers to co-create predictive models for crop yields or pest infestations. By contributing their knowledge of local conditions, farmers can help refine the models, making them more accurate and relevant to their specific context (Niles et al. 2021). This process not only improves the effectiveness of AI systems but also empowers local populations by valuing and incorporating their expertise into technological solutions.

17.13.2.2 Knowledge Co-production

Knowledge co-production is another collaborative model that integrates scientific knowledge with local expertise. It involves stakeholders from diverse backgrounds, including local communities, scientists, policymakers, and AI developers, working together to generate knowledge that addresses complex problems (Norström et al. 2020). This process recognizes the value of both formal scientific knowledge and experiential, place-based knowledge in addressing sustainability challenges.

In the context of AI, knowledge co-production can be applied to environmental monitoring and natural resource management. For example, AI systems can be used to analyze satellite data and predict deforestation risks, while local communities provide on-the-ground observations about changes in vegetation, wildlife, or water sources (Cochrane et al. 2019). By integrating both data sources, more accurate and context-specific solutions can be developed to prevent deforestation and promote sustainable land management practices.

17.13.3 Challenges in Integrating Local Knowledge and AI Systems

17.13.3.1 Data Standardization and Representation

A key challenge in integrating local knowledge with AI systems is the issue of data standardization. AI relies on structured, quantifiable data, while local knowledge is often qualitative, context-dependent, and difficult to formalize (Agrawal 2002). This discrepancy can create barriers to effective integration, as local knowledge may not easily fit into the data-driven frameworks used by AI technologies.

To overcome this challenge, researchers and developers must find ways to translate local knowledge into formats that are compatible with AI systems without losing the richness of the information. For example, participatory mapping tools can be used to gather spatial data from local communities about land use, which can then be incorporated into AI-driven geographic information systems (GIS) (McCall and Dunn 2012). By finding ways to digitize and systematize local knowledge, AI systems can leverage this information to make more informed predictions and decisions.

17.13.3.2 Power Dynamics and Knowledge Hierarchies

Another challenge in integrating local knowledge with AI systems is the potential for power imbalances between different knowledge systems. Historically, scientific and technological knowledge has been privileged over local and indigenous knowledge, leading to marginalization and the devaluation of traditional practices (Smith 2012). In the context of AI, there is a risk that local knowledge may be overshadowed by the perceived superiority of data-driven approaches.

Addressing this challenge requires a shift in the way knowledge is valued and integrated. Collaborative models must prioritize knowledge equity, ensuring that local knowledge is treated as equally important as scientific and technological expertise (Jasanoff 2004). This can be achieved through inclusive decision-making processes, transparent communication, and the co-design of AI systems that are responsive to local needs and values.

17.13.4 Case Studies of Successful Integration

17.13.4.1 AI for Climate-resilient Agriculture in Sub-Saharan Africa

In Sub-Saharan Africa, collaborative efforts between AI researchers and local farming communities have demonstrated the potential for integrating local knowledge into AI-driven agriculture. One such example is the use of AI to predict rainfall patterns and optimize planting times for smallholder farmers (Tesfaye et al. 2017). By incorporating local knowledge about traditional planting schedules and local soil conditions, AI systems were able to provide more accurate recommendations, helping farmers increase their crop yields and improve food security.

This collaboration has also helped foster a sense of ownership among local communities, as farmers were actively involved in the development and refinement of the AI models. By integrating their knowledge into the system, farmers felt more confident in the technology and were more likely to adopt its recommendations, leading to better outcomes for both food production and community resilience (Gebru et al. 2020).

17.13.4.2 Indigenous Knowledge and AI for Fire Management in Australia

In Australia, indigenous communities have long practiced traditional fire management, using controlled burns to prevent large wildfires and maintain healthy ecosystems. Recent collaborations between AI researchers and indigenous groups have sought to integrate these traditional practices into AI-based fire prediction models (Preece et al. 2021). By combining satellite imagery and environmental data with indigenous knowledge about fire behavior and local ecosystems, the AI models have become more accurate in predicting wildfire risks and informing fire management strategies.

This collaboration has also contributed to the recognition and validation of indigenous knowledge systems, demonstrating their relevance in addressing modern environmental challenges. Furthermore, it has empowered indigenous communities to take an active role in managing their lands using both traditional and technological tools, fostering greater sustainability and resilience (Sithole et al. 2021).

Consequently, the Integrating local knowledge with AI systems through collaborative models is essential for ensuring the effectiveness, relevance, and sustainability of AI-driven solutions. By involving local communities in the development and implementation of AI technologies, participatory approaches and knowledge co-production can bridge the gap between data-driven insights and contextual expertise. However, challenges such as data standardization and knowledge hierarchies must be addressed to ensure that local knowledge is valued and incorporated equitably. Successful examples of collaboration in agriculture and fire management demonstrate the potential for AI and local knowledge to work together in tackling complex global challenges, fostering greater resilience, sustainability, and community empowerment.

17.14 Conclusion

ML is rapidly transforming agriculture by introducing precision farming techniques that enhance efficiency, pest management, and resource allocation. However, the shift from traditional knowledge systems to data-driven models poses challenges, such as the potential erosion of local cultural practices and increased economic inequality between large-scale farms and smallholder farmers. While ML offers significant benefits, its widespread adoption is hindered by barriers like limited digital literacy, infrastructure, and access to technology in rural areas. There is also a need to ensure transparency through XAI systems, promoting trust in ML recommendations without undermining farmer autonomy. Additionally, ML's environmental impact is mixed, with concerns about monoculture and biodiversity loss in industrial farming systems, although it can also be leveraged to promote sustainable practices if integrated thoughtfully. Furthermore, issues of data ownership, algorithmic bias, and social inequities must be addressed to avoid exacerbating existing disparities, particularly for marginalized groups like women and low-income farmers. Therefore, balancing technological innovation with local knowledge, ensuring equitable access, and fostering inclusive adoption practices are essential for leveraging the full potential of ML in agriculture while promoting environmental sustainability, social equity, and long-term resilience in farming communities.

References

Adger, W.N. (2000). Social and ecological resilience: Are they related? *Progress in Human Geography* 24 (3): 347–364.

Adger, W.N. (2003). Building resilience to promote sustainability. *IHDP Update* 2 (1): 5–7.

Agarwal, B., Choudhury, N., and Sinha, S. (2021). Data, diversity, and development: The case of gender bias in agricultural data. *Journal of Development Studies* 57 (3): 453–467.

Agrawal, A. (2002). Indigenous knowledge and the politics of classification. *International Social Science Journal* 54 (173): 287–297.

Aker, J.C. and Mbiti, I.M. (2010). Mobile phones and economic development in Africa. *Journal of Economic Perspectives* 24 (3): 207–232.

Alonso, D. and Calderon, P. (2021). Impact of automation in agriculture: New technologies and employment. *Journal of Agricultural Science* 45 (2): 120–135.

Alston, M. (2015). Gender and agricultural sustainability: Evidence from the field. *Rural Society* 24 (2): 114–129.

Altieri, M.A. (1995). *Agroecology: The Science of Sustainable Agriculture* (2nd ed.). CRC Press.

Altieri, M.A. (1999). The ecological role of biodiversity in agroecosystems. *Agriculture, Ecosystems & Environment* 74 (1–3): 19–31.

Altieri, M.A. (2004). Linking ecologists and traditional farmers in the search for sustainable agriculture. *Frontiers in Ecology and the Environment* 2 (1): 35–42.

Altieri, M.A. and Toledo, V.M. (2011). The agroecological revolution in Latin America: Rescuing nature, ensuring food sovereignty, and empowering peasants. *Journal of Peasant Studies* 38 (3): 587–612.

Arnstein, S.R. (1969). A ladder of citizen participation. *Journal of the American Institute of Planners* 35 (4): 216–224.

Arora, S., Goel, A., and Gupta, P. (2022). Machine learning and gendered agricultural data: Challenges and opportunities. *Journal of Agricultural Informatics* 13 (1): 45–59.

Aydin, H., Saygin, C., and Caliskan, H. (2021). Machine learning for renewable energy optimization: A Review. *Energy Reports* 7: 94–106.

Barnes, A.P., Soto, I., Eory, V. et al. (2019). Understanding farmers' incentives to engage with knowledge exchange tools. *Land Use Policy* 83: 394–404.

Barocas, S. and Selbst, A.D. (2016). Big data's disparate impact. *California Law Review* 104 (3): 671–732.

Basso, B. and Antle, J. (2020). Digital agriculture to design sustainable agricultural systems. *Nature Sustainability* 3 (4): 254–256.

Basso, B., Cammarano, D., and L.Y. (2016). The impact of precision agriculture on rural economic development. *Agricultural Systems* 150: 113–123.

Beintema, N.M. and di Marco, D. (2019). Gender and agricultural research: Current status and future directions. *Agricultural Systems* 172: 1–8.

Bennett, A., Cole, A., and Murphy, R. (2020). *Advancing Equity in Agricultural Technology: Policy Recommendations for Inclusive ML Solutions*. Policy Brief, International Food Policy Research Institute. Retrieved from https://www.ifpri.org

Berkes, F. (2012). *Sacred Ecology* (3rd ed.). Routledge.

Berkes, F. and Folke, C. (1998). *Linking Social and Ecological Systems: Management Practices and Social Mechanisms for Building Resilience*. Cambridge University Press.

Bessen, J.E. (2019). *AI and Jobs: The Role of Demand*. NBER Working Paper No. 24235. National Bureau of Economic Research.

Binns, R., Veale, M., Van Kleek, M. et al. (2018). 'It's reducing a human being to a percentage': Perceptions of justice in algorithmic decisions. *Proceedings of the 2018 CHI Conference on Human Factors in Computing Systems*, 1–14. ACM.

Birner, R. (2018a). Developing and implementing successful sustainable intensification policies. *Global Food Security* 17: 1–9.

Birner, R. (2018b). Smallholder agriculture in developing countries: Challenges and opportunities. *The Journal of Agricultural Economics* 69 (1): 1–15.

Brondizio, E.S. and Moran, E.F. (2008). Human dimensions of climate change: The vulnerability of small farmers in the Amazon. *Philosophical Transactions of the Royal Society B: Biological Sciences* 363 (1498): 1803–1809.

Bronson, K. and Knezevic, I. (2016). Big data in food and agriculture. *Big Data & Society* 3 (1): 1–5.

Brynjolfsson, E. and McAfee, A. (2014). *The Second Machine Age: Work, Progress, and Prosperity in a Time of Brilliant Technologies*. W. W. Norton & Company.

Brynjolfsson, E. and McElheran, K. (2016). The rapid evolution of machine learning: Implications for business and the economy. In: *The Economics of Artificial Intelligence: An Agenda*, 187–220. University of Chicago Press.

Caffaro, F., Roccato, M., and Cremasco, M.M. (2021). Perceived barriers to the adoption of precision agriculture technologies in Italy. *Sustainability* 13 (2): 572.

Carbonell, I.M. (2016). The ethics of big data in agriculture. *Internet Policy Review* 5 (1).

Carolan, M. (2020). Automated agrifood futures: Robotics, labor, and the distributive politics of digital agriculture. *Journal of Peasant Studies* 47 (1): 184–207.

Carvalho, D.V., Pereira, E.M., and Cardoso, J.S. (2019). Machine learning interpretability: A survey on methods and metrics. *Electronics* 8 (8): 832.

Chisanga, B., Mwaanga, N., and Mulenga, M. (2020). AI and smallholder farmers: Challenges and opportunities. *Agricultural Economics* 51 (4): 567–577.

Chlingaryan, A., Sukkarieh, S., and Whelan, B. (2018). Machine learning approaches for crop yield prediction and nitrogen status estimation in precision agriculture: A review. *Computers and Electronics in Agriculture* 151: 61–69.

Chui, M., Manyika, J., and Miremadi, M. (2018). Where machines could replace humans—And where they can't (yet). *McKinsey Quarterly*. Retrieved from https://www.mckinsey.com

Cochrane, L. (2019). Strengthening farmer capacity in adapting to climate change and variability in Ethiopia. *Journal of Rural Studies* 65: 29–38.

Cochrane, M.A., Bowman, D.M., and Kolden, C.A. (2019). The integration of local knowledge into fire management strategies: Lessons from fire-prone regions. *Journal of Environmental Management* 234: 489–500.

Crossa, J., Pérez-Rodríguez, P., Cuevas, J. et al. (2017). Genomic selection in plant breeding: Methods, models, and perspectives. *Trends in Plant Science* 22 (11): 961–975.

Darnhofer, I. (2020). Farming resilience in the face of the unexpected: Exploring the strategic choices that position farms to balance change and stability. *Agriculture and Human Values* 37 (4): 923–939.

Daum, T. (2018). Agricultural robotics and their potential implications for rural employment. *Agricultural Systems* 164: 98–105.

Daum, T. and Birner, R. (2020). The neglected governance challenges of agricultural mechanization in Africa—Insights from Ghana. *Food Policy* 90: 101802.

Doshi-Velez, F., and Kim, B. (2017). Towards a rigorous science of interpretable machine learning. *arXiv preprint arXiv:1702.08608*.

Doss, C.R. (2018). Women and agricultural productivity: What have we learned? *Journal of Economic Perspectives* 32 (1): 73–92.

Eastwood, C., Klerkx, L., Ayre, M. et al. (2017). Managing socio-technical networks: Tools to support the co-evolution of agriculture and digital innovation. *Agricultural Systems* 155: 12–18.

FAO. (1996). *Rome Declaration on World Food Security and World Food Summit Plan of Action*. FAO.

FAO. (2011). The state of food and agriculture 2010–2011: Women in agriculture—Closing the gender gap for development. Food and Agriculture Organization of the United Nations. Retrieved from http://www.fao.org

FAO. (2020). The role of women in food security and nutrition: Gender perspectives on agricultural technology. Food and Agriculture Organization of the United Nations. Retrieved from http://www.fao.org

Floridi, L., Cowls, J., Beltrametti, M. et al. (2018). AI4People—An ethical framework for a good AI society: Opportunities, risks, principles, and recommendations. *Minds and Machines* 28 (4): 689–707.

Foley, J.A., DeFries, R., Asner, G.P. et al. (2005). Global consequences of land use. *Science* 309 (5734): 570–574.

Folke, C., Hahn, T., Olsson, P. et al. (2005). Adaptive governance of social-ecological systems. *Annual Review of Environment and Resources* 30: 441–473.

Gebbers, R. and Adamchuk, V.I. (2010). Precision agriculture and food security. *Science* 327 (5967): 828–831.

Gebru, B., Lee, J.A., and Tesfaye, K. (2020). AI-driven solutions for sustainable agriculture: Lessons from Sub-Saharan Africa. *Agricultural Systems* 182: 102846.

Godfray, H.C.J., Beddington, J.R., Crute, I.R., et al. (2010). Food security: The challenge of feeding 9 billion people. *Science* 327 (5967): 812–818.

Goh, J., Hwang, H., and Miller, R. (2017). Technological infrastructure and the digital divide. *Journal of Information Technology* 32 (1): 1–16.

Gordon, B. and Grant, D. (2021). The technological divide in agriculture: Issues and solutions. *Journal of Rural Studies* 84: 109–118.

Hargittai, E. (2020). The importance of digital literacy in the 21st century: Implications for education and training. *Information Society* 36 (1): 9–20.

Hargittai, E. and Hinnant, A. (2008). Digital inequality: Differences in young adults' use of the internet. *Sociological Inquiry* 78 (1): 56–78.

He, M. and Zhang, W. (2020). Agricultural innovation: STEM education and its impact on rural economies. *Technological Forecasting and Social Change* 155: 119930.

Hoffman, M. (2019). AI in agriculture: Challenges and opportunities. *Journal of Agricultural and Food Chemistry* 67 (15): 4053–4061.

Horrigan, L., Lawrence, R.S., and Walker, P. (2002). How sustainable agriculture can address the environmental and human health harms of industrial agriculture. *Environmental Health Perspectives* 110 (5): 445–456.

IAASTD. (2009). *Agriculture at a Crossroads: Global Report*. International Assessment of Agricultural Knowledge, Science and Technology for Development.

IPCC. (2014). *Climate Change 2014: Impacts, Adaptation, and Vulnerability*. Cambridge University Press.

Jasanoff, S. (2004). *States of Knowledge: The Co-production of Science and Social Order*. Routledge.

Jha, S., Doshi, R., Patel, V. et al. (2019). Artificial intelligence in agriculture: Challenges and prospects. *International Journal of Innovative Research in Technology* 6 (2): 101–107.

Jordan, M.I. and Mitchell, T.M. (2015). Machine learning: Trends, perspectives, and prospects. *Science* 349 (6245): 255–260.

Jose, S. (2009). Agroforestry for ecosystem services and environmental benefits: An overview. *Agroforestry Systems* 76 (1): 1–10.

Kamilaris, A. and Prenafeta-Boldú, F.X. (2018). Deep learning in agriculture: A survey. *Computers and Electronics in Agriculture* 147: 70–90.

Klerkx, L. and Begemann, S. (2020). Agricultural digitalization in the age of AI: Ethics, equity, and employment. *Agricultural Systems* 187: 102965.

Klerkx, L. and Rose, D. (2020). Innovation for sustainable agriculture: The role of digital technologies. *Current Opinion in Environmental Sustainability* 45: 26–32.

Klerkx, L., Jakku, E., and Labarthe, P. (2019). A review of social science on digital agriculture, smart farming and agriculture 4.0: New contributions and a future research agenda. *NJAS-Wageningen Journal of Life Sciences* 90: 100315.

Kotter, J.P. (1996). *Leading Change*. Harvard Business Review Press.

Kumar, P., Singh, S., and Sharma, A. (2021). Machine learning in pest and disease management: A case study. *Precision Agriculture* 22 (5): 1007–1020.

Kumar, R., Sharma, A., and Singh, A. (2020). Infrastructure challenges in rural ML adoption: A case study. *Journal of Rural Development* 39 (4): 1–16.

Lal, R. (2004). Soil carbon sequestration to mitigate climate change. *Geoderma* 123 (1): 1–22.

Lamb, D.W., Schneider, D., and Holland, K. (2008). An active crop canopy sensor for precise and spatially referenced control of variable rate fertilizer applications. *Precision Agriculture* 9 (6): 375–387.

Leach, M., Mearns, R., and Scoones, I. (1999). Environmental entitlements: Dynamics and institutions in community-based natural resource management. *World Development* 27 (2): 225–247.

Liakos, K.G., Busato, P., and Moshou, D. (2018). Machine learning in agriculture: A review. *Computers and Electronics in Agriculture* 151: 72–80.

Lindblom, J., Lundström, C., Ljung, M. et al. (2017). Promoting sustainable intensification in precision agriculture: Review of decision support systems development and strategies. *Precision Agriculture* 18 (3): 309–331.

Lowder, S.K., Skoet, J., and Raney, T. (2016). The number, size, and distribution of farms, smallholder farms, and family farms worldwide. *World Development* 87: 16–29.

Lowenberg-DeBoer, J., Miller, N., and Griffis, S. (2020). The economics of precision agriculture: Profitability, sustainability, and social impact. *Agricultural Systems* 182: 102834.

McCall, M.K. and Dunn, C.E. (2012). Geo-information tools for participatory spatial planning: Fulfilling the criteria for good governance? *Geoforum* 43 (1): 81–94.

McMichael, P. (2009). A food regime genealogy. *Journal of Peasant Studies* 36 (1): 139–169.

Miller, S.R., Tschang, F.T., and Chen, Y. (2018). Barriers to the adoption of machine learning technologies in rural areas. *Technology and Innovation Management Review* 8 (12): 21–32.

Mulla, D.J. (2013). Twenty-five years of remote sensing in precision agriculture: Key advances and remaining knowledge gaps. *Biosystems Engineering* 114 (4): 358–371.

Niles, M.T., Ahuja, R., and Farmer, J. (2021). Participatory AI: Co-developing climate resilience tools with smallholder farmers. *Journal of Rural Studies* 84: 1–11.

Norris, F.H., Stevens, S.P., Pfefferbaum, B. et al. (2008). Community resilience as a metaphor, theory, set of capacities, and strategy for disaster readiness. *American Journal of Community Psychology* 41 (1–2): 127–150.

Norström, A.V., Cvitanovic, C., and Löf, M.F. (2020). Principles for knowledge co-production in sustainability research. *Nature Sustainability* 3 (3): 182–190.

OECD. (2019). *The Future of Work: OECD Employment Outlook 2019*. OECD Publishing.

O'Neil, C. (2016). *Weapons of Math Destruction: How Big Data Increases Inequality and Threatens Democracy*. Crown Publishing Group.

Pasek, J., More, E., and Wilkins, C. (2018). The impact of machine learning on education. *Educational Data Mining Review* 6 (2): 45–60.

Patel, R. (2007). *Stuffed and Starved: The Hidden Battle for the World Food System*. Melville House.

Peterman, A., Quisumbing, A.R., and Behrman, J.A. (2020). A review of the evidence on gender differences in agricultural productivity. *World Development* 136: 105070.

Popkin, B.M., Adair, L.S., and Ng, S.W. (2012). Global nutrition transition and the pandemic of obesity in developing countries. *Nutrition Reviews* 70 (1): 3–21.

Preece, K., Mallon, E., and Sullivan, T. (2021). Indigenous knowledge integration into AI for fire prediction in Australia. *Ecological Applications* 31 (4): e2330.

Pretty, J. (1995). Participatory learning for sustainable agriculture. *World Development* 23 (8): 1247–1263.

Pretty, J. (2008). Agricultural sustainability: Concepts, principles and evidence. *Philosophical Transactions of the Royal Society B: Biological Sciences* 363 (1491): 447–465.

Putnam, R.D. (2000). *Bowling Alone: The Collapse and Revival of American Community*. Simon & Schuster.

Quisumbing, A.R., Meinzen-Dick, R., Raney, T.L., et al. (Eds.). (2014). *Gender in Agriculture: Closing the Knowledge Gap*. Springer.

Ragasa, C., Berhane, G., and Taffesse, A.S. (2019). The role of gender in the adoption of agricultural technologies: A review. *Journal of Agricultural Economics* 70 (1): 1–23.:10.1111/1477-9552.12295

Rahman, S. (2020). Machine learning in smallholder agriculture: Potentials, challenges, and future directions. *Food Policy* 93: 101847.

Rajkomar, A., Dean, J., and Kohane, I. (2019). Machine learning in medicine. *New England Journal of Medicine* 380 (14): 1347–1358.

Reichstein, B. (2020). Autonomous farming: The future of agriculture. *Journal of Precision Agriculture* 19 (4): 276–290.

Rigby, D. and Cáceres, D. (2001). Organic farming and the sustainability of agricultural systems. *Agricultural Systems* 68 (1): 21–40.

Rogers, E.M. (2003). *Diffusion of Innovations* (5th ed.). Free Press.

Rose, D.C., Wheeler, R., Winter, M. et al. (2021). Agricultural sustainability and the digital divide: The role of digital agriculture in addressing rural inequalities. *Agricultural Systems* 190: 103048.

Rotz, S., Gravely, E., Mosby, I. et al. (2019). Automated pastures and the digital divide: How agricultural technologies are shaping labour and rural communities. *Journal of Rural Studies* 68: 112–122.

Rudin, C. (2019). Stop explaining black-box machine learning models for high-stakes decisions and use interpretable models instead. *Nature Machine Intelligence* 1 (5): 206–215.

Salami, A., Kamara, A.B., and Brixiova, Z. (2020). Smallholder agriculture in East Africa: Trends, constraints, and opportunities. *African Development Bank Working Paper Series* 105: 1–24.

Samek, W., Wiegand, T., and Müller, K.R. (2017). Explainable artificial intelligence: Understanding, visualizing, and interpreting deep learning models. *arXiv preprint arXiv:1708.08296*.

Schimmelpfennig, D. (2018). Farm labor, machines, and AI: Emerging trends in agricultural employment. *Applied Economic Perspectives and Policy* 40 (1): 55–72.

Schneider, F. (2020). Adapting vocational training to the digitalization of agriculture. *Journal of Agribusiness Education* 23 (3): 177–192.

Singh, J., Nath, R., and Singla, J. (2020). Smart agriculture: Improving farming through IOT, AI, and automation. *IEEE Internet of Things Magazine* 3 (2): 28–34.

Sithole, B., Hunter-Xenie, H., and Maru, Y. (2021). The role of indigenous fire management in Australia's future: Combining ancient wisdom with modern science. *Fire Ecology* 17 (1): 4.

Smith, A. and Kopp, J. (2020). Data privacy and ownership in the digital age: Implications for smallholder farmers. *Information Society* 36 (2): 102–115.

Smith, A., Anderson, M., and Rainie, L. (2020). The state of digital literacy in rural communities. *Pew Research Center*. Retrieved from https://www.pewresearch.org

Srinivasan, S., Xu, H., and Collins, K. (2019). Gender, technology, and agriculture: An intersectional approach. *Agricultural Systems* 174: 14–23.

Tesfaye, A., Speelman, S., and Van Huylenbroeck, G. (2017). Climate-smart agriculture in smallholder farming communities in Sub-Saharan Africa. *Journal of Agricultural Economics* 68 (3): 759–780.

Tilman, D., Cassman, K.G., and Matson, P.A. et al. (2002). Agricultural sustainability and intensive production practices. *Nature* 418 (6898): 671–677.

Tittonell, P. and Giller, K.E. (2013). When yield gaps are poverty traps: The paradigm of ecological intensification in African smallholder agriculture. *Field Crops Research* 143: 76–90.

Tsiafouli, M.A., Thébault, E., Sgardelis, S.P. et al. (2015). Intensive agriculture reduces soil biodiversity across Europe. *Global Change Biology* 21 (2): 973–985.

United Nations. (2021). *The Impact of Artificial Intelligence on Work and Society*. United Nations Report.

van der Burg, S., Bogaardt, M.J., and Wolfert, S. (2019). Ethics of smart farming: Current questions and directions for responsible innovation towards the future. *NJAS-Wageningen Journal of Life Sciences* 90: 100289.

Van Dijk, J.A.G.M. (2020). *The Digital Divide: The Internet and Social Inequality in International Perspective*. SAGE Publications.

van Es, H.M. and Woodard, J. (2017). Innovation in precision agriculture: Machine learning in decision support. *Agricultural Systems* 153: 69–78.

Van Klompenburg, T., Kassahun, A., and Catal, C. (2020). Crop yield prediction using machine learning: A systematic literature review. *Computers and Electronics in Agriculture* 177: 105709.

Vanclay, F. (2004). Social principles for agricultural extension to assist in the promotion of natural resource management. *Australian Journal of Experimental Agriculture* 44 (3): 213–222.

Vankov, I., Todorova, L., and Ilieva, M. (2020). Challenges and perspectives for digital literacy in agriculture. *Agriculture* 10 (12): 593.

Wang, Y., Feng, L., and Xu, X. (2018). Precision irrigation: A review of AI-enabled irrigation technologies. *IEEE Access* 6: 14880–14897.

Warschauer, M. (2003). *Technology and Social Inclusion: Rethinking the Digital Divide*. MIT Press.

WCED. (1987). *Our Common Future: Report of the World Commission on Environment and Development*. Oxford University Press.

Wendling, Z.A., Emerson, J.W., de Sherbinin, A. et al. (2020). *2020 Environmental Performance Index*. Yale Center for Environmental Law & Policy.

Westermann, O., Ashby, J., and Pretty, J. (2018). Gender and sustainable agriculture: Bridging the divide. *World Development* 98: 236–247.

Wheeler, S.A., Zuo, A., and Bjornlund, H. (2018). Farmers' climate change beliefs and adaptation strategies for a water-scarce future in Australia. *Global Environmental Change* 48: 146–158.

Wolfert, S., Ge, L., and Verdouw, C. (2017). Big data in smart farming: A review. *Agricultural Systems* 153: 69–80.

Zeng, X., Liu, Z., and Hu, X. (2020). Economic barriers to accessing data for machine learning. *Journal of Economic Perspectives* 34 (3): 121–142.

Zimmerer, K.S., Carney, J.A., and Vanek, S.J. (2018). Sustainable smallholder intensification in global change? *World Development* 107: 286–305.

18

Ethical and Regulatory Considerations of Machine Learning in Modern Agriculture

Atef M. El-Sagheer[1], Mohamed M. M. Hamd[2]*

[1] *Agricultural Zoology and Nematology Department, Faculty of Agriculture, Al-Azhar University, Assiut, Egypt*
[2] *Foundations of Education, Faculty of Education, Assiut University, Assiut, Egypt*

*Corresponding author: atefelsagheer@azhar.edu.eg

18.1 Introduction

Machine learning (ML) is a rapidly advancing and powerful technology applied to many fields, with especially recent interest in the plant biology and agriculture sectors. It has the potential to affect agriculture by providing better decision accuracy, increased crop yield, efficient use of inputs, and more scientific application of disease management strategies. The general application of ML in agriculture includes predictive analytics, smart disease detection, resource optimization challenges, and pest management. However, their integration into the agricultural setup no doubt poses serious ethical, practical, and regulatory challenges that need consideration.

A major issue relates to the privacy and security of data. Agricultural ML systems rely on voluminous datasets involving sensitive information on farming operations, environmental conditions, or even farmers' personal data. Any breach, unauthorized access, adversarial attacks, and misuse are considered serious threats to the integrity of agricultural systems and farmers' economic wellbeing. Again, data ownership, data obfuscation, and data profiling raise the complexity of the dynamics in the ethical and responsible management of agricultural data. Simultaneously, the progress and deployment of ML technologies within the agricultural domain require the establishment of sound regulatory frameworks. General regulations like the European Union's General Data Protection Regulation and the FDA guidelines in the United States have a limit to the regulation they can provide; this often does not keep pace with the advancements that are happening with ML. That would require dynamic and adaptive regulation capable of weighing up data transparency, equity, and safety in the applications of ML while allowing technological innovation to reshape the sector.

This chapter shall look into the ethical dilemmas and regulatory challenges as a result of the deployment of ML in plant biology and agriculture. It brings forth main security vulnerabilities, provides practical solutions and assesses the deficiencies in regulations that must be resolved for ML technologies to be absorbed in agricultural frameworks in an ethical, secure, and sustainable way. This, in turn, would provide necessary insights for policymakers, business leaders, and scholars about the responsible development of ML in the agricultural sector.

18.2 Data Privacy and Security in Agricultural ML Systems

The integration of ML in agriculture is marking a revolution within the sector by offering precise decision-making, increasing crop yields, and enhancing disease management. The applications of ML within agriculture vary from predictive analytics to auto-detection of diseases to resource optimization. On the other hand, as the application of ML becomes more integral within agricultural practices, new challenges are arising regarding data security and privacy. Agricultural

Machine Learning for Plant Biology, First Edition. Edited by Jen-Tsung Chen.

data involves sensitive information about the methods of cultivation, ecological conditions, and personal details of farmers, which require protection from exploitation, unauthorized access, and potential abuses. In this context, this chapter aims to discuss the concerns on privacy and security associated with ML frameworks within the agricultural domain, present possible solutions, and highlight future challenges in the protection of agricultural data.

18.2.1 Agricultural Data and Its Sensitivity

Agricultural data encompasses a wide range of information, from farm-level data, such as soil health, weather conditions, crop yields, and pest and disease occurrence, to more personal data about the farmers themselves, including financial information, geographic locations, and business operations. As more ML systems rely on vast amounts of data for training and decision-making, safeguarding these datasets becomes critical.

Personal data collected from farmers could also be used for profiling, while agricultural data may also be used for competitive market benefit. For instance, companies could apply nonpublic agricultural data to realize price fluctuations and, therefore, undertake unfair competitive conduct. The third parties may apply such information to disrupt farming activities and, therefore, compromise food security.

18.2.2 Security Threats in Agricultural Machine Learning Systems

Several security challenges are prevalent in ML systems used in agriculture. These threats range from data breaches to adversarial attacks aimed at manipulating ML models.

18.2.2.1 Data Breaches

Agricultural data, whether collected by sensors, drones, or satellites, is often stored in cloud-based systems. Cloud services are vulnerable to cyber-attacks, potentially leading to the exposure of sensitive farm data. Data breaches can result in unauthorized parties gaining access to large-scale agricultural datasets, which can be used for unethical purposes.

18.2.2.2 Adversarial Attacks

In adversarial ML, the attacker intentionally perturbs the input data to deceive the model's predictions. In agriculture, for example, the attackers might directly interfere with the input data, which includes soil moisture levels, pest density, or weather forecasts to affect an automated decision-making system and cause huge losses to farmers.

18.2.2.3 Model Theft and Data Leakage

Models of ML themselves can be a target for theft. An adversary can reverse-engineer or replicate an ML model used in agriculture to gain access to proprietary insights. Data leakage during model training can also expose private agricultural data, leading to unauthorized use by third parties.

18.2.2.4 Unauthorized Access

Many systems of ML for agriculture require access to vast datasets collected from Internet of Things (IoT) devices and sensors. Without proper access control measures, unauthorized users may gain entry into these systems, compromising data integrity and security. Table 18.1 shows some of them.

Table 18.1 Key risk factors of unauthorized access in agricultural ML systems and mitigation strategies.

Risk Factor	Potential Impact	Mitigation Measures
Lack of authentication	Unauthorized access to sensitive data	Implement multi-factor authentication (MFA)
Weak encryption	Data breaches and manipulation	Use end-to-end encryption protocols
Unsecured IoT devices	System vulnerabilities exploited	Regular firmware updates and patches

18.2.3 Privacy Concerns in Agricultural Machine Learning Systems

18.2.3.1 Data Ownership and Consent

A critical issue is determining who owns the data collected by ML systems-farmers, governments, or the corporations providing the technology.

Farmers often lack control over the data generated on their farms, raising concerns over consent and data rights. ML systems must ensure that data collection is transparent and based on informed consent.

18.2.3.2 Data Obfuscation

Different levels of data obfuscation may be used in developing this system to reduce privacy concerns. It is still vulnerable to de-obfuscation, though, since anonymized data may again be re-identified by cross-referencing it with other sources of data. Translated to an agriculture application, sensitive information about a particular farm or location might still be identifiable.

18.2.3.3 Profiling and Surveillance

All this based on the extensive data collected and analyzed by ML systems could well characterize individual farmers on their agronomic practices, production, and other socio-economic parameters (Table 18.2). There is a possibility of unforeseen effects due to such developments, such as discrimination or monitoring against farmers, raising ethical challenges concerning privacy and autonomy.

18.2.4 Addressing Data Privacy and Security in Agricultural ML Systems

To mitigate the risks associated with data privacy and security in agricultural ML systems, several strategies can be adopted.

18.2.4.1 Secure Data Storage and Transmission

Agricultural data must be stored on secure cloud platforms. Encryption both at rest and in transit can ensure secured data against unauthorized access. End-to-end encryption makes sure that only the parties authorized can access the data throughout the lifecycle of the data.

18.2.4.2 Access Control and Authentication

Strict access control, such as multi-factor authentication, should be implemented to prevent unauthorized access to the agricultural ML system. Verification of authenticity and trust should be assured while allowing access to the data and ML models.

18.2.4.3 Federated Learning

The centralized model of ML has certain hindrances, especially on data privacy issues, which the federated learning decentralized approach tries to solve. In this model, the data stays on the farm (edge devices), while only the model updates are shared with the central server. This minimizes the chances of sensitive data exposure since data never leaves the local environment.

Table 18.2 Potential risks and ethical concerns of profiling and surveillance in agricultural ML systems.

Aspect	Potential Impact	Ethical Concern	References
Farmer profiling	Targeted services based on farm data	Risk of discrimination or exclusion	Doe et al. (2022)
Data surveillance	Continuous monitoring of farming activities	Privacy invasion and loss of autonomy	Smith and Lee (2021)
Algorithmic bias	Unfair resource allocation	Marginalization of smallholder farmers	Brown et al. (2023)
Third-party data access	Commercial exploitation of farmer data	Lack of informed consent	Jones et al. (2020)

18.2.4.4 Adversarial ML Defenses

Towards handling adversarial attacks, training ML agriculture models with adversarial examples is necessary to instill much-needed robustness. Techniques like differential privacy introduce noise into the data while guarding against adversarial manipulation; this can also reduce loss in model accuracy.

18.2.4.5 Ethical and Regulatory Frameworks

Governments and regulatory authorities should establish regulations related to data privacy and security in agricultural ML systems. The policies must respect ownership rights over data and grant farmers control over the usage of their data. The policy should be transparent regarding the collection, storage, and processing of data.

18.2.5 Future Directions and Challenges

With growing ML adoption in agriculture, several challenges have to be overcome to keep data privacy and security at the forefront. First, the equity in data access and security solutions between large-scale industrial farms and smallholder farmers needs to be guaranteed. Most of the small-scale farmers may not have the resources or the technical expertise to invest in advanced data security measures, making them more vulnerable to exploitation. Moreover, the challenge of balancing the need for large datasets that improve the performance of ML models with the ethical need to protect farmer privacy will continue to be a sensitive issue. Ultimately, international cooperation in establishing consistent ethical standards and security protocols across diverse geographical regions and agricultural sectors will be required.

In this light, the integration of ML into agriculture promises significant advances in productivity, sustainability, and resource management. At the same time, the collection of agricultural data by ML systems raises serious privacy and security concerns. The ethical, regulatory, and technical facets are involved in considering the whole gamut of these concerns. With increasing dependence on ML in the agricultural sector, the protection of sensitive information will be paramount to trusting a code of fairness and equitable access to these technologies.

18.3 Bias and Fairness in Artificial Intelligence Models for Plant Disease Prediction

Plant disease forecasting was revolutionized by ML models for early detection and better management of diseases, thus giving high yields in agriculture. However, there is growing concern over the biases and equity of ML in agriculture. These biases could be due to many reasons: nonrepresentative training datasets, poorly designed model architecture, and unequal technological access. These biases produce inequitable outcomes because there is a certain type of crop that may be affected, regions, or groups of farmers. This chapter discusses sources of bias in the ML model for plant disease prediction, consequences on agriculture, and mitigation strategies to remove bias in order to make agricultural outcomes nondiscriminatory.

18.3.1 Sources of Bias in ML for Plant Disease Prediction

Bias in ML models for plant disease prediction can originate from several areas, including the data used to train the models, algorithmic constraints, and even human factors during data labeling. Some common sources include the following.

18.3.1.1 Data Imbalance

Data imbalance occurs when the training data over-represents certain plant species, regions, or diseases, leading to biased models. For instance, a model trained primarily on data from high-income regions may underperform when predicting diseases in less developed areas (Sun et al. 2009a). If a model is trained disproportionately on data from major crops such as wheat or rice, it may not generalize well to other crops that are equally important in local economies but are underrepresented in the dataset (Kamiran and Calders 2012). As shown in Table 18.3.

18.3.1.2 Sampling Bias

Sampling bias arises when the data used for training does not adequately represent the entire population of interest. In agriculture, this could happen if the dataset predominantly comes from large commercial farms with advanced technological

Table 18.3 Impacts of data imbalance in agricultural ML systems.

Type of Imbalance	Example	Potential Consequences	References
Regional imbalance	Training data mainly from high-income regions	Poor accuracy in predicting diseases in low-income areas	Sun et al. (2009)
Crop-specific imbalance	Overrepresentation of wheat and rice data	Ineffective predictions for underrepresented crops	Kamiran and Calders (2012)
Disease imbalance	Limited data on rare plant diseases	Failure to detect or diagnose less common diseases	Zhao et al. (2018)
Climate zone imbalance	Focus on temperate climates, neglecting arid regions	Poor adaptability to diverse environmental conditions	Liu et al. (2020)

resources, while excluding data from smallholder farms that rely on traditional farming methods (Mehrabi et al. 2021b). As a result, the model may produce skewed predictions that favor large-scale farming practices.

18.3.1.3 Labeling Bias

Labeling bias occurs when the data used to train ML models is incorrectly or inconsistently labeled. In agricultural settings, plant disease data may be labeled by experts, who could introduce biases, especially in ambiguous cases. For instance, two experts might disagree on the exact stage of disease progression, leading to inconsistent labels that affect the model's learning (Goodfellow et al. 2016).

18.3.1.4 Algorithmic Bias

Bias can also stem from the algorithms themselves. Some ML algorithms tend to prioritize accuracy for the majority class while neglecting the minority class. In plant disease prediction, this could lead to high accuracy in predicting common diseases while failing to detect rare or emerging diseases (Watson and Mahler 2020).

18.3.1.5 Geographical and Climatic Bias

Plant disease manifestation is often influenced by local environmental factors such as climate and soil conditions. ML models trained on data from specific geographic regions may not perform well in areas with different environmental conditions. For instance, a model trained in temperate climates may not be reliable for predictions in tropical regions, where diseases can behave differently due to higher temperatures and humidity (Athey 2018).

18.3.2 Impact of Bias on Agricultural Communities

Bias in ML models can have significant consequences, particularly for vulnerable farming communities. The impacts include the following.

18.3.2.1 Disparities in Disease Management

ML models trained on data from well-resourced farms may perform poorly in low-income regions, where access to technology and farming practices differ (Mehrabi et al. 2021b). This disparity could result in smallholder farmers being unable to effectively manage diseases, exacerbating existing inequalities in agricultural productivity.

18.3.2.2 Inequity in Predictions for Minority Crops

Many ML models are designed for common crops such as maize, wheat, or rice, which receive significant research attention and funding. As a result, models may lack predictive accuracy for minority crops, which play a crucial role in the diets and economies of certain regions (Kamiran and Calders 2012). This can lead to crop losses, threatening food security and local economies.

18.3.2.3 Economic and Environmental Consequences

Inaccurate disease predictions can lead to misguided decisions, such as the overuse of pesticides or fertilizers, resulting in financial losses for farmers and environmental degradation. Biased models may also promote unsustainable agricultural practices, contributing to long-term ecological harm (Hino et al. 2018).

18.3.3 Strategies to Mitigate Bias in Prediction Models of Plant Disease

To address the issues of bias and ensure fairness in artificial intelligence (AI)-driven agriculture, several approaches can be employed.

18.3.3.1 Diversifying Training Data

To reduce bias, it is crucial to ensure that the training data reflects a wide variety of agricultural settings, including different crops, regions, and farming practices. Collaboration with local farmers, research institutions, and agricultural organizations can help collect data from diverse environments, ensuring that models generalize well across different contexts (Sun et al. 2009b).

18.3.3.2 Synthetic Data Generation

When real-world data is limited, synthetic data generation techniques such as data augmentation or generative adversarial networks (GANs) can be employed to create additional training examples for underrepresented crops or diseases. This approach helps balance datasets and improves the model's ability to predict diseases that might be rare or newly emerging (Goodfellow et al. 2016).

18.3.3.3 Fairness-aware Algorithms

Some algorithms can be explicitly designed to address bias and improve fairness. Fairness-aware algorithms attempt to ensure that the model's performance is equitable across different subgroups within the data (Chinta et al. 2024). In plant disease prediction, this might involve ensuring that predictions are accurate not only for major crops but also for minority crops or underrepresented regions (Kamiran and Calders 2012).

18.3.3.4 Transparency and Explainability

Models should be transparent and explainable to help identify and correct biases. Explainability tools can provide insights into how a model makes its predictions, which can highlight potential areas of bias (Hassija et al. 2024). For example, visualizing the parts of an image a model uses to diagnose a plant disease can help detect if the model is overly reliant on irrelevant features (Gunning and Aha 2019).

18.3.3.5 Continuous Monitoring and Audits

The ML models should be continuously monitored and regularly audited to identify emerging biases (Oyeniran et al. 2022). Over time, data distributions may change due to evolving environmental factors or agricultural practices, and models must be updated to remain fair and accurate (Wolfert et al. 2017). Ongoing audits can help ensure that biases do not develop or worsen after the initial deployment (Watson and Mahler 2020).

18.3.4 Challenges in Ensuring Fairness

Ensuring fairness in plant disease prediction remains challenging (Hino et al. 2018). Data collection in agriculture is often skewed towards commercially important crops and regions, leading to an unequal distribution of resources (Stevenson et al. 2020). Additionally, fairness is a complex concept that varies among different stakeholders in agriculture. Governments, farmers, agribusinesses, and consumers all have different perspectives on what constitutes fairness, making it difficult to define and implement a universally accepted standard (Mulligan et al. 2019).

Another significant challenge is the interdisciplinary nature of the problem. Addressing bias in agricultural ML models requires expertise from both data science and agricultural domains, making collaboration essential for developing effective solutions (Athey 2018; Bronson and Knezevic 2016).

It follows that bias in ML models for plant disease prediction poses significant challenges to equitable agricultural development (Joshi et al. 2021). Models trained on unrepresentative data can exacerbate existing inequalities, affecting the ability of smallholder farmers and minority crop producers to effectively manage diseases (Gebru et al. 2021). Mitigating bias requires a combination of technical solutions—such as fairness-aware algorithms and synthetic data generation—and collaborative efforts to collect more representative data (Barocas et al. 2019; Mehrabi et al. 2021a). Ensuring fairness in AI for agriculture is vital for creating sustainable and inclusive agricultural systems, benefiting all farmers regardless of their region or resources (Russell et al. 2020).

18.4 Environmental and Ecological Implications of AI in Agriculture

AI has rapidly gained traction as a transformative tool in agriculture, offering solutions to increase crop productivity, optimize resource use, and manage pests and diseases. However, while AI promises substantial benefits, its application in agriculture also raises significant environmental and ecological concerns. These implications are not just limited to the outcomes of AI-driven interventions but also extend to how AI systems are trained and integrated into existing agricultural practices.

18.4.1 AI-driven Precision Agriculture and Resource Optimization

One of the major applications of AI in agriculture is precision farming, which leverages AI tools, such as ML algorithms and satellite imagery, to monitor and manage crop health, soil conditions, and resource use (Zhang et al. 2019). By optimizing the application of water, fertilizers, and pesticides, precision agriculture can significantly reduce the environmental footprint of farming (Shamshiri et al. 2018).

18.4.1.1 Water Conservation

AI systems can predict crop water requirements by analyzing historical climate data, soil moisture, and plant health, enabling more efficient irrigation practices. Studies have shown that AI-assisted precision irrigation can reduce water use by 20–30% (McBratney et al. 2005). This is especially critical in regions facing water scarcity, such as California, where agriculture consumes about 80% of the state's water supply (Cooley et al. 2014).

18.4.1.2 Reduction of Chemical Inputs

AI-powered systems help farmers apply fertilizers and pesticides only where needed, reducing the overuse of these chemicals. Excessive pesticide use can lead to soil degradation, water contamination, and the loss of biodiversity (Popp et al. 2013). AI-based pest detection systems, like drone surveillance and image recognition algorithms, can accurately identify pest outbreaks, reducing the need for blanket pesticide applications and minimizing ecological harm (Mahlein 2016).

18.4.1.3 Energy Efficiency

AI can also optimize energy use in agricultural operations, such as greenhouses and farm machinery. AI-driven systems can control greenhouse environments more efficiently, adjusting lighting, heating, and ventilation to maintain optimal growing conditions while reducing energy consumption (Kaloxylos et al. 2014). Moreover, AI can guide the operation of autonomous tractors and harvesters, improving fuel efficiency and reducing carbon emissions (Jha et al. 2019).

18.4.2 Impact on Biodiversity and Ecosystem Services

While AI in agriculture promises to improve efficiency, its impact on biodiversity and ecosystem services presents both opportunities and risks.

18.4.2.1 Preservation of Biodiversity

AI systems that promote precision agriculture can help maintain biodiversity by reducing the environmental disturbances caused by traditional agricultural practices. For instance, AI can be used to monitor wildlife in agricultural areas, ensuring that farming activities do not disrupt local ecosystems (Correll et al. 2017). Moreover, AI-driven crop management systems that promote companion planting and crop rotation can enhance soil biodiversity and health, contributing to the resilience of farming ecosystems (Smith et al. 2020).

18.4.2.2 Risk of Monoculture Intensification

However, AI may also contribute to the intensification of monocultures-large-scale, single-crop farming systems-which can lead to the depletion of biodiversity. Monocultures, though efficient in terms of productivity, often require increased chemical inputs and are more susceptible to pests and diseases (Tilman et al. 2001). If AI is primarily utilized to optimize monoculture farming, it could exacerbate biodiversity loss and soil degradation, affecting ecosystem services such as pollination, water purification, and carbon sequestration (Bommarco et al. 2013).

18.4.2.3 Wildlife Habitat Displacement

The large-scale adoption of AI-driven autonomous machinery in agriculture, such as drones and robots, may also disturb wildlife habitats. Studies have shown that drone noise and movement can disrupt bird behavior, leading to habitat displacement and reduced breeding success (Vas et al. 2015). As AI technologies expand in agriculture, the potential for disrupting terrestrial ecosystems will need careful consideration.

18.4.3 Carbon Footprint of AI in Agriculture

The environmental impact of AI extends beyond its applications in agriculture to the carbon footprint associated with AI itself. Training complex AI models requires substantial computational power, contributing to greenhouse gas emissions. Recent studies have highlighted that the carbon emissions generated by training large AI models, such as deep learning algorithms, can be equivalent to the lifetime emissions of several cars (Strubell et al. 2019).

18.4.3.1 Energy Use in AI Training

The computational intensity of training AI models—especially those based on neural networks—can result in significant energy use. One study found that training a single deep learning model could emit as much as 626,155 pounds of CO_2, which is comparable to five times the lifetime emissions of an average American car (Strubell et al. 2019). This raises concerns about the sustainability of AI when applied at scale in agriculture.

18.4.3.2 Efforts to Reduce AI's Carbon Footprint

To mitigate these impacts, there is a growing movement towards developing more energy-efficient AI algorithms and utilizing renewable energy sources for AI infrastructure (Patterson et al. 2021). For instance, Google's Tensor Processing Units (TPUs), designed specifically for AI workloads, are more energy-efficient than general-purpose processors, reducing the environmental impact of training large models (Jouppi et al. 2017). Integrating such innovations into agricultural AI systems can help balance the environmental benefits of AI with its carbon costs.

18.4.4 Ecological Impacts of AI-powered Autonomous Systems

The introduction of AI-powered autonomous systems, such as robots, drones, and self-driving tractors, brings a new set of environmental challenges. These systems are often hailed for their potential to improve agricultural efficiency, but their ecological impacts need careful evaluation.

18.4.4.1 Soil Compaction

Autonomous tractors and robots can contribute to soil compaction, which impairs root growth and reduces soil biodiversity (Hamza and Anderson 2005). While autonomous systems can be programmed to minimize their environmental impact, such as by using lighter machinery, the overall effect of widespread use of autonomous agricultural vehicles on soil health remains a concern (Balafoutis et al. 2017).

18.4.4.2 E-waste and Resource Depletion

The rapid deployment of AI-driven hardware in agriculture, including sensors, drones, and robotics, will contribute to the growing problem of electronic waste (e-waste). E-waste contains hazardous materials, such as lead and mercury, which can leach into the environment and harm ecosystems (Balde et al. 2017). Additionally, the production of AI hardware depends on rare earth metals, the extraction of which has significant environmental and ecological impacts, including habitat destruction and water pollution (Massari and Ruberti 2013).

18.4.4.3 Opportunities for Regenerative Agriculture

On the positive side, AI-powered autonomous systems can also be used to promote regenerative agricultural practices. For example, robots that manage cover crops or apply organic fertilizers precisely can reduce the need for chemical inputs, improve soil health, and increase carbon sequestration (Lal 2004). Moreover, drones equipped with AI can monitor soil conditions and plant health, enabling farmers to adopt regenerative practices more effectively (Pereira et al. 2020).

18.4.5 Ethical and Ecological Governance of AI in Agriculture

As AI technologies become more integrated into agriculture, it is crucial to establish governance frameworks that prioritize environmental and ecological sustainability. These frameworks should address not only the environmental impact of AI itself but also its broader implications for agricultural systems and ecosystems.

18.4.5.1 Sustainable AI Development

Ensuring that AI technologies are developed with sustainability in mind requires collaboration between AI developers, environmental scientists, and policymakers. One potential solution is the adoption of sustainability benchmarks for AI, ensuring that AI-driven agricultural systems contribute to long-term ecological health rather than short-term gains (Berg et al. 2018).

18.4.5.2 Incorporating Ecological Indicators in AI Models

AI models used in agriculture should integrate ecological indicators, such as biodiversity metrics, soil health parameters, and water quality data, to ensure that the pursuit of productivity does not come at the expense of environmental sustainability (Bommarco et al. 2013). Such models can provide farmers with more holistic recommendations that balance productivity with ecological health.

18.4.5.3 Inclusive AI for Global Agriculture

Finally, it is essential to ensure that AI technologies are accessible to all farmers, particularly smallholder farmers in developing regions who may not have the same access to AI infrastructure as large-scale commercial operations. The unequal distribution of AI-driven technologies could exacerbate environmental degradation in regions that lack the resources to implement sustainable practices (Zeng et al. 2020).

From this, it can be concluded, the environmental and ecological implications of AI in agriculture are multifaceted, offering both significant opportunities and potential risks. While AI technologies can improve resource efficiency, reduce chemical inputs, and enhance regenerative agricultural practices, they also pose challenges related to biodiversity loss, carbon emissions, and e-waste. To maximize the environmental benefits of AI in agriculture while mitigating its negative impacts, it is crucial to adopt sustainable AI development practices, integrate ecological indicators into AI models, and promote inclusive access to AI-driven agricultural technologies.

18.5 Transparency and Explain Ability in AI-driven Agricultural Solutions

The rise of AI in agriculture is driving unprecedented changes in how farming is approached, with solutions ranging from predictive models for crop yield optimization to advanced pest control systems. However, alongside these innovations comes the growing demand for transparency and explain ability in AI systems. Farmers, policymakers, and other stakeholders need to understand how AI algorithms make decisions to trust and adopt these solutions fully. In agriculture, where decisions directly impact food security, the environment, and livelihoods, explainable AI (XAI) plays a critical role.

18.5.1 The Need for Transparency in AI-driven Agriculture

18.5.1.1 Building Trust Among Stakeholders

Transparency in AI systems is essential to build trust among farmers, agronomists, and policymakers. Since AI models are often perceived as "black boxes," stakeholders may be reluctant to adopt technologies they do not fully understand. Trust is vital in agriculture, where unpredictable environmental variables and high risks-such as crop failure-are involved. According to Doshi-Velez and Kim (2017), understanding how an AI model works helps users make informed decisions, which is especially crucial when high-stakes outcomes, such as food security and environmental sustainability, are on the line.

18.5.1.2 Ethical and Legal Implications

AI-driven agricultural solutions need to align with ethical standards, particularly concerning data privacy, ownership, and fairness. Transparency ensures that AI systems do not disproportionately benefit large-scale operations while disadvantaging smallholders. Moreover, from a legal perspective, governments are beginning to enforce regulations requiring explainability in AI systems. The European Union's General Data Protection Regulation (GDPR) mandates that AI-driven decisions affecting individuals, including those in agriculture, be explainable (Goodman and Flaxman 2017). Failure to adhere to these standards could lead to legal ramifications and hinder the widespread adoption of AI technologies.

18.5.2 Challenges in Achieving Explainability in Agricultural AI Systems

18.5.2.1 Complexity of AI Algorithms

Many AI systems used in agriculture, such as convolutional neural networks (CNNs) and reinforcement learning algorithms, are inherently complex, making them difficult to interpret. These models are often built to handle large datasets-like satellite imagery or sensor data-where millions of parameters are adjusted to optimize outcomes such as yield prediction or pest detection (Montavon et al. 2018). The challenge is translating the inner workings of these systems into an understandable format for nonexpert users, including farmers and agricultural workers.

18.5.2.2 Data Quality and Bias

One of the key barriers to transparency in AI models is the quality and bias inherent in the data used to train them. Agricultural AI systems rely on large datasets, often compiled from a variety of sources, including remote sensing, IoT devices, and historical agricultural records. However, if these datasets are biased-whether through geographic limitations or outdated information-the resulting models can yield inaccurate predictions (Mehrabi et al. 2021a). For instance, a pest management AI model trained predominantly on data from temperate regions might not perform well in tropical climates, leading to incorrect recommendations that could harm crops. Identifying and explaining these biases is critical to improving model transparency.

18.5.2.3 Trade-off Between Accuracy and Interpretability

A persistent issue in AI systems is the trade-off between model accuracy and interpretability. More interpretable models, like decision trees or linear regressions, often provide less accurate predictions compared to complex models such as deep learning networks (Rudin 2019). In agriculture, where precision is paramount, reducing the accuracy of AI systems for the sake of explainability can be problematic. This tension between accuracy and interpretability needs to be managed carefully to ensure both reliable predictions and understandable decision-making processes.

18.5.3 Strategies for Enhancing Transparency and Explainability

18.5.3.1 Post hoc Explainability Techniques

One of the most effective approaches to making AI systems more transparent is the use of post-hoc explain ability techniques, which aim to explain a model's decision after it has been made. Techniques such as Local Interpretable Model-agnostic Explanations (LIME) and Shapley Additive Explanations (SHAP) are commonly used to break down complex AI models into understandable components (Lundberg and Lee 2017). For instance, in a crop disease detection model, SHAP values can show which features-such as leaf texture or color-had the most significant influence on the model's prediction. These techniques can help farmers and agronomists had better understand how AI systems arrive at their conclusions, fostering trust in the technology.

18.5.3.2 Model-agnostic Interpretability Frameworks

Another approach is to design model-agnostic interpretability frameworks, which work independently of the underlying AI algorithm. These frameworks can be integrated into AI solutions used in agriculture, providing explanations in terms of crop characteristics, soil health, or environmental conditions, regardless of the specific model being employed. Ribeiro et al. (2016) argue that model-agnostic approaches offer flexibility, allowing for the explain ability of both simple and complex models across different agricultural applications, from irrigation management to pest detection.

18.5.3.3 Simplified User Interfaces for Farmers

Improving the transparency of AI systems in agriculture also involves designing user interfaces that clearly communicate the reasoning behind AI decisions. Many farmers may not be familiar with the technicalities of AI but need to understand the outcomes to make informed decisions. Simplified dashboards that show key metrics, model confidence levels, and visual explanations can help bridge the gap between complex AI systems and end-users (Tzachor et al. 2020). For instance, a pest detection AI system could provide a visual explanation of infected areas in a field and suggest targeted interventions, making the technology more accessible to farmers.

18.5.3.4 Collaboration Between AI Developers and Agronomists

Collaboration between AI developers and domain experts, such as agronomists and ecologists, is crucial for enhancing the transparency of AI systems. Domain experts can help identify which variables are most relevant to explainable models in agricultural contexts. Moreover, by working with agronomists, AI developers can better understand the needs and expectations of farmers, ensuring that AI tools provide useful and understandable recommendations (Clark et al. 2021). This interdisciplinary approach is essential for the development of AI solutions that are both accurate and explainable.

18.5.4 Ethical Implications of Transparency in AI-driven Agriculture

18.5.4.1 Equity and Accessibility

Ensuring transparency in AI-driven agricultural systems also has important ethical implications, particularly regarding equity and accessibility. Large-scale farms are more likely to have access to advanced AI tools, while smallholder farmers in developing countries may be left behind (World Bank 2021). By prioritizing transparency and explainability, AI solutions can become more accessible to a broader range of users, ensuring that all farmers, regardless of scale, can benefit from technological advancements.

18.5.4.2 Accountability in Decision-making

Transparency also enhances accountability in AI-driven agricultural decisions. When AI models are used to make critical decisions-such as when to irrigate crops or apply pesticides-understanding how these decisions were made is essential for assigning responsibility if something goes wrong (Lipton 2018). This is especially important in agriculture, where the consequences of incorrect decisions can be devastating for both the environment and food security.

18.5.5 Future Directions in Transparent AI for Agriculture

As AI continues to evolve, transparency and explainability will remain central to its successful integration into agriculture. Several promising directions could enhance transparency further.

18.5.5.1 Interpretable AI Models

Research into inherently interpretable AI models, such as hybrid models that combine deep learning with decision tree logic, could offer a balance between accuracy and explainability (Gilpin et al. 2018). These models would provide clear explanations for decision-making while maintaining high levels of predictive accuracy.

18.5.5.2 Regulation and Standardization

Governments and industry bodies are likely to introduce more robust regulations around the transparency of AI systems in agriculture. Standardization of transparency protocols across AI applications in agriculture could help ensure that explain ability becomes an integral part of AI development (Floridi et al. 2020).

18.5.5.3 Education and Training

Educating farmers and agricultural workers about AI technologies and their interpretability is another critical area for future development. Training programs that demystify AI and explain its decision-making processes could encourage broader adoption and more effective use of these technologies in agriculture (van Es and Woodard 2021).

On the whole, transparency and explain ability in AI-driven agricultural solutions are crucial for building trust, ensuring ethical implementation, and enhancing the accessibility of these technologies. While there are significant challenges, such as the complexity of AI models and the trade-off between accuracy and interpretability, several strategies-including post-hoc explain ability techniques, model-agnostic frameworks, and simplified user interfaces-can improve transparency. As the agricultural sector continues to integrate AI technologies, a strong focus on explainability will be essential to ensure that these solutions are both effective and ethical.

18.6 Balancing Innovation with Tradition: Ethical Challenges in Technological Adoption

The integration of new technologies into traditional practices poses significant ethical challenges, especially in sectors like agriculture, where time-honored methods are often deeply rooted in cultural and community identities. Technological innovation, driven by advancements such as AI, ML, and biotechnology, promises significant gains in productivity and sustainability (Table 18.4). However, these advancements often come with concerns over their impact on traditional farming practices, social equity, environmental sustainability, and the livelihoods of smallholder farmers.

18.6.1 Technological Innovation in Agriculture

18.6.1.1 Advancements in Agricultural Technology

Technological advancements in agriculture have transformed the way crops are grown, managed, and harvested. Precision agriculture, AI-powered crop monitoring, and biotechnology have become increasingly integrated into farming practices (Jha et al. 2019). These technologies offer numerous benefits, including increased yields, reduced resource use, and more efficient pest and disease management (Zhang et al. 2020). Despite these benefits, there are growing concerns about how these innovations may disrupt traditional farming methods that have been practiced for centuries, particularly in rural communities (Sumberg et al. 2017).

18.6.1.2 AI and Machine Learning in Crop Management

AI and ML technologies are revolutionizing crop management by providing predictive models that optimize yield, reduce waste, and ensure more sustainable resource use (Kamilaris and Prenafeta-Boldú 2018). For instance, ML algorithms can predict when to irrigate or apply fertilizers based on environmental data, helping farmers make data-driven decisions. However, these technologies are often inaccessible to smallholder farmers due to their high cost and technical complexity (Deichmann et al. 2016). The ethical challenge lies in ensuring that these innovations do not exacerbate inequalities between large-scale agribusinesses and small, traditional farms.

18.6.2 The Role of Tradition in Sustainable Farming

18.6.2.1 Traditional Farming Methods

Traditional farming practices, often passed down through generations, have been developed with a deep understanding of local ecosystems and environments. These methods, which include organic farming, crop rotation, and intercropping, are integral to the cultural and social fabric of many rural communities (Altieri 2018). Many of these practices are considered sustainable as they rely on natural processes and biodiversity to maintain soil health, manage pests, and produce food.

18.6.2.2 Cultural Significance of Farming Traditions

In many parts of the world, agriculture is not just an economic activity but also a way of life that is closely tied to cultural and social identities. The adoption of new technologies can threaten these cultural traditions, especially when modern

Table 18.4 Ethical challenges in balancing innovation with tradition in agriculture.

Ethical Concern	Description	Potential Impact	References
Displacement of traditional knowledge	AI and ML replacing indigenous farming techniques	Loss of cultural heritage and local expertise	Shiva (2016)
Social equity issues	Unequal access to advanced agricultural technologies	Widening gap between large and small farmers	Patel and Singh (2020)
Environmental sustainability	Intensive biotechnology applications altering ecosystems	Biodiversity loss and soil degradation	FAO (2019)
Economic disruption	High costs of technology adoption	Financial strain on smallholder farmers	World Bank (2021)

farming methods prioritize efficiency and scale over community-based approaches (Schultz et al. 2018). The ethical challenge is to balance the benefits of technological innovation with respect for these cultural traditions, ensuring that technology enhances rather than replaces traditional knowledge and practices.

18.6.3 Ethical Challenges in Technological Adoption

18.6.3.1 Equity and Access to Technology

One of the central ethical challenges in technological adoption is ensuring equitable access to new technologies. Large-scale agribusinesses are more likely to have the resources to invest in cutting-edge technologies, such as AI and biotechnology, whereas smallholder farmers often lack the financial means or technical knowledge to adopt these innovations (Klerkx et al. 2019). This creates a widening gap between those who can afford and benefit from new technologies and those who are left behind. According to Deichmann et al. (2016), more than 70% of farmers in low-income countries lack access to digital tools, limiting their ability to modernize their agricultural practices.

18.6.3.2 Environmental Sustainability vs. Technological Efficiency

While technology can enhance efficiency and productivity in agriculture, it can also raise concerns about environmental sustainability. Traditional farming practices are often aligned with sustainable principles, using natural methods to maintain soil health and biodiversity. In contrast, some modern technologies, such as genetically modified organisms (GMOs) or chemical-intensive farming may prioritize short-term gains over long-term sustainability (Bommarco et al. 2013). This raises ethical questions about the environmental impact of technological innovation, particularly in regions where ecosystems are fragile and vulnerable to degradation (Hamza and Anderson 2005).

18.6.3.3 The Threat to Smallholder Farmers

The rise of technological innovations in agriculture also poses a threat to the livelihoods of smallholder farmers. As large-scale farms adopt advanced technologies that enable them to produce more with fewer inputs, small farmers may find it increasingly difficult to compete. This trend has the potential to accelerate the consolidation of farmland and reduce the number of small-scale farming operations, leading to the erosion of rural communities and traditional farming knowledge (Scoones 2016). According to a report by the World Bank (2021), smallholder farmers make up 80% of the world's agricultural producers but often struggle to access the technological resources needed to remain competitive.

18.6.3.4 Technological Overload and Farmer Autonomy

Another ethical concern is the potential for technological overload, where farmers become overly reliant on AI-driven systems for decision-making, reducing their autonomy and expertise in managing their farms. AI systems, while powerful, are not infallible and can make incorrect predictions based on biased or incomplete data (Mehrabi et al. 2021a). When farmers rely too heavily on these systems, they may lose valuable traditional knowledge and skills that are critical to adapting to local conditions and unforeseen challenges. The balance between technological assistance and farmer autonomy is a key ethical issue in the adoption of AI-driven solutions in agriculture (Guston 2014).

18.6.4 Balancing Innovation and Tradition

18.6.4.1 Integrating Traditional Knowledge with Modern Technology

One approach to addressing the ethical challenges of technological adoption is to integrate traditional knowledge with modern technology. By combining AI-driven solutions with local expertise, farmers can benefit from technological advancements while maintaining their connection to traditional practices (Altieri 2018). For instance, AI systems can be designed to incorporate local knowledge about crop cycles, soil conditions, and pest management, providing a more holistic approach to farming that respects both innovation and tradition (Mazzucato 2018).

18.6.4.2 Participatory Approaches to Technology Development

A participatory approach to technology development, where farmers are actively involved in the design and implementation of new technologies, can help ensure that innovations meet the needs and values of local communities (van der Ploeg et al. 2019). By involving farmers in the development process, AI developers and agricultural researchers can create solutions that are culturally appropriate, accessible, and aligned with sustainable farming practices.

18.6.4.3 Policy and Regulation for Ethical Technology Adoption

Governments and international organizations play a crucial role in ensuring that technological adoption in agriculture is equitable and sustainable. Policies that promote access to technology for smallholder farmers, protect traditional knowledge, and regulate the environmental impact of new technologies are essential for addressing the ethical challenges of innovation in agriculture (Schutter 2010). According to the FAO (2020), national governments should focus on creating supportive frameworks that encourage the responsible and inclusive adoption of agricultural technologies.

In essence, the adoption of technological innovations in agriculture presents both opportunities and ethical challenges. While technologies such as AI, ML, and biotechnology offer the potential to improve productivity, sustainability, and resilience in farming, they also raise concerns about equity, environmental sustainability, and the erosion of traditional knowledge. Balancing innovation with tradition requires a multifaceted approach that integrates traditional knowledge with modern technology, promotes participatory development, and ensures equitable access to technological resources. By addressing these ethical challenges, we can create a future where technology enhances, rather than displaces, traditional farming practices.

18.7 Equity in Access to Machine Learning Technologies for Sustainable Agriculture

The integration of ML technologies into agriculture holds promise for enhancing productivity, efficiency, and sustainability. ML algorithms can optimize crop management, predict pest outbreaks, and improve resource allocation, thus contributing to more sustainable agricultural practices (Kamilaris and Prenafeta-Boldú 2018). However, the equitable distribution of these technologies remains a significant concern. Unequal access to ML technologies can exacerbate existing disparities between large-scale commercial farms and smallholder farmers, potentially undermining efforts to achieve global sustainability goals.

18.7.1 Challenges in Achieving Equity in Access to ML Technologies

18.7.1.1 Cost Barriers and Economic Disparities

One of the primary barriers to equitable access to ML technologies is the high cost associated with their implementation and maintenance. Advanced ML systems often require significant investment in hardware, software, and data infrastructure, which can be prohibitive for smallholder farmers and agricultural enterprises in developing regions (Deichmann et al. 2016). According to a study by He et al. (2021), the cost of acquiring and deploying precision agriculture technologies can be a substantial financial burden, particularly for small-scale and subsistence farmers. This economic disparity leads to unequal access to the benefits of ML technologies, reinforcing existing inequalities in agricultural productivity and income.

18.7.1.2 Lack of Technical Expertise

Effective utilization of ML technologies requires a level of technical expertise that many smallholder farmers and rural communities may lack. ML systems often involve complex algorithms and require a certain level of proficiency in data analysis and interpretation (Klerkx et al. 2019). A survey conducted by Lichtenberg et al. (2020) found that a significant gap exists in the availability of technical training and support for smallholder farmers, limiting their ability to implement and benefit from advanced technologies. This lack of expertise further compounds the challenge of achieving equity in ML technology access.

18.7.1.3 Data Availability and Quality

ML models rely heavily on data to make accurate predictions and recommendations. However, the availability and quality of agricultural data can vary widely across regions (Gou et al. 2021). In many developing countries, there is a lack of comprehensive and high-quality agricultural data, which hampers the effectiveness of ML technologies. Furthermore, data collection methods and standards can differ, leading to inconsistencies that affect the performance of ML models (Ray et al. 2021). The disparity in data availability and quality poses a significant challenge to ensuring that ML technologies can be effectively applied across diverse agricultural contexts.

18.7.1.4 Infrastructure Limitations

The deployment of ML technologies in agriculture often requires robust digital infrastructure, including internet connectivity and access to computational resources. In many rural and underserved areas, such infrastructure is lacking or inadequate (World Bank 2021). According to a report by the International Telecommunication Union (ITU 2020), regions with limited internet access and poor connectivity face substantial barriers to adopting digital technologies. These infrastructure limitations can prevent smallholder farmers from accessing and benefiting from ML technologies, further exacerbating inequities.

18.7.2 Strategies for Promoting Equity in Access to ML Technologies

18.7.2.1 Subsidies and Financial Support

To address the cost barriers associated with ML technologies, governments and international organizations can provide subsidies and financial support to smallholder farmers and rural enterprises. Such financial assistance can help reduce the initial investment required for technology adoption and make ML technologies more accessible (Deichmann et al. 2016). Programs that offer low-interest loans, grants, or subsidies for technology adoption can help bridge the economic gap and promote more equitable access to ML solutions.

18.7.2.2 Capacity Building and Training

Investing in capacity building and training programs is crucial for enabling smallholder farmers to effectively use ML technologies. Training initiatives should focus on equipping farmers with the skills needed to operate and interpret ML systems, including data analysis and decision-making (Klerkx et al. 2019). Extension services and educational programs that provide hands-on training and technical support can help overcome the expertise gap and facilitate the adoption of ML technologies in rural communities (Lichtenberg et al. 2020).

18.7.2.3 Improving Data Accessibility and Quality

Enhancing the availability and quality of agricultural data is essential for the effective application of ML technologies. Efforts to improve data collection methods, standardize data formats, and establish data-sharing platforms can help address data disparities (Gou et al. 2021). Collaborative initiatives between governments, research institutions, and private companies can facilitate the development of comprehensive agricultural datasets that are accessible to all stakeholders (Ray et al. 2021). Ensuring that data collection and management practices are inclusive and standardized can improve the effectiveness of ML models across diverse agricultural settings.

18.7.2.4 Developing Infrastructure and Connectivity

Addressing infrastructure limitations is crucial for ensuring equitable access to ML technologies. Investments in digital infrastructure, such as expanding internet connectivity and improving access to computational resources, can help overcome barriers to technology adoption (World Bank 2021). Public-private partnerships and development programs that focus on enhancing digital infrastructure in rural areas can support the broader deployment of ML technologies and reduce regional disparities (ITU 2020).

18.7.2.5 Promoting Open-source and Inclusive Technologies

The development and dissemination of open-source ML technologies can promote equity by making advanced tools and resources more accessible to a wider range of users. Open-source platforms and software provide free access to ML algorithms and data, allowing smallholder farmers and organizations with limited resources to benefit from technological advancements (Zhang et al. 2020). Inclusive technology design that considers the needs and contexts of diverse users can also help ensure that ML solutions are adaptable and relevant across different agricultural environments.

18.7.3 Case Studies and Examples

18.7.3.1 Precision Agriculture in India

In India, several initiatives have aimed to improve access to ML technologies for smallholder farmers. The Digital Green project, for example, uses digital platforms to provide farmers with tailored agricultural advice and training. By leveraging mobile technology and local knowledge, Digital Green helps bridge the gap between technological innovation and traditional farming practices (Gupta et al. 2019). This approach demonstrates how technology can be adapted to meet the needs of smallholder farmers and promote more equitable access to ML solutions.

18.7.3.2 Agricultural Data Platforms in Africa

In Africa, the Technologies of African Development Bank for African Agricultural Transformation (TAAT) initiative is working to improve data accessibility and quality for smallholder farmers. TAAT focuses on developing data-sharing platforms and improving data collection methods to support the adoption of ML technologies across the continent (African Development Bank 2021). By enhancing data infrastructure and accessibility, TAAT aims to address some of the key barriers to equitable access and promote sustainable agricultural practices.

18.7.3.3 Internet Connectivity Projects in Rural Areas

Efforts to improve internet connectivity in rural areas, such as the Google Project Loon initiative, have made significant strides in enhancing digital infrastructure. Project Loon uses high-altitude balloons to provide internet access in underserved regions, helping to overcome connectivity barriers that limit the adoption of ML technologies (Miller et al. 2020). Such initiatives highlight the importance of addressing infrastructure limitations to promote equitable access to digital tools and resources.

In light of the above, ensuring equity in access to ML technologies for sustainable agriculture is essential for addressing global challenges related to food security, environmental sustainability, and rural development. While there are significant challenges, including cost barriers, technical expertise gaps, data disparities, and infrastructure limitations, a range of strategies can help promote more equitable access. By investing in financial support, capacity building, data accessibility, infrastructure development, and inclusive technology design, stakeholders can work towards a more equitable and sustainable future for agriculture.

18.8 Human–AI Collaboration: Ethical Guidelines for Decision-making in Agriculture

The integration of AI into agricultural decision-making processes represents a significant shift towards data-driven farming. AI technologies, including ML and predictive analytics, offer the potential to enhance productivity, optimize resource use, and improve sustainability (Kamilaris and Prenafeta-Boldú 2018). However, as AI systems increasingly support or even make decisions in agriculture, ethical considerations around human-AI collaboration become paramount.

18.8.1 Ethical Challenges in Human–AI Collaboration

18.8.1.1 Transparency and Explainability

Transparency in AI systems involves ensuring that the decision-making process is understandable to users, particularly in high-stakes applications such as agriculture where decisions impact livelihoods and food security (Raji and Buolamwini 2019). Explainability is crucial for fostering trust and allowing users to understand how AI-generated recommendations are derived. Research highlights that opaque AI systems can lead to skepticism and reduced adoption among users (Gilpin et al. 2018). For example, AI models used for crop yield prediction must provide insights into how input variables influence predictions to be effectively used by farmers (Doshi-Velez and Kim 2017).

18.8.1.2 Accountability and Responsibility

Determining accountability in scenarios where AI systems are involved in decision-making is challenging. If an AI system makes a suboptimal recommendation that results in financial loss or environmental damage, it is essential to establish who responsible (Dignum 2019). Ethical guidelines must clarify the roles and responsibilities of AI developers, operators, and users. For instance, if an AI system advises on pesticide application and leads to environmental harm, accountability should be examined at multiple levels—from the developers who designed the system to the users who followed the recommendations (Floridi et al. 2018).

18.8.1.3 Bias and Fairness

AI systems can perpetuate or even exacerbate biases present in the training data, leading to unfair outcomes that disproportionately affect certain groups (Buolamwini and Gebru 2018). In agriculture, biases in AI models can manifest in various ways, such as uneven access to technology or skewed recommendations based on regional data disparities. Ensuring fairness requires implementing strategies to identify and mitigate biases in AI systems. For instance, algorithms used for loan approval or resource allocation in agriculture must be rigorously tested for fairness to prevent discrimination against smallholder farmers or marginalized communities (O'Neil 2016).

18.8.1.4 Human Autonomy and Decision-making

AI systems should support, rather than replace, human decision-making in agriculture. The risk of over-reliance on AI is that it may undermine human expertise and intuition, which are crucial for adapting to complex and dynamic agricultural environments (Shneiderman 2020). Ethical guidelines should emphasize the importance of maintaining human oversight and ensuring that AI systems are used as decision support tools rather than autonomous decision-makers. For example, while AI can provide insights on optimal planting times, final decisions should remain with the farmer who can consider local conditions and personal experience (Miller et al. 2020).

18.8.2 Ethical Guidelines for Human–AI Collaboration

18.8.2.1 Develop Transparent and Explainable AI Systems

To address transparency and explain ability, developers should prioritize creating AI systems that offer clear explanations of their decision-making processes. This includes using interpretable models or providing visualizations that help users understand how AI recommendations are generated (Carvalho et al. 2019). For instance, incorporating decision trees or rule-based systems can enhance transparency by showing the logic behind predictions or recommendations. Additionally, user-friendly interfaces and documentation can facilitate better understanding and trust in AI systems (Chen et al. 2018).

18.8.2.2 Establish Accountability Frameworks

Clear accountability frameworks are essential for managing the ethical implications of AI in agriculture. These frameworks should outline the responsibilities of AI developers, users, and other stakeholders (Dignum 2019). This involves creating mechanisms for reporting and addressing issues related to AI failures or ethical breaches. For example, setting up protocols for evaluating the impact of AI recommendations on environmental and economic outcomes can help ensure accountability (Binns et al. 2018).

18.8.2.3 Implement Bias Mitigation Strategies

To ensure fairness, AI developers should implement strategies to identify and mitigate biases in training data and algorithms. This can involve using diverse datasets that represent various agricultural contexts and regularly auditing AI systems for biased outcomes (Buolamwini and Gebru 2018). Additionally, engaging with stakeholders from diverse backgrounds during the development process can help address potential biases and ensure that AI systems are equitable and inclusive (Mehrabi et al. 2019).

18.8.2.4 Promote Human–AI Collaboration and Oversight

Ethical guidelines should emphasize the importance of human oversight in AI-assisted decision-making. AI systems should be designed to complement human judgment rather than replace it, providing insights and recommendations while allowing users to exercise their expertise and experience (Shneiderman 2020). Training programs for users should focus on integrating AI tools into their decision-making processes, ensuring that AI recommendations are evaluated in the context of local knowledge and conditions (Miller et al. 2020).

18.8.2.5 Foster Continuous Ethical Review and Improvement

Ongoing ethical review and improvement of AI systems are crucial for adapting to evolving challenges and ensuring responsible use. This includes establishing mechanisms for regular evaluation of AI systems' impact, incorporating feedback from users, and updating ethical guidelines as needed (Floridi et al. 2018). Collaborative efforts between researchers, developers, policymakers, and agricultural practitioners can facilitate continuous improvement and ensure that AI technologies align with ethical standards and societal values (Binns et al. 2018).

18.8.3 Case Studies and Examples

18.8.3.1 AI for Precision Agriculture in the United States

In the United States, AI-driven precision agriculture systems have been implemented to optimize resource use and increase crop yields. Projects like the AI for Earth initiative by Microsoft emphasize transparency and fairness by providing tools and resources to help farmers understand AI models and their recommendations (Microsoft 2021). These initiatives highlight the importance of developing AI systems that are both effective and accessible, addressing ethical concerns through transparency and user engagement.

18.8.3.2 AI-assisted Pest Management in India

In India, the use of AI for pest management has demonstrated the need for balancing human expertise with machine recommendations. The Plantix app, which uses AI to diagnose plant diseases and pests, involves local agricultural experts in verifying AI recommendations and providing context-specific advice (Tauseef et al. 2020). This approach underscores the value of human-AI collaboration, where AI supports decision-making while human expertise ensures the relevance and accuracy of recommendations.

18.8.3.3 Bias Mitigation in Agricultural Lending in Africa

In Africa, initiatives aimed at improving access to agricultural finance have highlighted the need for bias mitigation in AI systems. The use of AI to assess creditworthiness and allocate loans must consider the diverse financial situations of smallholder farmers. Efforts to develop fair lending algorithms include using inclusive datasets and engaging with local communities to ensure that AI systems do not reinforce existing biases (O'Neil 2016).

Overall, the human–AI collaboration in agricultural decision-making presents both opportunities and ethical challenges. By developing transparent and explainable AI systems, establishing accountability frameworks, implementing bias mitigation strategies, promoting human oversight, and fostering continuous ethical review, stakeholders can ensure that AI technologies are used responsibly and effectively. Balancing innovation with ethical considerations is essential for advancing sustainable agriculture while respecting human expertise and promoting equitable outcomes.

18.9 Regulatory Frameworks for ML in Agricultural Biotechnology

ML has become a transformative force in agricultural biotechnology, driving innovations in crop improvement, pest management, and precision agriculture (Kamilaris and Prenafeta-Boldú 2018). However, the rapid advancement of ML technologies necessitates robust regulatory frameworks to ensure their safe and ethical application in agriculture.

18.9.1 Current Regulatory Frameworks

18.9.1.1 Global and Regional Regulations

Regulatory frameworks for ML in agricultural biotechnology vary significantly across regions. For instance, the European Union (EU) has established comprehensive regulations under the General Data Protection Regulation (GDPR) and the European Food Safety Authority (EFSA) guidelines, which impact ML applications by emphasizing data protection, transparency, and safety (European Commission 2018 and EFSA 2021). In contrast, the United States relies on a combination of federal and state regulations, including the Food and Drug Administration (FDA) guidelines and the U.S. Department of Agriculture (USDA) regulations, which focus on the safety and efficacy of biotechnology products (FDA 2021 and USDA 2021).

18.9.1.2 Data Privacy and Security Regulations

Data privacy and security are critical concerns in ML applications, especially given the sensitive nature of agricultural data. The GDPR in the EU mandates stringent data protection measures, including obfuscation and secure storage, to safeguard personal and agricultural data (European Commission 2018). Similarly, the California Consumer Privacy Act (CCPA) in the United States provides rights related to data access and control, influencing how ML systems handle agricultural data (California Legislative Information 2020). These regulations aim to ensure that data used in ML models are handled ethically and securely.

18.9.1.3 Ethical and Safety Guidelines

Ethical considerations in ML for agricultural biotechnology are addressed through various safety and ethical guidelines. The OECD's guidelines for the safety of genetically modified organisms (GMOs) and the International Conference on Harmonisation of Technical Requirements for Registration of Pharmaceuticals for Human Use (ICH) provide frameworks for assessing the risks associated with biotechnology applications (OECD 2021; ICH 2021). These guidelines emphasize the importance of risk assessment, environmental impact, and the ethical use of biotechnological innovations.

18.9.2 Challenges and Gaps in Regulatory Frameworks

18.9.2.1 Rapid Technological Advancements

One of the primary challenges in regulating ML in agricultural biotechnology is the rapid pace of technological advancements. Existing regulatory frameworks often struggle to keep up with the evolving capabilities of ML technologies, leading to potential gaps in coverage and oversight (Binns et al. 2018). For instance, new ML techniques for genetic modification or pest management may not fit neatly within existing regulatory categories, necessitating updates and adaptations to regulatory standards (Shneiderman 2020).

18.9.2.2 Integration of ML into Existing Frameworks

Integrating ML technologies into existing regulatory frameworks can be complex due to the interdisciplinary nature of agricultural biotechnology. Traditional regulations may not fully address the unique aspects of ML, such as algorithmic transparency, explainability, and data handling (Doshi-Velez and Kim 2017). For example, while the FDA provides guidelines for GMOs, it may not have specific provisions for the use of ML in designing or assessing these organisms (FDA 2021).

18.9.2.3 Global Consistency and Harmonization

The lack of global consistency and harmonization in regulatory frameworks presents a challenge for international collaboration and trade. Different countries have varying approaches to regulating ML in agricultural biotechnology, which can create barriers to cross-border research and commercialization (Arai et al. 2019). Harmonizing regulations across regions can facilitate international cooperation and ensure that ML technologies are developed and deployed in a manner that is consistent with global standards (Cohen 2020).

18.9.3 Proposed Guidelines for Future Regulation

18.9.3.1 Dynamic and Adaptive Regulatory Frameworks

Future regulatory frameworks should be designed to be dynamic and adaptive to keep pace with technological advancements. This involves creating flexible guidelines that can be updated regularly to address new developments in ML technologies (Binns et al. 2018). For example, incorporating mechanisms for periodic review and revision of regulations can help ensure that they remain relevant and effective in the face of evolving ML capabilities (Floridi et al. 2018).

18.9.3.2 Enhanced Transparency and Explainability Requirements

To address concerns about algorithmic transparency and explainability, regulatory frameworks should include specific requirements for ML systems used in agricultural biotechnology. This includes mandating that ML models provide clear explanations of their decision-making processes and that users have access to information about how models are trained and validated (Gilpin et al. 2018). Transparency requirements can help build trust and ensure that ML systems are used responsibly and ethically.

18.9.3.3 Risk Assessment and Management Protocols

Developing comprehensive risk assessment and management protocols is essential for ensuring the safe use of ML technologies in agriculture. These protocols should address potential risks associated with data handling, algorithmic biases, and the environmental impact of biotechnology applications (OECD 2021). Risk management strategies should be integrated into regulatory frameworks to ensure that potential hazards are identified and mitigated effectively.

18.9.3.4 International Collaboration and Harmonization

Promoting international collaboration and harmonization of regulations can facilitate the development and deployment of ML technologies across borders. Establishing global standards and frameworks for regulating ML in agricultural biotechnology can help overcome barriers to international trade and research (Arai et al. 2019). Collaborative efforts among regulatory bodies, industry stakeholders, and international organizations can support the creation of consistent and effective regulatory practices (Cohen 2020).

18.9.4 Case Studies and Examples

18.9.4.1 EU Regulations for GMOs and ML

The EU's approach to regulating GMOs includes provisions for the use of ML technologies in assessing genetic modifications. The EFSA provides guidelines for evaluating the safety and efficacy of GMOs, including those developed with ML tools (EFSA 2021). This regulatory framework highlights the importance of integrating ML considerations into existing biotechnology regulations to ensure comprehensive oversight.

18.9.4.2 FDA's Approach to Biotechnology and AI

In the United States, the FDA's guidelines for biotechnology encompass ML applications used in drug development and genetic modification. The FDA has issued draft guidance on the use of AI in medical devices, which may be relevant for agricultural applications involving ML (FDA 2021). This approach demonstrates the need for adapting existing regulatory frameworks to address the specific challenges posed by ML technologies.

18.9.4.3 Global Harmonization Efforts

The OECD's initiatives to harmonize regulations for biotechnology and ML technologies highlight the importance of global collaboration. The OECD's guidelines for the safety of GMOs and its efforts to promote international dialogue on regulatory practices reflect ongoing efforts to create consistent and effective regulatory frameworks (OECD 2021).

Ultimately, the integration of ML into agricultural biotechnology presents significant opportunities and challenges. Developing robust regulatory frameworks is crucial for ensuring the safe and ethical application of ML technologies. By addressing challenges related to technological advancements, regulatory integration, and global harmonization, stakeholders can create guidelines that promote responsible innovation while safeguarding public health and the environment. Ongoing efforts to adapt and enhance regulatory frameworks will be essential for supporting the responsible development and deployment of ML technologies in agriculture.

18.10 Conclusion

The integration of ML technologies into plant biology and agriculture presents significant opportunities for enhancing productivity, sustainability, and precision in resource management. However, as these technologies become more prevalent, they bring forth critical challenges related to data privacy, security, fairness, and ethical governance. Safeguarding agricultural data from breaches and unauthorized access, ensuring transparency, and addressing biases in ML models are paramount for maintaining trust and equity among farming communities.

Moreover, existing regulatory frameworks, such as the GDPR and FDA guidelines, provide a foundation for addressing these concerns, but they require significant adaptation to keep pace with the rapid evolution of ML technologies. Dynamic and adaptive regulations, enhanced transparency, and comprehensive risk management protocols are essential for ensuring that ML systems are applied responsibly in agriculture.

Global collaboration will be crucial in harmonizing regulatory standards, enabling smallholder farmers to benefit equally from technological advancements, and preventing the exacerbation of existing inequalities. As ML continues to shape the future of agriculture, a multi-faceted approach, involving secure data practices, fairness in model design, and continuous regulatory improvements, will be necessary to harness its full potential while maintaining ethical oversight.

References

African Development Bank. (2021). *Technologies for African Agricultural Transformation (TAAT)*. African Development Bank. Retrieved from https://www.afdb.org/en/news-keywords/technologies-african-agricultural-transformation-taat

Altieri, M.A. (2018). *Agroecology: The Science of Sustainable Agriculture*. CRC Press.

Arai, Y., Sakurai, K., and Tanimura, K. (2019). Global regulatory perspectives on biotechnology and machine learning: A comparative analysis. *Journal of International Biotechnology Law* 16 (2): 135–150.

Athey, S. (2018). The impact of machine learning on economics. *The Economics Journal* 128 (610): 1373–1391.

Balafoutis, A., Fountas, S., Cavalaris, C. et al. (2017). Precision agriculture technologies positively contributing to GHG emissions mitigation, farm productivity and economics. *Sustainability* 9 (8): 1339.

Balde, C.P., Forti, V., Gray, V. et al. (2017). *The Global E-Waste Monitor 2017: Quantities, Flows, and Resources*. United Nations University.

Barocas, S., Hardt, M., and Narayanan, A. (2019). *Fairness and Machine Learning: Limitations and Opportunities*. The MIT Press.

Berg, A., Kunkel, B., and Schreiter, E. (2018). Sustainable AI: AI's role in achieving the SDGs. *Frontiers in Sustainability* 2: 12.

Binns, R., Veeramachaneni, K., and Louppe, G. (2018). The challenges of ethical AI: A review of fairness and accountability. *Proceedings of the 2018 AAAI/ACM Conference on AI, Ethics, and Society*, 43–49. New York, NY: Association for Computing Machinery.

Bommarco, R., Kleijn, D., and Potts, S.G. (2013). Ecological intensification: Harnessing ecosystem services for food security. *Trends in Ecology & Evolution* 28 (4): 230–238.

Bronson, K. and Knezevic, I. (2016). Big data in food and agriculture. *Big Data & Society* 3 (1): 2053951716648174.

Brown, T., Johnson, R., Miller, K. et al. (2023). AI bias and fairness in agriculture. *IEEE Transactions on AI Ethics* 15 (2): 112–130.

Buolamwini, J. and Gebru, T. (2018). Gender shades: Intersectional accuracy disparities in commercial gender classification. *Proceedings of the 2018 CHI Conference on Human Factors in Computing Systems*, 1–14, Association for Computing Machinery.

California Legislative Information. (2020). California Consumer Privacy Act (CCPA). Retrieved from https://leginfo.legislature.ca.gov/faces/billNavClient.xhtml?bill_id=201920200AB375

Carvalho, D.M., Oliveira, J.L., and Miranda, J.A. (2019). Explainable AI: A survey. *Proceedings of the 2019 International Conference on Artificial Intelligence*, 73–80, IEEE.

Chen, J., Song, L., and Wainwright, M.J. (2018). Learning explainable models with deep neural networks. *Proceedings of the 35th International Conference on Machine Learning*, 84–92, ML Research Press.

Chinta, S.V., Wang, Z., Yin, Z. et al. (2024). FairAIED: Navigating fairness, bias, and ethics in educational AI applications. *arXiv preprint arXiv:2407.18745*.

Clark, S., Shah, M., and Ghedira, K. (2021). Interdisciplinary collaboration in AI-driven agriculture. *Journal of Agricultural Systems* 190: 103087. https://doi.org/10.1016/j.agsy.2021.103087

Cohen, J.E. (2020). Harmonizing global regulations for agricultural biotechnology: Opportunities and challenges. *International Journal of Agricultural Sustainability* 18 (4): 375–387.

Cooley, H., Christian-Smith, J., and Gleick, P.H. (2014). *Water Use in California's Agriculture*. Pacific Institute.

Correll, M., Anderson, E., and Heer, J. (2017). Social and environmental challenges in real-time crop monitoring. *Nature Sustainability* 1 (4): 283–288.

Deichmann, U., Goyal, A., and Mishra, D. (2016). Will digital technologies transform agriculture in developing countries? *Agricultural Economics* 47 (S1): 21–33.

Dignum, V. (2019). Responsible artificial intelligence: How to develop ethical AI systems. *Proceedings of the 2019 AAAI/ACM Conference on AI, Ethics, and Society*, 62–70, Association for Computing Machinery.

Doe, J., Smith, A., White, P. et al. (2022). Ethical challenges in smart agriculture. *Agricultural Ethics Journal* 8 (4): 245–260.

Doshi-Velez, F. and Kim, B. (2017). Towards a rigorous science of interpretable machine learning. *Proceedings of the 2017 ICML Workshop on Human Interpretability in Machine Learning*, 1–6, arXiv.

EFSA. (2021). Guidelines for the risk assessment of genetically modified organisms. *European Food Safety Authority*. Retrieved from https://www.efsa.europa.eu/en/efsajournal/pub/4388

European Commission. (2018). General Data Protection Regulation (GDPR). Retrieved from https://ec.europa.eu/info/law/law-topic/data-protection_en

FDA. (2021). Guidance for industry: Use of artificial intelligence (AI) and machine learning in the development of medical devices. *U.S. Food and Drug Administration*. Retrieved from https://www.fda.gov/media/136290/download

Floridi, L., Cowls, J., Beltrametti, M. et al. (2018). AI4People: An ethical framework for a good AI society: Opportunities, risks, principles, and recommendations *Minds & Machines* 28: 689–707. https://doi.org/10.1007/s11023-018-9482-5.

Floridi, L., Cowls, J., and Taddeo, M. (2018). How to regulate artificial intelligence: The case for ethical guidelines. *Proceedings of the 2018 IEEE International Conference on Artificial Intelligence*, 205–214, IEEE.

Food and Agriculture Organization (FAO). (2019). *The State of Food and Agriculture: Sustainability in the Digital Age*, 1–150. United Nations.

Food and Agriculture Organization (FAO). (2020). *The State of Food and Agriculture: Overcoming Water Challenges in Agriculture*. FAO.

Gebru, T., Morgenstern, J., Vecchione, B., et al. (2021). Datasheets for datasets. *Communications of the ACM* 64 (12): 86–92.

Gilpin, L.H., Bau, D., and Yang, J. (2018). Explaining explanations: An overview of interpretability of machine learning. *Proceedings of the 2018 IEEE/ACM Conference on AI, Ethics, and Society*, 3–16.

Goodfellow, I., Bengio, Y., and Courville, A. (2016). *Deep Learning*. MIT Press.

Gou, J., Zhang, X., and Zhang, L. (2021). Big data and machine learning in precision agriculture: A review. *Computers and Electronics in Agriculture* 180: 105802.

Gunning, D. and Aha, D. (2019). DARPA's explainable artificial intelligence program. *AI Magazine* 40 (2): 44–58.

Gupta, A., Prakash, R., and Patel, V. (2019). Digital green: Enhancing agricultural extension through mobile technology. *Journal of Agricultural Education and Extension* 25 (3): 245–264.

Guston, D.H. (2014). Understanding 'anticipatory governance.' *Social Studies of Science* 44 (2): 218–242.

Hamza, M.A. and Anderson, W.K. (2005). Soil compaction in cropping systems: A review of the nature, causes and possible solutions. *Soil and Tillage Research* 82 (2): 121–145.

Hassija, V., Chamola, V., Mahapatra, A. et al. (2024). Interpreting black-box models: A review on explainable artificial intelligence. *Cognitive Computation* 16 (1): 45–74.

He, J., Zhang, L., and Xu, C. (2021). Economic impacts of precision agriculture technology adoption: Evidence from China. *Journal of Agricultural and Resource Economics* 46 (2): 321–337.

Hino, M., Benami, E., and Brooks, N. (2018). Machine learning for environmental monitoring. *Nature Sustainability* 1 (10): 583–588.

ICH. (2021). International Conference on Harmonisation of Technical Requirements For Registration of Pharmaceuticals For Human Use (ICH) guidelines. Retrieved from https://www.ich.org/page/ich-guidelines

ITU. (2020). Measuring digital development: Facts and figures 2020. International Telecommunication Union. Retrieved from https://www.itu.int/en/ITU-D/Statistics/Pages/publications/misr.aspx

Jha, K., Doshi, A., Patel, P. et al. (2019). A comprehensive review on automation in agriculture using artificial intelligence. *Artificial Intelligence in Agriculture* 2: 1–12.

Jones, P., Robinson, M., Clarke, J. et al. (2020). *The Role of Big Data in Agribusiness*, 50–210. Oxford University Press.

Joshi, P.K., Singh, N.P., and Sharma, P. (2021). Leveraging artificial intelligence for inclusive agricultural development: Challenges and opportunities. *AI in Agriculture* 5: 20–34.

Jouppi, N.P., Young, C., Patil, N. et al. (2017). In-datacenter performance analysis of a tensor processing unit. *Proceedings of the 44th Annual International Symposium on Computer Architecture*, 1–12.

Kamilaris, A. and Prenafeta-Boldú, F.X. (2018). Deep learning in agriculture: A survey. *Computers and Electronics in Agriculture* 147: 70–90.

Kamiran, F. and Calders, T. (2012). Data preprocessing techniques for classification without discrimination. *Knowledge and Information Systems* 33 (1): 1–33.

Klerkx, L., Jakku, E., and Labarthe, P. (2019). A review of social science on digital agriculture, smart farming and agricultural innovation systems. *Agricultural Systems* 171: 69–80.

Lichtenberg, E., Choi, C., and Zhao, Y. (2020). Assessing the knowledge gap in agricultural technology adoption among smallholder farmers. *Food Policy* 95: 101934.

Liu, Y., Chen, H., Zhang, W. et al. (2020). AI for global agriculture: Addressing regional bias. *Nature AI* 12 (3): 98–115.

Mazzucato, M. (2018). *The Entrepreneurial State: Debunking Public vs. Private Sector Myths*. Penguin Books.

Mehrabi, N., Morstatter, F., Saxena, N. et al. (2021a). A survey on bias and fairness in machine learning. *ACM Computing Surveys (CSUR)* 54 (6): 1–35.

Mehrabi, Z., McDowell, M.J., and Benton, T.G. (2021b). The environmental risks of large-scale agriculture. *Global Environmental Change* 68: 102279.

Microsoft. (2021). AI for Earth. Retrieved from https://www.microsoft.com/en-us/ai/ai-for-earth

Miller, J., Chen, J., and Li, C. (2020). Project Loon: Connectivity solutions for underserved regions. *IEEE Access* 8: 140143–140155.

Mulligan, D.K., Smith, J.M., and Koopman, C. (2019). Aligning fairness and efficiency in agricultural policy: A framework for stakeholders. *Journal of Agricultural Ethics* 32 (3): 445–461.

O'Neil, C. (2016). *Weapons of Math Destruction: How Big Data Increases Inequality and Threatens Democracy*. Crown Publishing Group.

OECD. (2021). OECD guidelines for the safety of genetically modified organisms. Organisation for Economic Co-operation and Development. Retrieved from https://www.oecd.org/env/biotrack

Oyeniran, C.O., Adewusi, A.O., Adeleke, A.G. et al. (2022). Ethical AI: Addressing bias in machine learning models and software applications. *Computer Science & IT Research Journal* 3 (3): 115–126.

Patel, R. and Singh, M. (2020). *Social Justice in Agricultural Innovation*, 30–180. Springer.

Raji, I.D. and Buolamwini, J. (2019). Actionable auditing: Investigating the impact of public accountability on algorithmic bias. *Proceedings of the 2019 ACM Conference on Fairness, Accountability, and Transparency*, 69–80, Association for Computing Machinery.

Ray, D., Sahu, S., and Zhang, X. (2021). Data quality in agricultural machine learning applications: A review. *Computers and Electronics in Agriculture* 182: 105881.

Russell, S., Dewey, D., and Tegmark, M. (2020). Research priorities for robust and beneficial artificial intelligence. *AI & Society* 35 (3): 459–466.

Schutter, O.D. (2010). The right to food. *United Nations General Assembly*, A/65/281.

Scoones, I. (2016). *Sustainable Livelihoods and Rural Development.* Practical Action Publishing.

Shamshiri, R.R., Kalantari, F., Ting, K.C., et al. (2018). Advances in greenhouse automation and controlled environment agriculture: A transition to plant factories and urban agriculture. *International Journal of Agricultural and Biological Engineering* 11 (1): 1–22.

Shiva, V. (2016). *Traditional Knowledge and Agricultural Sustainability*, 1–250. Oxford University Press.

Shneiderman, B. (2020). Human-centered AI. *ACM Transactions on Computer-Human Interaction* 27 (4): 1–28.

Smith, M. and Lee, R. (2021). *Data Privacy in Precision Farming*, 60–200. Springer.

Stevenson, J.R., Serraj, R., and Cassman, K.G. (2020). Evaluating and improving global farming systems data: Addressing gaps in scale, scope, and representativeness. *Agricultural Systems* 178: 102752.

Sumberg, J., Thompson, J., and Woodhouse, P. (2017). Why agronomy in the developing world has become contentious. *Agricultural Systems* 153: 1–10.

Sun, X., Wang, Y., Li, F. et al. (2009a). The effects of data imbalance in machine learning applications. *AI in Agriculture Journal* 5 (1): 33–50.

Sun, Y., Wong, A.K., and Kamel, M.S. (2009b). Classification of imbalanced data: A review. *International Journal of Pattern Recognition and Artificial Intelligence* 23 (4): 687–719.

Tauseef, M., Farid, M., and Owais, M. (2020). Plantix: AI-based plant disease diagnosis and management. *Journal of Agricultural Informatics* 11 (1): 1–15.

USDA. (2021). Biotechnology regulatory services: Overview. U.S. Department of Agriculture. Retrieved from https://www.aphis.usda.gov/aphis/ourfocus/biotechnology

Van der Ploeg, J.D., Ye, J., and Schneider, S. (2019). Rural development through the construction of new markets. *Journal of Peasant Studies* 46 (1): 133–154.

Watson, J. and Mahler, T. (2020). Addressing bias in artificial intelligence for sustainable development. *Nature Sustainability* 3 (9); 685–687.

Wolfert, S., Ge, L., Verdouw, C. et al. (2017). Big data in smart farming: A review. *Agricultural Systems* 153: 69–80.

World Bank. (2021). Economic impacts of agri-tech on small farmers. *World Development Report* (pp. 10–95).

Zhang, C. and Kovacs, J.M. (2019). The application of small unmanned aerial systems for precision agriculture: A review. *Precision Agriculture* 20 (6): 1287–1311. https://doi.org/10.1007/s11119-019-09676-x

Zhang, Y., He, Q., and Lu, G. (2020). AI-based solutions for crop disease identification in precision agriculture. *Journal of Agricultural Informatics* 11 (1): 23–30.

Zhao, L., Chen, B., Yang, D. et al. (2018). Deep learning for plant disease detection: Challenges and solutions. *IEEE Transactions on Agriculture* 27 (4): 187–205.

Index

Note: Page numbers in *italics* indicate figures and **bold** indicate tables.

Machine Learning for Plant Biology, First Edition. Edited by Jen-Tsung Chen.